STUDENT'S SOLUTIONS MANUAL

ELKA BLOCK
FRANK PURCELL

FINITE MATHEMATICS
TENTH EDITION

Margaret L. Lial
American River College
Raymond N. Greenwell
Hofstra University
Nathan P. Ritchey
Youngstown State University

PEARSON

Boston Columbus Indianapolis New York San Francisco Upper Saddle River
Amsterdam Cape Town Dubai London Madrid Milan Munich Paris Montreal Toronto
Delhi Mexico City Sao Paulo Sydney Hong Kong Seoul Singapore Taipei Tokyo

The author and publisher of this book have used their best efforts in preparing this book. These efforts include the development, research, and testing of the theories and programs to determine their effectiveness. The author and publisher make no warranty of any kind, expressed or implied, with regard to these programs or the documentation contained in this book. The author and publisher shall not be liable in any event for incidental or consequential damages in connection with, or arising out of, the furnishing, performance, or use of these programs.

Reproduced by Pearson from electronic files supplied by the author.

ISBN-13: 978-0-321-74867-6
ISBN-10: 0-321-74867-0

2 3 4 5 6 EBM 15 14 13 12

www.pearsonhighered.com

PEARSON

CONTENTS

CHAPTER 7   SETS AND PROBABILITY

CHAPTER 8   COUNTING PRINCIPLES

CHAPTER 9   STATISTICS

CHAPTER 10   MARKOV CHAINS

CHAPTER 11   GAME THEORY

ALGEBRA REFERENCE

R.1 Polynomials

Your Turn 1

$3(x^2 - 4x - 5) - 4(3x^2 - 5x - 7)$

$= 3x^2 - 12x - 15 - 12x^2 + 20x + 28$

$= -9x^2 + 8x + 13$

Your Turn 2

$(3y + 2)(4y^2 - 2y - 5)$

$= (3y)(4y^2 - 2y - 5) + (2)(4y^2 - 2y - 5)$

$= 12y^3 - 6y^2 - 15y + 8y^2 - 4y - 10$

$= 12y^3 + 2y^2 - 19y - 10$

R.1 Exercises

1. $(2x^2 - 6x + 11) + (-3x^2 + 7x - 2)$

$= 2x^2 - 6x + 11 - 3x^2 + 7x - 2$

$= (2 - 3)x^2 + (7 - 6)x + (11 - 2)$

$= -x^2 + x + 9$

3. $-6(2q^2 + 4q - 3) + 4(-q^2 + 7q - 3)$

$= (-12q^2 - 24q + 18)$

$\quad + (-4q^2 + 28q - 12)$

$= (-12q^2 - 4q^2)$

$\quad + (-24q + 28q) + (18 - 12)$

$= -16q^2 + 4q + 6$

5. $(0.613x^2 - 4.215x + 0.892) - 0.47(2x^2 - 3x + 5)$

$= 0.613x^2 - 4.215x + 0.892 - 0.94x^2$

$\qquad\qquad\qquad\qquad\qquad + 1.41x - 2.35$

$= -0.327x^2 - 2.805x - 1.458$

7. $-9m(2m^2 + 3m - 1)$

$= -9m(2m^2) - 9m(3m) - 9m(-1)$

$= -18m^3 - 27m^2 + 9m$

9. $(3t - 2y)(3t + 5y)$

$= (3t)(3t) + (3t)(5y) + (-2y)(3t) + (-2y)(5y)$

$= 9t^2 + 15ty - 6ty - 10y^2$

$= 9t^2 + 9ty - 10y^2$

11. $(2 - 3x)(2 + 3x)$

$= (2)(2) + (2)(3x) + (-3x)(2) + (-3x)(3x)$

$= 4 + 6x - 6x - 9x^2$

$= 4 - 9x^2$

13. $\left(\dfrac{2}{5}y + \dfrac{1}{8}z\right)\left(\dfrac{3}{5}y + \dfrac{1}{2}z\right)$

$= \left(\dfrac{2}{5}y\right)\left(\dfrac{3}{5}y\right) + \left(\dfrac{2}{5}y\right)\left(\dfrac{1}{2}z\right) + \left(\dfrac{1}{8}z\right)\left(\dfrac{3}{5}y\right)$

$\qquad\qquad\qquad\qquad + \left(\dfrac{1}{8}z\right)\left(\dfrac{1}{2}z\right)$

$= \dfrac{6}{25}y^2 + \dfrac{1}{5}yz + \dfrac{3}{40}yz + \dfrac{1}{16}z^2$

$= \dfrac{6}{25}y^2 + \left(\dfrac{8}{40} + \dfrac{3}{40}\right)yz + \dfrac{1}{16}z^2$

$= \dfrac{6}{25}y^2 + \dfrac{11}{40}yz + \dfrac{1}{16}z^2$

15. $(3p - 1)(9p^2 + 3p + 1)$

$= (3p - 1)(9p^2) + (3p - 1)(3p) + (3p - 1)(1)$

$= 3p(9p^2) - 1(9p^2) + 3p(3p)$

$\qquad\qquad\qquad\qquad - 1(3p) + 3p(1) - 1(1)$

$= 27p^3 - 9p^2 + 9p^2 - 3p + 3p - 1$

$= 27p^3 - 1$

17. $(2m + 1)(4m^2 - 2m + 1)$

$= 2m(4m^2 - 2m + 1) + 1(4m^2 - 2m + 1)$

$= 8m^3 - 4m^2 + 2m + 4m^2 - 2m + 1$

$= 8m^3 + 1$

19. $(x + y + z)(3x - 2y - z)$
$$= x(3x) + x(-2y) + x(-z) + y(3x) + y(-2y)$$
$$+ y(-z) + z(3x) + z(-2y) + z(-z)$$
$$= 3x^2 - 2xy - xz + 3xy - 2y^2 - yz + 3xz$$
$$- 2yz - z^2$$
$$= 3x^2 + xy + 2xz - 2y^2 - 3yz - z^2$$

21. $(x + 1)(x + 2)(x + 3)$
$$= [x(x + 2) + 1(x + 2)](x + 3)$$
$$= [x^2 + 2x + x + 2](x + 3)$$
$$= [x^2 + 3x + 2](x + 3)$$
$$= x^2(x + 3) + 3x(x + 3) + 2(x + 3)$$
$$= x^3 + 3x^2 + 3x^2 + 9x + 2x + 6$$
$$= x^3 + 6x^2 + 11x + 6$$

23. $(x + 2)^2 = (x + 2)(x + 2)$
$$= x(x + 2) + 2(x + 2)$$
$$= x^2 + 2x + 2x + 4$$
$$= x^2 + 4x + 4$$

25. $(x - 2y)^3$
$$= [(x - 2y)(x - 2y)](x - 2y)$$
$$= (x^2 - 2xy - 2xy + 4y^2)(x - 2y)$$
$$= (x^2 - 4xy + 4y^2)(x - 2y)$$
$$= (x^2 - 4xy + 4y^2)x$$
$$+ (x^2 - 4xy + 4y^2)(-2y)$$
$$= x^3 - 4x^2y + 4xy^2 - 2x^2y + 8xy^2 - 8y^3$$
$$= x^3 - 6x^2y + 12xy^2 - 8y^3$$

R.2 Factoring

Your Turn 1

Factor $4z^4 + 4z^3 + 18z^2$.
$$4z^4 + 4z^3 + 18z^2$$
$$= (2z^2) \cdot 2z^2 + (2z^2) \cdot 2z + (2z^2) \cdot 9$$
$$= (2z^2)(2z^2 + 2z + 9)$$

Your Turn 2

Factor $6a^2 + 5ab - 4b^2$.
$$6a^2 + 5ab - 4b^2 = (2a - b)(3a + 4b)$$

R.2 Exercises

1. $7a^3 + 14a^2 = 7a^2 \cdot a + 7a^2 \cdot 2$
$$= 7a^2(a + 2)$$

3. $13p^4q^2 - 39p^3q + 26p^2q^2$
$$= 13p^2q \cdot p^2q - 13p^2q \cdot 3p + 13p^2q \cdot 2q$$
$$= 13p^2q(p^2q - 3p + 2q)$$

5. $m^2 - 5m - 14 = (m - 7)(m + 2)$
since $(-7)(2) = -14$ and $-7 + 2 = -5$.

7. $z^2 + 9z + 20 = (z + 4)(z + 5)$
since $4 \cdot 5 = 20$ and $4 + 5 = 9$.

9. $a^2 - 6ab + 5b^2 = (a - b)(a - 5b)$
since $(-b)(-5b) = 5b^2$ and $-b + (-5b) = -6b$.

11. $y^2 - 4yz - 21z^2 = (y + 3z)(y - 7z)$
since $(3z)(-7z) = -21z^2$ and $3z + (-7z) = -4z$.

13. $3a^2 + 10a + 7$
The possible factors of $3a^2$ are $3a$ and a and the possible factors of 7 are 7 and 1. Try various combinations until one works.
$$3a^2 + 10a + 7 = (a + 1)(3a + 7)$$

15. $21m^2 + 13mn + 2n^2 = (7m + 2n)(3m + n)$

17. $3m^3 + 12m^2 + 9m = 3m(m^2 + 4m + 3)$
$$= 3m(m + 1)(m + 3)$$

19. $24a^4 + 10a^3b - 4a^2b^2$
$$= 2a^2(12a^2 + 5ab - 2b^2)$$
$$= 2a^2(4a - b)(3a + 2b)$$

21. $x^2 - 64 = x^2 - 8^2$
$$= (x + 8)(x - 8)$$

23. $10x^2 - 160 = 10(x^2 - 16)$
$$= 10(x^2 - 4^2)$$
$$= 10(x + 4)(x - 4)$$

25. $z^2 + 14zy + 49y^2 = z^2 + 2 \cdot 7zy + 7^2y^2$
$$= (z + 7y)^2$$

27. $9p^2 - 24p + 16 = (3p)^2 - 2 \cdot 3p \cdot 4 + 4^2$
$$= (3p - 4)^2$$

29. $27r^3 - 64s^3 = (3r)^3 - (4s)^3$

$$= (3r - 4s)(9r^2 + 12rs + 16s^2)$$

31. $x^4 - y^4 = (x^2)^2 - (y^2)^2$

$$= (x^2 + y^2)(x^2 - y^2)$$

$$= (x^2 + y^2)(x + y)(x - y)$$

R.3 Rational Expressions

Your Turn 1

Write in lowest terms $\dfrac{z^2 + 5z + 6}{2z^2 + 7z + 3}$.

$$\frac{z^2 + 5z + 6}{2z^2 + 7z + 3} = \frac{(z + 3)(z + 2)}{(z + 3)(2z + 1)}$$

$$= \frac{z + 2}{2z + 1}$$

Your Turn 2

Perform each of the following operations.

(a) $\dfrac{z^2 + 5z + 6}{2z^2 - 5z - 3} \cdot \dfrac{2z^2 - z - 1}{z^2 + 2z - 3}$

$$= \frac{(z + 2)(z + 3)}{(2z + 1)(z - 3)} \cdot \frac{(2z + 1)(z - 1)}{(z + 3)(z - 1)}$$

$$= \frac{(z + 2)(z + 3)\,(2z + 1)\,(z - 1)}{(2z + 1)(z - 3)(z + 3)\,(z - 1)}$$

$$= \frac{z + 2}{z - 3}$$

(b) $\dfrac{a - 3}{a^2 + 3a + 2} + \dfrac{5a}{a^2 - 4}$

$$= \frac{a - 3}{(a + 2)(a + 1)} + \frac{5a}{(a - 2)(a + 2)}$$

$$= \frac{a - 3}{(a + 2)(a + 1)} \cdot \frac{(a - 2)}{(a - 2)}$$

$$+ \frac{5a}{(a - 2)(a + 2)} \cdot \frac{(a + 1)}{(a + 1)}$$

$$= \frac{(a^2 - 5a + 6) + (5a^2 + 5a)}{(a - 2)(a + 2)(a + 1)}$$

$$= \frac{6a^2 + 6}{(a - 2)(a + 2)(a + 1)}$$

$$= \frac{6(a^2 + 1)}{(a - 2)(a + 2)(a + 1)}$$

R.3 Exercises

1. $\dfrac{5v^2}{35v} = \dfrac{5 \cdot v \cdot v}{5 \cdot 7 \cdot v} = \dfrac{v}{7}$

3. $\dfrac{8k + 16}{9k + 18} = \dfrac{8(k + 2)}{9(k + 2)} = \dfrac{8}{9}$

5. $\dfrac{4x^3 - 8x^2}{4x^2} = \dfrac{4x^2(x - 2)}{4x^2} = x - 2$

7. $\dfrac{m^2 - 4m + 4}{m^2 + m - 6} = \dfrac{(m - 2)(m - 2)}{(m - 2)(m + 3)}$

$$= \frac{m - 2}{m + 3}$$

9. $\dfrac{3x^2 + 3x - 6}{x^2 - 4} = \dfrac{3(x + 2)(x - 1)}{(x + 2)(x - 2)} = \dfrac{3(x - 1)}{x - 2}$

11. $\dfrac{m^4 - 16}{4m^2 - 16} = \dfrac{(m^2 + 4)(m + 2)(m - 2)}{4(m + 2)(m - 2)}$

$$= \frac{m^2 + 4}{4}$$

13. $\dfrac{9k^2}{25} \cdot \dfrac{5}{3k} = \dfrac{3 \cdot 3 \cdot 5k^2}{5 \cdot 5 \cdot 3k} = \dfrac{3k^2}{5k} = \dfrac{3k}{5}$

15. $\dfrac{3a + 3b}{4c} \cdot \dfrac{12}{5(a + b)} = \dfrac{3(a + b)}{4c} \cdot \dfrac{3 \cdot 4}{5(a + b)}$

$$= \frac{3 \cdot 3}{c \cdot 5}$$

$$= \frac{9}{5c}$$

17. $\dfrac{2k - 16}{6} \div \dfrac{4k - 32}{3} = \dfrac{2k - 16}{6} \cdot \dfrac{3}{4k - 32}$

$$= \frac{2(k - 8)}{6} \cdot \frac{3}{4(k - 8)}$$

$$= \frac{1}{4}$$

19. $\dfrac{4a + 12}{2a - 10} \div \dfrac{a^2 - 9}{a^2 - a - 20}$

$$= \frac{4(a + 3)}{2(a - 5)} \cdot \frac{(a - 5)(a + 4)}{(a - 3)(a + 3)}$$

$$= \frac{2(a + 4)}{a - 3}$$

21. $\dfrac{k^2 + 4k - 12}{k^2 + 10k + 24} \cdot \dfrac{k^2 + k - 12}{k^2 - 9}$

$$= \frac{(k + 6)(k - 2)}{(k + 6)(k + 4)} \cdot \frac{(k + 4)(k - 3)}{(k + 3)(k - 3)}$$

$$= \frac{k - 2}{k + 3}$$

23. $\dfrac{2m^2 - 5m - 12}{m^2 - 10m + 24} \div \dfrac{4m^2 - 9}{m^2 - 9m + 18}$

$\qquad = \dfrac{2m^2 - 5m - 12}{m^2 - 10m + 24} \cdot \dfrac{m^2 - 9m + 18}{4m^2 - 9}$

$\qquad = \dfrac{(2m + 3)(m - 4)(m - 6)(m - 3)}{(m - 6)(m - 4)(2m - 3)(2m + 3)}$

$\qquad = \dfrac{m - 3}{2m - 3}$

25. $\dfrac{a + 1}{2} - \dfrac{a - 1}{2} = \dfrac{(a + 1) - (a - 1)}{2}$

$\qquad\qquad\qquad = \dfrac{a + 1 - a + 1}{2}$

$\qquad\qquad\qquad = \dfrac{2}{2} = 1$

27. $\dfrac{6}{5y} - \dfrac{3}{2} = \dfrac{6 \cdot 2}{5y \cdot 2} - \dfrac{3 \cdot 5y}{2 \cdot 5y} = \dfrac{12 - 15y}{10y}$

29. $\dfrac{1}{m - 1} + \dfrac{2}{m} = \dfrac{m}{m}\left(\dfrac{1}{m - 1}\right) + \dfrac{m - 1}{m - 1}\left(\dfrac{2}{m}\right)$

$\qquad\qquad\quad = \dfrac{m + 2m - 2}{m(m - 1)}$

$\qquad\qquad\quad = \dfrac{3m - 2}{m(m - 1)}$

31. $\dfrac{8}{3(a - 1)} + \dfrac{2}{a - 1} = \dfrac{8}{3(a - 1)} + \dfrac{3}{3}\left(\dfrac{2}{a - 1}\right)$

$\qquad\qquad\qquad\quad = \dfrac{8 + 6}{3(a - 1)}$

$\qquad\qquad\qquad\quad = \dfrac{14}{3(a - 1)}$

33. $\dfrac{4}{x^2 + 4x + 3} + \dfrac{3}{x^2 - x - 2}$

$\qquad = \dfrac{4}{(x + 3)(x + 1)} + \dfrac{3}{(x - 2)(x + 1)}$

$\qquad = \dfrac{4(x - 2)}{(x - 2)(x + 3)(x + 1)}$

$\qquad\qquad + \dfrac{3(x + 3)}{(x - 2)(x + 3)(x + 1)}$

$\qquad = \dfrac{4(x - 2) + 3(x + 3)}{(x - 2)(x + 3)(x + 1)}$

$\qquad = \dfrac{4x - 8 + 3x + 9}{(x - 2)(x + 3)(x + 1)}$

$\qquad = \dfrac{7x + 1}{(x - 2)(x + 3)(x + 1)}$

35. $\dfrac{3k}{2k^2 + 3k - 2} - \dfrac{2k}{2k^2 - 7k + 3}$

$\qquad = \dfrac{3k}{(2k - 1)(k + 2)} - \dfrac{2k}{(2k - 1)(k - 3)}$

$\qquad = \left(\dfrac{k - 3}{k - 3}\right)\dfrac{3k}{(2k - 1)(k + 2)}$

$\qquad\qquad - \left(\dfrac{k + 2}{k + 2}\right)\dfrac{2k}{(2k - 1)(k - 3)}$

$\qquad = \dfrac{(3k^2 - 9k) - (2k^2 + 4k)}{(2k - 1)(k + 2)(k - 3)}$

$\qquad = \dfrac{k^2 - 13k}{(2k - 1)(k + 2)(k - 3)}$

$\qquad = \dfrac{k(k - 13)}{(2k - 1)(k + 2)(k - 3)}$

37. $\dfrac{2}{a + 2} + \dfrac{1}{a} + \dfrac{a - 1}{a^2 + 2a}$

$\qquad = \dfrac{2}{a + 2} + \dfrac{1}{a} + \dfrac{a - 1}{a(a + 2)}$

$\qquad = \left(\dfrac{a}{a}\right)\dfrac{2}{a + 2} + \left(\dfrac{a + 2}{a + 2}\right)\dfrac{1}{a} + \dfrac{a - 1}{a(a + 2)}$

$\qquad = \dfrac{2a + a + 2 + a - 1}{a(a + 2)}$

$\qquad = \dfrac{4a + 1}{a(a + 2)}$

R.4 Equations

Your Turn 1

Solve $3x - 7 = 4(5x + 2) - 7x$.

$3x - 7 = 20x + 8 - 7x$

$3x - 7 = 13x + 8$

$-10x = 15$

$\qquad x = -\dfrac{15}{10}$

$\qquad x = -\dfrac{3}{2}$

Your Turn 2

Solve $2m^2 + 7m = 15$.

$2m^2 + 7m - 15 = 0$

$(2m - 3)(m + 5) = 0$

$2m - 3 = 0 \quad \text{or} \quad m + 5 = 0$

$\qquad m = \dfrac{3}{2} \quad \text{or} \qquad m = -5$

Your Turn 3

Solve $z^2 + 6 = 8z$.

$z^2 - 8z + 6 = 0$

Use the quadratic formula with
$a = 1$, $b = -8$, and $c = 6$.

$$z = \frac{-(-8) \pm \sqrt{(-8)^2 - 4(1)(6)}}{2(1)}$$

$$= \frac{8 \pm \sqrt{64 - 24}}{2}$$

$$= \frac{8 \pm \sqrt{40}}{2}$$

$$= \frac{8 \pm \sqrt{4 \cdot 10}}{2}$$

$$= \frac{8 \pm 2\sqrt{10}}{2}$$

$$= 4 \pm \sqrt{10}$$

Your Turn 4

Solve $\dfrac{1}{x^2 - 4} + \dfrac{2}{x - 2} = \dfrac{1}{x}$.

$$\frac{1}{(x-2)(x+2)} + \frac{2}{x-2} = \frac{1}{x}$$

$$(x-2)(x+2)(x) \cdot \frac{1}{(x-2)(x+2)}$$

$$+ (x-2)(x+2)(x) \cdot \frac{2}{x-2}$$

$$= (x-2)(x+2)(x) \cdot \frac{1}{x}$$

$$x + 2x^2 + 4x = x^2 - 4$$

$$x^2 + 5x + 4 = 0$$

$$(x+1)(x+4) = 0$$

$$x = -1 \quad \text{or} \quad x = -4$$

Neither of these values makes a denominator equal to zero, so both are solutions.

R.4 Exercises

1. $2x + 8 = x - 4$

$x + 8 = -4$

$x = -12$

The solution is -12.

3. $0.2m - 0.5 = 0.1m + 0.7$

$10(0.2m - 0.5) = 10(0.1m + 0.7)$

$2m - 5 = m + 7$

$m - 5 = 7$

$m = 12$

The solution is 12.

5. $3r + 2 - 5(r + 1) = 6r + 4$

$3r + 2 - 5r - 5 = 6r + 4$

$-3 - 2r = 6r + 4$

$-3 = 8r + 4$

$-7 = 8r$

$-\dfrac{7}{8} = r$

The solution is $-\dfrac{7}{8}$.

7. $2[3m - 2(3 - m) - 4] = 6m - 4$

$2[3m - 6 + 2m - 4] = 6m - 4$

$2[5m - 10] = 6m - 4$

$10m - 20 = 6m - 4$

$4m - 20 = -4$

$4m = 16$

$m = 4$

The solution is 4.

9. $x^2 + 5x + 6 = 0$

$(x + 3)(x + 2) = 0$

$x + 3 = 0 \quad \text{or} \quad x + 2 = 0$

$x = -3 \quad \text{or} \quad x = -2$

The solutions are -3 and -2.

11. $m^2 = 14m - 49$

$m^2 - 14m + 49 = 0$

$(m)^2 - 2(7m) + (7)^2 = 0$

$(m - 7)^2 = 0$

$m - 7 = 0$

$m = 7$

The solution is 7.

13. $12x^2 - 5x = 2$

$12x^2 - 5x - 2 = 0$

$(4x + 1)(3x - 2) = 0$

$4x + 1 = 0 \quad \text{or } 3x - 2 = 0$

$4x = -1 \quad \text{or} \quad 3x = 2$

$x = -\dfrac{1}{4} \quad \text{or} \quad x = \dfrac{2}{3}$

The solutions are $-\dfrac{1}{4}$ and $\dfrac{2}{3}$.

15. $4x^2 - 36 = 0$

Divide both sides of the equation by 4.

$$x^2 - 9 = 0$$
$$(x + 3)(x - 3) = 0$$
$$x + 3 = 0 \quad \text{or} \quad x - 3 = 0$$
$$x = -3 \quad \text{or} \quad x = 3$$

The solutions are -3 and 3.

17. $12y^2 - 48y = 0$

$$12y(y) - 12y(4) = 0$$
$$12y(y - 4) = 0$$
$$12y = 0 \quad \text{or} \quad y - 4 = 0$$
$$y = 0 \quad \text{or} \quad y = 4$$

The solutions are 0 and 4.

19. $2m^2 - 4m = 3$

$$2m^2 - 4m - 3 = 0$$

$$m = \frac{-(-4) \pm \sqrt{(-4)^2 - 4(2)(-3)}}{2(2)}$$

$$= \frac{4 \pm \sqrt{40}}{4} = \frac{4 \pm \sqrt{4 \cdot 10}}{4}$$

$$= \frac{4 \pm \sqrt{4}\sqrt{10}}{4}$$

$$= \frac{4 \pm 2\sqrt{10}}{4} = \frac{2 \pm \sqrt{10}}{2}$$

The solutions are $\frac{2+\sqrt{10}}{2} \approx 2.5811$ and $\frac{2-\sqrt{10}}{2} \approx -0.5811$.

21. $k^2 - 10k = -20$

$$k^2 - 10k + 20 = 0$$

$$k = \frac{-(-10) \pm \sqrt{(-10)^2 - 4(1)(20)}}{2(1)}$$

$$k = \frac{10 \pm \sqrt{100 - 80}}{2}$$

$$k = \frac{10 \pm \sqrt{20}}{2}$$

$$k = \frac{10 \pm 2\sqrt{5}}{2}$$

$$k = \frac{2(5 \pm \sqrt{5})}{2}$$

$$k = 5 \pm \sqrt{5}$$

The solutions are $5 + \sqrt{5} \approx 7.2361$ and $5 - \sqrt{5} \approx 2.7639$.

23. $2r^2 - 7r + 5 = 0$

$$(2r - 5)(r - 1) = 0$$
$$r - 1 = 0 \quad \text{or} \quad r = 1$$
$$2r - 5 = 0 \quad \text{or} \quad r = \frac{5}{2}$$

The solutions are $\frac{5}{2}$ and 1.

25. $3k^2 + k = 6$

$$3k^2 + k - 6 = 0$$

$$k = \frac{-1 \pm \sqrt{1 - 4(3)(-6)}}{2(3)}$$

$$= \frac{-1 \pm \sqrt{73}}{6}$$

The solutions are $\frac{-1+\sqrt{73}}{6} \approx 1.2573$ and $\frac{-1-\sqrt{73}}{6} \approx -1.5907$.

27. $\dfrac{3x - 2}{7} = \dfrac{x + 2}{5}$

$$35\left(\frac{3x - 2}{7}\right) = 35\left(\frac{x + 2}{2}\right)$$
$$5(3x - 2) = 7(x + 2)$$
$$15x - 10 = 7x + 14$$
$$8x = 24$$
$$x = 3$$

The solution is $x = 3$.

29. $\dfrac{4}{x - 3} - \dfrac{8}{2x + 5} + \dfrac{3}{x - 3} = 0$

$$\frac{4}{x - 3} + \frac{3}{x - 3} - \frac{8}{2x + 5} = 0$$

$$\frac{7}{x - 3} - \frac{8}{2x + 5} = 0$$

Multiply both sides by $(x - 3)(2x + 5)$. Note that $x \neq 3$ and $x \neq \frac{5}{2}$.

$$(x - 3)(2x + 5)\left(\frac{7}{x - 3} - \frac{8}{2x + 5}\right) = (x - 3)(2x + 5)(0)$$

$$7(2x + 5) - 8(x - 3) = 0$$
$$14x + 35 - 8x + 24 = 0$$
$$6x + 59 = 0$$
$$6x = -59 \quad \text{or} \quad x = -\frac{59}{6}$$

Note: It is especially important to check solutions of equations that involve rational expressions. Here, a check shows that $-\frac{59}{6}$ is a solution.

31. $\dfrac{2m}{m-2} - \dfrac{6}{m} = \dfrac{12}{m^2 - 2m}$

$\dfrac{2m}{m-2} - \dfrac{6}{m} = \dfrac{12}{m(m-2)}$

Multiply both sides by $m(m-2)$. Note that $m \neq 0$ and $m \neq 2$.

$m(m-2)\left(\dfrac{2m}{m-2} - \dfrac{6}{m}\right)$

$\qquad\qquad = m(m-2)\left(\dfrac{12}{m(m-2)}\right)$

$m(2m) - 6(m-2) = 12$

$2m^2 - 6m + 12 = 12$

$2m^2 - 6m = 0$

$2m(m-3) = 0$

$2m = 0 \quad \text{or} \quad m - 3 = 0$

$m = 0 \quad \text{or} \qquad m = 3$

Since $m \neq 0$, 0 is not a solution. The solution is 3.

33. $\dfrac{1}{x-2} - \dfrac{3x}{x-1} = \dfrac{2x+1}{x^2 - 3x + 2}$

$\dfrac{1}{x-2} - \dfrac{3x}{x-1} = \dfrac{2x+1}{(x-2)(x-1)}$

Multiply both sides by $(x-2)(x-1)$. Note that $x \neq 2$ and $x \neq 1$.

$(x-2)(x-1)\left(\dfrac{1}{x-2} - \dfrac{3x}{x-1}\right)$

$\qquad = (x-2)(x-1) \cdot \left[\dfrac{2x+1}{(x-2)(x-1)}\right]$

$(x-2)(x-1)\left(\dfrac{1}{x-2}\right) - (x-2)(x-1) \cdot \left(\dfrac{3x}{x-1}\right)$

$\qquad = \dfrac{(x-2)(x-1)(2x+1)}{(x-2)(x-1)}$

$(x-1) - (x-2)(3x) = 2x + 1$

$x - 1 - 3x^2 + 6x = 2x + 1$

$-3x^2 + 7x - 1 = 2x + 1$

$-3x^2 + 5x - 2 = 0$

$3x^2 - 5x + 2 = 0$

$(3x-2)(x-1) = 0$

$3x - 2 = 0 \quad \text{or} \quad x - 1 = 0$

$x = \dfrac{2}{3} \quad \text{or} \qquad x = 1$

1 is not a solution since $x \neq 1$. The solution is $\frac{2}{3}$.

35. $\dfrac{5}{b+5} - \dfrac{4}{b^2 + 2b} = \dfrac{6}{b^2 + 7b + 10}$

$\dfrac{5}{b+5} - \dfrac{4}{b(b+2)} = \dfrac{6}{(b+5)(b+2)}$

Multiply both sides by $b(b+5)(b+2)$. Note that $b \neq 0$, $b \neq -5$, and $b \neq -2$.

$b(b+5)(b+2)\left(\dfrac{5}{b+5} - \dfrac{4}{b(b+2)}\right)$

$\qquad = b(b+5)(b+2)\left(\dfrac{6}{(b+5)(b+2)}\right)$

$5b(b+2) - 4(b+5) = 6b$

$5b^2 + 10b - 4b - 20 = 6b$

$5b^2 - 20 = 0$

$b^2 - 4 = 0$

$(b+2)(b-2) = 0$

$b + 2 = 0 \quad \text{or} \quad b - 2 = 0$

$b = -2 \quad \text{or} \qquad b = 2$

Since $b \neq -2$, -2 is not a solution. The solution is 2.

37. $\dfrac{4}{2x^2 + 3x - 9} + \dfrac{2}{2x^2 - x - 3}$

$\qquad\qquad\qquad = \dfrac{3}{x^2 + 4x + 3}$

$\dfrac{4}{(2x-3)(x+3)} + \dfrac{2}{(2x-3)(x+1)}$

$\qquad\qquad\qquad = \dfrac{3}{(x+3)(x+1)}$

Multiply both sides by $(2x-3)(x+3)(x+1)$. Note that $x \neq \frac{3}{2}$, $x \neq -3$, and $x \neq -1$.

$(2x-3)(x+3)(x+1)$

$\quad \cdot \left(\dfrac{4}{(2x-3)(x+3)} + \dfrac{2}{(2x-3)(x+1)}\right)$

$= (2x-3)(x+3)(x+1)\left(\dfrac{3}{(x+3)(x+1)}\right)$

$4(x+1) + 2(x+3) = 3(2x-3)$

$4x + 4 + 2x + 6 = 6x - 9$

$6x + 10 = 6x - 9$

$10 = -9$

This is a false statement. Therefore, there is no solution.

R.5 Inequalities

Your Turn 1

Solve $3z - 2 > 5z + 7$.

$$3z - 2 > 5z + 7$$
$$3z - 2 + 2 > 5z + 7 + 2$$
$$3z > 5z + 9$$
$$3z - 5z > 5z - 5z + 9$$
$$-2z > 9$$
$$\frac{-2z}{-2} < \frac{9}{-2}$$
$$z < -\frac{9}{2}$$

Your Turn 2

Solve $3y^2 \le 16y + 12$.

$$3y^2 \le 16y + 12$$
$$3y^2 - 16y - 12 \le 0$$

First solve the equation $3y^2 - 16y - 12 = 0$.

$$3y^2 - 16y - 12 = 0$$
$$(3y + 2)(y - 6) = 0$$
$$3y + 2 = 0 \quad \text{or} \quad y - 6 = 0$$
$$y = -\frac{2}{3} \quad \text{or} \quad y = 6$$

Determine three intervals on the number line and choose a test point in each interval.

$$
\begin{array}{ccc}
A & B & C \\
\end{array}
$$
$$-\frac{2}{3} \qquad 6$$

Choose -1 from interval A: $3(-1)^2 - 16(-1) - 12 > 0$

Choose 0 from interval B: $3(0)^2 - 16(0) - 12 < 0$

Choose 7 from interval C: $3(7)^2 - 16(7) - 12 > 0$

The numbers in interval B satisfy the inequality, and since the sign was less than or equal to, the boundary points of interval B are also part of the solution. The solution is $[-2/3, 6]$.

Your Turn 3

Solve $\dfrac{k^2 - 35}{k} \ge 2$.

First solve the corresponding equation $\dfrac{k^2 - 35}{k} = 2$.

$$\frac{k^2 - 35}{k} = 2$$
$$k^2 - 35 = 2k$$

$$k^2 - 2k - 35 = 0$$
$$(k - 7)(k + 5) = 0$$
$$k = 7 \quad \text{or} \quad k = -5$$

The denominator is 0 when $k = 0$, so there are four intervals to consider:

$$(-\infty, -5), (-5, 0), (0, 7), \text{ and } (7, \infty).$$

Choose a test point in each interval.

$$k = -8; \quad \frac{(-8)^2 - 35}{-8} < 2$$

$$k = -1; \quad \frac{(-1)^2 - 35}{-1} > 2$$

$$k = 5; \quad \frac{(-5)^2 - 35}{5} < 2$$

$$k = 10; \quad \frac{(10)^2 - 35}{10} > 2$$

The second and fourth intervals are part of the solution. Since the inequality is greater than or equal to, we can include the endpoints -5 and 7 but not the endpoint 0, which makes the denominator 0. The solution is $[-5, 0) \cup [7, \infty)$.

R.5 Exercises

1. $x < 4$

   Because the inequality symbol means "less than," the endpoint at 4 is not included. This inequality is written in interval notation as $(-\infty, 4)$. To graph this interval on a number line, place an open circle at 4 and draw a heavy arrow pointing to the left.

   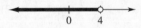

3. $1 \le x < 2$

   The endpoint at 1 is included, but the endpoint at 2 is not. This inequality is written in interval notation as $[1, 2)$. To graph this interval, place a closed circle at 1 and an open circle at 2; then draw a heavy line segment between them.

5. $-9 > x$

   This inequality may be rewritten as $x < -9$, and is written in interval notation as $(-\infty, -9)$. Note that the endpoint at -9 is not included. To graph this interval, place an open circle at -9 and draw a heavy arrow pointing to the left.

   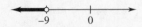

7. $[-7, -3]$

This represents all the numbers between -7 and -3, including both endpoints. This interval can be written as the inequality $-7 \le x \le -3$.

9. $(-\infty, -1]$

This represents all the numbers to the left of -1 on the number line and includes the endpoint. This interval can be written as the inequality $x \le -1$.

11. Notice that the endpoint -2 is included, but 6 is not. The interval shown in the graph can be written as the inequality $-2 \le x < 6$.

13. Notice that both endpoints are included. The interval shown in the graph can be written as $x \le -4$ or $x \ge 4$.

15.
$$6p + 7 \le 19$$
$$6p \le 12$$
$$\left(\frac{1}{6}\right)(6p) \le \left(\frac{1}{6}\right)(12)$$
$$p \le 2$$

The solution in interval notation is $(-\infty, 2]$.

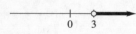

17.
$$m - (3m - 2) + 6 < 7m - 19$$
$$m - 3m + 2 + 6 < 7m - 19$$
$$-2m + 8 < 7m - 19$$
$$-9m + 8 < -19$$
$$-9m < -27$$
$$-\frac{1}{9}(-9m) > -\frac{1}{9}(-27)$$
$$m > 3$$

The solution is $(3, \infty)$.

19.
$$3p - 1 < 6p + 2(p - 1)$$
$$3p - 1 < 6p + 2p - 2$$
$$3p - 1 < 8p - 2$$
$$-5p - 1 < -2$$
$$-5p < -1$$
$$-\frac{1}{5}(-5p) > -\frac{1}{5}(-1)$$
$$p > \frac{1}{5}$$

The solution is $\left(\frac{1}{5}, \infty\right)$.

21.
$$-11 < y - 7 < -1$$
$$-11 + 7 < y - 7 + 7 < -1 + 7$$
$$-4 < y < 6$$

The solution is $(-4, 6)$.

23. $-2 < \dfrac{1 - 3k}{4} \le 4$
$$4(-2) < 4\left(\frac{1 - 3k}{4}\right) \le 4(4)$$
$$-8 < 1 - 3k \le 16$$
$$-9 < -3k \le 15$$
$$-\frac{1}{3}(-9) > -\frac{1}{3}(-3k) \ge -\frac{1}{3}(15)$$

Rewrite the inequalities in the proper order.
$$-5 \le k < 3$$

The solution is $[-5, 3)$.

25. $\dfrac{3}{5}(2p + 3) \ge \dfrac{1}{10}(5p + 1)$
$$10\left(\frac{3}{5}\right)(2p + 3) \ge 10\left(\frac{1}{10}\right)(5p + 1)$$
$$6(2p + 3) \ge 5p + 1$$
$$12p + 18 \ge 5p + 1$$
$$7p \ge -17$$
$$p \ge -\frac{17}{7}$$

The solution is $\left[-\frac{17}{7}, \infty\right)$.

27. $(m - 3)(m + 5) < 0$

Solve $(m - 3)(m + 5) = 0$.
$$(m - 3)(m + 5) = 0$$
$$m = 3 \quad \text{or} \quad m = -5$$

Intervals: $(-\infty, -5), (-5, 3), (3, \infty)$

For $(-\infty, -5)$, choose -6 to test for m.
$$(-6 - 3)(-6 + 5) = -9(-1) = 9 \not< 0$$

For $(-5, 3)$, choose 0.
$$(0 - 3)(0 + 5) = -3(5) = -15 < 0$$

For $(3, \infty)$, choose 4.
$$(4 - 3)(4 + 5) = 1(9) = 9 \not< 0$$

The solution is $(-5, 3)$.

29. $y^2 - 3y + 2 < 0$

$(y - 2)(y - 1) < 0$

Solve $(y - 2)(y - 1) = 0$.

$y = 2$ or $y = 1$

Intervals: $(-\infty, 1), (1, 2), (2, \infty)$

For $(-\infty, 1)$, choose $y = 0$.

$$0^2 - 3(0) + 2 = 2 \not< 0$$

For $(1, 2)$, choose $y = \frac{3}{2}$.

$$\left(\frac{3}{2}\right)^2 - 3\left(\frac{3}{2}\right) + 2 = \frac{9}{4} - \frac{9}{2} + 2$$

$$= \frac{9 - 18 + 8}{4}$$

$$= -\frac{1}{4} < 0$$

For $(2, \infty)$, choose 3.

$$3^2 - 3(3) + 2 = 2 \not< 0$$

The solution is $(1, 2)$.

31. $x^2 - 16 > 0$

Solve $x^2 - 16 = 0$.

$$x^2 - 16 = 0$$

$$(x + 4)(x - 4) = 0$$

$$x = -4 \quad \text{or} \quad x = 4$$

Intervals: $(-\infty, -4), \ (-4, 4), (4, \infty)$

For $(-\infty, -4)$, choose -5.

$$(-5)^2 - 16 = 9 > 0$$

For $(-4, 4)$, choose 0.

$$0^2 - 16 = -16 \not> 0$$

For $(4, \infty)$, choose 5.

$$5^2 - 16 = 9 > 0$$

The solution is $(-\infty, -4) \cup (4, \infty)$.

33. $x^2 - 4x \geq 5$

Solve $x^2 - 4x = 5$.

$$x^2 - 4x = 5$$

$$x^2 - 4x - 5 = 0$$

$$(x + 1)(x - 5) = 0$$

$x + 1 = 0$ or $x - 5 = 0$

$x = -1$ or $x = 5$

Intervals: $(-\infty, -1), (-1, 5), (5, \infty)$

For $(-\infty, -1)$, choose -2.

$$(-2)^2 - 4(-2) = 12 \geq 5$$

For $(-1, 5)$, choose 0.

$$0^2 - 4(0) = 0 \not\geq 5$$

For $(5, \infty)$, choose 6.

$$(6)^2 - 4(6) = 12 \geq 5$$

The solution is $(-\infty, -1] \cup [5, \infty)$.

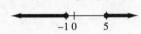

35. $3x^2 + 2x > 1$

Solve $3x^2 + 2x = 1$.

$$3x^2 + 2x = 1$$

$$3x^2 + 2x - 1 = 0$$

$$(3x - 1)(x + 1) = 0$$

$$x = \frac{1}{3} \quad \text{or} \quad x = -1$$

Intervals: $(-\infty, -1), \left(-1, \frac{1}{3}\right), \left(\frac{1}{3}, \infty\right)$

For $(-\infty, -1)$, choose -2.

$$3(-2)^2 + 2(-2) = 8 > 1$$

For $\left(-1, \frac{1}{3}\right)$, choose 0.

$$3(0)^2 + 2(0) = 0 \not> 1$$

For $\left(\frac{1}{3}, \infty\right)$, choose 1.

$$3(1)^2 + 2(1) = 5 > 1$$

The solution is $(-\infty, -1) \cup \left(\frac{1}{3}, \infty\right)$.

37. $9 - x^2 \leq 0$

Solve $9 - x^2 = 0$.

$$9 - x^2 = 0$$

$$(3 + x)(3 - x) = 0$$

$$x = -3 \quad \text{or} \quad x = 3$$

Intervals: $(-\infty, -3), (-3, 3), (3, \infty)$

For $(-\infty, -3)$, choose -4.

$$9 - (-4)^2 = -7 \leq 0$$

For $(-3, 3)$, choose 0.

$$9 - (0)^2 = 9 \not\leq 0$$

For $(3, \infty)$, choose 4.

$$9 - (4)^2 = -7 \le 0$$

The solution is $(-\infty, -3] \cup [3, \infty)$.

39. $x^3 - 4x \ge 0$

Solve $x^3 - 4x = 0$.

$$x^3 - 4x = 0$$
$$x(x^2 - 4) = 0$$
$$x(x + 2)(x - 2) = 0$$
$$x = 0, \quad \text{or} \quad x = -2, \quad \text{or} \quad x = 2$$

Intervals: $(-\infty, -2), (-2, 0), (0, 2), (2, \infty)$

For $(-\infty, -2)$, choose -3.

$$(-3)^3 - 4(-3) = -15 \not\ge 0$$

For $(-2, 0)$, choose -1.

$$(-1)^3 - 4(-1) = 3 \ge 0$$

For $(0, 2)$, choose 1.

$$(1)^3 - 4(1) = -3 \not\ge 0$$

For $(2, \infty)$, choose 3.

$$(3)^3 - 4(3) = 15 \ge 0$$

The solution is $[-2, 0] \cup [2, \infty)$.

41. $2x^3 - 14x^2 + 12x < 0$

Solve $2x^3 - 14x^2 + 12x = 0$.

$$2x^3 - 14x^2 + 12x = 0$$
$$2x(x^2 - 7x + 6) = 0$$
$$2x(x - 1)(x - 6) = 0$$
$$x = 0, \quad \text{or} \quad x = 1, \quad \text{or} \quad x = 6$$

Intervals: $(-\infty, 0), (0, 1), (1, 6), (6, \infty)$

For $(-\infty, 0)$, choose -1.

$$2(-1)^3 - 14(-1)^2 + 12(-1) = -28 < 0$$

For $(0, 1)$, choose $\frac{1}{2}$.

$$2\left(\frac{1}{2}\right)^3 - 14\left(\frac{1}{2}\right)^2 + 12\left(\frac{1}{2}\right) = \frac{11}{4} \not< 0$$

For $(1, 6)$, choose 2.

$$2(2)^3 - 14(2)^2 + 12(2) = -16 < 0$$

For $(6, \infty)$, choose 7.

$$2(7)^3 - 14(7)^2 + 12(7) = 84 \not< 0$$

The solution is $(-\infty, 0) \cup (1, 6)$.

43. $\dfrac{m - 3}{m + 5} \le 0$

Solve $\dfrac{m - 3}{m + 5} = 0$.

$$(m + 5)\frac{m - 3}{m + 5} = (m + 5)(0)$$
$$m - 3 = 0$$
$$m = 3$$

Set the denominator equal to 0 and solve.

$$m + 5 = 0$$
$$m = -5$$

Intervals: $(-\infty, -5), (-5, 3), (3, \infty)$

For $(-\infty, -5)$, choose -6.

$$\frac{-6 - 3}{-6 + 5} = 9 \not\le 0$$

For $(-5, 3)$, choose 0.

$$\frac{0 - 3}{0 + 5} = -\frac{3}{5} \le 0$$

For $(3, \infty)$, choose 4.

$$\frac{4 - 3}{4 + 5} = \frac{1}{9} \not\le 0$$

Although the $\le$ symbol is used, including -5 in the solution would cause the denominator to be zero.

The solution is $(-5, 3]$.

45. $\dfrac{k - 1}{k + 2} > 1$

Solve $\dfrac{k - 1}{k + 2} = 1$.

$$k - 1 = k + 2$$
$$-1 \ne 2$$

The equation has no solution. Solve $k + 2 = 0$.

$$k = -2$$

Intervals: $(-\infty, -2), (-2, \infty)$

For $(-\infty, -2)$, choose -3.

$$\frac{-3 - 1}{-3 + 2} = 4 > 1$$

For $(-2, \infty)$, choose 0.

$$\frac{0 - 1}{0 + 2} = -\frac{1}{2} \not> 1$$

The solution is $(-\infty, -2)$.

47. $\dfrac{2y + 3}{y - 5} \le 1$

Solve $\dfrac{2y + 3}{y - 5} = 1$.

$$2y + 3 = y - 5$$
$$y = -8$$

Solve $y - 5 = 0$.

$$y = 5$$

Intervals: $(-\infty, -8), (-8, 5), (5, \infty)$

For $(-\infty, -8)$, choose $y = -10$.

$$\frac{2(-10) + 3}{-10 - 5} = \frac{17}{15} \not\le 1$$

For $(-8, 5)$, choose $y = 0$.

$$\frac{2(0) + 3}{0 - 5} = -\frac{3}{5} \le 1$$

For $(5, \infty)$, choose $y = 6$.

$$\frac{2(6) + 3}{6 - 5} = \frac{15}{1} \not\le 1$$

The solution is $[-8, 5)$.

49. $\dfrac{2k}{k - 3} \le \dfrac{4}{k - 3}$

Solve $\dfrac{2k}{k - 3} = \dfrac{4}{k - 3}$.

$$\frac{2k}{k - 3} = \frac{4}{k - 3}$$

$$\frac{2k}{k - 3} - \frac{4}{k - 3} = 0$$

$$\frac{2k - 4}{k - 3} = 0$$

$$2k - 4 = 0$$

$$k = 2$$

Set the denominator equal to 0 and solve for k.

$$k - 3 = 0$$
$$k = 3$$

Intervals: $(-\infty, 2), (2, 3), (3, \infty)$

For $(-\infty, 2)$, choose 0.

$$\frac{2(0)}{0 - 3} = 0 \quad \text{and} \quad \frac{4}{0 - 3} = -\frac{4}{3}, \text{ so}$$

$$\frac{2(0)}{0 - 3} \not\le \frac{4}{0 - 3}.$$

For $(2, 3)$, choose $\frac{5}{2}$.

$$\frac{2\left(\frac{5}{2}\right)}{\frac{5}{2} - 3} = \frac{5}{-\frac{1}{2}} = -10$$

and $\dfrac{4}{\frac{5}{2} - 3} = \dfrac{4}{-\frac{1}{2}} = -8$, so

$$\frac{2\left(\frac{5}{2}\right)}{\frac{5}{2} - 3} \le \frac{4}{\frac{5}{2} - 3}.$$

For $(3, \infty)$, choose 4.

$$\frac{2(4)}{4 - 3} = 8 \quad \text{and} \quad \frac{4}{4 - 3} = 4, \text{ so}$$

$$\frac{2(4)}{4 - 3} \not\le \frac{4}{4 - 3}.$$

The solution is $[2, 3)$.

51. $\dfrac{2x}{x^2 - x - 6} \ge 0$

Solve $\dfrac{2x}{x^2 - x - 6} = 0$.

$$\frac{2x}{x^2 - x - 6} = 0$$

$$2x = 0$$

$$x = 0$$

Set the denominator equal to 0 and solve for x.

$$x^2 - x - 6 = 0$$
$$(x + 2)(x - 3) = 0$$
$$x + 2 = 0 \quad \text{or} \quad x - 3 = 0$$
$$x = -2 \quad \text{or} \qquad x = 3$$

Intervals: $(-\infty, -2), (-2, 0), (0, 3), (3, \infty)$

For $(-\infty, -2)$, choose -3.

$$\frac{2(-3)}{(-3)^2 - (-3) - 6} = -1 \not\ge 0$$

For $(-2, 0)$, choose -1.

$$\frac{2(-1)}{(-1)^2 - (-1) - 6} = \frac{1}{2} \ge 0$$

For $(0, 3)$, choose 2.

$$\frac{2(2)}{2^2 - 2 - 6} = -1 \not\ge 0$$

For $(3, \infty)$, choose 4.

$$\frac{2(4)}{4^2 - 4 - 6} = \frac{4}{3} \ge 0$$

The solution is $(-2, 0] \cup (3, \infty)$.

53. $\dfrac{z^2 + z}{z^2 - 1} \geq 3$

Solve

$$\frac{z^2 + z}{z^2 - 1} = 3.$$

$$z^2 + z = 3z^2 - 3$$

$$-2z^2 + z + 3 = 0$$

$$-1(2z^2 - z - 3) = 0$$

$$-1(z + 1)(2z - 3) = 0$$

$$z = -1 \quad \text{or} \quad z = \frac{3}{2}$$

Set $z^2 - 1 = 0$.

$$z^2 = 1$$

$$z = -1 \quad \text{or} \quad z = 1$$

Intervals: $(-\infty, -1), (-1, 1), \left(1, \frac{3}{2}\right), \left(\frac{3}{2}, \infty\right)$

For $(-\infty, -1)$, choose $x = -2$.

$$\frac{(-2)^2 + 3}{(-2)^2 - 1} = \frac{7}{3} \not\geq 3$$

For $(-1, 1)$, choose $x = 0$.

$$\frac{0^2 + 3}{0^2 - 1} = -3 \not\geq 3$$

For $\left(1, \frac{3}{2}\right)$, choose $x = \frac{3}{2}$.

$$\frac{\left(\frac{3}{2}\right)^2 + 3}{\left(\frac{3}{2}\right)^2 - 1} = \frac{21}{5} \geq 3$$

For $\left(\frac{3}{2}, \infty\right)$, choose $x = 2$.

$$\frac{2^2 + 3}{2^2 - 1} = \frac{7}{3} \not\geq 3$$

The solution is $\left(1, \frac{3}{2}\right]$.

R.6 Exponents

Your Turn 1

Simplify $\left(\dfrac{y^2 z^{-4}}{y^{-3} z^4}\right)^{-2}$.

$$\left(\frac{y^2 z^{-4}}{y^{-3} z^4}\right)^{-2} = \frac{\left(y^2 z^{-4}\right)^{-2}}{\left(y^{-3} z^4\right)^{-2}} = \frac{y^{(2)(-2)} z^{(-4)(-2)}}{y^{(-3)(-2)} z^{(4)(-2)}}$$

$$= \frac{y^{-4} z^8}{y^6 z^{-8}} = \frac{z^{8-(-8)}}{y^{6-(-4)}} = \frac{z^{16}}{y^{10}}$$

Your Turn 2

Factor $5z^{1/3} + 4z^{-2/3}$.

$$5z^{1/3} + 4z^{-2/3} = z^{-2/3}\left(5z^{(1/3)+(2/3)} + 4z^{(-2/3)+(2/3)}\right)$$

$$= z^{-2/3}(5z + 4)$$

R.6 Exercises

1. $8^{-2} = \dfrac{1}{8^2} = \dfrac{1}{64}$

3. $5^0 = 1$, by definition.

5. $-(-3)^{-2} = -\dfrac{1}{(-3)^2} = -\dfrac{1}{9}$

7. $\left(\dfrac{1}{6}\right)^{-2} = \dfrac{1}{\left(\frac{1}{6}\right)^2} = \dfrac{1}{\frac{1}{36}} = 36$

9. $\dfrac{4^{-2}}{4} = 4^{-2-1} = 4^{-3} = \dfrac{1}{4^3} = \dfrac{1}{64}$

11. $\dfrac{10^8 \cdot 10^{-10}}{10^4 \cdot 10^2}$

$$= \frac{10^{8+(-10)}}{10^{4+2}} = \frac{10^{-2}}{10^6}$$

$$= 10^{-2-6} = 10^{-8}$$

$$= \frac{1}{10^8}$$

13. $\dfrac{x^4 \cdot x^3}{x^5} = \dfrac{x^{4+3}}{x^5} = \dfrac{x^7}{x^5} = x^{7-5} = x^2$

15. $\dfrac{(4k^{-1})^2}{2k^{-5}} = \dfrac{4^2 k^{-2}}{2k^{-5}} = \dfrac{16k^{-2-(-5)}}{2}$

$$= 8k^{-2+5} = 8k^3$$

$$= 2^3 k^3$$

17. $\dfrac{3^{-1} \cdot x \cdot y^2}{x^{-4} \cdot y^5} = 3^{-1} \cdot x^{1-(-4)} \cdot y^{2-5}$

$$= 3^{-1} \cdot x^{1+4} \cdot y^{-3}$$

$$= \frac{1}{3} \cdot x^5 \cdot \frac{1}{y^3}$$

$$= \frac{x^5}{3y^3}$$

19. $\left(\dfrac{a^{-1}}{b^2}\right)^{-3} = \dfrac{(a^{-1})^{-3}}{(b^2)^{-3}} = \dfrac{a^{(-1)(-3)}}{b^{2(-3)}}$

$\qquad = \dfrac{a^3}{b^{-6}} = a^3 b^6$

21. $a^{-1} + b^{-1} = \dfrac{1}{a} + \dfrac{1}{b}$

$\qquad = \left(\dfrac{b}{b}\right)\left(\dfrac{1}{a}\right) + \left(\dfrac{a}{a}\right)\left(\dfrac{1}{b}\right)$

$\qquad = \dfrac{b}{ab} + \dfrac{a}{ab}$

$\qquad = \dfrac{b + a}{ab}$

$\qquad = \dfrac{a + b}{ab}$

23. $\dfrac{2n^{-1} - 2m^{-1}}{m + n^2} = \dfrac{\frac{2}{n} - \frac{2}{m}}{m + n^2}$

$\qquad = \dfrac{\frac{2}{n} \cdot \frac{m}{m} - \frac{2}{m} \cdot \frac{n}{n}}{(m + n^2)}$

$\qquad = \dfrac{2m - 2n}{mn(m + n^2)}$

$\qquad \text{or} \quad \dfrac{2(m - n)}{mn(m + n^2)}$

25. $(x^{-1} - y^{-1})^{-1} = \dfrac{1}{\frac{1}{x} - \frac{1}{y}}$

$\qquad = \dfrac{1}{\frac{1}{x} \cdot \frac{y}{y} - \frac{1}{y} \cdot \frac{x}{x}}$

$\qquad = \dfrac{1}{\frac{y}{xy} - \frac{x}{xy}}$

$\qquad = \dfrac{1}{\frac{y - x}{xy}}$

$\qquad = \dfrac{xy}{y - x}$

27. $121^{1/2} = (11^2)^{1/2} = 11^{2(1/2)} = 11^1 = 11$

29. $32^{2/5} = (32^{1/5})^2 = 2^2 = 4$

31. $\left(\dfrac{36}{144}\right)^{1/2} = \dfrac{36^{1/2}}{144^{1/2}} = \dfrac{6}{12} = \dfrac{1}{2}$

This can also be solved by reducing the fraction first.

$\left(\dfrac{36}{144}\right)^{1/2} = \left(\dfrac{1}{4}\right)^{1/2} = \dfrac{1^{1/2}}{4^{1/2}} = \dfrac{1}{2}$

33. $8^{-4/3} = (8^{1/3})^{-4} = 2^{-4} = \dfrac{1}{2^4} = \dfrac{1}{16}$

35. $\left(\dfrac{27}{64}\right)^{-1/3} = \dfrac{27^{-1/3}}{64^{-1/3}} = \dfrac{64^{1/3}}{27^{1/3}} = \dfrac{4}{3}$

37. $3^{2/3} \cdot 3^{4/3} = 3^{(2/3)+(4/3)} = 3^{6/3} = 3^2 = 9$

39. $\dfrac{4^{9/4} \cdot 4^{-7/4}}{4^{-10/4}} = 4^{9/4 - 7/4 - (-10/4)}$

$\qquad = 4^{12/4} = 4^3 = 64$

41. $\left(\dfrac{x^6 y^{-3}}{x^{-2} y^5}\right)^{1/2} = (x^{6-(-2)} y^{-3-5})^{1/2}$

$\qquad = (x^8 y^{-8})^{1/2}$

$\qquad = (x^8)^{1/2}(y^{-8})^{1/2}$

$\qquad = x^4 y^{-4}$

$\qquad = \dfrac{x^4}{y^4}$

43. $\dfrac{7^{-1/3} \cdot 7 r^{-3}}{7^{2/3} \cdot (r^{-2})^2} = \dfrac{7^{-1/3+1} r^{-3}}{7^{2/3} \cdot r^{-4}}$

$\qquad = 7^{-1/3 + 3/3 - 2/3} r^{-3-(-4)}$

$\qquad = 7^0 r^{-3+4} = 1 \cdot r^1 = r$

45. $\dfrac{3k^2 \cdot (4k^{-3})^{-1}}{4^{1/2} \cdot k^{7/2}}$

$\qquad = \dfrac{3k^2 \cdot 4^{-1} k^3}{2 \cdot k^{7/2}}$

$\qquad = 3 \cdot 2^{-1} \cdot 4^{-1} k^{2+3-(7/2)}$

$\qquad = \dfrac{3}{8} \cdot k^{3/2}$

$\qquad = \dfrac{3k^{3/2}}{8}$

47. $\dfrac{a^{4/3}}{a^{2/3}} \cdot \dfrac{b^{1/2}}{b^{-3/2}} = a^{4/3 - 2/3} b^{1/2 - (-3/2)}$

$\qquad = a^{2/3} b^2$

49. $\dfrac{k^{-3/5} \cdot h^{-1/3} \cdot t^{2/5}}{k^{-1/5} \cdot h^{-2/3} \cdot t^{1/5}}$

$\qquad = k^{-3/5 - (-1/5)} h^{-1/3 - (-2/3)} t^{2/5 - 1/5}$

$\qquad = k^{-3/5 + 1/5} h^{-1/3 + 2/3} t^{2/5 - 1/5}$

$\qquad = k^{-2/5} h^{1/3} t^{1/5} = \dfrac{h^{1/3} t^{1/5}}{k^{2/5}}$

51. $3x^3(x^2 + 3x)^2 - 15x(x^2 + 3x)^2$

$= 3x \cdot x^2(x^2 + 3x)^2 - 3x \cdot 5(x^2 + 3x)^2$

$= 3x(x^2 + 3x)^2(x^2 - 5)$

53. $10x^3(x^2 - 1)^{-1/2} - 5x(x^2 - 1)^{1/2}$

$= 5x \cdot 2x^2(x^2 - 1)^{-1/2} - 5x(x^2 - 1)^{-1/2}(x^2 - 1)^1$

$= 5x(x^2 - 1)^{-1/2}[2x^2 - (x^2 - 1)]$

$= 5x(x^2 - 1)^{-1/2}(x^2 + 1)$

55. $x(2x + 5)^2(x^2 - 4)^{-1/2} + 2(x^2 - 4)^{1/2}(2x + 5)$

$= (2x + 5)^2(x^2 - 4)^{-1/2}(x)$

$\quad\quad + (x^2 - 4)^1(x^2 - 4)^{-1/2}(2)(2x + 5)$

$= (2x + 5)(x^2 - 4)^{-1/2}$

$\quad\quad\quad\quad \cdot [(2x + 5)(x) + (x^2 - 4)(2)]$

$= (2x + 5)(x^2 - 4)^{-1/2} \cdot (2x^2 + 5x + 2x^2 - 8)$

$= (2x + 5)(x^2 - 4)^{-1/2}(4x^2 + 5x - 8)$

R.7 Radicals

Your Turn 1

Simplify $\sqrt{28x^9y^5}$.

$\sqrt{28x^9y^5} = \sqrt{4 \cdot x^8 \cdot y^4 \cdot 7xy}$

$= 2x^4y^2\sqrt{7xy}$

Your Turn 2

Rationalize the denominator in $\dfrac{5}{\sqrt{x} - \sqrt{y}}$.

$\dfrac{5}{\sqrt{x} - \sqrt{y}} = \dfrac{5}{\sqrt{x} - \sqrt{y}} \cdot \dfrac{\sqrt{x} + \sqrt{y}}{\sqrt{x} + \sqrt{y}}$

$= \dfrac{5\left(\sqrt{x} + \sqrt{y}\right)}{x - y}$

R.7 Exercises

1. $\sqrt[3]{125} = 5$ because $5^3 = 125$.

3. $\sqrt[5]{-3125} = -5$ because $(-5)^5 = -3125$.

5. $\sqrt{2000} = \sqrt{4 \cdot 100 \cdot 5}$

$= 2 \cdot 10\sqrt{5}$

$= 20\sqrt{5}$

7. $\sqrt{27} \cdot \sqrt{3} = \sqrt{27 \cdot 3} = \sqrt{81} = 9$

9. $7\sqrt{2} - 8\sqrt{18} + 4\sqrt{72}$

$= 7\sqrt{2} - 8\sqrt{9 \cdot 2} + 4\sqrt{36 \cdot 2}$

$= 7\sqrt{2} - 8(3)\sqrt{2} + 4(6)\sqrt{2}$

$= 7\sqrt{2} - 24\sqrt{2} + 24\sqrt{2}$

$= 7\sqrt{2}$

11. $4\sqrt{7} - \sqrt{28} + \sqrt{343}$

$= 4\sqrt{7} - \sqrt{4}\sqrt{7} + \sqrt{49}\sqrt{7}$

$= 4\sqrt{7} - 2\sqrt{7} + 7\sqrt{7}$

$= (4 - 2 + 7)\sqrt{7}$

$= 9\sqrt{7}$

13. $\sqrt[3]{2} - \sqrt[3]{16} + 2\sqrt[3]{54}$

$= \sqrt[3]{2} - (\sqrt[3]{8 \cdot 2}) + 2(\sqrt[3]{27 \cdot 2})$

$= \sqrt[3]{2} - \sqrt[3]{8}\sqrt[3]{2} + 2(\sqrt[3]{27}\sqrt[3]{2})$

$= \sqrt[3]{2} - 2\sqrt[3]{2} + 2(3\sqrt[3]{2})$

$= \sqrt[3]{2} - 2\sqrt[3]{2} + 6\sqrt[3]{2}$

$= 5\sqrt[3]{2}$

15. $\sqrt{2x^3y^2z^4} = \sqrt{x^2y^2z^4 \cdot 2x} = xyz^2\sqrt{2x}$

17. $\sqrt[3]{128x^3y^8z^9} = \sqrt[3]{64x^3y^6z^9 \cdot 2y^2}$

$= \sqrt[3]{64x^3y^6z^9}\sqrt[3]{2y^2}$

$= 4xy^2z^3\sqrt[3]{2y^2}$

19. $\sqrt{a^3b^5} - 2\sqrt{a^7b^3} + \sqrt{a^3b^9}$

$= \sqrt{a^2b^4ab} - 2\sqrt{a^6b^2ab} + \sqrt{a^2b^8ab}$

$= ab^2\sqrt{ab} - 2a^3b\sqrt{ab} + ab^4\sqrt{ab}$

$= (ab^2 - 2a^3b + ab^4)\sqrt{ab}$

$= ab\sqrt{ab}(b - 2a^2 + b^3)$

21. $\sqrt{a} \cdot \sqrt[3]{a} = a^{1/2} \cdot a^{1/3}$

$= a^{1/2 + (1/3)}$

$= a^{5/6} = \sqrt[6]{a^5}$

23. $\sqrt{16 - 8x + x^2}$

$= \sqrt{(4 - x)^2}$

$= |4 - x|$

25. $\sqrt{4 - 25z^2} = \sqrt{(2 + 5z)(2 - 5z)}$

This factorization does not produce a perfect square, so the expression $\sqrt{4 - 25z^2}$ cannot be simplified.

27. $\dfrac{5}{\sqrt{7}} = \dfrac{5}{\sqrt{7}} \cdot \dfrac{\sqrt{7}}{\sqrt{7}} = \dfrac{5\sqrt{7}}{7}$

29. $\dfrac{-3}{\sqrt{12}} = \dfrac{-3}{\sqrt{4 \cdot 3}}$

$\qquad = \dfrac{-3}{2\sqrt{3}} \cdot \dfrac{\sqrt{3}}{\sqrt{3}} = \dfrac{-3\sqrt{3}}{6} = -\dfrac{\sqrt{3}}{2}$

31. $\dfrac{3}{1 - \sqrt{2}} = \dfrac{3}{1 - \sqrt{2}} \cdot \dfrac{1 + \sqrt{2}}{1 + \sqrt{2}}$

$\qquad = \dfrac{3(1 + \sqrt{2})}{1 - 2}$

$\qquad = -3(1 + \sqrt{2})$

33. $\dfrac{6}{2 + \sqrt{2}} = \dfrac{6}{2 + \sqrt{2}} \cdot \dfrac{2 - \sqrt{2}}{2 - \sqrt{2}}$

$\qquad = \dfrac{6(2 - \sqrt{2})}{4 - 2\sqrt{2} + 2\sqrt{2} - \sqrt{4}}$

$\qquad = \dfrac{6(2 - \sqrt{2})}{4 - 2} = \dfrac{6(2 - \sqrt{2})}{2}$

$\qquad = 3(2 - \sqrt{2})$

35. $\dfrac{1}{\sqrt{r} - \sqrt{3}} = \dfrac{1}{\sqrt{r} - \sqrt{3}} \cdot \dfrac{\sqrt{r} + \sqrt{3}}{\sqrt{r} + \sqrt{3}}$

$\qquad = \dfrac{\sqrt{r} + \sqrt{3}}{r - 3}$

37. $\dfrac{y - 5}{\sqrt{y} - \sqrt{5}} = \dfrac{y - 5}{\sqrt{y} - \sqrt{5}} \cdot \dfrac{\sqrt{y} + \sqrt{5}}{\sqrt{y} + \sqrt{5}}$

$\qquad = \dfrac{(y - 5)(\sqrt{y} + \sqrt{5})}{y - 5}$

$\qquad = \sqrt{y} + \sqrt{5}$

39. $\dfrac{\sqrt{x} + \sqrt{x + 1}}{\sqrt{x} - \sqrt{x + 1}} = \dfrac{\sqrt{x} + \sqrt{x + 1}}{\sqrt{x} - \sqrt{x + 1}} \cdot \dfrac{\sqrt{x} + \sqrt{x + 1}}{\sqrt{x} + \sqrt{x + 1}}$

$\qquad = \dfrac{x + 2\sqrt{x(x + 1)} + (x + 1)}{x - (x + 1)}$

$\qquad = \dfrac{2x + 2\sqrt{x(x + 1)} + 1}{-1}$

$\qquad = -2x - 2\sqrt{x(x + 1)} - 1$

41. $\dfrac{1 + \sqrt{2}}{2} = \dfrac{\left(1 + \sqrt{2}\right)\left(1 - \sqrt{2}\right)}{2\left(1 - \sqrt{2}\right)}$

$\qquad = \dfrac{1 - 2}{2\left(1 - \sqrt{2}\right)}$

$\qquad = -\dfrac{1}{2\left(1 - \sqrt{2}\right)}$

43. $\dfrac{\sqrt{x} + \sqrt{x + 1}}{\sqrt{x} - \sqrt{x + 1}}$

$\qquad = \dfrac{\sqrt{x} + \sqrt{x + 1}}{\sqrt{x} - \sqrt{x + 1}} \cdot \dfrac{\sqrt{x} - \sqrt{x + 1}}{\sqrt{x} - \sqrt{x + 1}}$

$\qquad = \dfrac{x - (x + 1)}{x - 2\sqrt{x} \cdot \sqrt{x + 1} + (x + 1)}$

$\qquad = \dfrac{-1}{2x - 2\sqrt{x(x + 1)} + 1}$

LINEAR FUNCTIONS

1.1   Slopes and Equations of Lines

Your Turn 1

Find the slope of the line through $(1, 5)$ and $(4, 6)$.

Let $(x_1, y_1) = (1, 5)$ and $(x_2, y_2) = (4, 6)$.

$$m = \frac{6-5}{4-1} = \frac{1}{3}$$

Your Turn 2

Find the equation of the line with x-intercept -4 and y-intercept 6.

We know that $b = 6$ and that the line crosses the axes at $(-4, 0)$ and $(0, 6)$. Use these two intercepts to find the slope m.

$$m = \frac{6-0}{0-(-4)} = \frac{6}{4} = \frac{3}{2}$$

Thus the equation for the line in slope-intercept form is $y = \frac{3}{2}x + 6$.

Your Turn 3

Find the slope of the line whose equation is $8x + 3y = 5$.

Solve the equation for y.

$$8x + 3y = 5$$
$$3y = -8x + 5$$
$$y = -\frac{8}{3}x + \frac{5}{3}$$

The slope is $-8/3$.

Your Turn 4

Find the equation (in slope-intercept form) of the line through $(2, 9)$ and $(5, 3)$.

First find the slope.
$$m = \frac{3-9}{5-2} = \frac{-6}{3} = -2$$

Now use the point-slope form, with $(x_1, y_1) = (5, 3)$.

$$y - y_1 = m(x - x_1)$$
$$y - 3 = -2(x - 5)$$
$$y - 3 = -2x + 10$$
$$y = -2x + 13$$

Your Turn 5

Find (in slope-intercept form) the equation of the line that passes through the point $(4, 5)$ and is parallel to the line $3x - 6y = 7$.

First find the slope of the line $3x - 6y = 7$ by solving this equation for y.

$$3x - 6y = 7$$
$$6y = 3x - 7$$
$$y = \frac{3}{6}x - \frac{7}{6}$$
$$y = \frac{1}{2}x - \frac{7}{6}$$

Since the line we are to find is parallel to this line, it will also have slope $1/2$. Use the point-slope form with $(x_1, y_1) = (4, 5)$.

$$y - y_1 = m(x - x_1)$$
$$y - 5 = \frac{1}{2}(x - 4)$$
$$y - 5 = \frac{1}{2}x - 2$$
$$y = \frac{1}{2}x + 3$$

Your Turn 6

Find (in slope-intercept form) the equation of the line that passes through the point $(3, 2)$ and is perpendicular to the line $2x + 3y = 4$.

First find the slope of the line $2x + 3y = 4$ by solving this equation for y.

$$2x + 3y = 4$$
$$3y = -2x + 4$$
$$y = -\frac{2}{3}x + \frac{4}{3}$$

Since the line we are to find is perpendicular to a line with slope $-2/3$, , it will have slope $3/2$. (Note that $(-2/3)(3/2) = -1$.)

Use the point-slope form with $(x_1, y_1) = (3, 2)$.

$$y - y_1 = m(x - x_1)$$

$$y - 2 = \frac{3}{2}(x - 3)$$

$$y - 2 = \frac{3}{2}x - \frac{9}{2}$$

$$y = \frac{3}{2}x - \frac{5}{2}$$

1.1   Exercises

1. Find the slope of the line through $(4, 5)$ and $(-1, 2)$.

$$m = \frac{5 - 2}{4 - (-1)}$$

$$= \frac{3}{5}$$

3. Find the slope of the line through $(8, 4)$ and $(8, -7)$.

$$m = \frac{4 - (-7)}{8 - 8} = \frac{11}{0}$$

The slope is undefined; the line is vertical.

5. $y = x$

Using the slope-intercept form, $y = mx + b$, we see that the slope is 1.

7. $5x - 9y = 11$

Rewrite the equation in slope-intercept form.

$$9y = 5x - 11$$

$$y = \frac{5}{9}x - \frac{11}{9}$$

The slope is $\frac{5}{9}$.

9. $x = 5$

This is a vertical line. The slope is undefined.

11. $y = 8$

This is a horizontal line, which has a slope of 0.

13. Find the slope of a line parallel to $6x - 3y = 12$.

Rewrite the equation in slope-intercept form.

$$-3y = -6x + 12$$

$$y = 2x - 4$$

The slope is 2, so a parallel line will also have slope 2.

15. The line goes through $(1, 3)$, with slope $m = -2$. Use point-slope form.

$$y - 3 = -2(x - 1)$$

$$y = -2x + 2 + 3$$

$$y = -2x + 5$$

17. The line goes through $(-5, -7)$ with slope $m = 0$. Use point-slope form.

$$y - (-7) = 0[x - (-5)]$$

$$y + 7 = 0$$

$$y = -7$$

19. The line goes through $(4, 2)$ and $(1, 3)$. Find the slope, then use point-slope form with either of the two given points.

$$m = \frac{3 - 2}{1 - 4} = -\frac{1}{3}$$

$$y - 3 = -\frac{1}{3}(x - 1)$$

$$y = -\frac{1}{3}x + \frac{1}{3} + 3$$

$$y = -\frac{1}{3}x + \frac{10}{3}$$

21. The line goes through $\left(\frac{2}{3}, \frac{1}{2}\right)$ and $\left(\frac{1}{4}, -2\right)$.

$$m = \frac{-2 - \frac{1}{2}}{\frac{1}{4} - \frac{2}{3}} = \frac{-\frac{4}{2} - \frac{1}{2}}{\frac{3}{12} - \frac{8}{12}}$$

$$m = \frac{-\frac{5}{2}}{-\frac{5}{12}} = \frac{60}{10} = 6$$

$$y - (-2) = 6\left(x - \frac{1}{4}\right)$$

$$y + 2 = 6x - \frac{3}{2}$$

$$y = 6x - \frac{3}{2} - 2$$

$$y = 6x - \frac{3}{2} - \frac{4}{2}$$

$$y = 6x - \frac{7}{2}$$

23. The line goes through $(-8, 4)$ and $(-8, 6)$.

$$m = \frac{4 - 6}{-8 - (-8)} = \frac{-2}{0};$$

which is undefined.

This is a vertical line; the value of x is always -8. The equation of this line is $x = -8$.

25. The line has x-intercept -6 and y-intercept -3. Two points on the line are $(-6, 0)$ and $(0, -3)$. Find the slope; then use slope-intercept form.

$$m = \frac{-3 - 0}{0 - (-6)} = \frac{-3}{6} = -\frac{1}{2}$$

$$b = -3$$

$$y = -\frac{1}{2}x - 3$$
$$2y = -x - 6$$
$$x + 2y = -6$$

27. The vertical line through $(-6, 5)$ goes through the point $(-6, 0)$, so the equation is $x = -6$.

29. Write an equation of the line through $(-4, 6)$, parallel to $3x + 2y = 13$.

Rewrite the equation of the given line in slope-intercept form.
$$3x + 2y = 13$$
$$2y = -3x + 13$$
$$y = -\frac{3}{2}x + \frac{13}{2}$$

The slope is $-\frac{3}{2}$.

Use $m = -\frac{3}{2}$ and the point $(-4, 6)$ in the point-slope form.

$$y - 6 = -\frac{3}{2}[x - (-4)]$$
$$y = -\frac{3}{2}(x + 4) + 6$$
$$y = -\frac{3}{2}x - 6 + 6$$
$$y = -\frac{3}{2}x$$
$$2y = -3x$$
$$3x + 2y = 0$$

31. Write an equation of the line through $(3, -4)$, perpendicular to $x + y = 4$.

Rewrite the equation of the given line as
$$y = -x + 4.$$

The slope of this line is -1. To find the slope of a perpendicular line, solve
$$-1m = -1.$$
$$m = 1$$

Use $m = 1$ and $(3, -4)$ in the point-slope form.
$$y - (-4) = 1(x - 3)$$
$$y = x - 3 - 4$$
$$y = x - 7$$
$$x - y = 7$$

33. Write an equation of the line with y-intercept 4, perpendicular to $x + 5y = 7$.

Find the slope of the given line.

$$x + 5y = 7$$
$$5y = -x + 7$$
$$y = -\frac{1}{5}x + \frac{7}{5}$$

The slope is $-\frac{1}{5}$, so the slope of the perpendicular line will be 5. If the y-intercept is 4, then using the slope-intercept form we have
$$y = mx + b$$
$$y = 5x + 4, \quad \text{or} \quad 5x - y = -4$$

35. Do the points $(4, 3), (2, 0),$ and $(-18, -12)$ lie on the same line?

Find the slope between $(4, 3)$ and $(2, 0)$.
$$m = \frac{0 - 3}{2 - 4} = \frac{-3}{-2} = \frac{3}{2}$$

Find the slope between $(4, 3)$ and $(-18, -12)$.
$$m = \frac{-12 - 3}{-18 - 4} = \frac{-15}{-22} = \frac{15}{22}$$

Since these slopes are not the same, the points do not lie on the same line.

37. A parallelogram has 4 sides, with opposite sides parallel. The slope of the line through $(1, 3)$ and $(2, 1)$ is
$$m = \frac{3 - 1}{1 - 2}$$
$$= \frac{2}{-1}$$
$$= -2.$$

The slope of the line through $\left(-\frac{5}{2}, 2\right)$ and $\left(-\frac{7}{2}, 4\right)$ is
$$m = \frac{2 - 4}{-\frac{5}{2} - \left(-\frac{7}{2}\right)} = \frac{-2}{1} = -2.$$

Since these slopes are equal, these two sides are parallel.

The slope of the line through $\left(-\frac{7}{2}, 4\right)$ and $(1, 3)$ is
$$m = \frac{4 - 3}{-\frac{7}{2} - 1} = \frac{1}{-\frac{9}{2}} = -\frac{2}{9}.$$

Slope of the line through $\left(-\frac{5}{2}, 2\right)$ and $(2, 1)$ is
$$m = \frac{2 - 1}{-\frac{5}{2} - 2} = \frac{1}{-\frac{9}{2}} = -\frac{2}{9}.$$

Since these slopes are equal, these two sides are parallel.

Since both pairs of opposite sides are parallel, the quadrilateral is a parallelogram.

39. The line goes through $(0, 2)$ and $(-2, 0)$

$$m = \frac{2 - 0}{0 - (-2)} = \frac{2}{2} = 1$$

The correct choice is (a).

41. The line appears to go through $(0, 0)$ and $(-1, 4)$.

$$m = \frac{4 - 0}{-1 - 0} = \frac{4}{-1} = -4$$

43. (a) See the figure in the textbook.

Segment MN is drawn perpendicular to segment PQ. Recall that MQ is the length of segment MQ.

$$m_1 = \frac{\Delta y}{\Delta x} = \frac{MQ}{PQ}$$

From the diagram, we know that $PQ = 1$.

Thus, $m_1 = \frac{MQ}{1}$, so MQ has length m_1.

(b) $$m_2 = \frac{\Delta y}{\Delta x} = \frac{-QN}{PQ} = \frac{-QN}{1}$$

$$QN = -m_2$$

(c) Triangles MPQ, PNQ, and MNP are right triangles by construction. In triangles MPQ and MNP, angle M = angle M, and in the right triangles PNQ and MNP,

angle N = angle N.

Since all right angles are equal, and since triangles with two equal angles are similar, triangle MPQ is similar to triangle MNP and triangle PNQ is similar to triangle MNP.

Therefore, triangles MNQ and PNQ are similar to each other.

(d) Since corresponding sides in similar triangles are proportional,

$$MQ = k \cdot PQ \quad \text{and} \quad PQ = k \cdot QN.$$

$$\frac{MQ}{PQ} = \frac{k \cdot PQ}{k \cdot QN}$$

$$\frac{MQ}{PQ} = \frac{PQ}{QN}$$

From the diagram, we know that $PQ = 1$.

$$MQ = \frac{1}{QN}$$

From (a) and (b), $m_1 = MQ$ and $-m_2 = QN$.

Substituting, we get $m_1 = \frac{1}{-m_2}$.

Multiplying both sides by m_2, we have

$$m_1 m_2 = -1.$$

45. $y = x - 1$

Three ordered pairs that satisfy this equation are $(0, -1)$, $(1, 0)$, and $(4, 3)$. Plot these points and draw a line through them.

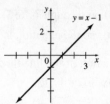

47. $y = -4x + 9$

Three ordered pairs that satisfy this equation are $(0, 9)$, $(1, 5)$, and $(2, 1)$. Plot these points and draw a line through them.

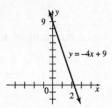

49. $2x - 3y = 12$

Find the intercepts.

If $y = 0$, then

$$2x - 3(0) = 12$$
$$2x = 12$$
$$x = 6$$

so the x-intercept is 6.

If $x = 0$, then

$$2(0) - 3y = 12$$
$$-3y = 12$$
$$y = -4$$

so the y-intercept is -4.

Plot the ordered pairs $(6, 0)$ and $(0, -4)$ and draw a line through these points. (A third point may be used as a check.)

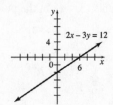

51. $3y - 7x = -21$

Find the intercepts.

If $y = 0$, then

$$3(0) + 7x = -21$$
$$-7x = -21$$
$$x = 3$$

so the x-intercept is 3.

If $x = 0$, then

$$3y - 7(0) = -21$$
$$3y = -21$$
$$y = -7$$

So the y-intercepts is -7.

Plot the ordered pairs $(3, 0)$ and $(0, -7)$ and draw a line through these points. (A third point may be used as a check.)

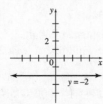

53. $y = -2$

The equation $y = -2$, or, equivalently, $y = 0x - 2$, always gives the same y-value, -2, for any value of x. The graph of this equation is the horizontal line with y-intercept -2.

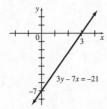

55. $x + 5 = 0$

This equation may be rewritten as $x = -5$. For any value of y, the x-value is -5. Because all ordered pairs that satisfy this equation have the same first number, this equation does not represent a function. The graph is the vertical line with x-intercept -5.

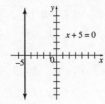

57. $y = 2x$

Three ordered pairs that satisfy this equation are $(0, 0)$, $(-2, -4)$, and $(2, 4)$. Use these points to draw the graph.

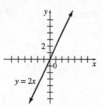

59. $x + 4y = 0$

If $y = 0$, then $x = 0$, so the x-intercept is 0. If $x = 0$, then $y = 0$, so the y-intercept is 0. Both intercepts give the same ordered pair, $(0, 0)$. To get a second point, choose some other value of x (or y). For example if $x = 4$, then

$$x + 4y = 0$$
$$4 + 4y = 0$$
$$4y = -4$$
$$y = -1,$$

giving the ordered pair $(4, -1)$. Graph the line through $(0, 0)$ and $(4, -1)$.

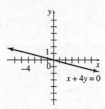

61. (a) The line goes through $(2, 27{,}000)$ and $(5, 63{,}000)$.

$$m = \frac{63{,}000 - 27{,}000}{5 - 2}$$
$$= 12{,}000$$
$$y - 27{,}000 = 12{,}000(x - 2)$$
$$y - 27{,}000 = 12{,}000x - 24{,}000$$
$$y = 12{,}000x + 3000$$

(b) Let $y = 100{,}000$; find x.

$$100{,}000 = 12{,}000x + 3000$$
$$97{,}000 = 12{,}000x$$
$$8.08 = x$$

Sales would surpass \$100,000 after 8 years, 1 month.

63. (a) The line goes through $(3, 100)$ and $(28, 215.3)$.

$$m = \frac{215.3 - 100}{28 - 3} \approx 4.612$$

Use the point $(3, 100)$ and the point-slope form.

$$y - 100 = 4.612(t - 3)$$
$$y = 4.612t - 13.836 + 100$$
$$y = 4.612t + 86.164$$

(b) The year 2000 corresponds to $t = 2000 - 1980 = 20$.

$$y = 4.612(20) + 86.164$$
$$y \approx 178.4$$

The predicted value is slightly more than the actual CPI of 172.2.

(c) The annual CPI is increasing at a rate of

65. (a) Let $x = $ age.

$$u = 0.85(220 - x) = 187 - 0.85x$$
$$l = 0.7(200 - x) = 154 - 0.7x$$

(b) $u = 187 - 0.85(20) = 170$
$l = 154 - 0.7(20) = 140$

The target heart rate zone is 140 to 170 beats per minute.

(c) $u = 187 - 0.85(40) = 153$
$l = 154 - 0.7(40) = 126$

The target heart rate zone is 126 to 153 beats per minute.

(d) $154 - 0.7x = 187 - 0.85(x + 36)$
$154 - 0.7x = 187 - 0.85x - 30.6$
$154 - 0.7x = 156.4 - 0.85x$
$0.15x = 2.4$
$x = 16$

The younger woman is 16; the older woman is $16 + 36 = 52$. $l = 0.7(220 - 16) \approx 143$ beats per minute.

67. Let $x = 0$ correspond to 1900. Then the "life expectancy from birth" line contains the points $(0, 46)$ and $(104, 77.8)$.

$$m = \frac{77.8 - 46}{104 - 0} = \frac{31.3}{102} = 0.306$$

Since $(0, 46)$ is one of the points, the line is given by the equation.

$$y = 0.306x + 46.$$

The "life expectancy from age 65" line contains the points $(0, 76)$ and $(104, 83.7)$.

$$m = \frac{83.7 - 76}{104 - 0} = \frac{7.7}{104} \approx 0.074$$

Since $(0, 76)$ is one of the points, the line is given by the equation

$$y = 0.07x + 76.$$

Set the two expressions for y equal to determine where the lines intersect. At this point, life expectancy should increase no further.

$$0.306x + 46 = 0.074x + 76$$
$$0.232x = 30$$
$$x \approx 129$$

Determine the y-value when $x = 129$. Use the first equation.

$$y = 0.306(129) + 46$$
$$= 39.474 + 46$$
$$= 85.474$$

Thus, the maximum life expectancy for humans is about 86 years.

69. (a) The line goes through $(9, 17.2)$ and $(18, 20.3)$.

$$m = \frac{20.3 - 17.2}{18 - 9} \approx 0.344$$

Use the point $(9, 17.2)$ and the point-slope form.

$$y - 17.2 = 0.344(t - 9)$$
$$y = 0.344t - 3.096 + 17.2$$
$$y = 0.344t + 14.1$$

(b) Let $y = 25$.

$$25 = 0.344t + 14.1$$
$$10.9 = 0.344t$$
$$32 \approx t$$

The percentage of adults without health insurance would be at least 25% in the year $1990 + 32 = 2022$.

71. (a) The line goes through $(50, 249{,}187)$ and $(108, 1{,}107{,}126)$.

$$m = \frac{1{,}107{,}126 - 249{,}187}{108 - 50}$$
$$\approx 14{,}792.05$$

Use the point $(50, 249{,}187)$ and the point-slope form.

$$y - 249{,}187 = 14{,}792.05(t - 50)$$
$$y = 14{,}792.05t - 739{,}602.5 + 249{,}187$$
$$y = 14{,}792.05t - 490{,}416$$

(b) The year 2015 corresponds to $t = 115$.

$$y = 14{,}792.05(115) - 490{,}416$$
$$y \approx 1{,}210{,}670$$

The number of immigrants admitted to the United States in 2015 will be about 1,210,670.

(c) The equation $y = 14,792.05t - 490,416$ has $-490,416$ for the y-intercept, indicating that the number of immigrants admitted in the year 1900 was $-490,416$. Realistically, the number of immigrants cannot be a negative value, so the equation cannot be used for valid predicted values.

73. (a) Plot the points $(15, 1600)$, $(200, 15,000)$, $(290, 24,000)$, and $(520, 40,000)$.

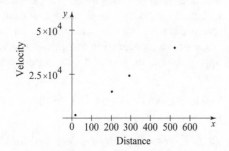

The points lie approximately on a line, so there appears to be a linear relationship between distance and time.

(b) The graph of any equation of the form $y = mx$ goes through the origin, so the line goes through $(520, 40,000)$ and $(0, 0)$.

$$m = \frac{40,000 - 0}{520 - 0} \approx 76.9$$
$$b = 0$$
$$y = 76.9x + 0$$
$$y = 76.9x$$

(c) Let $y = 60,000$; solve for x.

$$60,000 = 76.9x$$
$$780.23 \approx x$$

Hydra is about 780 megaparsecs from earth.

(d) $A = \dfrac{9.5 \times 10^{11}}{m}, m = 76.9$

$$A = \frac{9.5 \times 10^{11}}{76.9}$$
$$= 12.4 \text{ billion years}$$

75. (a)

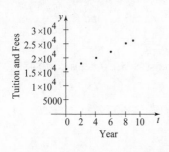

Yes, the data appear to lie roughly along a straight line.

(b) The line goes through $(0, 16,072)$ and $(9, 26,273)$.

$$m = \frac{26,273 - 16,072}{9 - 0} \approx 1133.4$$
$$b = 16,072$$
$$y = 1133.4t + 16,072$$

The slope 1133.4 indicates that tuition and fees have increased approximately $1133 per year.

(c) The year 2025 is too far in the future to rely on this equation to predict costs; too many other factors may influence these costs by then.

1.2 Linear Functions and Applications

Your Turn 1

For $g(x) = -4x + 5$, calculate $g(-5)$.

$$g(x) = -4x + 5$$
$$g(-5) = -4(-5) + 5$$
$$= 20 + 5$$
$$= 25$$

Your Turn 2

For the demand and supply functions given in Example 2, find the quantity of watermelon demanded and supplied at a price of $3.30 per watermelon.

$$p = D(q) = 9 - 0.75q$$
$$3.30 = 9 - 0.75q$$
$$0.75q = 5.7$$
$$q = \frac{5.7}{0.75} = 7.6$$

Since the quantity is in thousands, 7600 watermelon are demanded at a price of $3.30.

$$p = S(q) = 0.75q$$
$$3.30 = 0.75q$$
$$q = \frac{3.3}{0.75} = 4.4$$

Since the quantity is in thousands, 4400 watermelon are supplied at a price of $3.30.

Your Turn 3

Set the two price expressions equal and solve for the equilibrium quantity q.

$$10 - 0.85q = 0.4q$$
$$10 = 1.25q$$
$$q = \frac{10}{1.25} = 8$$

The equilibrium quantity is 8000 watermelon. Use either price expression to find the equilibrium price p.

$$p = 0.4q$$
$$p = 0.4(8) = 3.2$$

The equilibrium price is $3.20 per watermelon.

Your Turn 4

The marginal cost is the slope of the cost function $C(x)$, so this function has the form $C(x) = 15x + b$. To find b, use the fact that producing 80 batches costs $1930.

$$C(x) = 15x + b$$
$$C(80) = 15(80) + b$$
$$1930 = 1200 + b$$
$$b = 730$$

Thus the cost function is $C(x) = 15x + 730$.

Your Turn 5

The cost function is $C(x) = 35x + 250$ and the revenue function is $R(x) = 58x$. Thus the profit function is

$$P(x) = R(x) - C(x)$$
$$= 58x - (35x + 250)$$
$$= 23x - 250$$

The profit is to be $8030.

$$P(x) = 23x - 250$$
$$8030 = 23x - 250$$
$$23x = 8280$$
$$x = \frac{8280}{23} = 360$$

Sale of 360 units will produce $8030 profit.

1.2   Exercises

1. $f(2) = 7 - 5(2) = 7 - 10 = -3$

3. $f(-3) = 7 - 5(-3) = 7 + 15 = 22$

5. $g(1.5) = 2(1.5) - 3 = 3 - 3 = 0$

7. $g\left(-\frac{1}{2}\right) = 2\left(-\frac{1}{2}\right) - 3 = -1 - 3 = -4$

9. $f(t) = 7 - 5(t) = 7 - 5t$

11. This statement is true.

When we solve $y = f(x) = 0$, we are finding the value of x when $y = 0$, which is the x-intercept. When we evaluate $f(0)$, we are finding the value of y when $x = 0$, which is the y-intercept.

13. This statement is true.

Only a vertical line has an undefined slope, but a vertical line is not the graph of a function. Therefore, the slope of a linear function cannot be undefined.

15. The fixed cost is constant for a particular product and does not change as more items are made. The marginal cost is the rate of change of cost at a specific level of production and is equal to the slope of the cost function at that specific value; it approximates the cost of producing one additional item.

19. $10 is the fixed cost and $2.25 is the cost per hour.

Let $x =$ number of hours;

$R(x) =$ cost of renting a snowboard for x hours.

Thus,

$$R(x) = \text{fixed cost} + (\text{cost per hour}) \cdot (\text{number of hours})$$
$$R(x) = 10 + (2.25)(x)$$
$$= 2.25x + 10$$

21. $2 is the fixed cost and $0.75 is the cost per half-hour.

Let $x =$ the number of half-hours;

$C(x) =$ the cost of parking a car for x half-hours.

Thus,

$$C(x) = 2 + 0.75x$$
$$= 0.75x + 2$$

23. Fixed cost, $100; 50 items cost $1600 to produce.

Let $C(x) =$ cost of producing x items.

$C(x) = mx + b$, where b is the fixed cost.

$$C(x) = mx + 100$$

Now,

$C(x) = 1600$ when $x = 50$, so

$$1600 = m(50) + 100$$
$$1500 = 50m$$
$$30 = m.$$

Thus, $C(x) = 30x + 100$.

25. Marginal cost: $75; 50 items cost $4300.

$$C(x) = 75x + b$$

Now, $C(x) = 4300$ when $x = 50$.

$$4300 = 75(50) + b$$
$$550 = b$$

Thus, $C(x) = 75x + 550$.

27. $D(q) = 16 - 1.25q$

    (a) $D(0) = 16 - 1.25(0) = 16 - 0 = 16$

       When 0 watches are demanded, the price is $16.

    (b) $D(4) = 16 - 1.25(4) = 16 - 5 = 11$

       When 400 watches are demanded, the price is $11.

    (c) $D(8) = 16 - 1.25(8) = 16 - 10 = 6$

       When 800 watches are demanded, the price is $6.

    (d) Let $D(q) = 8$. Find q.

$$8 = 16 - 1.25q$$
$$\frac{5}{4}q = 8$$
$$q = 6.4$$

       When the price is $8, the number of watches demanded is 640.

    (e) Let $D(q) = 10$. Find q.

$$10 = 16 - 1.25q$$
$$\frac{5}{4}q = 6$$
$$q = 4.8$$

       When the price is $10, the number of watches demanded is 480.

    (f) Let $D(q) = 12$. Find q.

$$12 = 16 - 1.25q$$
$$\frac{5}{4}q = 4$$
$$q = 3.2$$

       When the price is $12, the number of watches demanded is 320.

    (g)

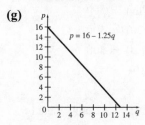

    (h) $S(q) = 0.75q$

       Let $S(q) = 0$. Find q.

$$0 = 0.75q$$
$$0 = q$$

       When the price is $0, the number of watches supplied is 0.

(i) Let $S(q) = 10$. Find q.

$$10 = 0.75q$$
$$\frac{40}{3} = q$$
$$q = 13.\overline{3}$$

When the price is $10, The number of watches supplied is about 1333.

(j) Let $S(q) = 20$. Find q.

$$20 = 0.75q$$
$$\frac{80}{3} = q$$
$$q = 26.\overline{6}$$

When the price is $20, the number of watches demanded is about 2667.

(k)

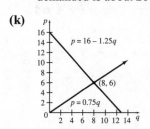

(l)
$$D(q) = S(q)$$
$$16 - 1.25q = 0.75q$$
$$16 = 2q$$
$$8 = q$$
$$S(8) = 0.75(8) = 6$$

The equilibrium quantity is 800 watches, and the equilibrium price is $6.

29. $p = S(q) = \dfrac{2}{5}q;\; p = D(q) = 100 - \dfrac{2}{5}q$

    (a)

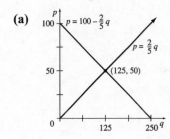

    (b) $S(q) = D(q)$

$$\frac{2}{5}q = 100 - \frac{2}{5}q$$
$$\frac{4}{5}q = 100$$
$$q = 125$$
$$S(125) = \frac{2}{5}(125) = 50$$

       The equilibrium quantity is 125, the equilibrium price is $50.

31. Use the supply function to find the equilibrium quantity that corresponds to the given equilibrium price of $4.50.

$$S(q) = p = 0.3q + 2.7$$
$$4.50 = 0.3q + 2.7$$
$$1.8 = 0.3q$$
$$6 = q$$

The line that represents the demand function goes through the given point $(2, 6.10)$ and the equilibrium point $(6, 4.50)$.

$$m = \frac{4.50 - 6.10}{6 - 2} = -0.4$$

Use point-slope form and the point $(2, 6.10)$.

$$D(q) - 6.10 = -0.4(q - 2)$$
$$D(q) = -0.4q + 0.8 + 6.10$$
$$D(q) = -0.4q + 6.9$$

33. (a) $C(x) = mx + b;\ m = 3.50;\ C(60) = 300$

$$C(x) = 3.50x + b$$

Find b.
$$300 = 3.50(60) + b$$
$$300 = 210 + b$$
$$90 = b$$
$$C(x) = 3.50x + 90$$

(b) $R(x) = 9x$
$$C(x) = R(x)$$
$$3.50x + 90 = 9x$$
$$90 = 5.5x$$
$$16.36 = x$$

Joanne must produce and sell 17 shirts.

(c) $P(x) = R(x) - C(x);\ P(x) = 500$

$$500 = 9x - (3.50x + 90)$$
$$500 = 5.5x - 90$$
$$590 = 5.5x$$
$$107.27 = x$$

To make a profit of $500, Joanne must produce and sell 108 shirts.

35. (a) Using the points $(100, 11.02)$ and $(400, 40.12)$,

$$m = \frac{40.12 - 11.02}{400 - 100}$$
$$= \frac{29.1}{300} = 0.097.$$

$$y - 11.02 = 0.097(x - 100)$$
$$y - 11.02 = 0.097x - 9.7$$
$$y = 0.097x + 1.32$$
$$C(x) = 0.097x + 1.32$$

(b) The fixed cost is given by the constant in $C(x)$. It is $1.32.

(c) $C(1000) = 0.097(1000) + 1.32$
$$= 97 + 1.32$$
$$= 98.32$$

The total cost of producing 1000 cups is $98.32.

(d) $C(1001) = 0.097(1001) + 1.32$
$$= 97.097 + 1.32$$
$$= 98.417$$

The total cost of producing 10001 cups is $98.42.

(e) Marginal cost $= 98.417 - 98.32$
$$= \$0.097 \quad \text{or} \quad 9.7\cent$$

(f) The marginal cost for *any* cup is the slope, $0.097 or 9.7¢. This means the cost of producing one additional cup of coffee would be 9.7¢.

37. $C(x) = 5x + 20;\ R(x) = 15x$

(a) $$C(x) = R(x)$$
$$5x + 20 = 15x$$
$$20 = 10x$$
$$2 = x$$

The break-even quantity is 2 units.

(b) $$P(x) = R(x) - C(x)$$
$$P(x) = 15x - (5x + 20)$$
$$P(100) = 15(100) - (5 \cdot 100 + 20)$$
$$= 1500 - 520$$
$$= 980$$

The profit from 100 units is $980.

(c) $$P(x) = 500$$
$$15x - (5x + 20) = 500$$
$$10x - 20 = 500$$
$$10x = 520$$
$$x = 52$$

For a profit of $500, 52 units must be produced.

39. $C(x) = 85x + 900$
$$R(x) = 105x$$

Set $C(x) = R(x)$ to find the break-even quantity.

$$85x + 900 = 105x$$
$$900 = 20x$$
$$45 = x$$

The break-even quantity is 45 units. You should decide not to produce since no more than 38 units can be sold.

$$P(x) = R(x) - C(x)$$
$$= 105x - (85x + 900)$$
$$= 20x - 900$$

The profit function is $P(x) = 20x - 900$.

41. $C(x) = 70x + 500$

$R(x) = 60x$

$$70x + 500 = 60x$$
$$10x = -500$$
$$x = -50$$

This represents a break-even quantity of -50 units. It is impossible to make a profit when the break-even quantity is negative. Cost will always be greater than revenue.

$$P(x) = R(x) - C(x)$$
$$= 60x - (70x + 500)$$
$$= -10x - 500$$

The profit function is $P(x) = -10x - 500$.

43. Since the fixed cost is $400, the cost function is $C(x) = mx + 100$, where m is the cost per unit. The revenue function is $R(x) = px$, where p is the price per unit.

The profit $P(x) = R(x) - C(x)$ is 0 at the given break-even quantity of 80.

$$P(x) = px - (mx + 400)$$
$$P(x) = px - mx - 400$$
$$P(x) = Mx - 400 \qquad \text{(Let } M = p - m.)$$
$$P(80) = M \cdot 80 - 400$$
$$0 = 80M - 400$$
$$400 = 80M$$
$$5 = M$$

So, the linear profit function is $P(x) = 5x - 400$, and the marginal profit is 5.

45. Use the formula derived in Example 7 in this section of the textbook.

$$F = \frac{9}{5}C + 32 \quad \text{or} \quad C = \frac{5}{9}(F - 32)$$

(a) $F = 58$; find C.

$$C = \frac{5}{9}(58 - 32)$$
$$C = \frac{5}{9}(26) = 14.4$$

The temperature is 14.4°C.

(b) $F = -20$; find C.

$$C = \frac{5}{9}(F - 32)$$
$$C = \frac{5}{9}(-20 - 32)$$
$$C = \frac{5}{9}(-52) = -28.9$$

The temperature is -28.9°C.

(c) $C = 50$; find F.

$$F = \frac{9}{5}C + 32$$
$$F = \frac{9}{5}(50) + 32$$
$$F = 90 + 32 = 122$$

The temperature is 122°F.

47. If the temperatures are numerically equal, then $F = C$.

$$F = \frac{9}{5}C + 32$$
$$C = \frac{9}{5}C + 32$$
$$-\frac{4}{5}C = 32$$
$$C = -40$$

The Celsius and Fahrenheit temperatures are numerically equal at -40°.

1.3   The Least Squares Line

Your Turn 1

Rather than recompute all the numbers in the solution table for Example 1, we record the changes to the totals. Note that we have $(90)(40.2) = 3618$, which replaces the next to last value in the xy column. Also note that we have $40.2^2 = 1616.04$, which replaces the next to last value in the y^2 column. The new totals are as follows:

$\Sigma x = 550 - 100 = 450$

$\Sigma y = 595.5 - 34.0 - 36.9 + 40.2 = 564.8$

$\Sigma xy = 28{,}135 - 3400 - 3321 + 3618 = 25{,}032$

$\Sigma x^2 = 38{,}500 - 10{,}000 = 28{,}500$

$\Sigma y^2 = 38{,}249.41 - 1156.00 - 1361.61 + 1616.04$

$\qquad = 37{,}347.84$

The number of data points n is now 9 rather than 10. Put the new column totals into the formulas for the slope and intercept.

$m = \dfrac{n\left(\Sigma xy\right) - \left(\Sigma x\right)\left(\Sigma y\right)}{n\left(\Sigma x^2\right) - \left(\Sigma x\right)^2}$

$m = \dfrac{9(25{,}032) - (450)(564.8)}{9(28{,}500) - (450)^2} \approx -0.534667$

$m \approx -0.535$

$b = \dfrac{\Sigma y - m(\Sigma x)}{n}$

$\quad = \dfrac{564.8 - (-0.534667)(450)}{9} \approx 89.5$

The least squares line is $Y = -0.535x + 89.5$.

Your Turn 2

Use the new column totals computed in Your Turn 1.

$r = \dfrac{n\left(\Sigma xy\right) - \left(\Sigma x\right)\left(\Sigma y\right)}{\sqrt{n\left(\Sigma x^2\right) - \left(\Sigma x\right)^2} \cdot \sqrt{n\left(\Sigma y^2\right) - \left(\Sigma y\right)^2}}$

$\quad = \dfrac{9(25{,}032) - (450)(564.8)}{\sqrt{9(28{,}500) - (450)^2} \cdot \sqrt{9(37{,}347.84) - (564.8)^2}}$

$\quad \approx -0.949$

1.3   Exercises

3. (a)

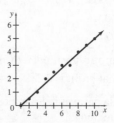

(b)

x	y	xy	x^2	y^2
1	0	0	1	0
2	0.5	1	4	0.25
3	1	3	9	1
4	2	8	16	4
5	2.5	12.5	25	6.25
6	3	18	36	9
7	3	21	49	9
8	4	32	64	16
9	4.5	40.5	81	20.25
10	5	50	100	25
55	25.5	186	385	90.75

$r = \dfrac{n(\Sigma xy) - (\Sigma x)(\Sigma y)}{\sqrt{n(\Sigma x^2) - (\Sigma x)^2} \cdot \sqrt{n(\Sigma y^2) - (\Sigma y)^2}}$

$\quad = \dfrac{10(186) - (55)(25.5)}{\sqrt{10(385) - (55)^2} \cdot \sqrt{10(90.75) - (25.5)^2}}$

$\quad \approx 0.993$

(c)   The least squares line is of the form $Y = mx + b$. First solve for m.

$m = \dfrac{n(\Sigma xy) - (\Sigma x)(\Sigma y)}{n(\Sigma x^2) - (\Sigma x)^2}$

$\quad = \dfrac{10(186) - (55)(25.5)}{10(385) - (55)^2}$

$\quad = 0.5545454545 \approx 0.555$

Now find b.

$b = \dfrac{\Sigma y - m(\Sigma x)}{n}$

$\quad = \dfrac{25.5 - 0.5545454545(55)}{10}$

$\quad = -0.5$

Thus, $Y = 0.555x - 0.5$.

(d)   Let $x = 11$. Find Y.

$Y = 0.55(11) - 0.5 = 5.6$

5.

x	y	xy	x^2	y^2
1	1	1	1	1
1	2	2	1	4
2	1	2	4	1
2	2	4	4	4
9	9	81	81	81
15	15	90	91	91

(a) $n = 5$

$$m = \frac{n(\sum xy) - (\sum x)(\sum y)}{n(\sum x^2) - (\sum x)^2}$$

$$= \frac{5(90) - (15)(15)}{5(91) - (15)^2}$$

$$= 0.9782608 \approx 0.9783$$

$$b = \frac{\sum y - m(\sum x)}{n}$$

$$= \frac{15 - (0.9782608)(15)}{5} \approx 0.0652$$

Thus, $Y = 0.9783x + 0.0652$.

$$r = \frac{n(\sum xy) - (\sum x)(\sum y)}{\sqrt{n(\sum x^2) - (\sum x)^2} \cdot \sqrt{n(\sum y^2) - (\sum y)^2}}$$

$$= \frac{5(90) - (15)(15)}{\sqrt{5(91) - (15)^2} \cdot \sqrt{5(91) - (15)^2}} \approx 0.9783$$

(b)

x	y	xy	x^2	y^2
1	1	1	1	1
1	2	2	1	4
2	1	2	4	1
2	2	4	4	4
6	6	9	10	10

$n = 4$

$$m = \frac{n(\sum xy) - (\sum x)(\sum y)}{n(\sum x^2) - (\sum x)^2}$$

$$= \frac{4(9) - (6)(6)}{4(10) - (6)^2} = 0$$

$$b = \frac{\sum y - m(\sum x)}{n} = \frac{6 - (0)(6)}{4} = 1.5$$

Thus, $Y = 0x + 1.5$, or $Y = 1.5$.

$$r = \frac{n(\sum xy) - (\sum x)(\sum y)}{\sqrt{n(\sum x^2) - (\sum x)^2} \cdot \sqrt{n(\sum y^2) - (\sum y)^2}}$$

$$= \frac{4(9) - (6)(6)}{\sqrt{4(10) - (6)^2} \cdot \sqrt{4(10) - (6)^2}}$$

$$= 0$$

(c)

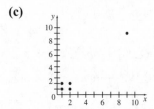

The point $(9, 9)$ is an outlier that has a strong effect on the least squares line and the correlation coefficient.

7.

x	y	xy	x^2	y^2
1	1	1	1	1
2	1	2	4	1
3	1	3	9	1
4	1.1	4.4	16	1.21
10	4.1	10.4	30	4.21

(a) $n = 4$

$$r = \frac{n(\sum xy) - (\sum x)(\sum y)}{\sqrt{n(\sum x^2) - (\sum x)^2} \cdot \sqrt{n(\sum y^2) - (\sum y)^2}}$$

$$= \frac{4(10.4) - (10)(4.1)}{\sqrt{4(30) - (10)^2} \cdot \sqrt{4(4.21) - (4.1)^2}}$$

$$= 0.7745966 \approx 0.7746$$

(b)

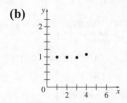

(c)   Yes; because the data points are either on or very close to the horizontal line $y = 1$, it seems that the data should have a strong linear relationship. The correlation coefficient does not describe well a linear relationship if the data points fit a horizontal line.

9. $nb + (\sum x)m = \sum y$

$(\sum x)b + (\sum x^2)m = \sum xy$

$nb + (\sum x)m = \sum y$

$nb = (\sum y) - (\sum x)m$

$b = \dfrac{\sum y - m(\sum x)}{n}$

$(\sum x)\left(\dfrac{\sum y - m(\sum x)}{n}\right) + (\sum x^2)m = \sum xy$

$(\sum x)[(\sum y) - m(\sum x)] + nm(\sum x^2) = n(\sum xy)$

$(\sum x)(\sum y) - m(\sum x)^2 + nm(\sum x^2) = n(\sum xy)$

$nm(\sum x^2) - m(\sum x)^2 = n(\sum xy) - (\sum x)(\sum y)$

$m[n(\sum x^2) - (\sum x)^2] = n(\sum xy) - (\sum x)(\sum y)$

$m = \dfrac{n(\sum xy) - (\sum x)(\sum y)}{n(\sum x^2) - (\sum x)^2}$

11. (a) $m = \dfrac{n(\sum xy) - (\sum x)(\sum y)}{n(\sum x^2) - (\sum x)^2}$

$= \dfrac{10(1810.095) - (235)(77.564)}{10(5605) - (235)^2}$

≈ -0.1534

$b = \dfrac{\sum y - m(\sum x)}{n}$

$= \dfrac{77.564 - (-0.1534)(235)}{10}$

≈ 11.36

Thus, $Y = -0.1534x + 11.36$.

(b)   The year 2020 corresponds to $x = 2020 - 1990 = 30$.

$Y = -0.1534(30) + 11.36 = 6.758$

If the trend continues linearly, there will be about 6760 banks in 2020.

(c)   Let $Y = 6$ (since Y is the number of banks in thousands) and find x.

$6 = -0.1534x + 11.36$

$-5.36 = -0.1534x$

$34.94 = x$

$35 \approx x$

The number of U.S. banks will drop below 6000 in the year $1990 + 35 = 2025$.

(d) $r = \dfrac{n(\sum xy) - (\sum x)(\sum y)}{\sqrt{n(\sum x^2) - (\sum x)^2} \cdot \sqrt{n(\sum y^2) - (\sum y)^2}}$

$= \dfrac{10(1810.095) - (235)(77.564)}{\sqrt{10(5605) - (235)^2} \cdot \sqrt{10(603.60324) - (77.564)^2}}$

≈ -0.9890

This means that the least squares line fits the data points very well. The negative sign indicates that the number of banks is decreasing as the years increase.

(e) $r = \dfrac{n(\sum xy) - (\sum x)(\sum y)}{\sqrt{n(\sum x^2) - (\sum x)^2} \cdot \sqrt{n(\sum y^2) - (\sum y)^2}}$

$= \dfrac{5(659) - (20)(126.2)}{\sqrt{5(120) - (20)^2} \cdot \sqrt{5(3784.82) - (126.2)^2}}$

≈ 0.9957

This means that the least squares line fits the data points extremely well.

13.

x	y	xy	x^2	y^2
4	2219.5	8878.0	16	4,926,180.25
5	2319.8	11,599.0	25	5,381,472.04
6	2415.0	14,490.0	36	5,832,225.00
7	2551.9	17,863.3	49	6,512,193.61
8	2592.1	20,786.8	64	6,718,982.41
30	12,098.3	73,567.1	190	29,371,053.31

(a) $m = \dfrac{n(\sum xy) - (\sum x)(\sum y)}{n(\sum x^2) - (\sum x)^2}$

$= \dfrac{5(73,567.1) - (30)(12,098.3)}{5(190) - (30)^2} \approx 97.73$

$b = \dfrac{\sum y - m(\sum x)}{n}$

$= \dfrac{12,098.3 - (97.73)(30)}{5} \approx 1833.3$

Thus, $Y = 97.73x + 1833.3$.

(b)   The total amount of consumer credit is increasing at a rate of about \$97.73 billion per year.

(c)   The year 2015 corresponds to $x = 15$.

$Y = 97.73(15) + 1833.3 = 3299.25$

If the trend continues linearly, the total amount of consumer credit will be about \$3299 billion in 2015.

(d) Let $Y = 4000$ and find x.

$$4000 = 97.73x + 1833.3$$
$$2166.7 = 97.73x$$
$$22.17 \approx x$$

The total debt will exceed \$4000 billion in the year $2000 + 23 = 2023$.

(e)
$$r = \frac{n(\sum xy) - (\sum x)(\sum y)}{\sqrt{n(\sum x^2) - (\sum x)^2} \cdot \sqrt{n(\sum y^2) - (\sum y)^2}}$$

$$= \frac{5(73,567.1) - (30)(12,098.3)}{\sqrt{5(190) - (30)^2} \cdot \sqrt{5(29,371,053.31) - (12,098.3)^2}}$$

$$\approx 0.9909$$

This means that the least squares line fits the data points extremely well.

15. (a)

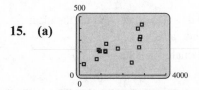

Yes, the data points lie in a linear pattern.

(b)

x	y	xy	x^2	y^2
206	95	19,570	42,436	9025
802	138	110,676	643,204	19,044
1771	228	403,788	3,136,441	51,984
1198	209	250,382	1,435,204	43,681
1238	269	333,022	1,532,644	72,361
2786	309	860,874	1,761,796	95,481
1207	202	243,814	1,456,849	40,804
892	217	193,564	795,664	47,089
2411	109	262,799	5,812,921	11,881
2885	434	1,252,090	8,323,225	188,356
2705	399	1,079,295	7,317,025	159,201
948	206	195,288	898,704	42,436
2762	239	660,118	7,628,644	57,121
2815	329	926,135	7,.924,255	108,241
24,626	3383	6,791,415	54,708,982	946,705

$$r = \frac{14(6,791,415) - (24,626)(3383)}{\sqrt{14(54,708,982) - 24,626^2} \cdot \sqrt{14(946,705) - 3383^2}}$$
$$\approx 0.693$$

There is a positive correlation between the price and the distance.

(c)
$$m = \frac{n(\sum xy) - (\sum x)(\sum y)}{n(\sum x^2) - (\sum x)^2}$$

$$m = \frac{14(6,791,415) - (24,626)(3383)}{14(54,708,982) - 24,626^2}$$

$$m = 0.0737999664 \approx 0.0738$$

$$b = \frac{\sum y - m(\sum x)}{n}$$

$$b = \frac{3383 - (0.0737999664)(24,626)}{14}$$

$$\approx 111.83$$

$$Y = 0.0738x + 111.83$$

The marginal cost is about 7.38 cents per mile.

(d) In 2000, the marginal cost was 2.43 cents per mile. It increased to 7.38 cents per mile by 2006.

(e) Phoenix is the outlier.

17. (a)

Yes, the points lie in a linear pattern.

(b) Using a calculator's STAT feature, the correlation coefficient is found to be $r \approx 0.959$. This indicates that the percentage of successful hunts does trend to increase with the size of the hunting party.

(c) $Y = 3.98x + 22.7$

19.

x	y	xy	x^2	y^2
0	17.4	0	0	302.76
4	17.7	70.8	16	313.29
8	16.9	135.2	64	285.61
12	16.2	194.4	144	262.44
16	15.8	252.8	256	249.64
40	84.0	653.2	480	1413.74

(a)
$$m = \frac{n(\sum xy) - (\sum x)(\sum y)}{n(\sum x^2) - (\sum x)^2}$$

$$= \frac{5(653.2) - (40)(84)}{5(480) - (40)^2} \approx -0.1175$$

$$b = \frac{\sum y - m(\sum x)}{n}$$

$$= \frac{84 - (-0.1175)(40)}{5} \approx 17.74$$

Thus, $Y = -0.1175x + 17.74$.

(b) The year 2020 corresponds to $x = 2020 - 1990 = 30$.

$$Y = -0.1175(30) + 17.74 = 14.215 \approx 14.2$$

If the trend continues linearly, the pupil-teacher ratio will be about 14.2 in 2020.

(c)
$$r = \frac{n(\sum xy) - (\sum x)(\sum y)}{\sqrt{n(\sum x^2) - (\sum x)^2} \cdot \sqrt{n(\sum y^2) - (\sum y)^2}}$$

$$= \frac{5(653.2) - (40)(84)}{\sqrt{5(480) - (40)^2} \cdot \sqrt{5(1413.74) - (84)^2}}$$

$$\approx -0.9326$$

The value indicates a strong linear correlation.

21.

x	y	xy	x^2	y^2
59	66	3894	3481	4356
62	71	4402	3844	5041
66	72	4752	4356	5184
68	73	4964	4624	5329
71	75	5325	5041	5625
67	63	4221	4489	3969
70	63	4410	4900	3969
71	67	4757	5041	4489
73	66	4818	5329	4356
75	66	4950	5625	4356
682	682	46,493	46,730	46,674

(a)
$$m = \frac{n(\sum xy) - (\sum x)(\sum y)}{n(\sum x^2) - (\sum x)^2}$$

$$= \frac{10(46,493) - (682)(682)}{10(46,730) - (682)^2} \approx -0.08915$$

$$b = \frac{\sum y - m(\sum x)}{n}$$

$$= \frac{682 - (-0.08915)(682)}{10} \approx 74.28$$

Thus, $Y = -0.08915x + 74.28$.

$$r = \frac{n(\sum xy) - (\sum x)(\sum y)}{\sqrt{n(\sum x^2) - (\sum x)^2} \cdot \sqrt{n(\sum y^2) - (\sum y)^2}}$$

$$= \frac{10(46,493) - (682)(682)}{\sqrt{10(46,730) - (682)^2} \cdot \sqrt{10(46,674) - (682)^2}}$$

$$\approx -0.1035$$

The taller the student, the shorter the ideal partner's height is.

(b) Data for female students:

x	y	xy	x^2	y^2
59	66	3894	3481	4356
62	71	4402	3844	5041
66	72	4752	4356	5184
68	73	4964	4624	5329
71	75	5325	5041	5625
326	357	23,337	21,346	25,535

$$m = \frac{n(\sum xy) - (\sum x)(\sum y)}{n(\sum x^2) - (\sum x)^2}$$

$$= \frac{5(23,337) - (326)(357)}{5(21,346) - (326)^2} \approx 0.6674$$

$$b = \frac{\sum y - m(\sum x)}{n}$$

$$= \frac{357 - (0.6674)(326)}{5} \approx 27.89$$

Thus, $Y = 0.6674x + 27.89$.

$$r = \frac{n(\sum xy) - (\sum x)(\sum y)}{\sqrt{n(\sum x^2) - (\sum x)^2} \cdot \sqrt{n(\sum y^2) - (\sum y)^2}}$$

$$= \frac{5(23,337) - (326)(357)}{\sqrt{5(21,346) - (326)^2} \cdot \sqrt{5(25,535) - (357)^2}}$$

$$\approx 0.9459$$

Data for male students:

x	y	xy	x^2	y^2
67	63	4221	4489	3969
70	63	4419	4900	3969
71	67	4757	5041	4489
73	66	4818	5329	4356
75	66	4950	5625	4356
356	325	23,156	25,384	21,139

$$m = \frac{n(\sum xy) - (\sum x)(\sum y)}{n(\sum x^2) - (\sum x)^2}$$

$$= \frac{5(23,156) - (356)(325)}{5(25,384) - (356)^2} \approx 0.4348$$

$$b = \frac{\sum y - m(\sum x)}{n}$$

$$= \frac{325 - (0.4348)(356)}{5} \approx 34.04$$

Thus, $Y = 0.4348x + 34.04$.

$$r = \frac{n(\sum xy) - (\sum x)(\sum y)}{\sqrt{n(\sum x^2) - (\sum x)^2} \cdot \sqrt{n(\sum y^2) - (\sum y)^2}}$$

$$= \frac{5(23,156) - (356)(325)}{\sqrt{5(25,384) - (356)^2} \cdot \sqrt{5(21,139) - (325)^2}}$$

$$\approx 0.7049$$

(c)

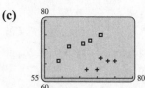

There is no linear relationship among all 10 data pairs. However, there is a linear relationship among the first five data pairs (female students) and a separate linear relationship among the second five data pairs (male students).

23. (a)

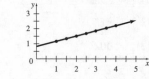

(b)

L	T	LT	L^2	T^2
1.0	1.11	1.11	1	1.2321
1.5	1.36	2.04	2.25	1.8496
2.0	1.57	3.14	4	2.4649
2.5	1.76	4.4	6.25	3.0976
3.0	1.92	5.76	9	3.6864
3.5	2.08	7.28	12.25	4.3264
4.0	2.22	8.88	16	4.9844
17.5	12.02	32.61	50.75	21.5854

$$m = \frac{n(\sum xy) - (\sum x)(\sum y)}{n(\sum x^2) - (\sum x)^2}$$

$$m = \frac{7(32.61) - (17.5)(12.02)}{7(50.75) - 17.5^2}$$

$$m = 0.3657142857$$

$$\approx 0.366$$

$$b = \frac{\sum T - m(\sum L)}{n}$$

$$b = \frac{12.02 - 0.3657142857(17.5)}{7}$$

$$\approx 0.803$$

$$Y = 0.366x + 0.803$$

The line seems to fit the data.

(c)

$$r = \frac{7(32.61) - (17.5)(12.02)}{\sqrt{7(50.75) - 17.5^2} \cdot \sqrt{7(21.5854) - 12.02^2}}$$

$$= 0.995,$$

which is a good fit and confirms the conclusion in part (b).

25. (a)

$$r = \frac{10(399.16) - (500)(20.668)}{\sqrt{10(33,250) - 500^2} \cdot \sqrt{10(91.927042) - (20.668)^2}}$$

$$= -0.995$$

Yes, there does appear to be a linear correlation.

(b) $$m = \frac{n(\sum xy) - (\sum x)(\sum y)}{n(\sum x^2) - (\sum x)^2}$$

$$m = \frac{10(399.16) - (500)(20.668)}{10(33,250) - 500^2}$$

$$m = -0.0768775758 \approx -0.0769$$

$$b = \frac{\sum y - m(\sum x)}{n}$$

$$b = \frac{20.668 - (-0.0768775758)(500)}{10}$$

$$\approx 5.91$$

$$Y = -0.0769x + 5.91$$

(c) Let $x = 50$

$$Y = -0.0769(50) + 5.91 \approx 2.07$$

The predicted number of points expected when a team is at the 50 yard line is 2.07 points.

27.

x	y
0.00	0.0
2.3167	11.5
3.7167	18.9
5.6000	27.8
7.0833	32.8
7.5000	36.0
8.5000	43.9
10.6000	51.5
11.9333	58.4
15.2333	71.8
17.8167	80.9
18.9667	85.2
20.8333	91.3
23.3833	100.5
153.4833	710.5

(a) Skaggs' average speed was $100.5/23.3833 \approx 4.298$ miles per hour.

(b)

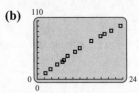

The data appear to lie approximately on a straight line.

(c) Using a graphing calculator,

$$Y = 4.317x + 3.419.$$

(d) Using a graphing calculator,

$$r \approx 0.9971$$

Yes, the least squares line is a very good fit to the data.

(e) A good value for Skaggs' average speed would be the slope of the least squares line, or

$$m = 4.317 \text{ miles per hour.}$$

This value is faster than the average speed found in part (a). The value 4.317 miles per hour is most likely the better value because it takes into account all 14 data pairs.

Chapter 1 Review Exercises

1. False; a line can have only one slant, so its slope is unique.

2. False; the equation $y = 3x + 4$ has slope 3.

3. True; the point $(3, -1)$ is on the line because $-1 = -2(3) + 5$ is a true statement.

4. False; the points $(2, 3)$ and $(2, 5)$ do not have the same y-coordinate.

5. True; the points $(4, 6)$ and $(5, 6)$ do have the same y-coordinate.

6. False; the x-intercept of the line $y = 8x + 9$ is $-\frac{9}{8}$.

7. True; $f(x) = \pi x + 4$ is a linear function because it is in the form $y = mx + b$, where m and b are real numbers.

8. False; $f(x) = 2x^2 + 3$ is not linear function because it isn't in the form $y = mx + b$, and it is a second-degree equation.

9. False; the line $y = 3x + 17$ has slope 3, and the line $y = -3x + 8$ has slope -3. Since $3 \cdot -3 \neq -1$, the lines cannot be perpendicular.

10. False; the line $4x + 3y = 8$ has slope $-\frac{4}{3}$, and the line $4x + y = 5$ has slope -4. Since the slopes are not equal, the lines cannot be parallel.

11. False; a correlation coefficient of zero indicates that there is no linear relationship among the data.

12. True; a correlation coefficient always will be a value between -1 and 1.

13. Marginal cost is the rate of change of the cost function; the fixed cost is the initial expenses before production begins.

15. Through $(-3, 7)$ and $(2, 12)$

$$m = \frac{12 - 7}{2 - (-3)} = \frac{5}{5} = 1$$

17. Through the origin and $(11, -2)$

$$m = \frac{-2 - 0}{11 - 0} = -\frac{2}{11}$$

19. $4x + 3y = 6$

$$3y = -4x + 6$$

$$y = -\frac{4}{3}x + 2$$

Therefore, the slope is $m = -\frac{4}{3}$.

21. $y + 4 = 9$

$$y = 5$$

$$y = 0x + 5$$

$$m = 0$$

23. $y = 5x + 4$

$$m = 5$$

25. Through $(5, -1)$; slope $\frac{2}{3}$

Use point-slope form.

$$y - (-1) = \frac{2}{3}(x - 5)$$

$$y + 1 = \frac{2}{3}(x - 5)$$

$$3(y + 1) = 2(x - 5)$$

$$3y + 3 = 2x - 10$$

$$3y = 2x - 13$$

$$y = \frac{2}{3}x - \frac{13}{3}$$

27. Through $(-6, 3)$ and $(2, -5)$

$$m = \frac{-5 - 3}{2 - (-6)} = \frac{-8}{8} = -1$$

Use point-slope form.

$$y - 3 = -1[x - (-6)]$$

$$y - 3 = -x - 6$$

$$y = -x - 3$$

29. Through $(2, -10)$, perpendicular to a line with undefined slope

A line with undefined slope is a vertical line. A line perpendicular to a vertical line is a horizontal line with equation of the form $y = k$. The desired line passed through $(2, -10)$, so $k = -10$. Thus, an equation of the desired line is $y = -10$.

31. Through $(3, -4)$ parallel to $4x - 2y = 9$

Solve $4x - 2y = 9$ for y.

$$-2y = -4x + 9$$

$$y = 2x - \frac{9}{2} \text{ so } m = 2$$

The desired line has the same slope. Use the point-slope form.

$$y - (-4) = 2(x - 3)$$

$$y + 4 = 2x - 6$$

$$y = 2x - 10$$

Rearrange.

$$2x - y = 10$$

33. Through $(-1, 4)$; undefined slope

Undefined slope means the line is vertical.

The equation of the vertical line through $(-1, 4)$ is $x = -1$.

35. Through $(3, -5)$, parallel to $y = 4$

Find the slope of the given line.

$y = 0x + 4$, so $m = 0$, and the required line will also have slope 0.

Use the point-slope from.

$$y - (-5) = 0(x - 3)$$

$$y + 5 = 0$$

$$y = -5$$

37. $y = 4x + 3$

Let $x = 0$: $\quad y = 4(0) + 3$

$$y = 3$$

Let $y = 0$: $\quad 0 = 4x + 3$

$$-3 = 4x$$

$$-\frac{3}{4} = x$$

Draw the line through $(0, 3)$ and $\left(-\frac{3}{4}, 0\right)$.

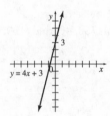

39. $3x - 5y = 15$

$$-5y = -3x + 15$$

$$y = \frac{3}{5}x - 3$$

When $x = 0$, $y = -3$; when $y = 0$, $x = 5$.

Draw the line through $(0, -3)$ and $(5, 0)$.

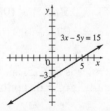

41. $x - 3 = 0$

   $\ x = 3$

   This is the vertical line through $(3, 0)$.

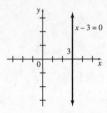

43. $y = 2x$

   When $x = 0$, $y = 0$.

   When $x = 1$, $y = 2$.

   Draw the line through $(0, 0)$ and $(1, 2)$.

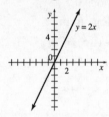

45. **(a)** $E = 352 + 42x$ (where x is in thousands)

   (b) $R = 130x$ (where x is in thousands)

   (c) $\quad R > E$

   $\quad 130x > 352 + 42x$

   $\quad\ \ 88x > 352$

   $\qquad\ x > 4$

   For a profit to be made, more than 4000 chips must be sold.

47. Using the points $(60, 40)$ and $(100, 60)$,

   $$m = \frac{60 - 40}{100 - 60} = \frac{20}{40} = 0.5.$$

   $$p - 40 = 0.5(q - 60)$$

   $$p - 40 = 0.5q - 30$$

   $$p = 0.5q + 10$$

   $$S(q) = 0.5q + 10$$

49. $S(q) = D(q)$

   $$0.5q + 10 = -0.5q + 72.50$$

   $$q = 62.5$$

   $$S(62.5) = 0.5(62.5) + 10$$

   $$= 31.25 + 10$$

   $$= 41.25$$

   The equilibrium price is \$41.25, and the equilibrium quantity is 62.5 diet pills.

51. Fixed cost is \$2000; 36 units cost \$8480.

   Two points on the line are $(0, 2000)$ and $(36, 8480)$, so

   $$m = \frac{8480 - 2000}{36 - 0} = \frac{6480}{36} = 180.$$

   Use point-slope form.

   $$y = 180x + 2000$$

   $$C(x) = 180x + 2000$$

53. Thirty units cost \$1500; 120 units cost \$5640. Two points on the line are $(30, 1500)$, $(120, 5640)$, so

   $$m = \frac{5640 - 1500}{120 - 30} = \frac{4140}{90} = 46.$$

   Use point-slope form.

   $$y - 1500 = 46(x - 30)$$

   $$y = 46x - 1380 + 1500$$

   $$y = 46x + 120$$

   $$C(x) = 46x + 120$$

55. **(a)** $C(x) = 3x + 160$; $R(x) = 7x$

   $$C(x) = R(x)$$

   $$3x + 160 = 7x$$

   $$160 = 4x$$

   $$40x = x$$

   The break-even quantity is 40 pounds.

   (b) $R(40) = 7 \cdot 40 = \$280$

   The revenue for 40 pounds is \$280.

57. Let y represent imports to China in billions of dollars. Using the points $(1, 19.1)$ and $(8, 69.7)$

   $$m = \frac{69.7 - 19.1}{8 - 1} = \frac{50.6}{7} \approx 7.23$$

   $$y - 19.1 = 7.23(t - 1)$$

   $$y - 19.1 = 7.23t - 7.23$$

   $$y = 7.23t + 11.9$$

59.

x	y
0	7500
5	12,000
10	16,000
15	20,450
20	24,900
25	28,400

   (a) Use the points $(0, 7500)$ and $(25, 28{,}400)$ to find the slope.

$$m = \frac{28{,}400 - 7500}{25 - 0} = 836$$

$$b = 7500$$

The linear equation for the average new car cost since 1980 is $y = 836x + 7500$.

(b) Use the points $(15, 20{,}450)$ and $(25, 28{,}400)$ to find the slope.

$$m = \frac{28{,}400 - 20{,}450}{25 - 15} = 795$$

$$y - 20{,}450 = 795(x - 15)$$

$$y - 20{,}450 = 795x - 11{,}925$$

$$y = 8525$$

The linear equation for the average new car cost since 1980 is $y = 795x + 8525$.

(c) Using a graphing calculator, the least square line is $Y = 843.7x + 7662$.

(d)

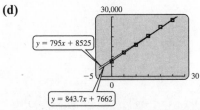

(e) The least squares lines best describes the data. Since the data seems to fit a straight line, a linear model describes the data very well.

(f) Using a graphing calculator, $r \approx 0.9995$.

61. (a)

x	y	xy	x^2	y^2
130	170	22,100	16,900	28,900
138	160	22,080	19,044	25,600
142	173	24,566	20,164	29,929
159	181	28,779	25,281	32,761
165	201	33,165	27,225	40,401
200	192	38,400	40,000	36,864
210	240	50,400	44,100	57,600
250	290	72,500	62,500	84,100
1394	1607	291,990	255,214	336,155

$$m = \frac{n(\sum xy) - (\sum x)(\sum y)}{n(\sum x^2) - (\sum x)^2}$$

$$m = \frac{8(291{,}990) - (1394)(1607)}{8(225{,}214) - 1394^2}$$

$$m = 0.9724399854 \approx 0.9724$$

$$b = \frac{\sum y - m(\sum x)}{n}$$

$$b = \frac{1607 - 0.9724(1394)}{8} \approx 31.43$$

$$Y = 0.9724x + 31.43$$

(b) Let $x = 190$; find Y.

$$Y = 0.9724(190) + 31.43$$

$$Y = 216.19 \approx 216$$

The cholesterol level for a person whose blood sugar level is 190 would be about 216.

(c) $r = \dfrac{8(291{,}990) - (1394)(1607)}{\sqrt{8(255{,}214) - 1394^2} \cdot \sqrt{8(336{,}155) - 1607^2}}$

$$= 0.933814 \approx 0.93$$

63. Using the points $(5, 55)$ and $(19, 72.1)$,

$$m = \frac{72.1 - 55}{19 - 5} = \frac{17.1}{14} \approx 1.22$$

$$y - 55 = 1.22(t - 5)$$

$$y - 55 = 1.22t - 6.1$$

$$y = 1.22t + 48.9$$

65. (a) Using a graphing calculator, $r = 0.6998$. The data seem to fit a line but the fit is not very good.

(b)

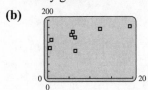

(c) Using a graphing calculator,

$$Y = 3.396x + 117.2$$

(d) The slope is 3.396 thousand (or 3396). On average, the governor's salary increases $3396 for each additional million in population.

Chapter 2

SYSTEMS OF LINEAR EQUATIONS AND MATRICES

2.1 Solution of Linear Systems by the Echelon Method

Your Turn 1

$$2x + 3y = 12 \quad (1)$$
$$3x - 4y = 1 \quad (2)$$

Use row transformations to eliminate x in equation (2).

$$2x + 3y = 12$$
$$3R_1 + (-2)R_2 \rightarrow R_2 \qquad 17y = 34$$

Now make the coefficient of the first term in each equation equal to 1.

$$\tfrac{1}{2}R_1 \rightarrow R_1 \qquad x + \frac{3}{2}y = 6$$
$$\tfrac{1}{17}R_2 \rightarrow R_2 \qquad y = 2$$

Back-substitute to solve for x.

$$x + \frac{3}{2}(2) = 6$$
$$x + 3 = 6$$
$$x = 3$$

The solution of the system is $(3, 2)$.

Your Turn 2

Let $x =$ flight time eastward

$y =$ difference in time zones

$$x + y = 16 \qquad (1)$$
$$x - y = 2 \qquad (2) \text{ (no wind)}$$

$$x + y = 16$$
$$R_1 + (-1)R_2 \rightarrow R_2 \qquad 2y = 14$$
$$y = 7$$
$$x + 7 = 16$$
$$x = 9$$

The flight time eastward is 9 hours and the difference in time zones is 7 hours.

Your Turn 3

Since the permissible values of z are 5, 6, 7, . . . , 16, the largest value of z is $z = 16$.

$$x = 2z - 9 = 2(16) - 9 = 23$$
$$y = 49 - 3z = 49 - 3(16) = 1$$

The solution with the largest number of spoons is 23 knives, 1 fork, and 16 spoons.

2.1 Exercises

In Exercises 1–16 and 19–28, check each solution by substituting it in the original equations of the system.

1.
$$x + y = 5 \quad (1)$$
$$2x - 2y = 2 \quad (2)$$

To eliminate x in equation (2), multiply equation (1) by -2 and add the result to equation (2). The new system is

$$x + y = 5 \quad (1)$$
$$-2R_1 + R_2 \rightarrow R_2 \qquad -4y = -8. \quad (3)$$

Now make the coefficient of the first term in each row equal 1. To accomplish this, multiply equation (3) by $-\tfrac{1}{4}$.

$$x + y = 5 \quad (1)$$
$$-\tfrac{1}{4}R_2 \rightarrow R_2 \qquad y = 2 \quad (4)$$

Substitute 2 for y in equation (1).

$$x + 2 = 5$$
$$y = 3$$

The solution is $(3, 2)$.

3.
$$3x - 2y = -3 \quad (1)$$
$$5x - y = 2 \quad (2)$$

To eliminate x in equation (2), multiply equation (1) by -5 and equation (2) by 3. Add the results. The new system is

$$3x - 2y = -3 \quad (1)$$
$$-5R_1 + 3R_2 \rightarrow R_2 \qquad 7y = 21. \quad (3)$$

Now make the coefficient of the first term in each row equal 1. To accomplish this, multiply equation (1) by $\tfrac{1}{3}$ and equation (3) by $\tfrac{1}{7}$.

$$\tfrac{1}{3}R_1 \rightarrow R_1 \qquad x - \frac{2}{3}y = -1 \quad (4)$$
$$\tfrac{1}{7}R_2 \rightarrow R_2 \qquad y = 3 \quad (5)$$

Back-substitution of 3 for y in equation (4) gives

$$x - \frac{2}{3}(3) = -1$$
$$x - 2 = -1$$
$$x = 1.$$

The solution is $(1, 3)$.

5. $3x + 2y = -6$ (*1*)

$5x - 2y = -10$ (*2*)

Eliminate x in equation (2) to get the system

$$3x + 2y = -6 \quad (1)$$

$5R_1 + (-3)R_2 \to R_2 \qquad 16y = 0. \quad (3)$

Make the coefficient of the first term in each equation equal 1.

$$\tfrac{1}{3}R_1 \to R_1 \quad x + \tfrac{2}{3}y = -2 \quad (4)$$

$$\tfrac{1}{16}R_2 \to R_2 \qquad y = 0 \quad (5)$$

Substitute 0 for y in equation (4) to get $x = -2$. The solution is $(-2, 0)$.

7. $6x - 2y = -4$ (*1*)

$3x + 4y = 8$ (*2*)

Eliminate x in equation (2).

$$6x - 2y = -4 \quad (1)$$

$-1R_1 + 2R_2 \to R_2 \qquad 10y = 20 \quad (3)$

Make the coefficient of the first term in each row equal 1.

$$\tfrac{1}{6}R_1 \to R_1 \quad x - \tfrac{1}{3}y = -\tfrac{2}{3} \quad (4)$$

$$\tfrac{1}{10}R_2 \to R_2 \qquad y = 2 \quad (5)$$

Substitute 2 for y in equation (4) to get $x = 0$. The solution is $(0, 2)$.

9. $5p + 11q = -7$ (*1*)

$3p - 8q = 25$ (*2*)

Eliminate p in equation (2).

$$5p + 11q = -7 \quad (1)$$

$-3R_1 + 5R_2 \to R_2 \qquad -73q = 146 \quad (3)$

Make the coefficient of the first term in each row equal 1.

$$\tfrac{1}{5}R_1 \to R_1 \quad p + \tfrac{11}{5}q = -\tfrac{7}{5} \quad (4)$$

$$-\tfrac{1}{73}R_2 \to R_2 \qquad q = -1 \quad (5)$$

Substitute -2 for q in equation (4) to get $p = 3$. The solution is $(3, -2)$.

11. $6x + 7y = -2$ (*1*)

$7x - 6y = 26$ (*2*)

Eliminate x in equation (2).

$$6x + 7y = -2 \quad (1)$$

$7R_1 + (-6)R_2 \to R_2 \qquad 85y = -170 \quad (3)$

Make the coefficient of the first term in each equation equal 1.

$$\tfrac{1}{6}R_1 \to R_1 \quad x + \tfrac{7}{6}y = -\tfrac{1}{3} \quad (4)$$

$$\tfrac{1}{85}R_2 \to R_2 \qquad y = -2 \quad (5)$$

Substitute -2 for y in equation (4) to get $x = 2$. The solution is $(2, -2)$.

13. $3x + 2y = 5$ (*1*)

$6x + 4y = 8$ (*2*)

Eliminate x in equation (2).

$$3x + 2y = 5 \quad (1)$$

$-2R_1 + R_2 \to R_2 \qquad 0 = -2 \quad (3)$

Equation (3) is a false statement.

The system is inconsistent and has no solution.

15. $3x - 2y = -4$ (*1*)

$-6x + 4y = 8$ (*2*)

Eliminate x in equation (2).

$$3x - 2y = -4 \quad (1)$$

$2R_1 + R_2 \to R_2 \qquad 0 = 0 \quad (3)$

The true statement in equation (3) indicates that there are an infinite number of solutions for the system. Solve equation (1) for x.

$$3x - 2y = -4 \qquad (1)$$
$$3x = 2y - 4$$
$$x = \frac{2y - 4}{3} \quad (4)$$

For each value of y, equation (4) indicates that $x = \frac{2y-4}{3}$, and all ordered pairs of the form $\left(\frac{2y-4}{3}, y\right)$ are solutions.

17. $x - \dfrac{3y}{2} = \dfrac{5}{2}$ (*1*)

$\dfrac{4x}{3} + \dfrac{2y}{3} = 6$ (*2*)

Rewrite the equations without fractions.

$2R_1 \to R_1 \quad 2x - 3y = 5 \quad (3)$

$3R_2 \to R_2 \quad 4x + 2y = 18 \quad (4)$

Eliminate x in equation (4).

$$2x - 3y = 5 \quad (3)$$

$-2R_1 + R_2 \to R_2 \qquad 8y = 8 \quad (5)$

Make the coefficient of the first term in each equation equal 1.

$$\tfrac{1}{2}R_1 \to R_1 \quad x - \tfrac{3}{2}y = \tfrac{5}{2} \quad (6)$$

$$\tfrac{1}{8}R_2 \to R_2 \qquad y = 1 \quad (7)$$

Substitute 1 for y in equation (6) to get $x = 4$. The solution is $(4, 1)$.

19. $\dfrac{x}{2} + y = \dfrac{3}{2}$ (1)

$\dfrac{x}{3} + y = \dfrac{1}{3}$ (2)

Rewrite the equations without fractions.

$2R_1 \rightarrow R_1 \quad x + 2y = 3 \quad (3)$

$3R_2 \rightarrow R_2 \quad x + 3y = 1 \quad (4)$

Eliminate x in equation (4).

$x + 2y = \quad 3 \quad (3)$

$-1R_1 + R_2 \rightarrow R_2 \quad\quad y = -2 \quad (5)$

Substitute -2 for y in equation (3) to get $x = 7$.
The solution is $(7, -2)$.

21. An inconsistent system has *no* solutions.

23. $x + 2y + 3z = 90 \quad (1)$

$3y + 4z = 36 \quad (2)$

Let z be the parameter and solve equation (2) for y in terms of z.

$3y + 4z = 36$

$3y = 36 - 4z$

$y = 12 - \dfrac{4}{3}z$

Substitute this expression for y in equation (1) to solve for x in terms of z.

$x + 2\left(12 - \dfrac{4}{3}z\right) + 3z = 90$

$x + 24 - \dfrac{8}{3}z + 3z = 90$

$x + \dfrac{1}{3}z = 66$

$x = 66 - \dfrac{1}{3}z$

Thus the solutions are $\left(66 - \dfrac{1}{3}z, 12 - \dfrac{4}{3}z, z\right)$,

where z is any real number. Since the solutions have to be nonnegative integers, set $66 - \dfrac{1}{3}z \geq 0$.
Solving for z gives $z \leq 198$.

Since y must be nonnegative, we have $12 - \dfrac{4}{3}z \geq 0$. Solving for z gives $z \leq 9$.

Since z must be a multiple of 3 for x and y to be integers, the permissible values of z are 0, 3, 6, and 9, which gives 4 solutions.

25. $3x + 2y + 4z = 80 \quad (1)$

$y - 3z = 10 \quad (2)$

Let z be the parameter and solve equation (2) for y in terms of z.

$y - 3z = 10$

$y = 3z + 10$

Substitute this expression for y in equation (1) to solve for x in terms of z.

$3x + 2(3z + 10) + 4z = 80$

$3x + 6z + 20 + 4z = 80$

$3x + 10z = 60$

$3x = 60 - 10z$

$x = 20 - \dfrac{10}{3}z$

Thus the solutions are $\left(20 - \dfrac{10}{3}z, 3z + 10, z\right)$,

where z is any real number. Since the solutions have to be nonnegative integers, set
$20 - \dfrac{10}{3}z \geq 0$. Solving for z gives $z \leq 6$.

Since y must be nonnegative, we have
$3z + 10 \geq 0$. Solving for z gives $z \geq -10\frac{1}{3}$.

Since z must be a multiple of 3 for x to be an integer, the permissible values of z are 0, 3, and, 6, which gives 3 solutions.

29. $2x + 3y - z = 1 \quad (1)$

$3x + 5y + z = 3 \quad (2)$

Eliminate x in equation (2).

$2x + 3y - \quad z = 1 \quad (1)$

$-3R_1 + 2R_2 \rightarrow R_2 \quad\quad y + 5z = 3 \quad (3)$

Since there are only two equations, it is not possible to continue with the echelon method as in the previous exercises involving systems with three equations and three variables. To complete the solution, make the coefficient of the first term in the each equation equal 1.

$\frac{1}{2}R_1 \rightarrow R_1 \quad x + \dfrac{3}{2}y + \dfrac{1}{2}z = \dfrac{1}{2} \quad (4)$

$y + 5z = 3 \quad (3)$

Solve equation (3) for y in terms of the parameter z.

$y + 5z = 3$

$y = 3 - 5z$

Substitute this expression for y in equation (4) to solve for x in terms of the parameter z.

$x + \dfrac{3}{2}(3 - 5z) - \dfrac{1}{2}z = \dfrac{1}{2}$

$x + \dfrac{9}{2} - \dfrac{15}{2}z - \dfrac{1}{2}z = \dfrac{1}{2}$

$x - 8z = -4$

$x = 8z - 4$

The solution is $(8z - 4, 3 - 5z, z)$.

31. $x + 2y + 3z = 11$ (1)

     $2x - y + z = 2$ (2)

$$-2R_1 + R_2 \rightarrow R_2 \quad \begin{array}{l} x + 2y + 3z = 11 \quad (1) \\ -5y - 5z = -20 \quad (3) \end{array}$$

$$-\tfrac{1}{5}R_2 \rightarrow R_2 \quad \begin{array}{l} x + 2y + 3z = 11 \quad (1) \\ y + z = 4 \quad (4) \end{array}$$

Since there are only two equations, it is not possible to continue with the echelon method. To complete the solution, solve equation (4) for y in terms of the parameter z.

$$y = 4 - z$$

Now substitute $4 - z$ for y in equation (1) and solve for x in terms of z.

$$x + 2(4 - z) + 3z = 11$$
$$x + 8 - 2z + 3z = 11$$
$$x = 3 - z$$

The solution is $(3 - z, \, 4 - z, \, z)$.

33. $nb + (\sum x)m = \sum y$ (1)

     $(\sum x)b + (\sum x^2)m = \sum xy$ (2)

Multiply equation (1) by $\frac{1}{n}$.

$$b + \frac{\sum x}{n}m = \frac{\sum y}{n} \quad (3)$$

$$(\sum x)b + (\sum x^2)m = \sum xy \quad (2)$$

Eliminate b from equation (2).

$$b + \frac{\sum x}{n}m = \frac{\sum y}{n} \quad (3)$$

$(-\sum x)R_1 + R_2 \rightarrow R_2$

$$\left[-\frac{(\sum x)^2}{n} + \sum x^2 \right]m = \frac{-(\sum x)(\sum y)}{n} + \sum xy \quad (4)$$

Multiply equation (4) by $\dfrac{1}{-\dfrac{(\sum x)^2}{n} + \sum x^2}$.

$$b + \frac{\sum x}{n}m = \frac{\sum y}{n} \quad (3)$$

$$m = \left[\frac{-(\sum x)(\sum y)}{n} + \sum xy \right]\left[\frac{1}{-\dfrac{(\sum x)^2}{n} + \sum x^2} \right] \quad (5)$$

Simplify the right side of equation (5).

$$m = \left[\frac{-(\sum x)(\sum y) + n\sum xy}{n} \right]\left[\frac{n}{-(\sum x)^2 + n(\sum x^2)} \right]$$

$$m = \frac{n\sum xy - (\sum x)(\sum y)}{n(\sum x^2) - (\sum x)^2}$$

From equation (3) we have

$$b = \frac{\sum y}{n} - \frac{\sum x}{n}m$$

$$b = \frac{\sum y - m(\sum x)}{n}.$$

35. Let $x =$ the cost per pound of rice, and

     $y =$ the cost per pound of potatoes.

The system to be solved is

     $20x + 10y = 16.20$ (1)

     $30x + 12y = 23.04.$ (2)

Multiply equation (1) by $\frac{1}{20}$.

$$\tfrac{1}{20}R_1 \rightarrow R_1 \quad \begin{array}{l} x + 0.5y = 0.81 \quad (3) \\ 30x + 12y = 23.04 \quad (2) \end{array}$$

Eliminate x in equation (2).

$$-30R_1 + R_2 \rightarrow R_2 \quad \begin{array}{l} x + 0.5y = 0.81 \quad (3) \\ -3y = -1.26 \quad (4) \end{array}$$

Multiply equation (4) by $-\frac{1}{3}$.

$$-\tfrac{1}{3}R_2 \rightarrow R_2 \quad \begin{array}{l} x + 0.5y = 0.81 \quad (3) \\ y = 0.42 \quad (5) \end{array}$$

Substitute 0.42 for y in equation (3).

$$x + 0.5(0.42) = 0.81$$
$$x + 0.21 = 0.81$$
$$x = 0.60$$

The cost of 10 pounds of rice and 50 pounds of potatoes is

$$10(0.60) + 50(0.42) = 27,$$

that is, $27.

37. Let $x =$ the number of seats on the main floor, and

     $y =$ the number of seats in the balcony.

The system to be solved is

     $8x + 5y = 4200$ (1)

     $0.25(8x) + 0.40(5y) = 1200.$ (2)

Make the coefficient of the first term in equation (1) equal 1.

$$\tfrac{1}{8}R_1 \rightarrow R_1 \quad \begin{array}{l} x + \frac{5}{8}y = 525 \quad (3) \\ 2x + 2y = 1200 \quad (2) \end{array}$$

Eliminate x in equation (2).

$$x + \frac{5}{8}y = 525 \quad (3)$$

$$-2R_1 + R_2 \rightarrow R_2 \qquad \frac{6}{8}y = 150 \quad (4)$$

Make the coefficient of the first term in equation (4) equal 1.

$$x + \frac{5}{8}y = 525 \quad (3)$$

$$\frac{8}{6}R_2 \rightarrow R_2 \qquad y = 200 \quad (5)$$

Substitute 200 for y in equation (3).

$$x + \frac{5}{8}(200) = 525$$

$$x + 125 = 525$$

$$x = 400$$

There are 400 main floor seats and 200 balcony seats.

39. Let $x =$ the number of model 201 to make each day, and

$y =$ the number of model 301 to make each day.

The system to be solved is

$$2x + 3y = 34 \quad (1)$$
$$18x + 27y = 335. \quad (2)$$

Make the coefficient of the first term in equation (1) equal 1.

$$\frac{1}{2}R_1 \rightarrow R_1 \qquad x + \frac{3}{2}y = 17 \quad (3)$$
$$18x + 27y = 335 \quad (2)$$

Eliminate x in equation (2).

$$x + \frac{3}{2}y = 17 \quad (3)$$

$$-18R_1 + R_2 \rightarrow R_2 \qquad 0 = 29 \quad (4)$$

Since equation (4) is false, the system is inconsistent. Therefore, this situation is impossible.

41. Let $x =$ the number of buffets produced each week,

$y =$ the number of chairs produced each week.

$z =$ the number of tables produced each week.

Make a table.

	Buffet	Chair	Table	Totals
Construction	30	10	10	350
Finishing	10	10	30	150

The system to be solved is

$$30x + 10y + 10z = 350 \quad (1)$$
$$10x + 10y + 30z = 150. \quad (2)$$

Make the coefficient of the first term in equation (1) equal 1.

$$\frac{1}{30}R_1 \rightarrow R_1 \quad x + \frac{1}{3}y + \frac{1}{3}z = \frac{35}{3} \quad (3)$$
$$10x + 10y + 30z = 150 \quad (2)$$

Eliminate x from equation (2).

$$x + \frac{1}{3}y + \frac{1}{3}z = \frac{35}{3} \quad (3)$$

$$-10R_1 + R_2 \rightarrow R_2 \qquad \frac{20}{3}y + \frac{80}{3}z = \frac{100}{3} \quad (4)$$

Solve equation (4) for y. Multiply by 3.

$$20y + 80z = 100$$
$$y + 4z = 5$$
$$y = 5 - 4z$$

Substitute $5 - 4z$ for y in equation (1) and solve for x.

$$30x + 10(5 - 4z) + 10z = 350$$
$$30x + 50 - 40z + 10z = 350$$
$$30x = 300 + 30z$$
$$x = 10 + z$$

The solution is $(10 + z, 5 - 4z, z)$. All variables must be nonnegative integers. Therefore,

$$5 - 4z \geq 0$$
$$5 \geq 4z$$
$$z \leq \frac{5}{4},$$

so $z = 0$ or $z = 1$. (Any larger value of z would cause y to be negative, which would make no sense in the problem.) If $z = 0$, then the solution is $(10, 5, 0)$. If $z = 1$, then the solution is $(11, 1, 1)$. Therefore, the company should make either 10 buffets, 5 chairs, and no tables or 11 buffets, 1 chair, and 1 table each week.

43. Let $x =$ the number of long-sleeve blouses,

$y =$ the number of short-sleeve blouses, and

$z =$ the number of sleeveless blouses.

Make a table.

	Long Sleeve	Short Sleeve	Sleeveless	Totals
Cutting	1.5	1	0.5	380
Sewing	1.2	0.9	0.6	330

The system to be solved is

$$1.5x + y + 0.5z = 380 \quad (1)$$
$$1.2x + 0.9y + 0.6z = 330. \quad (2)$$

Simplify the equations. Multiply equation (1) by 2 and equation (2) by $\frac{10}{3}$.

$$3x + 2y + z = 760 \quad (3)$$
$$4x + 3y + 2z = 1100 \quad (4)$$

Make the leading coefficient of equation (3) equal 1.

$$\tfrac{1}{3}R_1 \to R_1 \quad x + \frac{2}{3}y + \frac{1}{3}z = \frac{760}{3} \quad (5)$$
$$4x + 3y + 2z = 1100 \quad (4)$$

Eliminate x from equation (4).

$$x + \frac{2}{3}y + \frac{1}{3}z = \frac{760}{3} \quad (5)$$
$$-4R_1 + R_2 \to R_2 \quad \frac{1}{3}y + \frac{2}{3}z = \frac{260}{3} \quad (6)$$

Make the leading coefficient of equation (6) equal 1.

$$x + \frac{2}{3}y + \frac{1}{3}z = \frac{760}{3} \quad (5)$$
$$3R_2 \to R_2 \quad y + 2z = 260 \quad (7)$$

From equation (7), $y = 260 - 2z$. Substitute this into equation (5).

$$x + \frac{2}{3}(260 - 2z) + \frac{1}{3}z = \frac{760}{3}$$
$$x + \frac{520}{3} - \frac{4}{3}z + \frac{1}{3}z = \frac{760}{3}$$
$$x - z = \frac{240}{3}$$
$$x = z + 80$$

The solution is $(z + 80, 260 - 2z, z)$. In this problem x, y, and z must be nonnegative, so

$$260 - 2z \geq 0$$
$$-2z \geq -260$$
$$z \leq 130.$$

Therefore, the plant should make $z = 80$ long-sleeve blouses, $260 - 2z$ short-sleeve blouses, and z sleeveless blouses with $0 \leq z \leq 130$.

45. **(a)** For the first equation, the first sighting in 2000 was on day $y = 759 - 0.338(2000) = 83$, or during the eighty-third day of the year. Since 2000 was a leap year, the eighty-third day fell on March 23.

For the second equation, the first sighting in 2000 was on day $y = 1637 - 0.779(2000) = 79$, or during the seventy-ninth day of the year. Since 2000 was a leap year, the seventy-ninth day fell on March 19.

(b) $y = 759 - 0.338x \quad (1)$

$y = 1637 - 0.779x \quad (2)$

Rewrite equations so that variables are on the left side and constant term is on the right side.

$$0.338x + y = 759 \quad (3)$$
$$0.779x + y = 1637 \quad (4)$$

Eliminate y from equation (4).

$$0.338x + y = 759 \quad (3)$$
$$-1R_1 + R_2 \to R_2 \quad 0.441x = 878 \quad (5)$$

Make leading coefficient for equation (5) equal 1.

The two estimates agree in the year closest to $x = \frac{878}{0.441} \approx 1990.93$, so they agree in 1991.

The estimated number of days into the year when a robin can be expected is

$$0.338\left(\frac{878}{0.441}\right) + y = 759$$
$$y \approx 86.$$

47. Let $x =$ number of field goals, and
$y =$ number of foul shots.

Then
$$x + y = 64 \quad (1)$$
$$2x + y = 100 \quad (2).$$

Eliminate x in equation (2).

$$x + y = 64 \quad (1)$$
$$-2R_1 + R_2 \to R_2 \quad -y = -28 \quad (3)$$

Make the coefficients of the first term of each equation equal 1.

$$x + y = 64 \quad (1)$$
$$-1R_2 \to R_2 \quad y = 28 \quad (4)$$

Substitute 28 for y in equation (1) to get $x = 36$. Wilt Chamberlain made 36 field goals and 28 foul shots.

2.2 Solution of Linear Systems by the Gauss-Jordan Method

Your Turn 1

$$4x + 5y = 10$$
$$7x + 8y = 19$$

Write the augmented matrix and transform the matrix.

$$\begin{bmatrix} 4 & 5 & | & 10 \\ 7 & 8 & | & 19 \end{bmatrix}$$

$$\tfrac{1}{4}R_1 \to R_1 \begin{bmatrix} 1 & \frac{5}{4} & | & \frac{5}{2} \\ 7 & 8 & | & 19 \end{bmatrix}$$

$$-7R_1 + R_2 \to R_2 \begin{bmatrix} 1 & \frac{5}{4} & | & \frac{5}{2} \\ 0 & -\frac{3}{4} & | & \frac{3}{2} \end{bmatrix}$$

$$-\frac{4}{3}R_2 \rightarrow R_2 \begin{bmatrix} 1 & \frac{5}{4} & \frac{5}{2} \\ 0 & 1 & -2 \end{bmatrix}$$

$$-\frac{5}{4}R_2 + R_1 \rightarrow R_1 \begin{bmatrix} 1 & 0 & 5 \\ 0 & 1 & -2 \end{bmatrix}$$

The solution is $x = 5$ and $y = -2$, or $(5, -2)$.

Your Turn 2

$$x + 2y + 3z = 2$$
$$2x + 2y - 3z = 27$$
$$3x + 2y + 5z = 10$$

Write the augmented matrix and transform the matrix.

$$\begin{bmatrix} 1 & 2 & 3 & 2 \\ 2 & 2 & -3 & 27 \\ 3 & 2 & 5 & 10 \end{bmatrix}$$

$$\begin{matrix} -2R_1 + R_2 \rightarrow R_2 \\ -3R_1 + R_3 \rightarrow R_3 \end{matrix} \begin{bmatrix} 1 & 2 & 3 & 2 \\ 0 & -2 & -9 & 23 \\ 0 & -4 & -4 & 4 \end{bmatrix}$$

$$-\frac{1}{2}R_2 \rightarrow R_2 \begin{bmatrix} 1 & 2 & 3 & 2 \\ 0 & 1 & \frac{9}{2} & -\frac{23}{2} \\ 0 & -4 & -4 & 4 \end{bmatrix}$$

$$\begin{matrix} -2R_2 + R_1 \rightarrow R_1 \\ \\ 4R_2 + R_3 \rightarrow R_3 \end{matrix} \begin{bmatrix} 1 & 0 & -6 & 25 \\ 0 & 1 & \frac{9}{2} & -\frac{23}{2} \\ 0 & 0 & 14 & -42 \end{bmatrix}$$

$$\frac{1}{14}R_3 \rightarrow R_3 \begin{bmatrix} 1 & 0 & -6 & 25 \\ 0 & 1 & \frac{9}{2} & -\frac{23}{2} \\ 0 & 0 & 1 & -3 \end{bmatrix}$$

$$\begin{matrix} 6R_3 + R_1 \rightarrow R_1 \\ -\frac{9}{2}R_3 + R_2 \rightarrow R_2 \end{matrix} \begin{bmatrix} 1 & 0 & 0 & 7 \\ 0 & 1 & 0 & 2 \\ 0 & 0 & 1 & -3 \end{bmatrix}$$

The solution is $x = 7, y = 2$, and $z = -3$, or $(7, 2, -3)$.

Your Turn 3

$$2x - 2y + 3z - 4w = 6$$
$$3x + 2y + 5z - 3w = 7$$
$$4x + y + 2z - 2w = 8$$

Write the augmented matrix and transform the matrix.

$$\begin{bmatrix} 2 & -2 & 3 & -4 & 6 \\ 3 & 2 & 5 & -3 & 7 \\ 4 & 1 & 2 & -2 & 8 \end{bmatrix}$$

$$\frac{1}{2}R_1 \rightarrow R_1 \begin{bmatrix} 1 & -1 & \frac{3}{2} & -2 & 3 \\ 3 & 2 & 5 & -3 & 7 \\ 4 & 1 & 2 & -2 & 8 \end{bmatrix}$$

$$\begin{matrix} -3R_1 + R_2 \rightarrow R_2 \\ -4R_1 + R_3 \rightarrow R_3 \end{matrix} \begin{bmatrix} 1 & -1 & \frac{3}{2} & -2 & 3 \\ 0 & 5 & \frac{1}{2} & 3 & -2 \\ 0 & 5 & -4 & 6 & -4 \end{bmatrix}$$

$$\frac{1}{5}R_2 \rightarrow R_2 \begin{bmatrix} 1 & -1 & \frac{3}{2} & -2 & 3 \\ 0 & 1 & \frac{1}{10} & \frac{3}{5} & -\frac{2}{5} \\ 0 & 5 & -4 & 6 & -4 \end{bmatrix}$$

$$\begin{matrix} R_2 + R_1 \rightarrow R_1 \\ \\ -5R_2 + R_3 \rightarrow R_3 \end{matrix} \begin{bmatrix} 1 & 0 & \frac{8}{5} & -\frac{7}{5} & \frac{13}{5} \\ 0 & 1 & \frac{1}{10} & \frac{3}{5} & -\frac{2}{5} \\ 0 & 0 & -\frac{9}{2} & 3 & -2 \end{bmatrix}$$

$$-\frac{2}{9}R_3 \rightarrow R_3 \begin{bmatrix} 1 & 0 & \frac{8}{5} & -\frac{7}{5} & \frac{13}{5} \\ 0 & 1 & \frac{1}{10} & \frac{3}{5} & -\frac{2}{5} \\ 0 & 0 & 1 & -\frac{2}{3} & \frac{4}{9} \end{bmatrix}$$

$$\begin{matrix} -\frac{8}{5}R_3 + R_1 \rightarrow R_1 \\ -\frac{1}{10}R_3 + R_2 \rightarrow R_2 \end{matrix} \begin{bmatrix} 1 & 0 & 0 & -\frac{1}{3} & \frac{17}{9} \\ 0 & 1 & 0 & \frac{2}{3} & -\frac{4}{9} \\ 0 & 0 & 1 & -\frac{2}{3} & \frac{4}{9} \end{bmatrix}$$

We cannot change the values in column 4 without changing the form of the other three columns. So, let w be the parameter. The last matrix gives these equations.

$$x - \frac{1}{3}w = \frac{17}{9}, \quad \text{or} \quad x = \frac{17}{9} + \frac{1}{3}w$$

$$y + \frac{2}{3}w = -\frac{4}{9}, \quad \text{or} \quad y = -\frac{4}{9} - \frac{2}{3}w$$

$$z - \frac{2}{3}w = \frac{4}{9}, \quad \text{or} \quad z = \frac{4}{9} + \frac{2}{3}w$$

The solution is $\left(\frac{17}{9} + \frac{1}{3}w, -\frac{4}{9} - \frac{2}{3}w, \frac{4}{9} + \frac{2}{3}w, w \right)$, where w is a real number.

2.2 Exercises

1. $3x + y = 6$
$2x + 5y = 15$

The equations are already in proper form. The augmented matrix obtained from the coefficients and the constants is

$$\begin{bmatrix} 3 & 1 & | & 6 \\ 2 & 5 & | & 15 \end{bmatrix}.$$

3. $2x + y + z = 3$
$3x - 4y + 2z = -7$
$x + y + z = 2$

leads to the augmented matrix

$$\begin{bmatrix} 2 & 1 & 1 & | & 3 \\ 3 & -4 & 2 & | & -7 \\ 1 & 1 & 1 & | & 2 \end{bmatrix}.$$

5. We are given the augmented matrix

$$\begin{bmatrix} 1 & 0 & | & 2 \\ 0 & 1 & | & 3 \end{bmatrix}.$$

This is equivalent to the system of equations

$$x = 2$$
$$y = 3,$$

or $x = 2, y = 3$.

7. $\begin{bmatrix} 1 & 0 & 0 & | & 4 \\ 0 & 1 & 0 & | & -5 \\ 0 & 0 & 1 & | & 1 \end{bmatrix}$

The system associated with this matrix is

$$x = 4$$
$$y = -5$$
$$z = 1,$$

or $x = 4, \ y = -5, \ z = 1$.

9. *Row operations* on a matrix correspond to transformations of a system of equations.

11. $\begin{bmatrix} 3 & 7 & 4 & | & 10 \\ 1 & 2 & 3 & | & 6 \\ 0 & 4 & 5 & | & 11 \end{bmatrix}$

Find $R_1 + (-3)R_2$.

In row 2, column 1,

$$3 + (-3)1 = 0.$$

In row 2, column 2,

$$7 + (-3)2 = 1.$$

In row 2, column 3,

$$4 + (-3)3 = -5.$$

In row 2, column 4,

$$10 + (-3)6 = -8.$$

Replace R_2 with these values. The new matrix is

$$\begin{bmatrix} 3 & 7 & 4 & | & 10 \\ 0 & 1 & -5 & | & -8 \\ 0 & 4 & 5 & | & 11 \end{bmatrix}.$$

13. $\begin{bmatrix} 1 & 6 & 4 & | & 7 \\ 0 & 3 & 2 & | & 5 \\ 0 & 5 & 3 & | & 7 \end{bmatrix}$

Find $(-2)R_2 + R_1 \to R_1$

$$\begin{bmatrix} (-2)0+1 & (-2)3+6 & (-2)2+4 & | & (-2)5+7 \\ 0 & 3 & 2 & | & 5 \\ 0 & 5 & 3 & | & 7 \end{bmatrix}$$

$$= \begin{bmatrix} 1 & 0 & 0 & | & -3 \\ 0 & 3 & 2 & | & 5 \\ 0 & 5 & 3 & | & 7 \end{bmatrix}$$

15. $\begin{bmatrix} 3 & 0 & 0 & | & 18 \\ 0 & 5 & 0 & | & 9 \\ 0 & 0 & 4 & | & 8 \end{bmatrix}$

$$\tfrac{1}{3}R_1 \to R_1 \begin{bmatrix} \tfrac{1}{3}(3) & \tfrac{1}{3}(0) & \tfrac{1}{3}(0) & | & \tfrac{1}{3}(18) \\ 0 & 5 & 0 & | & 9 \\ 0 & 0 & 4 & | & 8 \end{bmatrix} = \begin{bmatrix} 1 & 0 & 0 & | & 6 \\ 0 & 5 & 0 & | & 9 \\ 0 & 0 & 4 & | & 8 \end{bmatrix}$$

17. $x + y = 5$
$3x + 2y = 12$

Write the augmented matrix and use row operations.

$$\begin{bmatrix} 1 & 1 & | & 5 \\ 3 & 2 & | & 12 \end{bmatrix}$$

$$-3R_1 + R_2 \to R_2 \begin{bmatrix} 1 & 1 & | & 5 \\ 0 & -1 & | & -3 \end{bmatrix}$$

$$-1R_2 \to R_2 \begin{bmatrix} 1 & 1 & | & 5 \\ 0 & 1 & | & 3 \end{bmatrix}$$

$$-1R_2 + R_1 \to R_1 \begin{bmatrix} 1 & 0 & | & 2 \\ 0 & 1 & | & 3 \end{bmatrix}$$

The solution is $(2, 3)$.

19. $x + y = 7$
$4x + 3y = 22$

Write the augmented matrix and use row operations.

$$\begin{bmatrix} 1 & 1 & | & 7 \\ 4 & 3 & | & 22 \end{bmatrix}$$

$$-4R_1 + R_2 \rightarrow R_2 \begin{bmatrix} 1 & 1 & | & 7 \\ 0 & -1 & | & -6 \end{bmatrix}$$

$$-1R_2 \rightarrow R_2 \begin{bmatrix} 1 & 1 & | & 7 \\ 0 & 1 & | & 6 \end{bmatrix}$$

$$-1R_2 + R_1 \rightarrow R_1 \begin{bmatrix} 1 & 0 & | & 1 \\ 0 & 1 & | & 6 \end{bmatrix}$$

The solution is $(1, 6)$.

21. $2x - 3y = 2$
$4x - 6y = 1$

Write the augmented matrix and use row operations.

$$\begin{bmatrix} 2 & -3 & | & 2 \\ 4 & -6 & | & 1 \end{bmatrix}$$

$$-2R_1 + R_2 \rightarrow R_2 \begin{bmatrix} 2 & -3 & | & 2 \\ 0 & 0 & | & -3 \end{bmatrix}$$

The system associated with the last matrix is

$$2x - 3y = 2$$
$$0x + 0y = -3.$$

Since the second equation, $0 = -3$, is false, the system is inconsistent and therefore has no solution.

23. $6x - 3y = 1$
$-12x + 6y = -2$

Write the augmented matrix of the system and use row operations.

$$\begin{bmatrix} 6 & -3 & | & 1 \\ -12 & 6 & | & -2 \end{bmatrix}$$

$$2R_1 + R_2 \rightarrow R_2 \begin{bmatrix} 6 & -3 & | & 1 \\ 0 & 0 & | & 0 \end{bmatrix}$$

$$\tfrac{1}{6}R_1 \rightarrow R_1 \begin{bmatrix} 1 & -\tfrac{1}{2} & | & \tfrac{1}{6} \\ 0 & 0 & | & 0 \end{bmatrix}$$

This is as far as we can go with the Gauss-Jordan method. To complete the solution, write the equation that corresponds to the first row of the matrix.

$$x - \frac{1}{2}y = \frac{1}{6}$$

Solve this equation for x in terms of y.

$$x = \frac{1}{2}y + \frac{1}{6} = \frac{3y + 1}{6}$$

The solution is $\left(\frac{3y+1}{6}, y \right)$, y any real number.

25. $y = x - 3$
$y = 1 + z$
$z = 4 - x$

First write the system in proper form.

$$-x + y \quad\quad = -3$$
$$y - z = 1$$
$$x \quad\quad + z = 4$$

Write the augmented matrix and use row operations.

$$\begin{bmatrix} -1 & 1 & 0 & | & -3 \\ 0 & 1 & -1 & | & 1 \\ 1 & 0 & 1 & | & 4 \end{bmatrix}$$

$$-1R_1 \rightarrow R_1 \begin{bmatrix} 1 & -1 & 0 & | & 3 \\ 0 & 1 & -1 & | & 1 \\ 1 & 0 & 1 & | & 4 \end{bmatrix}$$

$$-1R_1 + R_3 \rightarrow R_3 \begin{bmatrix} 1 & -1 & 0 & | & 3 \\ 0 & 1 & -1 & | & 1 \\ 0 & 1 & 1 & | & 1 \end{bmatrix}$$

$$R_2 + R_1 \rightarrow R_1 \begin{bmatrix} 1 & 0 & -1 & | & 4 \\ 0 & 1 & -1 & | & 1 \\ 0 & 0 & 2 & | & 0 \end{bmatrix}$$
$$-1R_2 + R_3 \rightarrow R_3$$

$$R_3 + 2R_1 \rightarrow R_1 \begin{bmatrix} 2 & 0 & 0 & | & 8 \\ 0 & 2 & 0 & | & 2 \\ 0 & 0 & 2 & | & 0 \end{bmatrix}$$
$$R_3 + 2R_2 \rightarrow R_2$$

$$\tfrac{1}{2}R_1 \rightarrow R_1 \begin{bmatrix} 1 & 0 & 0 & | & 4 \\ 0 & 1 & 0 & | & 1 \\ 0 & 0 & 1 & | & 0 \end{bmatrix}$$
$$\tfrac{1}{2}R_2 \rightarrow R_2$$
$$\tfrac{1}{2}R_3 \rightarrow R_3$$

The solution is $(4, 1, 0)$.

27. $2x - 2y = -5$
$2y + z = 0$
$2x + z = -7$

Write the augmented matrix and use row operations.

$$\begin{bmatrix} 2 & -2 & 0 & | & -5 \\ 0 & 2 & 1 & | & 0 \\ 2 & 0 & 1 & | & -7 \end{bmatrix}$$

$$-1R_1 + R_3 \rightarrow R_3 \begin{bmatrix} 2 & -2 & 0 & | & -5 \\ 0 & 2 & 1 & | & 0 \\ 0 & 2 & 1 & | & -2 \end{bmatrix}$$

$$R_2 + R_1 \rightarrow R_1 \begin{bmatrix} 2 & 0 & 1 & | & -5 \\ 0 & 2 & 1 & | & 0 \\ 0 & 0 & 0 & | & -2 \end{bmatrix}$$
$$-1R_2 + R_3 \rightarrow R_3$$

This matrix corresponds to the system of equations

$$2x + z = -5$$
$$2y + z = 0$$
$$0 = -2.$$

This false statement $0 = -2$ indicates that the system is inconsistent and therefore has no solution.

29. $4x + 4y - 4z = 24$
$2x - y + z = -9$
$x - 2y + 3z = 1$

Write the augmented matrix and use row operations.

$$\begin{bmatrix} 4 & 4 & -4 & | & 24 \\ 2 & -1 & 1 & | & -9 \\ 1 & -2 & 3 & | & 1 \end{bmatrix}$$

$$\begin{matrix} \\ R_1 + (-2)R_2 \rightarrow R_2 \\ R_1 + (-4)R_3 \rightarrow R_3 \end{matrix} \begin{bmatrix} 4 & 4 & -4 & | & 24 \\ 0 & 6 & -6 & | & 42 \\ 0 & 12 & -16 & | & 20 \end{bmatrix}$$

$$\begin{matrix} 2R_2 + (-3)R_1 \rightarrow R_1 \\ \\ -2R_2 + R_3 \rightarrow R_3 \end{matrix} \begin{bmatrix} -12 & 0 & 0 & | & 12 \\ 0 & 6 & -6 & | & 42 \\ 0 & 0 & -4 & | & -64 \end{bmatrix}$$

$$\begin{matrix} \\ -3R_3 + 2R_2 \rightarrow R_2 \\ \\ \end{matrix} \begin{bmatrix} -12 & 0 & 0 & | & 12 \\ 0 & 12 & 0 & | & 276 \\ 0 & 0 & -4 & | & -64 \end{bmatrix}$$

$$\begin{matrix} -\frac{1}{12}R_1 \rightarrow R_1 \\ \frac{1}{12}R_2 \rightarrow R_2 \\ -\frac{1}{4}R_3 \rightarrow R_3 \end{matrix} \begin{bmatrix} 1 & 0 & 0 & | & -1 \\ 0 & 1 & 0 & | & 23 \\ 0 & 0 & 1 & | & 16 \end{bmatrix}$$

The solution is $(-1, 23, 16)$.

31. $3x + 5y - z = 0$
$4x - y + 2z = 1$
$7x + 4y + z = 1$

Write the augmented matrix and use row operations.

$$\begin{bmatrix} 3 & 5 & -1 & | & 0 \\ 4 & -1 & 2 & | & 1 \\ 7 & 4 & 1 & | & 1 \end{bmatrix}$$

$$\begin{matrix} \\ 4R_1 + (-3)R_2 \rightarrow R_2 \\ 7R_1 + (-3)R_3 \rightarrow R_3 \end{matrix} \begin{bmatrix} 3 & 5 & -1 & | & 0 \\ 0 & 23 & -10 & | & -3 \\ 0 & 23 & -10 & | & -3 \end{bmatrix}$$

$$\begin{matrix} 23R_1 + (-5)R_2 \rightarrow R_1 \\ \\ R_2 + (-1)R_3 \rightarrow R_3 \end{matrix} \begin{bmatrix} 69 & 0 & 27 & | & 15 \\ 0 & 23 & -10 & | & -3 \\ 0 & 0 & 0 & | & 0 \end{bmatrix}$$

$$\begin{matrix} \frac{1}{69}R_1 \rightarrow R_1 \\ \frac{1}{23}R_2 \rightarrow R_2 \\ \\ \end{matrix} \begin{bmatrix} 1 & 0 & \frac{9}{23} & | & \frac{5}{23} \\ 0 & 1 & -\frac{10}{23} & | & -\frac{3}{23} \\ 0 & 0 & 0 & | & 0 \end{bmatrix}$$

The row of zeros indicates dependent equations. Solve the first two equations respectively for x and y in terms of z to obtain

$$x = -\frac{9}{23}z + \frac{5}{23} = \frac{-9z + 5}{23}$$

and

$$y = \frac{10}{23}z - \frac{3}{23} = \frac{10z - 3}{23}.$$

The solution is $\left(\frac{-9z + 5}{23}, \frac{10z - 3}{23}, z \right)$.

33. $5x - 4y + 2z = 6$
$5x + 3y - z = 11$
$15x - 5y + 3z = 23$

Write the augmented matrix and use row operations.

$$\begin{bmatrix} 5 & -4 & 2 & | & 6 \\ 5 & 3 & -1 & | & 11 \\ 15 & -5 & 3 & | & 23 \end{bmatrix}$$

$$\begin{matrix} \\ -1R_1 + R_2 \rightarrow R_2 \\ -3R_1 + R_3 \rightarrow R_3 \end{matrix} \begin{bmatrix} 5 & -4 & 2 & | & 6 \\ 0 & 7 & -3 & | & 5 \\ 0 & 7 & -3 & | & 5 \end{bmatrix}$$

$$\begin{matrix} 4R_2 + 7R_1 \rightarrow R_1 \\ \\ -1R_2 + R_3 \rightarrow R_3 \end{matrix} \begin{bmatrix} 35 & 0 & 2 & | & 62 \\ 0 & 7 & -3 & | & 5 \\ 0 & 0 & 0 & | & 0 \end{bmatrix}$$

$$\begin{matrix} \frac{1}{35}R_1 \rightarrow R_1 \\ \frac{1}{7}R_2 \rightarrow R_2 \\ \\ \end{matrix} \begin{bmatrix} 1 & 0 & \frac{2}{35} & | & \frac{62}{35} \\ 0 & 1 & -\frac{3}{7} & | & \frac{5}{7} \\ 0 & 0 & 0 & | & 0 \end{bmatrix}$$

The row of zeros indicates dependent equations. Solve the first two equations respectively for x and y in terms of z to obtain

$$x = -\frac{2}{35}z + \frac{62}{35} = \frac{-2z + 62}{35}$$

and

$$y = \frac{3}{7}z + \frac{5}{7} = \frac{3z + 5}{7}.$$

The solution is $\left(\frac{-2z + 62}{35}, \frac{3z + 5}{7}, z \right)$.

35. $2x + 3y + z = 9$
$4x + 6y + 2z = 18$
$-\frac{1}{2}x - \frac{3}{4}y - \frac{1}{4}z = -\frac{9}{4}$

Write the augmented matrix and use row operations.

$$\begin{bmatrix} 2 & 3 & 1 & 9 \\ 4 & 6 & 2 & 18 \\ -\frac{1}{2} & -\frac{3}{4} & -\frac{1}{4} & -\frac{9}{4} \end{bmatrix}$$

$$\begin{array}{c} -2R_1 + R_2 \rightarrow R_2 \\ \frac{1}{4}R_1 + R_3 \rightarrow R_3 \end{array} \begin{bmatrix} 2 & 3 & 1 & 9 \\ 0 & 0 & 0 & 0 \\ 0 & 0 & 0 & 0 \end{bmatrix}$$

The rows of zeros indicate dependent equations. Since the equation involves x, y, and z, let y and z be parameters. Solve the equation for x to obtain $x = \dfrac{9 - 3y - z}{2}$.

The solution is $\left(\dfrac{9z - 3y - z}{2}, y, z \right)$, where y and z are any real numbers.

37. $\quad x + 2y \qquad - w = \quad 3$
$\quad 2x + \qquad 4z + 2w = -6$
$\quad x + 2y - z \qquad = \quad 6$
$\quad 2x - \quad y + z + \quad w = -3$

Write the augmented matrix and use row operations.

$$\begin{bmatrix} 1 & 2 & 0 & -1 & 3 \\ 2 & 0 & 4 & 2 & -6 \\ 1 & 2 & -1 & 0 & 6 \\ 2 & -1 & 1 & 1 & -3 \end{bmatrix}$$

$$\begin{array}{c} -2R_1 + R_2 \rightarrow R_2 \\ -1R_1 + R_3 \rightarrow R_3 \\ -2R_1 + R_4 \rightarrow R_4 \end{array} \begin{bmatrix} 1 & 2 & 0 & -1 & 3 \\ 0 & -4 & 4 & 4 & -12 \\ 0 & 0 & -1 & 1 & 3 \\ 0 & -5 & 1 & 3 & -9 \end{bmatrix}$$

$$\begin{array}{c} R_2 + 2R_1 \rightarrow R_1 \\ \\ \\ -5R_2 + 4R_4 \rightarrow R_4 \end{array} \begin{bmatrix} 2 & 0 & 4 & 2 & -6 \\ 0 & -4 & 4 & 4 & -12 \\ 0 & 0 & -1 & 1 & 3 \\ 0 & 0 & -16 & -8 & 24 \end{bmatrix}$$

$$\begin{array}{c} 4R_3 + R_1 \rightarrow R_1 \\ 4R_3 + R_2 \rightarrow R_2 \\ \\ 16R_3 + (-1)R_4 \rightarrow R_4 \end{array} \begin{bmatrix} 2 & 0 & 0 & 6 & 6 \\ 0 & -4 & 0 & 8 & 0 \\ 0 & 0 & -1 & 1 & 3 \\ 0 & 0 & 0 & 24 & 24 \end{bmatrix}$$

$$\begin{array}{c} R_4 + (-4)R_1 \rightarrow R_1 \\ R_4 + (-3)R_2 \rightarrow R_2 \\ R_4 + (-24)R_3 \rightarrow R_3 \end{array} \begin{bmatrix} -8 & 0 & 0 & 0 & 0 \\ 0 & 12 & 0 & 0 & 24 \\ 0 & 0 & 24 & 0 & -48 \\ 0 & 0 & 0 & 24 & 24 \end{bmatrix}$$

$$\begin{array}{c} -\frac{1}{8}R_1 \rightarrow R_1 \\ \frac{1}{12}R_2 \rightarrow R_2 \\ \frac{1}{24}R_3 \rightarrow R_3 \\ \frac{1}{24}R_4 \rightarrow R_4 \end{array} \begin{bmatrix} 1 & 0 & 0 & 0 & 0 \\ 0 & 1 & 0 & 0 & 2 \\ 0 & 0 & 1 & 0 & -2 \\ 0 & 0 & 0 & 1 & 1 \end{bmatrix}$$

The solution is $x = 0$, $y = 2$, $z = -2$, $w = 1$, or $(0, 2, -2, 1)$.

39. $\quad x + y - z + 2w = -20$
$\quad 2x - \quad y + z + \quad w = \quad 11$
$\quad 3x - 2y + z - 2w = \quad 27$

$$\begin{bmatrix} 1 & 1 & -1 & 2 & -20 \\ 2 & -1 & 1 & 1 & 11 \\ 3 & -2 & 1 & -2 & 27 \end{bmatrix}$$

$$\begin{array}{c} -2R_1 + R_2 \rightarrow R_2 \\ -3R_1 + R_3 \rightarrow R_3 \end{array} \begin{bmatrix} 1 & 1 & -1 & 2 & -20 \\ 0 & -3 & 3 & -3 & 51 \\ 0 & -5 & 4 & -8 & 87 \end{bmatrix}$$

$$-\frac{1}{3}R_2 \rightarrow R_2 \begin{bmatrix} 1 & 1 & -1 & 2 & -20 \\ 0 & 1 & -1 & 1 & -17 \\ 0 & -5 & 4 & -8 & 87 \end{bmatrix}$$

$$\begin{array}{c} -1R_2 + R_1 \rightarrow R_1 \\ \\ 5R_2 + R_3 \rightarrow R_3 \end{array} \begin{bmatrix} 1 & 0 & 0 & 1 & -3 \\ 0 & 1 & -1 & 1 & -17 \\ 0 & 0 & -1 & -3 & 2 \end{bmatrix}$$

$$-1R_3 \rightarrow R_3 \begin{bmatrix} 1 & 0 & 0 & 1 & -3 \\ 0 & 1 & -1 & 1 & -17 \\ 0 & 0 & 1 & 3 & -2 \end{bmatrix}$$

$$R_3 + R_2 \rightarrow R_2 \begin{bmatrix} 1 & 0 & 0 & 1 & -3 \\ 0 & 1 & 0 & 4 & -19 \\ 0 & 0 & 1 & 3 & -2 \end{bmatrix}$$

This is as far as we can go using row operations. To complete the solution, write the equations that correspond to the matrix.

$$x + \quad w = \quad -3$$
$$y + 4w = -19$$
$$z + 3w = \quad -2$$

Let w be the parameter and express x, y, and z in terms of w. From the equations above, $x = -w - 3$, $y = -4w - 19$, and $z = -3w - 2$.

The solution is $(-w - 3, \ -4w - 19, \ -3w - 2, \ w)$, where w is any real number.

41.
$$10.47x + 3.52y + 2.58z - 6.42w = 218.65$$
$$8.62x - 4.93y - 1.75z + 2.83w = 157.03$$
$$4.92x + 6.83y - 2.97z + 2.65w = 462.3$$
$$2.86x + 19.10y - 6.24z - 8.73w = 398.4$$

Write the augmented matrix of the system.

$$\begin{bmatrix} 10.47 & 3.52 & 2.58 & -6.42 & | & 218.65 \\ 8.62 & -4.93 & -1.75 & 2.83 & | & 157.03 \\ 4.92 & 6.83 & -2.97 & 2.65 & | & 462.3 \\ 2.86 & 19.10 & -6.24 & -8.73 & | & 398.4 \end{bmatrix}$$

This exercise should be solved by graphing calculator or computer methods. The solution, which may vary slightly, is $x \approx 28.9436$, $y \approx 36.6326$, $z \approx 9.6390$, and $w \approx 37.1036$, or

$$(28.9436, 36.6326, 9.6390, 37.1036).$$

43. Insert the given values, introduce variables, and the table is as follows.

$\frac{3}{8}$	a	b
c	d	$\frac{1}{4}$
e	f	g

From this, we obtain the following system of equations.

$$a + b \qquad\qquad\qquad + \tfrac{3}{8} = 1$$
$$c + d \qquad\qquad + \tfrac{1}{4} = 1$$
$$e + f + g = 1$$
$$c \qquad + e \qquad + \tfrac{3}{8} = 1$$
$$a \qquad + d \qquad + f = 1$$
$$b \qquad\qquad + g + \tfrac{1}{4} = 1$$
$$d \qquad\qquad + g + \tfrac{3}{8} = 1$$
$$b \qquad + d + e = 1$$

The augmented matrix and the final form after row operations are as follows.

$$\begin{bmatrix} 1 & 1 & 0 & 0 & 0 & 0 & 0 & | & \frac{5}{8} \\ 0 & 0 & 1 & 1 & 0 & 0 & 0 & | & \frac{3}{4} \\ 0 & 0 & 0 & 0 & 1 & 1 & 1 & | & 1 \\ 0 & 0 & 1 & 0 & 1 & 0 & 0 & | & \frac{5}{8} \\ 1 & 0 & 0 & 1 & 0 & 1 & 0 & | & 1 \\ 0 & 1 & 0 & 0 & 0 & 0 & 1 & | & \frac{3}{4} \\ 0 & 0 & 0 & 1 & 0 & 0 & 1 & | & \frac{5}{8} \\ 0 & 1 & 0 & 1 & 1 & 0 & 0 & | & 1 \end{bmatrix} \rightarrow \begin{bmatrix} 1 & 0 & 0 & 0 & 0 & 0 & 0 & | & \frac{1}{6} \\ 0 & 1 & 0 & 0 & 0 & 0 & 0 & | & \frac{11}{24} \\ 0 & 0 & 1 & 0 & 0 & 0 & 0 & | & \frac{5}{12} \\ 0 & 0 & 0 & 1 & 0 & 0 & 0 & | & \frac{1}{3} \\ 0 & 0 & 0 & 0 & 1 & 0 & 0 & | & \frac{5}{24} \\ 0 & 0 & 0 & 0 & 0 & 1 & 0 & | & \frac{1}{2} \\ 0 & 0 & 0 & 0 & 0 & 0 & 1 & | & \frac{7}{12} \\ 0 & 0 & 0 & 0 & 0 & 0 & 0 & | & 0 \end{bmatrix}$$

The solution to the system is read from the last column.

$$a = \tfrac{1}{6}, b = \tfrac{11}{24}, c = \tfrac{5}{12}, d = \tfrac{1}{3},$$
$$e = \tfrac{5}{24}, f = \tfrac{1}{2}, \text{ and } g = \tfrac{7}{24}$$

So the magic square is:

$\frac{3}{8}$	$\frac{1}{6}$	$\frac{11}{24}$
$\frac{5}{12}$	$\frac{1}{3}$	$\frac{1}{4}$
$\frac{5}{24}$	$\frac{1}{2}$	$\frac{7}{24}$

45. Let x = amount invested in U.S. savings bonds, y = amount invested in mutual funds, and z = amount invested in a money market account.

Since the total amount invested was \$10,000, $x + y + z = 10,000$.

Katherine invested twice as much in mutual funds as in savings bonds, so $y = 2x$.

The total return on her investments was \$470, so $0.025x + 0.06y + 0.045z = 470$.

The system to be solved is

$$x + y + z = 10,000 \quad (1)$$
$$2x - y = 0 \quad (2)$$
$$0.025x + 0.06y + 0.045z = 470 \quad (3).$$

Multiply equation (3) by 1000.

$$x + y + z = 10,000 \quad (1)$$
$$2x - y = 0 \quad (2)$$
$$1000R_3 \rightarrow R_3 \quad 25x + 60y + 45z = 470,000 \quad (4)$$

Eliminate x in equations (2) and (4).

$$x + y + z = 10,000 \quad (1)$$
$$-2R_1 + R_2 \rightarrow R_2 \quad -3y - 2z = -20,000 \quad (5)$$
$$-25R_1 + R_3 \rightarrow R_3 \quad 35y + 20z = 220,000 \quad (6)$$

Eliminate y in equation (6).

$$x + y + z = 10,000 \quad (1)$$
$$-3y - 2z = -20,000 \quad (5)$$
$$35R_2 + 3R_3 \rightarrow R_3 \quad -10z = -40,000 \quad (7)$$

Make each leading coefficient equal 1.

$$x + y + z = 10,000 \quad (1)$$
$$-\tfrac{1}{3}R_2 \rightarrow R_2 \quad y + \tfrac{2}{3}z = \frac{20,000}{3} \quad (8)$$
$$-\tfrac{1}{10}R_3 \rightarrow R_3 \quad z = 4000 \quad (9)$$

Substitute 4000 for z in equation (8) to get $y = 4000$. Finally, substitute 4000 for z and 4000 for y in equation (1) to get $x = 2000$. Ms. Chong invested \$2000 in U.S. savings bonds, \$4000 in mutual funds, and \$4000 in a money market account.

47. Let x = the number of chairs produced each week,
y = the number of cabinets produced each week, and
z = the number of buffets produced each week.

Make a table to organize the information.

	Chair	Cabinet	Buffet	Totals
Cutting	0.2	0.5	0.3	1950
Assembly	0.3	0.4	0.1	1490
Finishing	0.1	0.6	0.4	2160

The system to be solved is

$$0.2x + 0.5y + 0.3z = 1950$$
$$0.3x + 0.4y + 0.1z = 1490$$
$$0.1x + 0.6y + 0.4z = 2160.$$

Write the augmented matrix of the system

$$\begin{bmatrix} 0.2 & 0.5 & 0.3 & | & 1950 \\ 0.3 & 0.4 & 0.1 & | & 1490 \\ 0.1 & 0.6 & 0.4 & | & 2160 \end{bmatrix}$$

$$\begin{matrix} 10R_1 \to R_1 \\ 10R_2 \to R_2 \\ 10R_3 \to R_3 \end{matrix} \begin{bmatrix} 2 & 5 & 3 & | & 19{,}500 \\ 3 & 4 & 1 & | & 14{,}900 \\ 1 & 6 & 4 & | & 21{,}600 \end{bmatrix}$$

Interchange rows 1 and 3.

$$\begin{bmatrix} 1 & 6 & 4 & | & 21{,}600 \\ 3 & 4 & 1 & | & 14{,}900 \\ 2 & 5 & 3 & | & 19{,}500 \end{bmatrix}$$

$$\begin{matrix} -3R_1 + R_2 \to R_2 \\ -2R_1 + R_3 \to R_3 \end{matrix} \begin{bmatrix} 1 & 6 & 4 & | & 21{,}600 \\ 0 & -14 & -11 & | & -49{,}900 \\ 0 & -7 & -5 & | & -23{,}700 \end{bmatrix}$$

$$-\frac{1}{14}R_2 \to R_2 \begin{bmatrix} 1 & 6 & 4 & | & 21{,}600 \\ 0 & 1 & \frac{11}{14} & | & \frac{24{,}950}{7} \\ 0 & -7 & -5 & | & -23{,}700 \end{bmatrix}$$

$$\begin{matrix} -6R_2 + R_1 \to R_1 \\ \\ 7R_2 + R_3 \to R_3 \end{matrix} \begin{bmatrix} 1 & 0 & -\frac{5}{7} & | & \frac{1500}{7} \\ 0 & 1 & \frac{11}{14} & | & \frac{24{,}950}{7} \\ 0 & 0 & \frac{1}{2} & | & 1250 \end{bmatrix}$$

$$2R_3 \to R_3 \begin{bmatrix} 1 & 0 & -\frac{5}{7} & | & \frac{1500}{7} \\ 0 & 1 & \frac{11}{14} & | & \frac{24{,}950}{7} \\ 0 & 0 & 1 & | & 2500 \end{bmatrix}$$

$$\begin{matrix} \frac{5}{7}R_3 + R_1 \to R_1 \\ -\frac{11}{14}R_3 + R_2 \to R_2 \end{matrix} \begin{bmatrix} 1 & 0 & 0 & | & 2000 \\ 0 & 1 & 0 & | & 1600 \\ 0 & 0 & 1 & | & 2500 \end{bmatrix}$$

The solution is (2000, 1600, 2500). Therefore, 2000 chairs, 1600 cabinets, and 2500 buffets should be produced.

49. (a) Let x be the number of trucks used, y be the number of vans, and z be the number of SUVs. We first obtain the equations given here.

$$2x + 3y + 3z = 25$$
$$2x + 4y + 5z = 33$$
$$3x + 2y + z = 22$$

Write the augmented matrix and use row operations.

$$\begin{bmatrix} 2 & 3 & 3 & | & 25 \\ 2 & 4 & 5 & | & 33 \\ 3 & 2 & 1 & | & 22 \end{bmatrix}$$

$$\begin{matrix} -1R_1 + R_2 \to R_2 \\ -3R_1 + 2R_3 \to R_3 \end{matrix} \begin{bmatrix} 2 & 3 & 3 & | & 25 \\ 0 & 1 & 2 & | & 8 \\ 0 & -5 & -7 & | & -31 \end{bmatrix}$$

$$\begin{matrix} -3R_2 + R_1 \to R_1 \\ 5R_2 + R_3 \to R_3 \end{matrix} \begin{bmatrix} 2 & 0 & -3 & | & 1 \\ 0 & 1 & 2 & | & 8 \\ 0 & 0 & 3 & | & 9 \end{bmatrix}$$

$$\begin{matrix} R_3 + R_1 \to R_1 \\ -2R_3 + 3R_2 \to R_2 \end{matrix} \begin{bmatrix} 2 & 0 & 0 & | & 10 \\ 0 & 3 & 0 & | & 6 \\ 0 & 0 & 3 & | & 9 \end{bmatrix}$$

$$\begin{matrix} \frac{1}{2}R_1 \to R_1 \\ \frac{1}{3}R_2 \to R_2 \\ \frac{1}{3}R_3 \to R_3 \end{matrix} \begin{bmatrix} 1 & 0 & 0 & | & 5 \\ 0 & 1 & 0 & | & 2 \\ 0 & 0 & 1 & | & 3 \end{bmatrix}$$

Read the solution from the last column of the matrix. The solution is 5 trucks, 2 vans, and 3 SUVs.

(b) The system of equations is now

$$2x + 3y + 3z = 25$$
$$2x + 4y + 5z = 33.$$

Write the augmented matrix and use row operations.

$$\begin{bmatrix} 2 & 3 & 3 & | & 25 \\ 2 & 4 & 5 & | & 33 \end{bmatrix}$$

$$-R_1 + R_2 \to R_2 \begin{bmatrix} 2 & 3 & 3 & | & 25 \\ 0 & 1 & 2 & | & 8 \end{bmatrix}$$

$-3R_2 + R_1 \to R_1 \begin{bmatrix} 2 & 0 & -3 & | & 1 \\ 0 & 1 & 2 & | & 8 \end{bmatrix}$

Obtain a one in row 1, column 1.

$\frac{1}{2}R_1 \to R_1 \begin{bmatrix} 1 & 0 & -\frac{3}{2} & | & \frac{1}{2} \\ 0 & 1 & 2 & | & 8 \end{bmatrix}$

The last row indicates multiple solutions are possible. The remaining equations are

$$x - \frac{3}{2}z = \frac{1}{2} \quad \text{and} \quad y + 2z = 8.$$

Solving these for x and y, we have

$$x = \frac{3}{2}z + \frac{1}{2} \quad \text{and} \quad y = -2z + 8.$$

The form of the solution is

$$\left(\frac{3}{2}z + \frac{1}{2}, -2z + 8, z\right).$$

Since the solutions must be whole numbers,

$$\frac{3}{2}z + \frac{1}{2} \geq 0 \quad \text{and} \quad -2z + 8 \geq 0$$
$$\frac{3}{2}z \geq -\frac{1}{2} \qquad\qquad -2z \geq -8$$
$$z \geq -\frac{1}{3} \qquad\qquad\qquad z \leq 4$$

Thus, there are 4 possible solutions but each must be checked to determine if they produce whole numbers for x and y.

When $z = 0, \left(\frac{1}{2}, 8, 0\right)$ which is not realistic.

When $z = 1, (2, 6, 1)$.

When $z = 2, \left(\frac{7}{2}, 4, 2\right)$ which is not realistic.

When $z = 3, (5, 2, 3)$.

When $z = 4, \left(\frac{13}{2}, 0, 4\right)$ which is not realistic.

The company has 2 options. Either use 2 trucks, 6 vans, and 1 SUV or use 5 trucks, 2 vans, and 3 SUVs.

51. Let $x =$ the amount borrowed at 8%,
$y =$ the amount borrowed at 9%, and
$z =$ the amount borrowed at 10%.

(a) The system to be solved is

$$x + y + z = 25{,}000$$
$$0.08x + 0.09y + 0.10z = 2190$$
$$y = z + 1000$$

Multiply the second equation by 100 and rewrite the equations in standard form.

$$x + y + z = 25{,}000$$
$$8x + 9y + 10z = 219{,}000$$
$$y - z = 1000.$$

Write the augmented matrix and use row operations to solve

$$\begin{bmatrix} 1 & 1 & 1 & | & 25{,}000 \\ 8 & 9 & 10 & | & 219{,}000 \\ 0 & 1 & -1 & | & 1000 \end{bmatrix}$$

$-8R_1 + R_2 \to R_2 \begin{bmatrix} 1 & 1 & 1 & | & 25{,}000 \\ 0 & 1 & 2 & | & 19{,}000 \\ 0 & 1 & -1 & | & 1000 \end{bmatrix}$

$\begin{matrix} -1R_2 + R_1 \to R_1 \\ \\ -1R_2 + R_3 \to R_3 \end{matrix} \begin{bmatrix} 1 & 0 & -1 & | & 6000 \\ 0 & 1 & 2 & | & 19{,}000 \\ 0 & 0 & -3 & | & -18{,}000 \end{bmatrix}$

$\begin{matrix} -1R_3 + 3R_1 \to R_1 \\ 2R_3 + 3R_2 \to R_2 \end{matrix} \begin{bmatrix} 3 & 0 & 0 & | & 36{,}000 \\ 0 & 3 & 0 & | & 21{,}000 \\ 0 & 0 & -3 & | & -18{,}000 \end{bmatrix}$

$\begin{matrix} \frac{1}{3}R_1 \to R_1 \\ \frac{1}{3}R_2 \to R_2 \\ \frac{1}{3}R_3 \to R_3 \end{matrix} \begin{bmatrix} 1 & 0 & 0 & | & 12{,}000 \\ 0 & 1 & 0 & | & 7000 \\ 0 & 0 & 1 & | & 6000 \end{bmatrix}$

The solution is $(12{,}000, 7000, 6000)$. The company borrowed \$12,000 at 8%, \$7000 at 9%, and \$6000 at 10%.

(b) If the condition is dropped, the initial augmented matrix and solution is found as before.

$$\begin{bmatrix} 1 & 1 & 1 & | & 25{,}000 \\ 8 & 9 & 10 & | & 219{,}000 \end{bmatrix}$$

$-8R_1 + R_2 \to R_2 \begin{bmatrix} 1 & 1 & 1 & | & 25{,}000 \\ 0 & 1 & 2 & | & 19{,}000 \end{bmatrix}$

$-1R_2 + R_1 \to R_1 \begin{bmatrix} 1 & 0 & -1 & | & 6000 \\ 0 & 1 & 2 & | & 19{,}000 \end{bmatrix}$

This gives the system of equations

$$x = z + 6000$$
$$y = -2x + 19{,}000$$

Since all values must be nonnegative,

$$z + 6000 \geq 0 \quad \text{and} \quad -2z + 19{,}000 \geq 0$$
$$z \geq -6000 \qquad\qquad\qquad z \leq 9500.$$

The second inequality produces the condition that the amount borrowed at 10% must be less than or equal to \$9500. If \$5000 is borrowed at 10%, $z = 5000$, and

$$x = 500 + 6000 = 11{,}000$$
$$y = -2(5000) + 19{,}000 = 9000.$$

This means \$11,000 is borrowed at 8% and \$9000 is borrowed at 9%.

(c) The original conditions resulted in \$12,000 borrowed at 8%. So, if the bank sets a maximum of \$10,000 at the 8% rate, no solution is possible.

(d) The total interest would be

$$0.08(10{,}000) + 0.09(8000) + 0.10(7000)$$
$$= 800 + 720 + 700$$
$$= 2220$$

or \$2220, which is not the \$2190 interest as specified as one of the conditions of the problem.

53. Let x_1 = the number of units from first supplier for Roseville,

x_2 = the number of units from first supplier for Akron,

x_3 = the number of units from second supplier for Roseville, and

x_4 = the number of units from second supplier for Akron.

Roseville needs 40 units so

$$x_1 + x_3 = 40.$$

Akron needs 75 units so

$$x_2 + x_4 = 75.$$

The manufacturer orders 75 units from the first supplier so

$$x_1 + x_2 = 75.$$

The total cost is \$10,750 so

$$70x_1 + 90x_2 + 80x_3 + 120x_4 = 10{,}750.$$

The system to be solved is

$$
\begin{aligned}
x_1 \qquad\quad + x_3 \qquad\quad &= 40 \\
x_2 \qquad\quad + x_4 &= 75 \\
x_1 + \quad x_2 \qquad\qquad &= 75 \\
70x_1 + 90x_2 + 80x_3 + 120x_4 &= 10{,}750.
\end{aligned}
$$

Write augmented matrix and use row operations.

$$
\left[\begin{array}{cccc|c}
1 & 0 & 1 & 0 & 40 \\
0 & 1 & 0 & 1 & 75 \\
1 & 1 & 0 & 0 & 75 \\
70 & 90 & 80 & 120 & 10{,}750
\end{array}\right]
$$

$$
\begin{array}{c}
 \\
 \\
-1R_1 + R_3 \to R_3 \\
-70R_1 + R_4 \to R_4
\end{array}
\left[\begin{array}{cccc|c}
1 & 0 & 1 & 0 & 40 \\
0 & 1 & 0 & 1 & 75 \\
0 & 1 & -1 & 0 & 35 \\
0 & 90 & 10 & 120 & 7950
\end{array}\right]
$$

$$
\begin{array}{c}
 \\
 \\
-1R_2 + R_3 \to R_3 \\
-90R_2 + R_4 \to R_4
\end{array}
\left[\begin{array}{cccc|c}
1 & 0 & 1 & 0 & 40 \\
0 & 1 & 0 & 1 & 75 \\
0 & 0 & -1 & -1 & -40 \\
0 & 0 & 10 & 30 & 1200
\end{array}\right]
$$

$$
\begin{array}{c}
 \\
 \\
-1R_3 \to R_3 \\
10R_3 + R_4 \to R_4
\end{array}
\left[\begin{array}{cccc|c}
1 & 0 & 1 & 0 & 40 \\
0 & 1 & 0 & 1 & 75 \\
0 & 0 & 1 & 1 & 40 \\
0 & 0 & 0 & 20 & 800
\end{array}\right]
$$

$$
\frac{1}{20}R_4 \to R_4
\left[\begin{array}{cccc|c}
1 & 0 & 1 & 0 & 40 \\
0 & 1 & 0 & 1 & 75 \\
0 & 0 & 1 & 1 & 40 \\
0 & 0 & 0 & 1 & 40
\end{array}\right]
$$

$$
\begin{array}{c}
 \\
-1R_4 + R_2 \to R_2 \\
-1R_4 + R_3 \to R_3 \\

\end{array}
\left[\begin{array}{cccc|c}
1 & 0 & 1 & 0 & 40 \\
0 & 1 & 0 & 0 & 35 \\
0 & 0 & 1 & 0 & 0 \\
0 & 0 & 0 & 1 & 40
\end{array}\right]
$$

$$
-1R_3 + R_1 \to R_1
\left[\begin{array}{cccc|c}
1 & 0 & 0 & 0 & 40 \\
0 & 1 & 0 & 0 & 35 \\
0 & 0 & 1 & 0 & 0 \\
0 & 0 & 0 & 1 & 40
\end{array}\right]
$$

The solution of the system is $x_1 = 40$, $x_2 = 35$, $x_3 = 0$, $x_4 = 40$, or (40, 35, 0, 40). The first supplier should send 40 units to Roseville and 35 units to Akron. The second supplier should send 0 units to Roseville and 40 units to Akron.

55. Let x = the number of two-person tents,

y = the number of four-person tents, and

z = the number of six-person tents that were ordered.

(a) The problem is to solve the following system of equations.

$$
\begin{aligned}
2x + 4y + 6z &= 200 \\
40x + 64y + 88z &= 3200 \\
129x + 179y + 229z &= 8950
\end{aligned}
$$

Write the augmented matrix and use row operations to solve.

$$
\left[\begin{array}{ccc|c}
2 & 4 & 6 & 200 \\
40 & 64 & 88 & 3200 \\
129 & 179 & 229 & 8950
\end{array}\right]
$$

$$
\begin{array}{c}
 \\
20R_1 + (-1)R_2 \to R_2 \\
129R_1 + (-2)R_3 \to R_3
\end{array}
\left[\begin{array}{ccc|c}
2 & 4 & 6 & 200 \\
0 & 16 & 32 & 800 \\
0 & 158 & 316 & 7900
\end{array}\right]
$$

$R_2 + (-4)R_1 \to R_1$ $\begin{bmatrix} -8 & 0 & 8 & | & 0 \\ 0 & 16 & 32 & | & 800 \\ 0 & 0 & 0 & | & 0 \end{bmatrix}$

$79R_2 + (-8)R_3 \to R_3$

$-\frac{1}{8}R_1 \to R_1$ $\begin{bmatrix} 1 & 0 & -1 & | & 0 \\ 0 & 1 & 2 & | & 50 \\ 0 & 0 & 0 & | & 0 \end{bmatrix}$

$\frac{1}{16}R_2 \to R_2$

Since the last row is all zeros, there is more than one solution. Let z be the parameter. The matrix gives

$$x - z = 0$$
$$y + 2z = 50.$$

Solving these equations for x and y, the solution is $(z, -2z + 50, z)$. The numbers in the solution must be nonnegative integers. Therefore,

$$y \geq 0$$
$$-2z + 50 \geq 0$$
$$z \leq 25.$$

Thus, $z \in \{0, 1, 2, 3, \ldots, 25\}$ In other words, depending on the number of six-person tents, there are 26 solutions to this problem.

(b) The number of four-person tents is given by the value of the variable y. Since $y = -2z + 50$, the most four-person tents will result when z is as small as possible, or 0. When this occurs, $y = -2(0) + 50 = 50$. And since $x = z$, the solution with the most four-person tents is 0 two-person tents, 50 four-person tents, and 0 six-person tents.

(c) The number of two-person tents is given by the value of the variable x. Since $x = z$, the most two-person tents will result when y is as small as possible, or 0. When this occurs,

$$2x + 4y + 6z = 200$$
$$2(z) + 4(0) + 6(z) = 200$$
$$8z = 200$$
$$z = 25.$$

The solution with the most two-person tents is 25 two-person tents, 0 four-person tents, and 25 six-person tents.

57. Let $x =$ the number of grams of group A,
  $y =$ the number of grams of group B, and
  $z =$ the number of grams of group C.

(a) The system to be solved is

$$x + y + z = 400 \quad (1)$$
$$x = \frac{1}{3}y \quad (2)$$
$$x + z = 2y. \quad (3)$$

Rewrite equations (2) and (3) in proper form and multiply both sides of equation (2) by 3.

$$x + y + z = 400$$
$$3x - y = 0$$
$$x - 2y + z = 0$$

Write the augmented matrix.

$$\begin{bmatrix} 1 & 1 & 1 & | & 400 \\ 3 & -1 & 0 & | & 0 \\ 1 & -2 & 1 & | & 0 \end{bmatrix}$$

$-3R_1 + R_2 \to R_2$ $\begin{bmatrix} 1 & 1 & 1 & | & 400 \\ 0 & -4 & -3 & | & -1200 \\ 0 & -3 & 0 & | & -400 \end{bmatrix}$

$-1R_1 + R_3 \to R_3$

$-\frac{1}{3}R_3 \to R_3$ $\begin{bmatrix} 1 & 1 & 1 & | & 400 \\ 0 & -4 & -3 & | & -1200 \\ 0 & 1 & 0 & | & \frac{400}{3} \end{bmatrix}$

Interchange rows 2 and 3.

$$\begin{bmatrix} 1 & 1 & 1 & | & 400 \\ 0 & 1 & 0 & | & \frac{400}{3} \\ 0 & -4 & -3 & | & -1200 \end{bmatrix}$$

$-1R_2 + R_1 \to R_1$ $\begin{bmatrix} 1 & 0 & 1 & | & \frac{800}{3} \\ 0 & 1 & 0 & | & \frac{400}{3} \\ 0 & 0 & -3 & | & -\frac{2000}{3} \end{bmatrix}$

$4R_2 + R_3 \to R_3$

$-\frac{1}{3}R_3 \to R_3$ $\begin{bmatrix} 1 & 0 & 1 & | & \frac{800}{3} \\ 0 & 1 & 0 & | & \frac{400}{3} \\ 0 & 0 & 1 & | & \frac{2000}{9} \end{bmatrix}$

$-1R_3 + R_1 \to R_1$ $\begin{bmatrix} 1 & 0 & 0 & | & \frac{400}{9} \\ 0 & 1 & 0 & | & \frac{400}{3} \\ 0 & 0 & 1 & | & \frac{2000}{9} \end{bmatrix}$

The solution is $\left(\frac{400}{9}, \frac{400}{3}, \frac{2000}{9} \right)$. Include $\frac{400}{9}$ g of group A, $\frac{400}{3}$ g of group B, and $\frac{2000}{9}$ g of group C.

(b) If the requirement that the diet include one-third as much of A as of B is dropped, refer to the first two rows of the fifth augmented matrix in part (a).

$$\begin{bmatrix} 1 & 0 & 1 & | & \frac{800}{3} \\ 0 & 1 & 0 & | & \frac{400}{3} \end{bmatrix}$$

This gives

$$x = \frac{800}{3} - z$$

$$y = \frac{400}{3}.$$

Therefore, for any positive number z of grams of group C, there should be z grams less than $\frac{800}{3}$ g of group A and $\frac{400}{3}$ g of group B.

(c) Since there was a unique solution for the original problem, by adding an additional condition. the only possible solution would be the one from part (a). However, by substituting those values of A, B, and C for x, y, and z in the equation for the additional condition, $0.02x + 0.02y + 0.03z = 8.00$, the values do not work. Thus, no solution is possible.

59. Let $x =$ the number of species A,
$y =$ the number of species B, and
$z =$ the number of species C.

Use a chart to organize the information.

		Species			
		A	B	C	Totals
Food	I	1.32	2.1	0.86	490
	II	2.9	0.95	1.52	897
	III	1.75	0.6	2.01	653

The system to be solved is

$$1.32x + 2.1y + 0.86z = 490$$
$$2.9x + 0.95y + 1.52z = 897$$
$$1.75x + 0.6y + 2.01z = 653.$$

Use graphing calculator or computer methods to solve this system. The solution, which may vary slightly, is to stock about 244 fish of species A, 39 fish of species B, and 101 fish of species C.

61. **(a)** Bulls:
The number of white ones was one half plus one third the number of black greater than the brown.

$$X = \left(\frac{1}{2} + \frac{1}{3}\right)Y + T$$

$$X = \frac{5}{6}Y + T$$

$$6X - 5Y = 6T$$

The number of the black, one quarter plus one fifth the number of the spotted greater than the brown.

$$Y = \left(\frac{1}{4} + \frac{1}{5}\right)Z + T$$

$$Y = \frac{9}{20}Z + T$$

$$20Y = 9Z + 20T$$

$$20Y - 9Z = 20T$$

The number of the spotted, one sixth and one seventh the number of the white greater than the brown.

$$Z = \left(\frac{1}{6} + \frac{1}{7}\right)X + T$$

$$Z = \frac{13}{42}X + T$$

$$42Z = 13X + 42T$$

$$42Z - 13X = 42T$$

So the system of equations for the bulls is

$$6X - 5Y = 6T$$
$$20Y - 9Z = 20T$$
$$42Z - 13X = 42T.$$

Cows:
The number of white ones was one third plus one quarter of the total black cattle.

$$x = \left(\frac{1}{3} + \frac{1}{4}\right)(Y + y)$$

$$x = \frac{7}{12}(Y + y)$$

$$12x = 7Y + 7y$$

$$12x - 7y = 7Y$$

The number of the black, one quarter plus one fifth the total of the spotted cattle.

$$y = \left(\frac{1}{4} + \frac{1}{5}\right)(Z + z)$$

$$y = \frac{9}{20}(Z + z)$$

$$20y = 9Z + 9z$$

$$20y - 9z = 9Z$$

The number of the spotted, one fifth plus one sixth the total of the brown cattle.

$$z = \left(\frac{1}{5} + \frac{1}{6}\right)(T + t)$$

$$z = \frac{11}{30}(T + t)$$

$$30z = 11T + 11t$$

$$30z - 11t = 11T$$

The number of the brown, one sixth plus one seventh the total of the white cattle.

$$t = \left(\frac{1}{6} + \frac{1}{7}\right)(X + x)$$

$$t = \frac{13}{42}(X + x)$$

$$42t = 13X + 13x$$

$$42t - 13x = 13X$$

So the system of equations for the cows is

$$12x - 7y = 7Y$$
$$20y - 9z = 9Z$$
$$30z - 11t = 11T$$
$$-13x + 42t = 13X$$

(b) For $T = 4,149,387$, the 3×3 system to be solved is

$$6X - 5Y \qquad = 24,896,322$$
$$20Y - 9Z = 82,987,740$$
$$-13X \qquad + 42Z = 174,274,254$$

Write the augmented matrix of the system.

$$\begin{bmatrix} 6 & -5 & 0 & | & 24,896,322 \\ 0 & 20 & -9 & | & 82,987,740 \\ -13 & 0 & 42 & | & 174,274,254 \end{bmatrix}$$

This exercise should be solved by graphing calculator or computer methods. The solution is $X = 10,366,482$ white bulls, $Y = 7,460,514$ black bulls, and $Z = 7,358,060$ spotted bulls.

For $X = 10,366,482$, $Y = 7,460,514$, and $Z = 7,358,060$, the 4×4 system to be solved is

$$12x - 7y = 52,223,598$$
$$20y - 9z = 66,222,540$$
$$30z - 11t = 45,643,257$$
$$-13x + 42t = 134,764,266$$

Write the augmented matrix of the system.

$$\begin{bmatrix} 12 & -7 & 0 & 0 & | & 52,223,598 \\ 0 & 20 & -9 & 0 & | & 66,222,540 \\ 0 & 0 & 30 & -11 & | & 45,643,257 \\ -13 & 0 & 0 & 42 & | & 134,764,266 \end{bmatrix}$$

This exercise should be solved by graphing calculator or computer methods. The solution is $x = 7,206,360$ white cows, $y = 4,893,246$ black cows, $z = 3,515,820$ spotted cows, and $t = 5,439,213$ brown cows.

63. (a) The system to be solved is
$$0 = 200,000 - 0.5r - 0.3b$$
$$0 = 350,000 - 0.5r - 0.7b.$$

First, write the system in proper form.
$$0.5r + 0.3b = 200,000$$
$$0.5r + 0.7b = 350,000$$

Write the augmented matrix and use row operations.

$$\begin{bmatrix} 0.5 & 0.3 & | & 200,000 \\ 0.5 & 0.7 & | & 350,000 \end{bmatrix}$$

$$\begin{array}{c} 10R_1 \to R_1 \\ 10R_2 \to R_2 \end{array} \begin{bmatrix} 5 & 3 & | & 2,000,000 \\ 5 & 7 & | & 3,500,000 \end{bmatrix}$$

$$-1R_1 + R_2 \to R_2 \begin{bmatrix} 5 & 3 & | & 2,000,000 \\ 0 & 4 & | & 1,500,000 \end{bmatrix}$$

$$-\tfrac{3}{4}R_2 + R_1 \to R_1 \begin{bmatrix} 5 & 0 & | & 875,000 \\ 0 & 4 & | & 1,500,000 \end{bmatrix}$$

$$\begin{array}{c} \tfrac{1}{5}R_1 \to R_1 \\ \tfrac{1}{4}R_2 \to R_2 \end{array} \begin{bmatrix} 1 & 0 & | & 175,000 \\ 0 & 1 & | & 375,000 \end{bmatrix}$$

The solution is $(175,000,\ 375,000)$. When the rate of increase for each is zero, there are 175,000 soldiers in the Red Army and 375,000 soldiers in the Blue Army.

65. Let $x =$ the number of calories in each gram of fat
$y =$ the number of calories in each gram of carbohydrates
$z =$ the number of calories in each gram of protein

We want to solve the following system.
$$10x + 36y + 2z = 240$$
$$14x + 37y + 3z = 280$$
$$20x + 23y + 11z = 295$$

Write the augmented matrix and transform the matrix.

$$\begin{bmatrix} 10 & 36 & 2 & | & 240 \\ 14 & 37 & 3 & | & 280 \\ 20 & 23 & 11 & | & 295 \end{bmatrix}$$

$$\tfrac{1}{10}R_1 \to R_1 \begin{bmatrix} 1 & 3.6 & 0.2 & | & 24 \\ 14 & 37 & 3 & | & 280 \\ 20 & 23 & 11 & | & 295 \end{bmatrix}$$

$$\begin{array}{c} -14R_1 + R_2 \to R_2 \\ -20R_1 + R_3 \to R_3 \end{array} \begin{bmatrix} 1 & 3.6 & 0.2 & | & 24 \\ 0 & -13.4 & 0.2 & | & -56 \\ 0 & -49 & 7 & | & -185 \end{bmatrix}$$

$$-\tfrac{1}{13.4}R_2 \to R_2 \begin{bmatrix} 1 & 3.6 & 0.2 & | & 24 \\ 0 & 1 & -0.014925 & | & 4.179104 \\ 0 & -49 & 7 & | & -185 \end{bmatrix}$$

$$\begin{array}{c} -3.6R_2 + R_1 \to R_1 \\ 49R_2 + R_3 \to R_3 \end{array} \begin{bmatrix} 1 & 0 & 0.25373 & | & 8.95522 \\ 0 & 1 & -0.014925 & | & 4.179104 \\ 0 & 0 & 6.268675 & | & 19.7761 \end{bmatrix}$$

$$\frac{1}{6.268675}R_3 \to R_3 \begin{bmatrix} 1 & 0 & 0.25373 & | & 8.95522 \\ 0 & 1 & -0.014925 & | & 4.179104 \\ 0 & 0 & 1 & | & 3.15475 \end{bmatrix}$$

$$\begin{matrix} -0.25373R_3 + R_1 \to R_1 \\ 0.014925R_3 + R_2 \to R_2 \end{matrix} \begin{bmatrix} 1 & 0 & 0 & | & 8.15476 \\ 0 & 1 & 0 & | & 4.22619 \\ 0 & 0 & 1 & | & 3.15475 \end{bmatrix}$$

The solution is $(8.15, 4.23, 3.15)$. There are 8.15 calories in a gram of fat, 4.23 calories in a gram of carbohydrates, and 3.15 calories in a gram of protein.

67. Let $x =$ the number of balls,
$y =$ the number of dolls, and
$z =$ the number of cars.

(a) The system to be solved is

$$\begin{aligned} x + y + z &= 100 \\ 2x + 3y + 4z &= 295 \\ 12x + 16y + 18z &= 1542. \end{aligned}$$

Write the augmented matrix of the system.

$$\begin{bmatrix} 1 & 1 & 1 & | & 100 \\ 2 & 3 & 4 & | & 295 \\ 12 & 16 & 18 & | & 1542 \end{bmatrix}$$

$$\begin{matrix} -2R_1 + R_2 \to R_2 \\ -12R_1 + R_3 \to R_3 \end{matrix} \begin{bmatrix} 1 & 1 & 1 & | & 100 \\ 0 & 1 & 2 & | & 95 \\ 0 & 4 & 6 & | & 342 \end{bmatrix}$$

$$\begin{matrix} -1R_2 + R_1 \to R_1 \\ \\ -4R_2 + R_3 \to R_3 \end{matrix} \begin{bmatrix} 1 & 0 & -1 & | & 5 \\ 0 & 1 & 2 & | & 95 \\ 0 & 0 & -2 & | & -38 \end{bmatrix}$$

$$-\frac{1}{2}R_3 \to R_3 \begin{bmatrix} 1 & 0 & -1 & | & 5 \\ 0 & 1 & 2 & | & 95 \\ 0 & 0 & 1 & | & 19 \end{bmatrix}$$

$$\begin{matrix} R_3 + R_1 \to R_1 \\ -2R_3 + R_2 \to R_2 \end{matrix} \begin{bmatrix} 1 & 0 & 0 & | & 24 \\ 0 & 1 & 0 & | & 57 \\ 0 & 0 & 1 & | & 19 \end{bmatrix}$$

The solution is $(24, 57, 19)$. There were 24 balls, 57 dolls, and 19 cars.

(b) The augmented matrix becomes

$$\begin{bmatrix} 1 & 1 & 1 & | & 100 \\ 2 & 3 & 4 & | & 295 \\ 11 & 15 & 19 & | & 1542 \end{bmatrix}.$$

$$\begin{matrix} -2R_1 + R_2 \to R_2 \\ -11R_1 + R_3 \to R_3 \end{matrix} \begin{bmatrix} 1 & 1 & 1 & | & 100 \\ 0 & 1 & 2 & | & 95 \\ 0 & 4 & 8 & | & 442 \end{bmatrix}$$

$$\begin{matrix} -1R_2 + R_1 \to R_1 \\ \\ -4R_2 + R_3 \to R_3 \end{matrix} \begin{bmatrix} 1 & 0 & -1 & | & 5 \\ 0 & 1 & 2 & | & 95 \\ 0 & 0 & 0 & | & 62 \end{bmatrix}$$

Since row 3 yields a false statement, $0 = 62$, there is no solution.

(c) The augmented matrix becomes

$$\begin{bmatrix} 1 & 1 & 1 & | & 100 \\ 2 & 3 & 4 & | & 295 \\ 11 & 15 & 19 & | & 1480 \end{bmatrix}.$$

$$\begin{matrix} -2R_1 + R_2 \to R_2 \\ -11R_1 + R_3 \to R_3 \end{matrix} \begin{bmatrix} 1 & 1 & 1 & | & 100 \\ 0 & 1 & 2 & | & 95 \\ 0 & 4 & 8 & | & 380 \end{bmatrix}$$

$$\begin{matrix} -1R_2 + R_1 \to R_1 \\ \\ -4R_2 + R_3 \to R_3 \end{matrix} \begin{bmatrix} 1 & 0 & -1 & | & 5 \\ 0 & 1 & 2 & | & 95 \\ 0 & 0 & 0 & | & 0 \end{bmatrix}$$

Since the last row is all zeros, there are infinitely many solutions Let z be the parameter. The matrix gives

$$\begin{aligned} x - z &= 5 \\ y + 2z &= 95. \end{aligned}$$

Solving these equations for x and y, the solution is $(5 + z, 95 - 2z, z)$. The numbers in the solution must be nonnegative integers. Therefore,

$$\begin{aligned} 95 - 2z &\geq 0 \\ -2z &\geq -95 \\ z &\leq 47.5. \end{aligned}$$

Thus, $z \in \{0, 1, 2, 3, ..., 47\}$. There are 48 possible solutions.

(d) For the smallest number of cars, $z = 0$, the solution is $(5, 95, 0)$. This means 5 balls, 95 dolls, and no cars.

(e) For the largest number of cars, $z = 47$, the solution is $(52, 1, 47)$. This means 52 balls, 1 doll, and 47 cars.

69. (a) $x_{11} + x_{12} + x_{21} = 1$

$x_{11} + x_{12} + x_{22} = 1$

$x_{11} + x_{21} + x_{22} = 1$

$x_{12} + x_{21} + x_{22} = 1$

Write the augmented matrix of the system.

$$\begin{bmatrix} 1 & 1 & 1 & 0 & | & 1 \\ 1 & 1 & 0 & 1 & | & 1 \\ 1 & 0 & 1 & 1 & | & 1 \\ 0 & 1 & 1 & 1 & | & 1 \end{bmatrix}$$

$$\begin{matrix} \\ \\ -1R_1 + R_2 \to R_2 \\ -1R_1 + R_3 \to R_3 \end{matrix} \begin{bmatrix} 1 & 1 & 1 & 0 & | & 1 \\ 0 & 0 & -1 & 1 & | & 0 \\ 0 & -1 & 0 & 1 & | & 0 \\ 0 & 1 & 1 & 1 & | & 1 \end{bmatrix}$$

Since $-1 = 1$ modulo 2, replace -1 with 1.

$$\begin{bmatrix} 1 & 1 & 1 & 0 & | & 1 \\ 0 & 0 & 1 & 1 & | & 0 \\ 0 & 1 & 0 & 1 & | & 0 \\ 0 & 1 & 1 & 1 & | & 1 \end{bmatrix}$$

Interchange rows 2 and 3.

$$\begin{bmatrix} 1 & 1 & 1 & 0 & | & 1 \\ 0 & 1 & 0 & 1 & | & 0 \\ 0 & 0 & 1 & 1 & | & 0 \\ 0 & 1 & 1 & 1 & | & 1 \end{bmatrix}$$

$$\begin{matrix} -1R_2 + R_1 \to R_1 \\ \\ \\ -R_2 + R_4 \to R_4 \end{matrix} \begin{bmatrix} 1 & 0 & 1 & -1 & | & 1 \\ 0 & 1 & 0 & 1 & | & 0 \\ 0 & 0 & 1 & 1 & | & 0 \\ 0 & 0 & 1 & 0 & | & 1 \end{bmatrix}$$

Again, replace -1 with 1.

$$\begin{bmatrix} 1 & 0 & 1 & 1 & | & 1 \\ 0 & 1 & 0 & 1 & | & 0 \\ 0 & 0 & 1 & 1 & | & 0 \\ 0 & 0 & 1 & 0 & | & 1 \end{bmatrix}$$

$$\begin{matrix} -1R_3 + R_1 \to R_1 \\ \\ \\ -1R_3 + R_4 \to R_4 \end{matrix} \begin{bmatrix} 1 & 0 & 0 & 0 & | & 1 \\ 0 & 1 & 0 & 1 & | & 0 \\ 0 & 0 & 1 & 1 & | & 0 \\ 0 & 0 & 0 & -1 & | & 1 \end{bmatrix}$$

Replace -1 with 1.

$$\begin{bmatrix} 1 & 0 & 0 & 0 & | & 1 \\ 0 & 1 & 0 & 1 & | & 0 \\ 0 & 0 & 1 & 1 & | & 0 \\ 0 & 0 & 0 & 1 & | & 1 \end{bmatrix}$$

$$\begin{matrix} \\ -1R_4 + R_2 \to R_2 \\ -1R_4 + R_3 \to R_3 \\ \\ \end{matrix} \begin{bmatrix} 1 & 0 & 0 & 0 & | & 1 \\ 0 & 1 & 0 & 0 & | & -1 \\ 0 & 0 & 1 & 0 & | & -1 \\ 0 & 0 & 0 & 1 & | & 1 \end{bmatrix}$$

Finally, replace -1 with 1.

$$\begin{bmatrix} 1 & 0 & 0 & 0 & | & 1 \\ 0 & 1 & 0 & 0 & | & 1 \\ 0 & 0 & 1 & 0 & | & 1 \\ 0 & 0 & 0 & 1 & | & 1 \end{bmatrix}$$

The solution $(1, 1, 1, 1)$ corresponds to $x_{11} = 1$, $x_{12} = 1$, $x_{21} = 1$, and $x_{22} = 1$. Since 1 indicates that a button is pushed, the strategy required to turn all the lights out is to push every button one time.

(b) $x_{11} + x_{12} + x_{21} = 0$

$x_{11} + x_{12} + x_{22} = 1$

$x_{11} + x_{21} + x_{22} = 1$

$x_{12} + x_{21} + x_{22} = 0$

Write the augmented matrix of the system.

$$\begin{bmatrix} 1 & 1 & 1 & 0 & | & 0 \\ 1 & 1 & 0 & 1 & | & 1 \\ 1 & 0 & 1 & 1 & | & 1 \\ 0 & 1 & 1 & 1 & | & 0 \end{bmatrix}$$

$$\begin{matrix} \\ \\ -1R_1 + R_2 \to R_2 \\ -1R_1 + R_3 \to R_3 \end{matrix} \begin{bmatrix} 1 & 1 & 1 & 0 & | & 0 \\ 0 & 0 & -1 & 1 & | & 1 \\ 0 & -1 & 0 & 1 & | & 1 \\ 0 & 1 & 1 & 1 & | & 0 \end{bmatrix}$$

Replace -1 with 1.

$$\begin{bmatrix} 1 & 1 & 1 & 0 & | & 0 \\ 0 & 0 & 1 & 1 & | & 1 \\ 0 & 1 & 0 & 1 & | & 1 \\ 0 & 1 & 1 & 1 & | & 0 \end{bmatrix}$$

Interchange rows 2 and 3.

$$\begin{bmatrix} 1 & 1 & 1 & 0 & | & 0 \\ 0 & 1 & 0 & 1 & | & 1 \\ 0 & 0 & 1 & 1 & | & 1 \\ 0 & 1 & 1 & 1 & | & 0 \end{bmatrix}$$

$$-1R_2 + R_1 \to R_1 \quad \begin{bmatrix} 1 & 0 & 1 & -1 & -1 \\ 0 & 1 & 0 & 1 & 1 \\ 0 & 0 & 1 & 1 & 1 \\ 0 & 0 & 1 & 0 & -1 \end{bmatrix}$$

$$-1R_2 + R_4 \to R_4$$

Replace -1 with 1.

$$\begin{bmatrix} 1 & 0 & 1 & 1 & 1 \\ 0 & 1 & 0 & 1 & 1 \\ 0 & 0 & 1 & 1 & 1 \\ 0 & 0 & 1 & 0 & 1 \end{bmatrix}$$

$$-1R_3 + R_1 \to R_1 \quad \begin{bmatrix} 1 & 0 & 0 & 0 & 0 \\ 0 & 1 & 0 & 1 & 1 \\ 0 & 0 & 1 & 1 & 1 \\ 0 & 0 & 0 & -1 & 0 \end{bmatrix}$$

$$-1R_3 + R_4 \to R_4$$

Replace -1 with 1.

$$\begin{bmatrix} 1 & 0 & 0 & 0 & 0 \\ 0 & 1 & 0 & 1 & 1 \\ 0 & 0 & 1 & 1 & 1 \\ 0 & 0 & 0 & 1 & 0 \end{bmatrix}$$

$$-1R_4 + R_2 \to R_2 \quad \begin{bmatrix} 1 & 0 & 0 & 0 & 0 \\ 0 & 1 & 0 & 0 & 1 \\ 0 & 0 & 1 & 0 & 1 \\ 0 & 0 & 0 & 1 & 0 \end{bmatrix}$$

$$-1R_4 + R_3 \to R_3$$

The solution $(0, 1, 1, 0)$ corresponds to $x_{11} = 0$, $x_{12} = 1$, $x_{21} = 1$, and $x_{22} = 0$. Since 1 indicates that a button is pushed and 0 indicates that it is not, the strategy required to turn all the lights out is to push the button in the first row, second column, and push the button in the second row first column.

2.3 Addition and Subtraction of Matrices

Your Turn 1

(a) It is not possible to add a 2×4 matrix and a 2×3 matrix.

(b)

$$\begin{bmatrix} 3 & 4 & 5 \\ 1 & 2 & 3 \end{bmatrix} + \begin{bmatrix} 1 & -2 & 4 \\ -2 & -4 & 8 \end{bmatrix} = \begin{bmatrix} 4 & 2 & 9 \\ -1 & -2 & 11 \end{bmatrix}$$

Your Turn 2

$$\begin{bmatrix} 3 & 4 & 5 \\ 1 & 2 & 3 \end{bmatrix} - \begin{bmatrix} 1 & -2 & 4 \\ -2 & -4 & 8 \end{bmatrix} = \begin{bmatrix} 2 & 6 & 1 \\ 3 & 6 & -5 \end{bmatrix}$$

2.3 Exercises

1. $\begin{bmatrix} 1 & 3 \\ 5 & 7 \end{bmatrix} = \begin{bmatrix} 1 & 5 \\ 3 & 7 \end{bmatrix}$

This statement is false, since not all corresponding elements are equal.

3. $\begin{bmatrix} x \\ y \end{bmatrix} = \begin{bmatrix} -2 \\ 8 \end{bmatrix}$ if $x = -2$ and $y = 8$.

This statement is true. The matrices are the same size and corresponding elements are equal.

5. $\begin{bmatrix} 1 & 9 & -4 \\ 3 & 7 & 2 \\ -1 & 1 & 0 \end{bmatrix}$ is a square matrix.

This statement is true. The matrix has 3 rows and 3 columns.

7. $\begin{bmatrix} -4 & 8 \\ 2 & 3 \end{bmatrix}$ is a 2×2 square matrix.

9. $\begin{bmatrix} -6 & 8 & 0 & 0 \\ 4 & 1 & 9 & 2 \\ 3 & -5 & 7 & 1 \end{bmatrix}$ is a 3×4 matrix.

11. $\begin{bmatrix} -7 \\ 5 \end{bmatrix}$ is a 2×1 column matrix.

13. Undefined

15. $\begin{bmatrix} 3 & 4 \\ -8 & 1 \end{bmatrix} = \begin{bmatrix} 3 & x \\ y & z \end{bmatrix}$

Corresponding elements must be equal for the matrices to be equal. Therefore, $x = 4$, $y = -8$, and $z = 1$.

17. $\begin{bmatrix} s - 4 & t + 2 \\ -5 & 7 \end{bmatrix} = \begin{bmatrix} 6 & 2 \\ -5 & r \end{bmatrix}$

Corresponding elements must be equal

$$s - 4 = 6 \qquad t + 2 = 2 \qquad r = 7.$$
$$s = 10 \qquad t = 0$$

Thus, $s = 10$, $t = 0$, and $r = 7$.

19. $\begin{bmatrix} a + 2 & 3b & 4c \\ d & 7f & 8 \end{bmatrix} + \begin{bmatrix} -7 & 2b & 6 \\ -3d & -6 & -2 \end{bmatrix} = \begin{bmatrix} 15 & 25 & 6 \\ -8 & 1 & 6 \end{bmatrix}$

Add the two matrices on the left side to obtain

$$\begin{bmatrix} a + 2 & 3b & 4c \\ d & 7f & 8 \end{bmatrix} + \begin{bmatrix} -7 & 2b & 6 \\ -3d & -6 & -2 \end{bmatrix}$$

$$= \begin{bmatrix} (a+2)+(-7) & 3b+2b & 4c+6 \\ d+(-3d) & 7f+(-6) & 8+(-2) \end{bmatrix}$$

$$= \begin{bmatrix} a-5 & 5b & 4c+6 \\ -2d & 7f-6 & 6 \end{bmatrix}$$

Corresponding elements of this matrix and the matrix on the right side of the original equation must be equal.

$$a-5 = 15 \qquad 5b = 25 \qquad 4c+6 = 6$$
$$a = 20 \qquad b = 5 \qquad c = 0$$

$$-2d = -8 \qquad 7f-6 = 1$$
$$d = 4 \qquad f = 1$$

Thus, $a = 20$, $b = 5$, $c = 0$, $d = 4$, and $f = 1$.

21. $\begin{bmatrix} 2 & 4 & 5 & -7 \\ 6 & -3 & 12 & 0 \end{bmatrix} + \begin{bmatrix} 8 & 0 & -10 & 1 \\ -2 & 8 & -9 & 11 \end{bmatrix}$

$$= \begin{bmatrix} 2+8 & 4+0 & 5+(-10) & -7+1 \\ 6+(-2) & -3+8 & 12+(-9) & 0+11 \end{bmatrix}$$

$$= \begin{bmatrix} 10 & 4 & -5 & -6 \\ 4 & 5 & 3 & 11 \end{bmatrix}$$

23. $\begin{bmatrix} 1 & 3 & -2 \\ 4 & 7 & 1 \end{bmatrix} + \begin{bmatrix} 3 & 0 \\ 6 & 4 \\ -5 & 2 \end{bmatrix}$

These matrices cannot be added since the first matrix has size 2×3, while the second has size 3×2. Only matrices that are the same size can be added.

25. The matrices have the same size, so the subtraction can be done. Let A and B represent the given matrices.

$A - B =$

$$= \begin{bmatrix} 2-1 & 8-3 & 12-6 & 0-9 \\ 7-2 & 4-(-3) & -1-(-3) & 5-4 \\ 1-8 & 2-0 & 0-(-2) & 10-17 \end{bmatrix}$$

$$= \begin{bmatrix} 1 & 5 & 6 & -9 \\ 5 & 7 & 2 & 1 \\ -7 & 2 & 2 & -7 \end{bmatrix}$$

27. $\begin{bmatrix} 2 & 3 \\ -2 & 4 \end{bmatrix} + \begin{bmatrix} 4 & 3 \\ 7 & 8 \end{bmatrix} - \begin{bmatrix} 3 & 2 \\ 1 & 4 \end{bmatrix}$

$$= \begin{bmatrix} 2+4-3 & 3+3-2 \\ -2+7-1 & 4+8-4 \end{bmatrix} = \begin{bmatrix} 3 & 4 \\ 4 & 8 \end{bmatrix}$$

29. $\begin{bmatrix} 2 & -1 \\ 0 & 13 \end{bmatrix} - \begin{bmatrix} 4 & 8 \\ -5 & 7 \end{bmatrix} + \begin{bmatrix} 12 & 7 \\ 5 & 3 \end{bmatrix}$

$$= \begin{bmatrix} 2-4+12 & -1-8+7 \\ 0-(-5)+5 & 13-7+3 \end{bmatrix} = \begin{bmatrix} 10 & -2 \\ 10 & 9 \end{bmatrix}$$

31. $\begin{bmatrix} -4x+2y & -3x+y \\ 6x-3y & 2x-5y \end{bmatrix} + \begin{bmatrix} -8x+6y & 2x \\ 3y-5x & 6x+4y \end{bmatrix}$

$$= \begin{bmatrix} (-4x+2y)+(-8x+6y) & (-3x+y)+2x \\ (6x-3y)+(3y-5x) & (2x-5y)+(6x+4y) \end{bmatrix}$$

$$= \begin{bmatrix} -12x+8y & -x+y \\ x & 8x-y \end{bmatrix}$$

33. $O - X = \begin{bmatrix} 0 & 0 \\ 0 & 0 \end{bmatrix} - \begin{bmatrix} x & y \\ z & w \end{bmatrix}$

$$= \begin{bmatrix} 0-x & 0-y \\ 0-z & 0-w \end{bmatrix} = \begin{bmatrix} -x & -y \\ -z & -w \end{bmatrix}$$

35. Show that $X + (T + P) = (X + T) + P$.

On the left side, the sum $T + P$ is obtained first, and then

$$X + (T + P).$$

This gives the matrix

$$\begin{bmatrix} x+(r+m) & y+(s+n) \\ z+(t+p) & w+(u+q) \end{bmatrix}.$$

For the right side, first the sum $X + T$ is obtained, and then

$$(X + T) + P.$$

This gives the matrix

$$\begin{bmatrix} (x+r)+m & (y+s)+n \\ (z+t)+p & (w+u)+q \end{bmatrix}.$$

Comparing corresponding elements, we see that they are equal by the associative property of addition of real numbers. Thus,

$$X + (T + P) = (X + T) + P.$$

37. Show that $P + O = P$.

$$P + O = \begin{bmatrix} m & n \\ p & q \end{bmatrix} + \begin{bmatrix} 0 & 0 \\ 0 & 0 \end{bmatrix} = \begin{bmatrix} m+0 & n+0 \\ p+0 & q+0 \end{bmatrix}$$

$$= \begin{bmatrix} m & n \\ p & q \end{bmatrix} = P$$

Thus, $P + O = P$.

39. **(a)** The production cost matrix for Chicago is

$$\begin{array}{cc} & \text{Phones} \quad \text{Calculators} \\ \begin{array}{c} \text{Material} \\ \text{Labor} \end{array} & \begin{bmatrix} 4.05 & 7.01 \\ 3.27 & 3.51 \end{bmatrix}. \end{array}$$

The production cost matrix for Seattle is

$$\begin{array}{cc} & \text{Phones} \quad \text{Calculators} \\ \begin{array}{c} \text{Material} \\ \text{Labor} \end{array} & \begin{bmatrix} 4.40 & 6.90 \\ 3.54 & 3.76 \end{bmatrix}. \end{array}$$

(b) The new production cost matrix for Chicago is

$$\begin{array}{cc} & \text{Phones} \qquad \text{Calculators} \\ \begin{array}{c} \text{Material} \\ \text{Labor} \end{array} & \begin{bmatrix} 4.05 + 0.37 & 7.01 + 0.42 \\ 3.27 + 0.11 & 3.51 + 0.11 \end{bmatrix} \end{array}$$

or $\begin{bmatrix} 4.42 & 7.43 \\ 3.38 & 3.62 \end{bmatrix}$.

41. **(a)** There are four food groups and three meals. To represent the data by a 3×4 matrix, we must use the rows to correspond to the meals, breakfast, lunch, and dinner, and the columns to correspond to the four food groups. Thus, we obtain the matrix

$$\begin{bmatrix} 2 & 1 & 2 & 1 \\ 3 & 2 & 2 & 1 \\ 4 & 3 & 2 & 1 \end{bmatrix}.$$

(b) There are four food groups. These will correspond to the four rows. There are three components in each food group: fat, carbohydrates, and protein. These will correspond to the three columns. The matrix is

$$\begin{bmatrix} 5 & 0 & 7 \\ 0 & 10 & 1 \\ 0 & 15 & 2 \\ 10 & 12 & 8 \end{bmatrix}.$$

(c) The matrix is

$$\begin{bmatrix} 8 \\ 4 \\ 5 \end{bmatrix}.$$

43.

$$\begin{array}{cc} & \text{Obtained Pain Relief} \\ & \text{Yes} \quad \text{No} \\ \begin{array}{c} \text{Painfree} \\ \text{Placebo} \end{array} & \begin{bmatrix} 22 & 3 \\ 8 & 17 \end{bmatrix} \end{array}$$

(a) Of the 25 patients who took the placebo, 8 got relief.

(b) Of the 25 patients who took Painfree, 3 got no relief.

(c)

$$\begin{bmatrix} 22 & 3 \\ 8 & 17 \end{bmatrix} + \begin{bmatrix} 21 & 4 \\ 6 & 19 \end{bmatrix} + \begin{bmatrix} 19 & 6 \\ 10 & 15 \end{bmatrix} + \begin{bmatrix} 23 & 2 \\ 3 & 22 \end{bmatrix} = \begin{bmatrix} 85 & 15 \\ 27 & 73 \end{bmatrix}$$

(d) Yes, it appears that Painfree is effective. Of the 100 patients who took the medication, 85% got relief.

45. **(a)** The matrix for the life expectancy of African Americans is

$$\begin{array}{cc} & \text{M} \qquad \text{F} \\ \begin{array}{c} 1970 \\ 1980 \\ 1990 \\ 2000 \end{array} & \begin{bmatrix} 60.0 & 68.3 \\ 63.8 & 72.5 \\ 64.5 & 73.6 \\ 68.3 & 75.2 \end{bmatrix} \end{array}$$

(b) The matrix for the life expectancy of White Americans is

$$\begin{array}{cc} & \text{M} \qquad \text{F} \\ \begin{array}{c} 1970 \\ 1980 \\ 1990 \\ 2000 \end{array} & \begin{bmatrix} 68.0 & 75.6 \\ 70.7 & 78.1 \\ 72.7 & 79.4 \\ 74.9 & 80.1 \end{bmatrix} \end{array}$$

(c) The matrix showing the difference between the life expectancy between the two groups is

$$\begin{bmatrix} 60.0 & 68.3 \\ 63.8 & 72.5 \\ 64.5 & 73.6 \\ 68.3 & 75.2 \end{bmatrix} - \begin{bmatrix} 68.0 & 75.6 \\ 70.7 & 78.1 \\ 72.7 & 79.4 \\ 74.9 & 80.1 \end{bmatrix} = \begin{bmatrix} -8.0 & -7.3 \\ -6.9 & -5.6 \\ -8.2 & -5.8 \\ -6.6 & -4.9 \end{bmatrix}$$

47. **(a)** The matrix for the educational attainment of African Americans is

	4 Years of High School or More	4 Years of College or More
1980	51.2	7.9
1985	59.8	11.1
1990	66.2	11.3
1995	73.8	13.2
2000	78.5	16.5
2008	83.0	19.6

(b) The matrix for the educational attainment of Hispanic Americans is

	4 Years of High School or More	4 Years of College or More
1980	45.3	7.9
1985	47.9	8.5
1990	50.8	9.2
1995	53.4	9.3
2000	57.0	10.6
2008	62.3	13.3

(c) The matrix showing the difference in the educational attainment between African and Hispanic Americans is

$$\begin{bmatrix} 51.2 & 7.9 \\ 59.8 & 11.1 \\ 66.2 & 11.3 \\ 73.8 & 13.2 \\ 78.5 & 16.5 \\ 83.0 & 19.6 \end{bmatrix} - \begin{bmatrix} 45.3 & 7.9 \\ 47.9 & 8.5 \\ 50.8 & 9.2 \\ 53.4 & 9.3 \\ 57.0 & 10.6 \\ 62.3 & 13.3 \end{bmatrix} = \begin{bmatrix} 5.9 & 0 \\ 11.9 & 2.6 \\ 15.4 & 2.1 \\ 20.4 & 3.9 \\ 21.5 & 5.9 \\ 20.7 & 6.3 \end{bmatrix}$$

2.4 Multiplication of Matrices

Your Turn 1

$$AB = \begin{bmatrix} 3 & 4 \\ 1 & 2 \end{bmatrix}\begin{bmatrix} 1 & -2 \\ -2 & -4 \end{bmatrix}$$

$$= \begin{bmatrix} 3(1) + 4(-2) & 3(-2) + 4(-4) \\ 1(1) + 2(-2) & 1(-2) + 2(-4) \end{bmatrix}$$

$$= \begin{bmatrix} -5 & -22 \\ -3 & -10 \end{bmatrix}$$

Your Turn 2

$$AB = \begin{bmatrix} 3 & 5 & -1 \\ 2 & 4 & -2 \end{bmatrix}\begin{bmatrix} 3 & -4 \\ -5 & -3 \end{bmatrix}$$

AB does not exist because a 2×3 matrix cannot be multiplied by a 2×2 matrix.

$$BA = \begin{bmatrix} 3 & -4 \\ -5 & -3 \end{bmatrix}\begin{bmatrix} 3 & 5 & -1 \\ 2 & 4 & -2 \end{bmatrix}$$

$$= \begin{bmatrix} 3(3) - 4(2) & 3(5) - 4(4) & 3(-1) - 4(-2) \\ -5(3) - 3(2) & -5(5) - 3(4) & -5(-1) - 3(-2) \end{bmatrix}$$

$$= \begin{bmatrix} 1 & -1 & 5 \\ -21 & -37 & 11 \end{bmatrix}$$

2.4 Exercises

In Exercises 1-6, let

$$A = \begin{bmatrix} -2 & 4 \\ 0 & 3 \end{bmatrix} \text{ and } B = \begin{bmatrix} -6 & 2 \\ 4 & 0 \end{bmatrix}.$$

1. $2A = 2\begin{bmatrix} -2 & 4 \\ 0 & 3 \end{bmatrix} = \begin{bmatrix} -4 & 8 \\ 0 & 6 \end{bmatrix}$

3. $-6A = -6\begin{bmatrix} -2 & 4 \\ 0 & 3 \end{bmatrix} = \begin{bmatrix} 12 & -24 \\ 0 & -18 \end{bmatrix}$

5. $-4A + 5B = -4\begin{bmatrix} -2 & 4 \\ 0 & 3 \end{bmatrix} + 5\begin{bmatrix} -6 & 2 \\ 4 & 0 \end{bmatrix}$

$$= \begin{bmatrix} 8 & -16 \\ 0 & -12 \end{bmatrix} + \begin{bmatrix} -30 & 10 \\ 20 & 0 \end{bmatrix}$$

$$= \begin{bmatrix} -22 & -6 \\ 20 & -12 \end{bmatrix}$$

7. Matrix A size   Matrix B size
    $2 \times \underline{\mathbf{2}}$ $\underline{\mathbf{2}} \times 2$

The number of columns of A is the same as the number of rows of B, so the product AB exists. The size of the matrix AB is 2×2.

Matrix B size   Matrix A size
    $2 \times \underline{\mathbf{2}}$ $\underline{\mathbf{2}} \times 2$

Since the number of columns of B is the same as the number of rows of A, the product BA also exists and has size 2×2.

9. Matrix A size   Matrix B size
    $3 \times \underline{\mathbf{4}}$ $\underline{\mathbf{4}} \times 4$

Since matrix A has 4 columns and matrix B has 4 rows, the product AB exists and has size 3×4.

Matrix B size   Matrix A size
    $4 \times \underline{\mathbf{4}}$ $\underline{\mathbf{3}} \times 4$

Since B has 4 columns and A has 3 rows, the product BA does not exist.

11. Matrix A size   Matrix B size
    $4 \times \underline{\mathbf{2}}$ $\underline{\mathbf{3}} \times 4$

The number of columns of A is not the same as the number of rows of B, so the product AB does not exist.

Matrix B size   Matrix A size
    $3 \times \underline{\mathbf{4}}$ $\underline{\mathbf{4}} \times 2$

The number of columns of B is the same as the number of rows of A, so the product BA exists and has size 3×2.

13. To find the product matrix AB, the number of *columns* of A must be the same as the number of *rows* of B.

15. Call the first matrix A and the second matrix B. The product matrix AB will have size 2×1.

Step 1: Multiply the elements of the first row of A by the corresponding elements of the column of B and add.

$$\begin{bmatrix} 2 & -1 \\ 5 & 8 \end{bmatrix} \begin{bmatrix} 3 \\ -2 \end{bmatrix} \qquad 2(3) + (-1)(-2) = 8$$

Therefore, 8 is the first row entry of the product matrix AB.

Step 2: Multiply the elements of the second row of A by the corresponding elements of the column of B and add.

$$\begin{bmatrix} 2 & -1 \\ 5 & 8 \end{bmatrix} \begin{bmatrix} 3 \\ -2 \end{bmatrix} \qquad 5(3) + 8(-2) = -1$$

The second row entry of the product is -1.

Step 3: Write the product using the two entries found above.

$$AB = \begin{bmatrix} 2 & -1 \\ 5 & 8 \end{bmatrix} \begin{bmatrix} 3 \\ -2 \end{bmatrix} = \begin{bmatrix} 8 \\ -1 \end{bmatrix}$$

17. $\begin{bmatrix} 2 & -1 & 7 \\ -3 & 0 & -4 \end{bmatrix} \begin{bmatrix} 5 \\ 10 \\ 2 \end{bmatrix}$

$$= \begin{bmatrix} 2 \cdot 5 + (-1) \cdot 10 + 7 \cdot 2 \\ (-3) \cdot 5 + 0 \cdot 10 + (-4) \cdot 2 \end{bmatrix}$$

$$= \begin{bmatrix} 14 \\ -23 \end{bmatrix}$$

19. $\begin{bmatrix} 2 & -1 \\ 3 & 6 \end{bmatrix} \begin{bmatrix} -1 & 0 & 4 \\ 5 & -2 & 0 \end{bmatrix}$

$$= \begin{bmatrix} 2 \cdot (-1) + (-1) \cdot 5 & 2 \cdot 0 + (-1) \cdot (-2) & 2 \cdot 4 + (-1) \cdot 0 \\ 3 \cdot (-1) + 6 \cdot 5 & 3 \cdot 0 + 6 \cdot (-2) & 3 \cdot 4 + 6 \cdot 0 \end{bmatrix}$$

$$= \begin{bmatrix} -7 & 2 & 8 \\ 27 & -12 & 12 \end{bmatrix}$$

21. $\begin{bmatrix} 2 & 2 & -1 \\ 3 & 0 & 1 \end{bmatrix} \begin{bmatrix} 0 & 2 \\ -1 & 4 \\ 0 & 2 \end{bmatrix}$

$$= \begin{bmatrix} 2 \cdot 0 + 2(-1) + (-1)0 & 2 \cdot 2 + 2 \cdot 4 + (-1)2 \\ 3 \cdot 0 + 0(-1) + 1(0) & 3 \cdot 2 + 0 \cdot 4 + 1 \cdot 2 \end{bmatrix}$$

$$= \begin{bmatrix} -2 & 10 \\ 0 & 8 \end{bmatrix}$$

23. $\begin{bmatrix} 1 & 2 \\ 3 & 4 \end{bmatrix} \begin{bmatrix} -1 & 5 \\ 7 & 0 \end{bmatrix}$

$$= \begin{bmatrix} 1(-1) + 2 \cdot 7 & 1 \cdot 5 + 2 \cdot 0 \\ 3(-1) + 4 \cdot 7 & 3 \cdot 5 + 4 \cdot 0 \end{bmatrix}$$

$$= \begin{bmatrix} 13 & 5 \\ 25 & 15 \end{bmatrix}$$

25. $\begin{bmatrix} -2 & -3 & 7 \\ 1 & 5 & 6 \end{bmatrix} \begin{bmatrix} 1 \\ 2 \\ 3 \end{bmatrix} = \begin{bmatrix} -2(1) + (-3)2 + 7 \cdot 3 \\ 1 \cdot 1 + 5 \cdot 2 + 6 \cdot 3 \end{bmatrix}$

$$= \begin{bmatrix} 13 \\ 29 \end{bmatrix}$$

27. $\left(\begin{bmatrix} 2 & 1 \\ -3 & -6 \\ 4 & 0 \end{bmatrix} \begin{bmatrix} 1 & -2 \\ 2 & -1 \end{bmatrix} \right) \begin{bmatrix} 3 \\ 1 \end{bmatrix} = \begin{bmatrix} 4 & -5 \\ -15 & 12 \\ 4 & -8 \end{bmatrix} \begin{bmatrix} 3 \\ 1 \end{bmatrix}$

$$= \begin{bmatrix} 7 \\ -33 \\ 4 \end{bmatrix}$$

29. $\begin{bmatrix} 2 & -2 \\ 1 & -1 \end{bmatrix} \left(\begin{bmatrix} 4 & 3 \\ 1 & 2 \end{bmatrix} + \begin{bmatrix} 7 & 0 \\ -1 & 5 \end{bmatrix} \right)$

$$= \begin{bmatrix} 2 & -2 \\ 1 & -1 \end{bmatrix} \begin{bmatrix} 11 & 3 \\ 0 & 7 \end{bmatrix}$$

$$= \begin{bmatrix} 22 & -8 \\ 11 & -4 \end{bmatrix}$$

31. **(a)** $AB = \begin{bmatrix} -2 & 4 \\ 1 & 3 \end{bmatrix} \begin{bmatrix} -2 & 1 \\ 3 & 6 \end{bmatrix} = \begin{bmatrix} 16 & 22 \\ 7 & 19 \end{bmatrix}$

(b) $BA = \begin{bmatrix} -2 & 1 \\ 3 & 6 \end{bmatrix} \begin{bmatrix} -2 & 4 \\ 1 & 3 \end{bmatrix} = \begin{bmatrix} 5 & -5 \\ 0 & 30 \end{bmatrix}$

(c) No, AB and BA are not equal here.

(d) No, AB does not always equal BA.

33. Verify that $P(X + T) = PX + PT$.

Find $P(X + T)$ and $PX + PT$ separately and compare their values to see if they are the same.

$$P(X + T) = \begin{bmatrix} m & n \\ p & q \end{bmatrix} \left(\begin{bmatrix} x & y \\ z & w \end{bmatrix} + \begin{bmatrix} r & s \\ t & u \end{bmatrix} \right) = \begin{bmatrix} m & n \\ p & q \end{bmatrix} \left(\begin{bmatrix} x + r & y + s \\ z + t & w + u \end{bmatrix} \right)$$

$$= \begin{bmatrix} m(x + r) + n(z + t) & m(y + s) + n(w + u) \\ p(x + r) + q(z + t) & p(y + s) + q(w + u) \end{bmatrix} = \begin{bmatrix} mx + mr + nz + nt & my + ms + nw + nu \\ px + pr + qz + qt & py + ps + qw + qu \end{bmatrix}$$

$$PX + PT = \begin{bmatrix} m & n \\ p & q \end{bmatrix} \begin{bmatrix} x & y \\ z & w \end{bmatrix} + \begin{bmatrix} m & n \\ p & q \end{bmatrix} \begin{bmatrix} r & s \\ t & u \end{bmatrix} = \begin{bmatrix} mx + nz & my + nw \\ px + qz & py + qw \end{bmatrix} + \begin{bmatrix} mr + nt & ms + nu \\ pr + qt & ps + qu \end{bmatrix}$$

$$= \begin{bmatrix} (mx + nz) + (mr + nt) & (my + nw) + (ms + nu) \\ (px + qz) + (pr + qt) & (py + qw) + (ps + qu) \end{bmatrix} = \begin{bmatrix} mx + nz + mr + nt & my + nw + ms + nu \\ px + qz + pr + qt & py + qw + ps + qu \end{bmatrix}$$

$$= \begin{bmatrix} mx + mr + nz + nt & my + ms + nw + nu \\ px + pr + qz + qt & py + ps + qw + qu \end{bmatrix}$$

Observe that the two results are identical. Thus, $P(X + T) = PX + PT$.

35. Verify that $(k + h)P = kP + hP$ for any real numbers k and h.

$$(k + h)P = (k + h) \begin{bmatrix} m & n \\ p & q \end{bmatrix}$$

$$= \begin{bmatrix} (k + h)m & (k + h)n \\ (k + h)p & (k + h)q \end{bmatrix}$$

$$= \begin{bmatrix} km + hm & kn + hn \\ kp + hp & kq + hq \end{bmatrix}$$

$$= \begin{bmatrix} km & kn \\ kp & kq \end{bmatrix} + \begin{bmatrix} hm & hn \\ hp & hq \end{bmatrix}$$

$$= k \begin{bmatrix} m & n \\ p & q \end{bmatrix} + h \begin{bmatrix} m & n \\ p & q \end{bmatrix}$$

$$= kP + hP$$

Thus, $(k + h)P = kP + hP$ for any real numbers k and h.

37.
$$\begin{bmatrix} 2 & 3 & 1 \\ 1 & -4 & 5 \end{bmatrix} \begin{bmatrix} x_1 \\ x_2 \\ x_3 \end{bmatrix} = \begin{bmatrix} 2x_1 + 3x_2 + x_3 \\ x_1 - 4x_2 + 5x_3 \end{bmatrix},$$

and $\begin{bmatrix} 2x_1 + 3x_2 + x_3 \\ x_1 - 4x_2 + 5x_3 \end{bmatrix} = \begin{bmatrix} 5 \\ 8 \end{bmatrix}$.

This is equivalent to

$$2x_1 + 3x_2 + x_3 = 5$$
$$x_1 - 4x_2 + 5x_3 = 8$$

since corresponding elements of equal matrices must be equal. Reversing this, observe that the

given system of linear equations can be written as the matrix equation

$$\begin{bmatrix} 2 & 3 & 1 \\ 1 & -4 & 5 \end{bmatrix} \begin{bmatrix} x_1 \\ x_2 \\ x_3 \end{bmatrix} = \begin{bmatrix} 5 \\ 8 \end{bmatrix}.$$

39. **(a)** Use a graphing calculator or a computer to find the product matrix. The answer is

$$AC = \begin{bmatrix} 6 & 106 & 158 & 222 & 28 \\ 120 & 139 & 64 & 75 & 115 \\ -146 & -2 & 184 & 144 & -129 \\ 106 & 94 & 24 & 116 & 110 \end{bmatrix}.$$

(b) CA does not exist.

(c) AC and CA are clearly not equal, since CA does not even exist.

41. Use a graphing calculator or computer to find the matrix products and sums. The answers are as follows.

(a) $C + D = \begin{bmatrix} -1 & 5 & 9 & 13 & -1 \\ 7 & 17 & 2 & -10 & 6 \\ 18 & 9 & -12 & 12 & 22 \\ 9 & 4 & 18 & 10 & -3 \\ 1 & 6 & 10 & 28 & 5 \end{bmatrix}$

(b) $(C + D)B = \begin{bmatrix} -2 & -9 & 90 & 77 \\ -42 & -63 & 127 & 62 \\ 413 & 76 & 180 & -56 \\ -29 & -44 & 198 & 85 \\ 137 & 20 & 162 & 103 \end{bmatrix}$

(c) $CB = \begin{bmatrix} -56 & -1 & 1 & 45 \\ -156 & -119 & 76 & 122 \\ 315 & 86 & 118 & -91 \\ -17 & -17 & 116 & 51 \\ 118 & 19 & 125 & 77 \end{bmatrix}$

(d) $DB = \begin{bmatrix} 54 & -8 & 89 & 32 \\ 114 & 56 & 51 & -60 \\ 98 & -10 & 62 & 35 \\ -12 & -27 & 82 & 34 \\ 19 & 1 & 37 & 26 \end{bmatrix}$

(e) $CB + DB = \begin{bmatrix} -2 & -9 & 90 & 77 \\ -42 & -63 & 127 & 62 \\ 413 & 76 & 180 & -56 \\ -29 & -44 & 198 & 85 \\ 137 & 20 & 162 & 103 \end{bmatrix}$

(f) Yes, $(C + D)B$ and $CB + DB$ are equal, as can be seen by observing that the answers to parts (b) and (e) are identical.

43. (a) $\begin{bmatrix} 10 & 4 & 3 & 5 & 6 \\ 7 & 2 & 2 & 3 & 8 \\ 4 & 5 & 1 & 0 & 10 \\ 0 & 3 & 4 & 5 & 5 \end{bmatrix} \begin{bmatrix} 2 & 3 \\ 1 & 1 \\ 4 & 3 \\ 3 & 3 \\ 1 & 2 \end{bmatrix}$

$= \begin{array}{c} \text{Dept. 1} \\ \text{Dept. 2} \\ \text{Dept. 3} \\ \text{Dept. 4} \end{array} \begin{array}{cc} A & B \\ \begin{bmatrix} 57 & 70 \\ 41 & 54 \\ 27 & 40 \\ 39 & 40 \end{bmatrix} \end{array}$

(b) The total cost to buy from supplier A is $57 + 41 + 27 + 39 = \$164$, and the total cost to buy from supplier B is $70 + 54 + 40 + 40 = \$204$. The company should make the purchase from supplier A, since \$164 is a lower total cost than \$204.

45. (a) To find the average, add the matrices. Then multiply the resulting matrix by $\frac{1}{3}$. (Multiplying by $\frac{1}{3}$ is the same as dividing by 3.)

$\frac{1}{3}\left(\begin{bmatrix} 4.27 & 6.94 \\ 3.45 & 3.65 \end{bmatrix} + \begin{bmatrix} 4.05 & 7.01 \\ 3.27 & 3.51 \end{bmatrix} + \begin{bmatrix} 4.40 & 6.90 \\ 3.54 & 3.76 \end{bmatrix}\right)$

$= \frac{1}{3}\begin{bmatrix} 12.72 & 20.85 \\ 10.26 & 10.92 \end{bmatrix} = \begin{bmatrix} 4.24 & 6.95 \\ 3.42 & 3.64 \end{bmatrix}$

(b) To find the new average, add the new matrix for the Chicago plant and the matrix for the Seattle plant. Since there are only two matrices now, multiply the resulting matrix by $\frac{1}{2}$ to get the average. (Multiplying by $\frac{1}{2}$ is the same as dividing by 2.)

$\frac{1}{2}\left(\begin{bmatrix} 4.42 & 7.43 \\ 3.38 & 3.62 \end{bmatrix} + \begin{bmatrix} 4.40 & 6.90 \\ 3.54 & 3.76 \end{bmatrix}\right)$

$= \frac{1}{2}\begin{bmatrix} 8.82 & 14.33 \\ 6.92 & 7.38 \end{bmatrix} = \begin{bmatrix} 4.41 & 7.17 \\ 3.46 & 3.69 \end{bmatrix}$

47. (a) $P = \begin{array}{c} \text{Sal's} \\ \text{Fred's} \end{array} \begin{array}{ccc} \text{Sh} & \text{Sa} & \text{B} \\ \begin{bmatrix} 80 & 40 & 120 \\ 60 & 30 & 150 \end{bmatrix} \end{array}$

(b) $F = \begin{array}{c} \text{Sh} \\ \text{Sa} \\ \text{B} \end{array} \begin{array}{cc} \text{CA} & \text{AR} \\ \begin{bmatrix} \frac{1}{2} & \frac{1}{5} \\ \frac{1}{4} & \frac{1}{5} \\ \frac{1}{4} & \frac{3}{5} \end{bmatrix} \end{array}$

(c) $PF = \begin{bmatrix} 80 & 40 & 120 \\ 60 & 30 & 150 \end{bmatrix} \begin{bmatrix} \frac{1}{2} & \frac{1}{5} \\ \frac{1}{4} & \frac{1}{5} \\ \frac{1}{4} & \frac{3}{5} \end{bmatrix}$

$= \begin{bmatrix} 80\left(\frac{1}{2}\right) + 40\left(\frac{1}{4}\right) + 120\left(\frac{1}{4}\right) & 80\left(\frac{1}{5}\right) + 40\left(\frac{1}{5}\right) + 120\left(\frac{3}{5}\right) \\ 60\left(\frac{1}{2}\right) + 30\left(\frac{1}{4}\right) + 150\left(\frac{1}{4}\right) & 60\left(\frac{1}{5}\right) + 30\left(\frac{1}{5}\right) + 150\left(\frac{3}{5}\right) \end{bmatrix}$

$= \begin{bmatrix} 80 & 96 \\ 75 & 108 \end{bmatrix}$

The rows give the average price per pair of footwear sold by each store, and the columns give the state.

(d) The average price of footwear at a Fred's outlet in Arizona is \$108.

49. (a)

$$XY = \begin{bmatrix} 2 & 1 & 2 & 1 \\ 3 & 2 & 2 & 1 \\ 4 & 3 & 2 & 1 \end{bmatrix} \begin{bmatrix} 5 & 0 & 7 \\ 0 & 10 & 1 \\ 0 & 15 & 2 \\ 10 & 12 & 8 \end{bmatrix} = \begin{bmatrix} 20 & 52 & 27 \\ 25 & 62 & 35 \\ 30 & 72 & 43 \end{bmatrix}$$

The rows give the amounts of fat, carbohydrates, and protein, respectively, in each of the daily meals.

(b)
$$YZ = \begin{bmatrix} 5 & 0 & 7 \\ 0 & 10 & 1 \\ 0 & 15 & 2 \\ 10 & 12 & 8 \end{bmatrix} \begin{bmatrix} 8 \\ 4 \\ 5 \end{bmatrix} = \begin{bmatrix} 75 \\ 45 \\ 70 \\ 168 \end{bmatrix}$$

The rows give the number of calories in one exchange of each of the food groups.

(c) Use the matrices found for XY and YZ from parts (a) and (b).

$$(XY)Z = \begin{bmatrix} 20 & 52 & 27 \\ 25 & 62 & 35 \\ 30 & 72 & 43 \end{bmatrix} \begin{bmatrix} 8 \\ 4 \\ 5 \end{bmatrix} = \begin{bmatrix} 503 \\ 623 \\ 743 \end{bmatrix}$$

$$X(YZ) = \begin{bmatrix} 2 & 1 & 2 & 1 \\ 3 & 2 & 2 & 1 \\ 4 & 3 & 2 & 1 \end{bmatrix} \begin{bmatrix} 75 \\ 45 \\ 70 \\ 168 \end{bmatrix} = \begin{bmatrix} 503 \\ 623 \\ 743 \end{bmatrix}$$

The rows give the number of calories in each meal.

51.
$$\frac{1}{6}\begin{bmatrix} 60.0 & 68.3 \\ 63.8 & 72.5 \\ 64.5 & 73.6 \\ 68.3 & 75.2 \end{bmatrix} + \frac{5}{6}\begin{bmatrix} 68.0 & 75.6 \\ 70.7 & 78.1 \\ 72.7 & 79.4 \\ 74.9 & 80.1 \end{bmatrix}$$

$$= \frac{1}{6}\left(\begin{bmatrix} 60.0 & 68.3 \\ 63.8 & 72.5 \\ 64.5 & 73.6 \\ 68.3 & 75.2 \end{bmatrix} + 5\begin{bmatrix} 68.0 & 75.6 \\ 70.7 & 78.1 \\ 72.7 & 79.4 \\ 74.9 & 80.1 \end{bmatrix}\right)$$

$$= \frac{1}{6}\left(\begin{bmatrix} 60.0 & 68.3 \\ 63.8 & 72.5 \\ 64.5 & 73.6 \\ 68.3 & 75.2 \end{bmatrix} + \begin{bmatrix} 340.0 & 378.0 \\ 353.5 & 390.5 \\ 363.5 & 397.0 \\ 374.5 & 400.5 \end{bmatrix}\right)$$

$$= \frac{1}{6}\begin{bmatrix} 400.0 & 446.3 \\ 417.3 & 463.0 \\ 428.0 & 470.6 \\ 442.8 & 475.7 \end{bmatrix} = \begin{bmatrix} 66.7 & 74.4 \\ 69.6 & 77.2 \\ 71.3 & 78.4 \\ 73.8 & 79.3 \end{bmatrix}$$

53. (a) The matrices are

$$A = \begin{bmatrix} 0.0346 & 0.0118 \\ 0.0174 & 0.0073 \\ 0.0189 & 0.0059 \\ 0.0135 & 0.0083 \\ 0.0099 & 0.0103 \end{bmatrix}$$

$$B = \begin{bmatrix} 361 & 2038 & 286 & 227 & 460 \\ 473 & 2494 & 362 & 252 & 484 \\ 627 & 2978 & 443 & 278 & 499 \\ 803 & 3435 & 524 & 314 & 511 \\ 1013 & 3824 & 591 & 344 & 522 \end{bmatrix}$$

(b) The total number of births and deaths each year is found by multiplying matrix B by matrix A.

$$BA = \begin{bmatrix} 361 & 2038 & 286 & 227 & 460 \\ 473 & 2494 & 362 & 252 & 484 \\ 627 & 2978 & 443 & 278 & 499 \\ 803 & 3435 & 524 & 314 & 511 \\ 1013 & 3824 & 591 & 344 & 522 \end{bmatrix}\begin{bmatrix} 0.0346 & 0.0118 \\ 0.0174 & 0.0073 \\ 0.0189 & 0.0059 \\ 0.0135 & 0.0083 \\ 0.0099 & 0.0103 \end{bmatrix}$$

$$= \begin{array}{c} \\ 1970 \\ 1980 \\ 1990 \\ 2000 \\ 2010 \end{array}\begin{bmatrix} \text{Births} & \text{Deaths} \\ 60.98 & 27.45 \\ 74.80 & 33.00 \\ 90.58 & 39.20 \\ 106.75 & 45.51 \\ 122.57 & 51.59 \end{bmatrix}$$

2.5 Matrix Inverses

Your Turn 1

Use row operations to transform the augmented matrix so that the identity matrix is the first three columns.

$$A = \begin{bmatrix} 2 & 3 & 1 & | & 1 & 0 & 0 \\ 1 & -2 & -1 & | & 0 & 1 & 0 \\ 3 & 3 & 2 & | & 0 & 0 & 1 \end{bmatrix}$$

$$R_1 \leftrightarrow R_2 \begin{bmatrix} 1 & -2 & -1 & | & 0 & 1 & 0 \\ 2 & 3 & 1 & | & 1 & 0 & 0 \\ 3 & 3 & 2 & | & 0 & 0 & 1 \end{bmatrix}$$

$$\begin{array}{c} \\ -2R_1 + R_2 \to R_2 \\ -3R_1 + R_3 \to R_3 \end{array}\begin{bmatrix} 1 & -2 & -1 & | & 0 & 1 & 0 \\ 0 & 7 & 3 & | & 1 & -2 & 0 \\ 0 & 9 & 5 & | & 0 & -3 & 1 \end{bmatrix}$$

$$\frac{1}{7}R_2 \rightarrow R_2 \begin{bmatrix} 1 & -2 & -1 & 0 & 1 & 0 \\ 0 & 1 & \frac{3}{7} & \frac{1}{7} & -\frac{2}{7} & 0 \\ 0 & 9 & 5 & 0 & -3 & 1 \end{bmatrix}$$

$$2R_2+R_1 \rightarrow R_1 \begin{bmatrix} 1 & 0 & -\frac{1}{7} & \frac{2}{7} & \frac{3}{7} & 0 \\ 0 & 1 & \frac{3}{7} & \frac{1}{7} & -\frac{2}{7} & 0 \\ -9R_2+R_3 \rightarrow R_3 & 0 & 0 & \frac{8}{7} & -\frac{9}{7} & -\frac{3}{7} & 1 \end{bmatrix}$$

$$\frac{7}{8}R_3 \rightarrow R_3 \begin{bmatrix} 1 & 0 & -\frac{1}{7} & \frac{2}{7} & \frac{3}{7} & 0 \\ 0 & 1 & \frac{3}{7} & \frac{1}{7} & -\frac{2}{7} & 0 \\ 0 & 0 & 1 & -\frac{9}{8} & -\frac{3}{8} & \frac{7}{8} \end{bmatrix}$$

$$\frac{1}{7}R_3+R_1 \rightarrow R_1 \\ -\frac{3}{7}R_3+R_2 \rightarrow R_2 \begin{bmatrix} 1 & 0 & 0 & \frac{1}{8} & \frac{3}{8} & \frac{1}{8} \\ 0 & 1 & 0 & \frac{5}{8} & -\frac{1}{8} & -\frac{3}{8} \\ 0 & 0 & 1 & -\frac{9}{8} & -\frac{3}{8} & \frac{7}{8} \end{bmatrix}$$

Thus, $A^{-1} = \begin{bmatrix} \frac{1}{8} & \frac{3}{8} & \frac{1}{8} \\ \frac{5}{8} & -\frac{1}{8} & -\frac{3}{8} \\ -\frac{9}{8} & -\frac{3}{8} & \frac{7}{8} \end{bmatrix}$.

Your Turn 2

Solve the linear system

$$5x + 4y = 23$$
$$4x - 3y = 6 .$$

Let $A = \begin{bmatrix} 5 & 4 \\ 4 & -3 \end{bmatrix}$ be the coefficient matrix,

$B = \begin{bmatrix} 23 \\ 6 \end{bmatrix}$, and $X = \begin{bmatrix} x \\ y \end{bmatrix}$.

First find A^{-1}.

$$\begin{bmatrix} 5 & 4 & 1 & 0 \\ 4 & -3 & 0 & 1 \end{bmatrix}$$

$$\frac{1}{5}R_1 \rightarrow R_1 \begin{bmatrix} 1 & \frac{4}{5} & \frac{1}{5} & 0 \\ 4 & -3 & 0 & 1 \end{bmatrix}$$

$$-4R_1+R_2 \rightarrow R_2 \begin{bmatrix} 1 & \frac{4}{5} & \frac{1}{5} & 0 \\ 0 & -\frac{31}{5} & -\frac{4}{5} & 1 \end{bmatrix}$$

$$-\frac{5}{31}R_2 \rightarrow R_2 \begin{bmatrix} 1 & \frac{4}{5} & \frac{1}{5} & 0 \\ 0 & 1 & \frac{4}{31} & -\frac{5}{31} \end{bmatrix}$$

$$-\frac{4}{5}R_2+R_1 \rightarrow R_1 \begin{bmatrix} 1 & 0 & \frac{3}{31} & \frac{4}{31} \\ 0 & 1 & \frac{4}{31} & -\frac{5}{31} \end{bmatrix}$$

$$A^{-1} = \begin{bmatrix} \frac{3}{31} & \frac{4}{31} \\ \frac{4}{31} & -\frac{5}{31} \end{bmatrix}$$

$$X = A^{-1}B = \begin{bmatrix} \frac{3}{31} & \frac{4}{31} \\ \frac{4}{31} & -\frac{5}{31} \end{bmatrix} \begin{bmatrix} 23 \\ 6 \end{bmatrix} = \begin{bmatrix} 3 \\ 2 \end{bmatrix}$$

The solution to the system is $(3, 2)$.

Your Turn 3

(a) The word "behold" gives two 3×1 matrices:

$$\begin{bmatrix} 2 \\ 5 \\ 8 \end{bmatrix} \quad \text{and} \quad \begin{bmatrix} 15 \\ 12 \\ 4 \end{bmatrix}$$

Use the coding matrix $A = \begin{bmatrix} 1 & 3 & 4 \\ 2 & 1 & 3 \\ 4 & 2 & 1 \end{bmatrix}$ to

find the product of A with each column matrix.

$$\begin{bmatrix} 1 & 3 & 4 \\ 2 & 1 & 3 \\ 4 & 2 & 1 \end{bmatrix} \begin{bmatrix} 2 \\ 5 \\ 8 \end{bmatrix} = \begin{bmatrix} 49 \\ 33 \\ 26 \end{bmatrix} \quad \text{and} \quad \begin{bmatrix} 1 & 3 & 4 \\ 2 & 1 & 3 \\ 4 & 2 & 1 \end{bmatrix} \begin{bmatrix} 15 \\ 12 \\ 4 \end{bmatrix} = \begin{bmatrix} 67 \\ 54 \\ 88 \end{bmatrix}$$

The coded message is 49, 33, 26, 67, 54, 88.

(b) Use the inverse of the coding matrix to decode the message 96, 87, 74, 141, 117, 114.

$$A^{-1} = \begin{bmatrix} -0.2 & 0.2 & 0.2 \\ 0.4 & -0.6 & 0.2 \\ 0 & 0.4 & -0.2 \end{bmatrix}, \begin{bmatrix} 96 \\ 87 \\ 74 \end{bmatrix}, \text{ and } \begin{bmatrix} 141 \\ 117 \\ 114 \end{bmatrix}$$

$$\begin{bmatrix} -0.2 & 0.2 & 0.2 \\ 0.4 & -0.6 & 0.2 \\ 0 & 0.4 & -0.2 \end{bmatrix} \begin{bmatrix} 96 \\ 87 \\ 74 \end{bmatrix} = \begin{bmatrix} 13 \\ 1 \\ 20 \end{bmatrix}$$

$$\begin{bmatrix} -0.2 & 0.2 & 0.2 \\ 0.4 & -0.6 & 0.2 \\ 0 & 0.4 & -0.2 \end{bmatrix} \begin{bmatrix} 141 \\ 117 \\ 114 \end{bmatrix} = \begin{bmatrix} 18 \\ 9 \\ 24 \end{bmatrix}$$

The message is the word "matrix."

2.5   Exercises

1.
$$\begin{bmatrix} 2 & 1 \\ 5 & 3 \end{bmatrix} \begin{bmatrix} 3 & -1 \\ -5 & 2 \end{bmatrix} = \begin{bmatrix} 6-5 & -2+2 \\ 15-15 & -5+6 \end{bmatrix}$$

$$= \begin{bmatrix} 1 & 0 \\ 0 & 1 \end{bmatrix} = I$$

$$\begin{bmatrix} 3 & -1 \\ -5 & 2 \end{bmatrix} \begin{bmatrix} 2 & 1 \\ 5 & 3 \end{bmatrix} = \begin{bmatrix} 6-5 & 3-3 \\ -10+10 & -5+6 \end{bmatrix}$$

$$= \begin{bmatrix} 1 & 0 \\ 0 & 1 \end{bmatrix} = I$$

Since the products obtained by multiplying the matrices in either order are both the 2×2 identity matrix, the given matrices are inverses of each other.

3.
$$\begin{bmatrix} 2 & 6 \\ 2 & 4 \end{bmatrix} \begin{bmatrix} -1 & 2 \\ 2 & -4 \end{bmatrix} = \begin{bmatrix} 10 & -20 \\ 6 & -12 \end{bmatrix} \neq I$$

No, the matrices are not inverses of each other since their product matrix is not I.

5.
$$\begin{bmatrix} 2 & 0 & 1 \\ 1 & 1 & 2 \\ 0 & 1 & 0 \end{bmatrix} \begin{bmatrix} 1 & 1 & -1 \\ 0 & 1 & 0 \\ -1 & -2 & 2 \end{bmatrix}$$

$$= \begin{bmatrix} 2+0-1 & 2+0-2 & -2+0+2 \\ 1+0-2 & 1+1-4 & -1+0+4 \\ 0+0+0 & 0+1+0 & 0+0+0 \end{bmatrix}$$

$$= \begin{bmatrix} 1 & 0 & 0 \\ -1 & -2 & 3 \\ 0 & 1 & 0 \end{bmatrix} \neq I$$

No, the matrices are not inverses of each other since their product matrix is not I.

7.
$$\begin{bmatrix} 1 & 3 & 3 \\ 1 & 4 & 3 \\ 1 & 3 & 4 \end{bmatrix} \begin{bmatrix} 7 & -3 & -3 \\ -1 & 1 & 0 \\ -1 & 0 & 1 \end{bmatrix} = \begin{bmatrix} 1 & 0 & 0 \\ 0 & 1 & 0 \\ 0 & 0 & 1 \end{bmatrix} = I$$

$$\begin{bmatrix} 7 & -3 & -3 \\ -1 & 1 & 0 \\ -1 & 0 & 1 \end{bmatrix} \begin{bmatrix} 1 & 3 & 3 \\ 1 & 4 & 3 \\ 1 & 3 & 4 \end{bmatrix} = \begin{bmatrix} 1 & 0 & 0 \\ 0 & 1 & 0 \\ 0 & 0 & 1 \end{bmatrix} = I$$

Yes, these matrices are inverses of each other.

9. No, a matrix with a row of all zeros does not have an inverse; the row of all zeros makes it impossible to get all the 1's in the main diagonal of the identity matrix.

11. Let $A = \begin{bmatrix} 1 & -1 \\ 2 & 0 \end{bmatrix}$.

Form the augmented matrix $[A|I]$.

$$[A|I] = \begin{bmatrix} 1 & -1 & 1 & 0 \\ 2 & 0 & 0 & 1 \end{bmatrix}$$

Perform row operations on $[A|I]$ to get a matrix of the form $[I|B]$.

$$\begin{bmatrix} 1 & -1 & 1 & 0 \\ 2 & 0 & 0 & 1 \end{bmatrix}$$

$-2R_1 + R_2 \to R_2$
$$\begin{bmatrix} 1 & -1 & 1 & 0 \\ 0 & 2 & -2 & 1 \end{bmatrix}$$

$2R_1 + R_2 \to R_1$
$$\begin{bmatrix} 2 & 0 & 0 & 1 \\ 0 & 2 & -2 & 1 \end{bmatrix}$$

$\frac{1}{2}R_1 \to R_1$
$\frac{1}{2}R_2 \to R_2$
$$\begin{bmatrix} 1 & 0 & 0 & \frac{1}{2} \\ 0 & 1 & -1 & \frac{1}{2} \end{bmatrix} = [I|B]$$

The matrix B in the last transformation is the desired multiplicative inverse.

$$A^{-1} = \begin{bmatrix} 0 & \frac{1}{2} \\ -1 & \frac{1}{2} \end{bmatrix}$$

This answer may be checked by showing that $AA^{-1} = I$ and $A^{-1}A = I$.

13. Let $A = \begin{bmatrix} 3 & -1 \\ -5 & 2 \end{bmatrix}$.

$$[A|I] = \begin{bmatrix} 3 & -1 & 1 & 0 \\ -5 & 2 & 0 & 1 \end{bmatrix}$$

$5R_1 + 3R_2 \to R_2$
$$\begin{bmatrix} 3 & -1 & 1 & 0 \\ 0 & 1 & 5 & 3 \end{bmatrix}$$

$R_1 + R_2 \to R_1$
$$\begin{bmatrix} 3 & 0 & 6 & 3 \\ 0 & 1 & 5 & 3 \end{bmatrix}$$

$\frac{1}{3}R_1 \to R_1$
$$\begin{bmatrix} 1 & 0 & 2 & 1 \\ 0 & 1 & 5 & 3 \end{bmatrix} = [I|B]$$

The desired inverse is

$$A^{-1} = \begin{bmatrix} 2 & 1 \\ 5 & 3 \end{bmatrix}.$$

15. Let $A = \begin{bmatrix} 1 & -3 \\ -2 & 6 \end{bmatrix}$.

$[A|I] = \begin{bmatrix} 1 & -3 & | & 1 & 0 \\ -2 & 6 & | & 0 & 1 \end{bmatrix}$

$2R_1 + R_2 \rightarrow R_2 \begin{bmatrix} 1 & -3 & | & 1 & 0 \\ 0 & 0 & | & 2 & 1 \end{bmatrix}$

Because the last row has all zeros to the left of the vertical bar, there is no way to complete the desired transformation. A has no inverse.

17. Let $A = \begin{bmatrix} 1 & 0 & 0 \\ 0 & -1 & 0 \\ 1 & 0 & 1 \end{bmatrix}$.

$[A|I] = \begin{bmatrix} 1 & 0 & 0 & | & 1 & 0 & 0 \\ 0 & -1 & 0 & | & 0 & 1 & 0 \\ 1 & 0 & 1 & | & 0 & 0 & 1 \end{bmatrix}$

$-1R_1 + R_3 \rightarrow R_3 \begin{bmatrix} 1 & 0 & 0 & | & 1 & 0 & 0 \\ 0 & -1 & 0 & | & 0 & 1 & 0 \\ 0 & 0 & 1 & | & -1 & 0 & 1 \end{bmatrix}$

$-1R_2 \rightarrow R_2 \begin{bmatrix} 1 & 0 & 0 & | & 1 & 0 & 0 \\ 0 & 1 & 0 & | & 0 & -1 & 0 \\ 0 & 0 & 1 & | & -1 & 0 & 1 \end{bmatrix}$

$A^{-1} = \begin{bmatrix} 1 & 0 & 0 \\ 0 & -1 & 0 \\ -1 & 0 & 1 \end{bmatrix}$

19. Let $A = \begin{bmatrix} -1 & -1 & -1 \\ 4 & 5 & 0 \\ 0 & 1 & -3 \end{bmatrix}$.

$[A|I] = \begin{bmatrix} -1 & -1 & -1 & | & 1 & 0 & 0 \\ 4 & 5 & 0 & | & 0 & 1 & 0 \\ 0 & 1 & -3 & | & 0 & 0 & 1 \end{bmatrix}$

$4R_1 + R_2 \rightarrow R_2 \begin{bmatrix} -1 & -1 & -1 & | & 1 & 0 & 0 \\ 0 & 1 & -4 & | & 4 & 1 & 0 \\ 0 & 1 & -3 & | & 0 & 0 & 1 \end{bmatrix}$

$R_2 + R_1 \rightarrow R_1 \begin{bmatrix} -1 & 0 & -5 & | & 5 & 1 & 0 \\ 0 & 1 & -4 & | & 4 & 1 & 0 \\ 0 & 0 & 1 & | & -4 & -1 & 1 \end{bmatrix}$
$-1R_2 + R_3 \rightarrow R_3$

$5R_3 + R_1 \rightarrow R_1 \begin{bmatrix} -1 & 0 & 0 & | & -15 & -4 & 5 \\ 0 & 1 & 0 & | & -12 & -3 & 4 \\ 0 & 0 & 1 & | & -4 & -1 & 1 \end{bmatrix}$
$4R_3 + R_2 \rightarrow R_2$

$-1R_1 \rightarrow R_1 \begin{bmatrix} 1 & 0 & 0 & | & 15 & 4 & -5 \\ 0 & 1 & 0 & | & -12 & -3 & 4 \\ 0 & 0 & 1 & | & -4 & -1 & 1 \end{bmatrix}$

$A^{-1} = \begin{bmatrix} 15 & 4 & -5 \\ -12 & -3 & 4 \\ -4 & -1 & 1 \end{bmatrix}$

21. Let $A = \begin{bmatrix} 1 & 2 & 3 \\ -3 & -2 & -1 \\ -1 & 0 & 1 \end{bmatrix}$.

$[A|I] = \begin{bmatrix} 1 & 2 & 3 & | & 1 & 0 & 0 \\ -3 & -2 & -1 & | & 0 & 1 & 0 \\ -1 & 0 & 1 & | & 0 & 0 & 1 \end{bmatrix}$

$3R_1 + R_2 \rightarrow R_2 \begin{bmatrix} 1 & 2 & 3 & | & 1 & 0 & 0 \\ 0 & 4 & 8 & | & 3 & 1 & 0 \\ 0 & 2 & 4 & | & 1 & 0 & 1 \end{bmatrix}$
$R_1 + R_3 \rightarrow R_3$

$R_2 + (-2R_1) \rightarrow R_1 \begin{bmatrix} -2 & 0 & 2 & | & 1 & 1 & 0 \\ 0 & 4 & 8 & | & 3 & 1 & 0 \\ 0 & 0 & 0 & | & 1 & 1 & -2 \end{bmatrix}$
$R_2 + (-2R_3) \rightarrow R_3$

Because the last row has all zeros to the left of the vertical bar, there is no way to complete the desired transformation. A has no inverse.

23. Find the inverse of $A = \begin{bmatrix} 1 & 3 & -2 \\ 2 & 7 & -3 \\ 3 & 8 & -5 \end{bmatrix}$, if it exists.

$[A|I] = \begin{bmatrix} 1 & 3 & -2 & | & 1 & 0 & 0 \\ 2 & 7 & -3 & | & 0 & 1 & 0 \\ 3 & 8 & -5 & | & 0 & 0 & 1 \end{bmatrix}$

$-2R_1 + R_2 \rightarrow R_2 \begin{bmatrix} 1 & 3 & -2 & | & 1 & 0 & 0 \\ 0 & 1 & 1 & | & -2 & 1 & 0 \\ 0 & -1 & 1 & | & -3 & 0 & 1 \end{bmatrix}$
$-3R_1 + R_3 \rightarrow R_3$

$-3R_2 + R_1 \rightarrow R_1 \begin{bmatrix} 1 & 0 & -5 & | & 7 & -3 & 0 \\ 0 & 1 & 1 & | & -2 & 1 & 0 \\ 0 & 0 & 2 & | & -5 & 1 & 1 \end{bmatrix}$
$R_2 + R_3 \rightarrow R_3$

$5R_3 + 2R_1 \rightarrow R_1 \begin{bmatrix} 2 & 0 & 0 & | & -11 & -1 & 5 \\ 0 & 2 & 0 & | & 1 & 1 & -1 \\ 0 & 0 & 2 & | & -5 & 1 & 1 \end{bmatrix}$
$-1R_3 + 2R_2 \rightarrow R_2$

$\frac{1}{2}R_1 \to R_1$
$\frac{1}{2}R_2 \to R_2$
$\frac{1}{2}R_3 \to R_3$
$\begin{bmatrix} 1 & 0 & 0 & -\frac{11}{2} & -\frac{1}{2} & \frac{5}{2} \\ 0 & 1 & 0 & \frac{1}{2} & \frac{1}{2} & -\frac{1}{2} \\ 0 & 0 & 1 & -\frac{5}{2} & \frac{1}{2} & \frac{1}{2} \end{bmatrix}$

$$A^{-1} = \begin{bmatrix} -\frac{11}{2} & -\frac{1}{2} & \frac{5}{2} \\ \frac{1}{2} & \frac{1}{2} & -\frac{1}{2} \\ -\frac{5}{2} & \frac{1}{2} & \frac{1}{2} \end{bmatrix}$$

25. Let $A = \begin{bmatrix} 1 & -2 & 3 & 0 \\ 0 & 1 & -1 & 1 \\ -2 & 2 & -2 & 4 \\ 0 & 2 & -3 & 1 \end{bmatrix}$.

$[A|I] = \begin{bmatrix} 1 & -2 & 3 & 0 & 1 & 0 & 0 & 0 \\ 0 & 1 & -1 & 1 & 0 & 1 & 0 & 0 \\ -2 & 2 & -2 & 4 & 0 & 0 & 1 & 0 \\ 0 & 2 & -3 & 1 & 0 & 0 & 0 & 1 \end{bmatrix}$

$2R_1 + R_3 \to R_3$
$\begin{bmatrix} 1 & -2 & 3 & 0 & 1 & 0 & 0 & 0 \\ 0 & 1 & -1 & 1 & 0 & 1 & 0 & 0 \\ 0 & -2 & 4 & 4 & 2 & 0 & 1 & 0 \\ 0 & 2 & -3 & 1 & 0 & 0 & 0 & 1 \end{bmatrix}$

$2R_2 + R_1 \to R_1$

$2R_2 + R_3 \to R_3$
$-2R_2 + R_4 \to R_4$
$\begin{bmatrix} 1 & 0 & 1 & 2 & 1 & 2 & 0 & 0 \\ 0 & 1 & -1 & 1 & 0 & 1 & 0 & 0 \\ 0 & 0 & 2 & 6 & 2 & 2 & 1 & 0 \\ 0 & 0 & -1 & -1 & 0 & -2 & 0 & 1 \end{bmatrix}$

$R_3 + (-2)R_1 \to R_1$
$R_3 + 2R_2 \to R_2$

$R_3 + 2R_4 \to R_4$
$\begin{bmatrix} -2 & 0 & 0 & 2 & 0 & -2 & 1 & 0 \\ 0 & 2 & 0 & 8 & 2 & 4 & 1 & 0 \\ 0 & 0 & 2 & 6 & 2 & 2 & 1 & 0 \\ 0 & 0 & 0 & 4 & 2 & -2 & 1 & 2 \end{bmatrix}$

$R_4 + (-2)R_1 \to R_1$
$-2R_4 + R_2 \to R_2$
$-3R_4 + 2R_3 \to R_3$
$\begin{bmatrix} 4 & 0 & 0 & 0 & 2 & 2 & -1 & 2 \\ 0 & 2 & 0 & 0 & -2 & 8 & -1 & -4 \\ 0 & 0 & 4 & 0 & -2 & 10 & -1 & -6 \\ 0 & 0 & 0 & 4 & 2 & -2 & 1 & 2 \end{bmatrix}$

$\frac{1}{4}R_1 \to R_1$
$\frac{1}{2}R_2 \to R_2$
$\frac{1}{4}R_3 \to R_3$
$\frac{1}{4}R_4 \to R_4$
$\begin{bmatrix} 1 & 0 & 0 & 0 & \frac{1}{2} & \frac{1}{2} & -\frac{1}{4} & \frac{1}{2} \\ 0 & 1 & 0 & 0 & -1 & 4 & -\frac{1}{2} & -2 \\ 0 & 0 & 1 & 0 & -\frac{1}{2} & \frac{5}{2} & -\frac{1}{4} & -\frac{3}{2} \\ 0 & 0 & 0 & 1 & \frac{1}{2} & -\frac{1}{2} & \frac{1}{4} & \frac{1}{2} \end{bmatrix}$

$$A^{-1} = \begin{bmatrix} \frac{1}{2} & \frac{1}{2} & -\frac{1}{4} & \frac{1}{2} \\ -1 & 4 & -\frac{1}{2} & -2 \\ -\frac{1}{2} & \frac{5}{2} & -\frac{1}{4} & -\frac{3}{2} \\ \frac{1}{2} & -\frac{1}{2} & \frac{1}{4} & \frac{1}{2} \end{bmatrix}$$

27. $2x + 5y = 15$
$\quad x + 4y = 9$

First, write the system in matrix form.

$$\begin{bmatrix} 2 & 5 \\ 1 & 4 \end{bmatrix}\begin{bmatrix} x \\ y \end{bmatrix} = \begin{bmatrix} 15 \\ 9 \end{bmatrix}$$

Let $A = \begin{bmatrix} 2 & 5 \\ 1 & 4 \end{bmatrix}$, $X = \begin{bmatrix} x \\ y \end{bmatrix}$, and $B = \begin{bmatrix} 15 \\ 9 \end{bmatrix}$.

The system in matrix form is $AX = B$. We wish to find $X = A^{-1}AX = A^{-1}B$. Use row operations to find A^{-1}.

$$[A|I] = \begin{bmatrix} 2 & 5 & 1 & 0 \\ 1 & 4 & 0 & 1 \end{bmatrix}$$

$-1R_1 + 2R_2 \to R_2$ $\begin{bmatrix} 2 & 5 & 1 & 0 \\ 0 & 3 & -1 & 2 \end{bmatrix}$

$-5R_2 + 3R_1 \to R_1$ $\begin{bmatrix} 6 & 0 & 8 & -10 \\ 0 & 3 & -1 & 2 \end{bmatrix}$

$\frac{1}{6}R_1 \to R_1$
$\frac{1}{3}R_2 \to R_2$
$\begin{bmatrix} 1 & 0 & \frac{4}{3} & -\frac{5}{3} \\ 0 & 1 & -\frac{1}{3} & \frac{2}{3} \end{bmatrix}$

$$A^{-1} = \begin{bmatrix} \frac{4}{3} & -\frac{5}{3} \\ -\frac{1}{3} & \frac{2}{3} \end{bmatrix} = \frac{1}{3}\begin{bmatrix} 4 & -5 \\ -1 & 2 \end{bmatrix}$$

Next find the product $A^{-1}B$.

$$X = A^{-1}B = \begin{bmatrix} \frac{4}{3} & -\frac{5}{3} \\ -\frac{1}{3} & \frac{2}{3} \end{bmatrix}\begin{bmatrix} 15 \\ 9 \end{bmatrix}$$

$$= \frac{1}{3}\begin{bmatrix} 4 & -5 \\ -1 & 2 \end{bmatrix}\begin{bmatrix} 15 \\ 9 \end{bmatrix}$$

$$= \frac{1}{3}\begin{bmatrix} 15 \\ 3 \end{bmatrix} = \begin{bmatrix} 5 \\ 1 \end{bmatrix}$$

Thus, the solution is $(5, 1)$.

29. $2x + y = 5$
$5x + 3y = 13$

Let $A = \begin{bmatrix} 2 & 1 \\ 5 & 3 \end{bmatrix}$, $X = \begin{bmatrix} x \\ y \end{bmatrix}$, $B = \begin{bmatrix} 5 \\ 13 \end{bmatrix}$.

Use row operations to obtain

$$A^{-1} = \begin{bmatrix} 3 & -1 \\ -5 & 2 \end{bmatrix}.$$

$$X = A^{-1}B = \begin{bmatrix} 3 & -1 \\ -5 & 2 \end{bmatrix}\begin{bmatrix} 5 \\ 13 \end{bmatrix} = \begin{bmatrix} 2 \\ 1 \end{bmatrix}$$

The solution is $(2, 1)$.

31. $3x - 2y = 3$
$7x - 5y = 0$

First, write the system in matrix form.

$$\begin{bmatrix} 3 & -2 \\ 7 & -5 \end{bmatrix}\begin{bmatrix} x \\ y \end{bmatrix} = \begin{bmatrix} 3 \\ 0 \end{bmatrix}$$

Let $A = \begin{bmatrix} 3 & -2 \\ 7 & -5 \end{bmatrix}$, $X = \begin{bmatrix} x \\ y \end{bmatrix}$, and $B = \begin{bmatrix} 3 \\ 0 \end{bmatrix}$.

The system is in matrix form $AX = B$. We wish to find $X = A^{-1}$ $AX = A^{-1}B$. Use row operations to find A^{-1}.

$$[A|I] = \begin{bmatrix} 3 & -2 & | & 1 & 0 \\ 7 & -5 & | & 0 & 1 \end{bmatrix}$$

$-7R_1 + 3R_2 \to R_2 \quad \begin{bmatrix} 3 & -2 & | & 1 & 0 \\ 0 & -1 & | & -7 & 3 \end{bmatrix}$

$-2R_2 + R_1 \to R_1 \quad \begin{bmatrix} 3 & 0 & | & 15 & -6 \\ 0 & -1 & | & -7 & 3 \end{bmatrix}$

$\frac{1}{3}R_1 \to R_1 \quad \begin{bmatrix} 1 & 0 & | & 5 & -2 \\ 0 & 1 & | & 7 & -3 \end{bmatrix}$
$-1R_2 \to R_2$

$$A^{-1} = \begin{bmatrix} 5 & -2 \\ 7 & -3 \end{bmatrix}$$

Next find the product $A^{-1}B$.

$$X = A^{-1}B = \begin{bmatrix} 5 & -2 \\ 7 & -3 \end{bmatrix}\begin{bmatrix} 3 \\ 0 \end{bmatrix} = \begin{bmatrix} 15 \\ 21 \end{bmatrix}$$

Thus, the solution is $(15, 21)$.

33. $-x - 8y = 12$
$3x + 24y = -36$

Let $A = \begin{bmatrix} -1 & -8 \\ 3 & 24 \end{bmatrix}$, $X = \begin{bmatrix} x \\ y \end{bmatrix}$, $B = \begin{bmatrix} 12 \\ -36 \end{bmatrix}$.

Using row operations on $[A|I]$ leads to the matrix

$$\begin{bmatrix} 1 & 8 & | & -1 & 0 \\ 0 & 0 & | & 3 & 1 \end{bmatrix},$$

but the zeros in the second row indicate that matrix A does not have an inverse. We cannot complete the solution by this method.

Since the second equation is a multiple of the first, the equations are dependent. Solve the first equation of the system for x.

$$-x - 8y = 12$$
$$-x = 8y + 12$$
$$x = -8y - 12$$

The solution is $(-8y - 12, y)$, where y is any real number.

35. $-x - y - z = 1$
$4x + 5y \quad = -2$
$y - 3z = 3$

has coefficient matrix

$$A = \begin{bmatrix} -1 & -1 & -1 \\ 4 & 5 & 0 \\ 0 & 1 & -3 \end{bmatrix}.$$

In Exercise 19, it was found that

$$A^{-1} = \begin{bmatrix} -1 & -1 & -1 \\ 4 & 5 & 0 \\ 0 & 1 & 3 \end{bmatrix}^{-1}$$

$$= \begin{bmatrix} 15 & 4 & -5 \\ -12 & -3 & 4 \\ -4 & -1 & 1 \end{bmatrix}.$$

Since $X = A^{-1}B$,

$$\begin{bmatrix} x \\ y \\ z \end{bmatrix} = \begin{bmatrix} 15 & 4 & -5 \\ -12 & -3 & 4 \\ -4 & -1 & 1 \end{bmatrix}\begin{bmatrix} 1 \\ -2 \\ 3 \end{bmatrix} = \begin{bmatrix} -8 \\ 6 \\ 1 \end{bmatrix}.$$

The solution is $(-8, 6, 1)$.

37.
$$x + 3y - 2z = 4$$
$$2x + 7y - 3z = 8$$
$$3x + 8y - 5z = -4$$

has coefficient matrix

$$A = \begin{bmatrix} 1 & 3 & -2 \\ 2 & 7 & -3 \\ 3 & 8 & -5 \end{bmatrix}.$$

In Exercise 23, it was calculated that

$$A^{-1} = \begin{bmatrix} 1 & 3 & -2 \\ 2 & 7 & -3 \\ 3 & 8 & -5 \end{bmatrix}^{-1} = \begin{bmatrix} -\frac{11}{2} & -\frac{1}{2} & \frac{5}{2} \\ \frac{1}{2} & \frac{1}{2} & -\frac{1}{2} \\ -\frac{5}{2} & \frac{1}{2} & \frac{1}{2} \end{bmatrix}$$

$$= \frac{1}{2} \begin{bmatrix} -11 & -1 & 5 \\ 1 & 1 & -1 \\ -5 & 1 & 1 \end{bmatrix}.$$

Since $X = A^{-1}B$.

$$\begin{bmatrix} x \\ y \\ z \end{bmatrix} = \frac{1}{2} \begin{bmatrix} -11 & -1 & 5 \\ 1 & 1 & -1 \\ -5 & 1 & 1 \end{bmatrix} \begin{bmatrix} 4 \\ 8 \\ -4 \end{bmatrix}$$

$$= \frac{1}{2} \begin{bmatrix} -72 \\ 16 \\ -16 \end{bmatrix} = \begin{bmatrix} -36 \\ 8 \\ -8 \end{bmatrix}$$

39.
$$2x - 2y = 5$$
$$4y + 8z = 7$$
$$x + 2z = 1$$

has coefficient matrix

$$A = \begin{bmatrix} 2 & -2 & 0 \\ 0 & 4 & 8 \\ 1 & 0 & 2 \end{bmatrix}.$$

However, using row operations on $\begin{bmatrix} A|I \end{bmatrix}$ shows that A does not have an inverse, so another method must be used.

Try the Gauss-Jordan method. The augmented matrix is

$$\begin{bmatrix} 2 & -2 & 0 & 5 \\ 0 & 4 & 8 & 7 \\ 1 & 0 & 2 & 1 \end{bmatrix}.$$

After several row operations, we obtain the matrix

$$\begin{bmatrix} 1 & 0 & 2 & \frac{17}{4} \\ 0 & 1 & 2 & \frac{7}{4} \\ 0 & 0 & 0 & 13 \end{bmatrix}.$$

The bottom row of this matrix shows that the system has no solution, since $0 = 13$ is a false statement.

41.
$$x - 2y + 3z \qquad = 4$$
$$y - z + w = -8$$
$$-2x + 2y - 2z + 4w = 12$$
$$2y - 3z + w = -4$$

has coefficient matrix

$$A = \begin{bmatrix} 1 & -2 & 3 & 0 \\ 0 & 1 & -1 & 1 \\ -2 & 2 & -2 & 4 \\ 0 & 2 & -3 & 1 \end{bmatrix}.$$

In Exercise 25, it was found that

$$A^{-1} = \begin{bmatrix} \frac{1}{2} & \frac{1}{2} & -\frac{1}{4} & \frac{1}{2} \\ -1 & 4 & -\frac{1}{2} & -2 \\ -\frac{1}{2} & \frac{5}{2} & -\frac{1}{4} & -\frac{3}{2} \\ \frac{1}{2} & -\frac{1}{2} & \frac{1}{4} & \frac{1}{2} \end{bmatrix}.$$

Since $X = A^{-1}B$,

$$\begin{bmatrix} x \\ y \\ z \\ w \end{bmatrix} = \begin{bmatrix} \frac{1}{2} & \frac{1}{2} & -\frac{1}{4} & \frac{1}{2} \\ -1 & 4 & -\frac{1}{2} & -2 \\ -\frac{1}{2} & \frac{5}{2} & -\frac{1}{4} & -\frac{3}{2} \\ \frac{1}{2} & -\frac{1}{2} & \frac{1}{4} & \frac{1}{2} \end{bmatrix} \begin{bmatrix} 4 \\ -8 \\ 12 \\ -4 \end{bmatrix} = \begin{bmatrix} -7 \\ -34 \\ -19 \\ 7 \end{bmatrix}.$$

The solution is $(-7, -34, -19, 7)$.

In Exercises 43–48, let $A = \begin{bmatrix} a & b \\ c & d \end{bmatrix}$.

43. $IA = \begin{bmatrix} 1 & 0 \\ 0 & 1 \end{bmatrix} \begin{bmatrix} a & b \\ c & d \end{bmatrix} = \begin{bmatrix} a & b \\ c & d \end{bmatrix} = A$

Thus, $IA = A$.

45. $A \cdot O = \begin{bmatrix} a & b \\ c & d \end{bmatrix} \begin{bmatrix} 0 & 0 \\ 0 & 0 \end{bmatrix} = \begin{bmatrix} 0 & 0 \\ 0 & 0 \end{bmatrix} = O$

Thus, $A \cdot O = O$.

47. In Exercise 46, it was found that

$$A^{-1} = \frac{1}{ad - bc}\begin{bmatrix} d & -b \\ -c & a \end{bmatrix}.$$

$$A^{-1}A = \left(\frac{1}{ad - bc}\begin{bmatrix} d & -b \\ -c & a \end{bmatrix}\right)\begin{bmatrix} a & b \\ c & d \end{bmatrix}$$

$$= \frac{1}{ad - bc}\left(\begin{bmatrix} d & -b \\ -c & a \end{bmatrix}\begin{bmatrix} a & b \\ c & d \end{bmatrix}\right)$$

$$= \frac{1}{ad - bc}\begin{bmatrix} ad - bc & 0 \\ 0 & ad - bc \end{bmatrix}$$

$$= \begin{bmatrix} 1 & 0 \\ 0 & 1 \end{bmatrix} = I$$

Thus, $A^{-1}A = I$.

49.
$$AB = O$$
$$A^{-1}(AB) = A^{-1} \cdot O$$
$$(A^{-1}A)B = O$$
$$I \cdot B = O$$
$$B = O$$

Thus, if $AB = O$ and A^{-1} exists, then $B = O$.

51. This exercise should be solved by graphing calculator or computer methods. The solution, which may vary slightly, is

$$C^{-1} = \begin{bmatrix} -0.0477 & -0.0230 & 0.0292 & 0.0895 & -0.0402 \\ 0.0921 & 0.0150 & 0.0321 & 0.0209 & -0.0276 \\ -0.0678 & 0.0315 & -0.0404 & 0.0326 & 0.0373 \\ 0.0171 & -0.0248 & 0.0069 & -0.0003 & 0.0246 \\ -0.0208 & 0.0740 & 0.0096 & -0.1018 & 0.0646 \end{bmatrix}.$$

(Entries are rounded to 4 places.)

53. This exercise should be solved by graphing calculator or computer methods. The solution, which may vary slightly, is

$$D^{-1} = \begin{bmatrix} 0.0394 & 0.0880 & 0.0033 & 0.0530 & -0.1499 \\ -0.1492 & 0.0289 & 0.0187 & 0.1033 & 0.1668 \\ -0.1330 & -0.0543 & 0.0356 & 0.1768 & 0.1055 \\ 0.1407 & 0.0175 & -0.0453 & -0.1344 & 0.0655 \\ 0.0102 & -0.0653 & 0.0993 & 0.0085 & -0.0388 \end{bmatrix}.$$

(Entries are rounded to 4 places.)

55. This exercise should be solved by graphing calculator or computer methods. The solution may vary slightly. The answer is, yes, $D^{-1}C^{-1} = (CD)^{-1}$.

57. This exercise should be solved by graphing calculator or computer methods. The solution, which may vary slightly, is

$$\begin{bmatrix} 1.51482 \\ 0.053479 \\ -0.637242 \\ 0.462629 \end{bmatrix}.$$

59. (a) The matrix is $B = \begin{bmatrix} 72 \\ 48 \\ 60 \end{bmatrix}$.

(b) The matrix equation is

$$\begin{bmatrix} 2 & 4 & 2 \\ 2 & 1 & 2 \\ 2 & 1 & 3 \end{bmatrix}\begin{bmatrix} x_1 \\ x_2 \\ x_3 \end{bmatrix} = \begin{bmatrix} 72 \\ 48 \\ 60 \end{bmatrix}.$$

(c) To solve the system, begin by using row operations to find A^{-1}.

$$[A|I] = \begin{bmatrix} 2 & 4 & 2 & | & 1 & 0 & 0 \\ 2 & 1 & 2 & | & 0 & 1 & 0 \\ 2 & 1 & 3 & | & 0 & 0 & 1 \end{bmatrix}$$

$$\begin{matrix} \\ R_1 + (-1)R_2 \to R_2 \\ R_1 + (-1)R_3 \to R_3 \end{matrix}\begin{bmatrix} 2 & 4 & 2 & | & 1 & 0 & 0 \\ 0 & 3 & 0 & | & 1 & -1 & 0 \\ 0 & 3 & -1 & | & 1 & 0 & -1 \end{bmatrix}$$

$$\begin{matrix} -4R_2 + 3R_1 \to R_1 \\ \\ R_2 + (-1)R_3 \to R_3 \end{matrix}\begin{bmatrix} 6 & 0 & 6 & | & -1 & 4 & 0 \\ 0 & 3 & 0 & | & 1 & -1 & 0 \\ 0 & 0 & 1 & | & 0 & -1 & 1 \end{bmatrix}$$

$$\begin{matrix} -6R_3 + R_1 \to R_1 \\ \\ \end{matrix}\begin{bmatrix} 6 & 0 & 0 & | & -1 & 10 & -6 \\ 0 & 3 & 0 & | & 1 & -1 & 0 \\ 0 & 0 & 1 & | & 0 & -1 & 1 \end{bmatrix}$$

$$\begin{matrix} \frac{1}{6}R_1 \to R_1 \\ \frac{1}{3}R_2 \to R_2 \\ \\ \end{matrix}\begin{bmatrix} 1 & 0 & 0 & | & -\frac{1}{6} & \frac{5}{3} & -1 \\ 0 & 1 & 0 & | & \frac{1}{3} & -\frac{1}{3} & 0 \\ 0 & 0 & 1 & | & 0 & -1 & 1 \end{bmatrix}$$

The inverse matrix is

$$A^{-1} = \begin{bmatrix} -\frac{1}{6} & \frac{5}{3} & -1 \\ \frac{1}{3} & -\frac{1}{3} & 0 \\ 0 & -1 & 1 \end{bmatrix}.$$

Since $X = A^{-1}B$,

$$\begin{bmatrix} x_1 \\ x_2 \\ x_3 \end{bmatrix} = \begin{bmatrix} -\frac{1}{6} & \frac{5}{3} & -1 \\ \frac{1}{3} & -\frac{1}{3} & 0 \\ 0 & -1 & 1 \end{bmatrix} \begin{bmatrix} 72 \\ 48 \\ 60 \end{bmatrix} = \begin{bmatrix} 8 \\ 8 \\ 12 \end{bmatrix}.$$

There are 8 daily orders for type I, 8 for type II, and 12 for type III.

61. Let $x =$ the amount invested in AAA bonds,
$y =$ the amount invested in A bonds, and
$z =$ amount invested in B bonds.

(a) The total investment is $x + y + z = 25,000$.
The annual return is $0.06x + 0.065y + 0.08z = 1650$. Since twice as much is invested in AAA bonds as in B bonds, $x = 2z$.

The system to be solved is

$$\begin{array}{rcl} x + \quad y + \quad z &=& 25,000 \\ 0.06x + 0.065y + 0.08z &=& 1650 \\ x \quad\quad\quad - \quad 2z &=& 0 \end{array}$$

Let $A = \begin{bmatrix} 1 & 1 & 1 \\ 0.06 & 0.065 & 0.08 \\ 1 & 0 & -2 \end{bmatrix}$, $B = \begin{bmatrix} 25,000 \\ 1650 \\ 0 \end{bmatrix}$,

and $X = \begin{bmatrix} x \\ y \\ z \end{bmatrix}$.

Use a graphing calculator to obtain

$$A^{-1} = \begin{bmatrix} -26 & 400 & 3 \\ 40 & -600 & -4 \\ -13 & 200 & 1 \end{bmatrix}.$$

Use a graphing calculator again to solve the matrix equation $X = A^{-1}B$.

$$\begin{bmatrix} x \\ y \\ z \end{bmatrix} = \begin{bmatrix} -26 & 400 & 3 \\ 40 & -600 & -4 \\ -13 & 200 & 1 \end{bmatrix} \begin{bmatrix} 25,000 \\ 1650 \\ 0 \end{bmatrix}$$

$$= \begin{bmatrix} 10,000 \\ 10,000 \\ 5000 \end{bmatrix}$$

$10,000 should be invested at 6% in AAA bonds, $10,000 at 6.5% in A bonds, and $5000 at 8% in B bonds.

(b) The matrix of constants is changed to

$$B = \begin{bmatrix} 30,000 \\ 1985 \\ 0 \end{bmatrix}.$$

$$\begin{bmatrix} x \\ y \\ z \end{bmatrix} = \begin{bmatrix} -26 & 400 & 3 \\ 40 & -600 & -4 \\ -13 & 200 & 1 \end{bmatrix} \begin{bmatrix} 30,000 \\ 1985 \\ 0 \end{bmatrix}$$

$$= \begin{bmatrix} 14,000 \\ 9,000 \\ 7000 \end{bmatrix}$$

$14,000 should be invested at 6% in AAA bonds, $9000 at 6.5% in A bonds, and $7000 at 8% in B bonds.

(c) The matrix of constants is changed to

$$B = \begin{bmatrix} 40,000 \\ 2660 \\ 0 \end{bmatrix}.$$

$$\begin{bmatrix} x \\ y \\ z \end{bmatrix} = \begin{bmatrix} -26 & 400 & 3 \\ 40 & -600 & -4 \\ -13 & 200 & 1 \end{bmatrix} \begin{bmatrix} 40,000 \\ 2660 \\ 0 \end{bmatrix}$$

$$= \begin{bmatrix} 24,000 \\ 4000 \\ 12,000 \end{bmatrix}$$

$24,000 should be invested at 6% in AAA bonds, $4000 at 6.5% in A bonds, and $12,000 at 8% in B bonds.

63. Let $x =$ the number of Super Vim tablets,
$y =$ the number of Multitab tablets, and
$z =$ the number of Mighty Mix tablets.
The total number of vitamins is
$$x + y + z.$$
The total amount of niacin is
$$15x + 20y + 25z.$$
The total amount of Vitamin E is
$$12x + 15y + 35z.$$

(a) The system to be solved is
$$\begin{array}{rcl} x + \quad y + \quad z &=& 225 \\ 15x + 20y + 25z &=& 4750 \\ 12x + 15y + 35z &=& 5225. \end{array}$$

Let $A = \begin{bmatrix} 1 & 1 & 1 \\ 15 & 20 & 25 \\ 20 & 15 & 35 \end{bmatrix}$, $X = \begin{bmatrix} x \\ y \\ z \end{bmatrix}$, $B = \begin{bmatrix} 225 \\ 4750 \\ 5225 \end{bmatrix}$.

Thus, $AX = B$ and

$$\begin{bmatrix} 1 & 1 & 1 \\ 15 & 20 & 25 \\ 12 & 15 & 35 \end{bmatrix} \begin{bmatrix} x \\ y \\ z \end{bmatrix} = \begin{bmatrix} 225 \\ 4750 \\ 5225 \end{bmatrix}.$$

Use row operations to obtain the inverse of the coefficient matrix.

$$A^{-1} = \begin{bmatrix} \dfrac{65}{17} & -\dfrac{4}{17} & \dfrac{1}{17} \\ -\dfrac{45}{17} & \dfrac{23}{85} & -\dfrac{2}{17} \\ -\dfrac{3}{17} & -\dfrac{3}{85} & \dfrac{1}{17} \end{bmatrix}$$

Since $X = A^{-1}B$,

$$\begin{bmatrix} x \\ y \\ z \end{bmatrix} = \begin{bmatrix} \dfrac{65}{17} & -\dfrac{4}{17} & \dfrac{1}{17} \\ -\dfrac{45}{17} & \dfrac{23}{85} & -\dfrac{2}{17} \\ -\dfrac{3}{17} & -\dfrac{3}{85} & \dfrac{1}{17} \end{bmatrix}\begin{bmatrix} 225 \\ 4750 \\ 5225 \end{bmatrix} = \begin{bmatrix} 50 \\ 75 \\ 100 \end{bmatrix}.$$

There are 50 Super Vim tablets, 75 Multitab tablets, and 100 Mighty Mix tablets.

(b) The matrix of constants is changed to

$$B = \begin{bmatrix} 185 \\ 3625 \\ 3750 \end{bmatrix}.$$

$$\begin{bmatrix} x \\ y \\ z \end{bmatrix} = \begin{bmatrix} \dfrac{65}{17} & \dfrac{4}{17} & \dfrac{1}{17} \\ -\dfrac{45}{17} & \dfrac{23}{85} & -\dfrac{2}{17} \\ -\dfrac{3}{17} & -\dfrac{3}{85} & \dfrac{1}{17} \end{bmatrix}\begin{bmatrix} 185 \\ 3625 \\ 3750 \end{bmatrix} = \begin{bmatrix} 75 \\ 50 \\ 60 \end{bmatrix}$$

There are 75 Super Vim tablets, 50 Multitab tablets, and 60 Mighty Mix tablets.

(c) The matrix of constants is changed to

$$B = \begin{bmatrix} 230 \\ 4450 \\ 4210 \end{bmatrix}.$$

$$\begin{bmatrix} x \\ y \\ z \end{bmatrix} = \begin{bmatrix} \dfrac{65}{17} & \dfrac{4}{17} & \dfrac{1}{17} \\ -\dfrac{45}{17} & \dfrac{23}{85} & -\dfrac{2}{17} \\ -\dfrac{3}{17} & -\dfrac{3}{85} & \dfrac{1}{17} \end{bmatrix}\begin{bmatrix} 230 \\ 4450 \\ 4210 \end{bmatrix} = \begin{bmatrix} 80 \\ 100 \\ 50 \end{bmatrix}$$

There are 80 Super Vim tablets, 100 Multitab tablets, and 50 Mighty Mix tablets.

65. (a) First, divide the letters and spaces of the message into groups of three, writing each group as a column vector.

$$\begin{bmatrix} T \\ o \\ (\text{space}) \end{bmatrix}, \begin{bmatrix} b \\ e \\ (\text{space}) \end{bmatrix}, \begin{bmatrix} o \\ r \\ (\text{space}) \end{bmatrix}, \begin{bmatrix} n \\ o \\ t \end{bmatrix}, \begin{bmatrix} (\text{space}) \\ t \\ o \end{bmatrix}, \begin{bmatrix} (\text{space}) \\ b \\ e \end{bmatrix}$$

Next, convert each letter into a number, assigning 1 to A, 2 to B, and so on, with the number 27 used to represent each space between words.

$$\begin{bmatrix} 20 \\ 15 \\ 27 \end{bmatrix}, \begin{bmatrix} 2 \\ 5 \\ 27 \end{bmatrix}, \begin{bmatrix} 15 \\ 18 \\ 27 \end{bmatrix}, \begin{bmatrix} 14 \\ 15 \\ 20 \end{bmatrix}, \begin{bmatrix} 27 \\ 20 \\ 15 \end{bmatrix}, \begin{bmatrix} 27 \\ 2 \\ 5 \end{bmatrix}$$

Now, find the product of the coding matrix B and each column vector. This produces a new set of vectors, which represents the coded message.

$$\begin{bmatrix} 262 \\ -161 \\ -12 \end{bmatrix}, \begin{bmatrix} 186 \\ -103 \\ -22 \end{bmatrix}, \begin{bmatrix} 264 \\ -168 \\ -9 \end{bmatrix}, \begin{bmatrix} 208 \\ -134 \\ -5 \end{bmatrix}, \begin{bmatrix} 224 \\ -152 \\ 5 \end{bmatrix}, \begin{bmatrix} 92 \\ -50 \\ -3 \end{bmatrix}$$

This message will be transmitted as 262, −161, −12, 186, −103, −22, 264, −168, −9, 208, −134, −5, 224, −152, 5, 92, −50, −3.

(b) Use row operations or a graphing calculator to find the inverse of the coding matrix B.

$$B^{-1} = \begin{bmatrix} 1.75 & 2.5 & 3 \\ -0.25 & -0.5 & 0 \\ -0.25 & -0.5 & -1 \end{bmatrix}$$

(c) First, divide the coded message into groups of three numbers and form each group into a column vector.

$$\begin{bmatrix} 116 \\ -60 \\ -15 \end{bmatrix}, \begin{bmatrix} 294 \\ -197 \\ -2 \end{bmatrix}, \begin{bmatrix} 148 \\ -92 \\ -9 \end{bmatrix}, \begin{bmatrix} 96 \\ -64 \\ -4 \end{bmatrix}, \begin{bmatrix} 264 \\ -182 \\ -2 \end{bmatrix}$$

Next, find the product of the decoding matric B^{-1} and each of the column vectors. This produces a new set of vectors, which represents the decoded message.

$$\begin{bmatrix} 8 \\ 1 \\ 16 \end{bmatrix}, \begin{bmatrix} 16 \\ 25 \\ 27 \end{bmatrix}, \begin{bmatrix} 2 \\ 9 \\ 18 \end{bmatrix}, \begin{bmatrix} 20 \\ 8 \\ 4 \end{bmatrix}, \begin{bmatrix} 1 \\ 25 \\ 27 \end{bmatrix}$$

Last, convert each number into a letter, assigning A to 1, B to 2, and so on, with the number 27 used to represent a space between words. The decoded message is HAPPY BIRTHDAY.

2.6 Input-Output Models

Your Turn 1

$$X = (I - A)^{-1}D$$

$$X = \begin{bmatrix} 1.395 & 0.496 & 0.589 \\ 0.837 & 1.364 & 0.620 \\ 0.558 & 0.465 & 1.302 \end{bmatrix} \begin{bmatrix} 322 \\ 447 \\ 133 \end{bmatrix}$$

$$= \begin{bmatrix} 749.239 \\ 961.682 \\ 560.697 \end{bmatrix}$$

The productions of 749 units of agriculture, 962 units of manufacturing, and 561 units of transportation are required to satisfy the demands of 322, 447, and 133 units, respectively.

Your Turn 2

$$A = \begin{bmatrix} \frac{1}{2} & \frac{1}{4} & \frac{1}{6} \\ 0 & \frac{1}{4} & \frac{1}{6} \\ \frac{1}{2} & \frac{1}{2} & \frac{2}{3} \end{bmatrix} \qquad I - A = \begin{bmatrix} \frac{1}{2} & -\frac{1}{4} & -\frac{1}{6} \\ 0 & \frac{3}{4} & -\frac{1}{6} \\ -\frac{1}{2} & -\frac{1}{2} & \frac{1}{3} \end{bmatrix}$$

$$(I - A)X = \begin{bmatrix} \frac{1}{2}x_1 & -\frac{1}{4}x_2 & -\frac{1}{6}x_3 \\ 0x_1 & \frac{3}{4}x_2 & -\frac{1}{6}x_3 \\ -\frac{1}{2}x_1 & -\frac{1}{2}x_2 & \frac{1}{3}x_3 \end{bmatrix} = \begin{bmatrix} 0 \\ 0 \\ 0 \end{bmatrix}$$

We get the following system.

$$\frac{1}{2}x_1 - \frac{1}{4}x_2 - \frac{1}{6}x_3 = 0$$

$$\frac{3}{4}x_2 - \frac{1}{6}x_3 = 0$$

$$-\frac{1}{2}x_1 - \frac{1}{2}x_2 + \frac{1}{3}x_3 = 0$$

Clearing fractions gives the following system.

$$6x_1 - 3x_2 - 2x_3 = 0$$
$$9x_2 - 2x_3 = 0$$
$$-3x_1 - 3x_2 + 2x_3 = 0$$

$$9x_2 - 2x_3 = 0 \qquad\qquad 6x_1 - 3\left(\frac{2}{9}x_3\right) - 2x_3 = 0$$
$$9x_2 = 2x_3$$
$$\qquad\qquad\qquad 6x_1 - \frac{2}{3}x_3 - 2x_3 = 0$$
$$x_2 = \frac{2}{9}x_3$$

$$6x_1 - \frac{8}{3}x_3 = 0$$

$$6x_1 = \frac{8}{3}x_3$$

$$x_1 = \frac{4}{9}x_3$$

The solution is $\left(\frac{4}{9}x_3, \frac{2}{9}x_3, x_3\right)$, where x_3 is any real number. For $x_3 = 9$, the solution is $(4, 2, 9)$. So, the production of the three commodities should be in the ratio 4:2:9.

2.6 Exercises

1. $A = \begin{bmatrix} 0.8 & 0.2 \\ 0.2 & 0.7 \end{bmatrix}, D = \begin{bmatrix} 2 \\ 3 \end{bmatrix}$

To find the production matrix, first calculate $I - A$.

$$I - A = \begin{bmatrix} 1 & 0 \\ 0 & 1 \end{bmatrix} - \begin{bmatrix} 0.8 & 0.2 \\ 0.2 & 0.7 \end{bmatrix} = \begin{bmatrix} 0.2 & -0.2 \\ -0.2 & 0.3 \end{bmatrix}$$

Using row operations, find the inverse of $I - A$.

$$[I - A | I] = \begin{bmatrix} 0.2 & -0.2 & | & 1 & 0 \\ -0.2 & 0.3 & | & 0 & 1 \end{bmatrix}$$

$$\begin{matrix} 10R_1 \to R_1 \\ 10R_2 \to R_2 \end{matrix} \begin{bmatrix} 2 & -2 & | & 10 & 0 \\ -2 & 3 & | & 0 & 10 \end{bmatrix}$$

$$R_1 + R_2 \to R_2 \begin{bmatrix} 2 & -2 & | & 10 & 0 \\ 0 & 1 & | & 10 & 10 \end{bmatrix}$$

$$2R_2 + R_1 \to R_1 \begin{bmatrix} 2 & 0 & | & 30 & 20 \\ 0 & 1 & | & 10 & 10 \end{bmatrix}$$

$$\frac{1}{2}R_1 \to R_1 \begin{bmatrix} 1 & 0 & | & 15 & 10 \\ 0 & 1 & | & 10 & 10 \end{bmatrix}$$

$$(I - A)^{-1} = \begin{bmatrix} 15 & 10 \\ 10 & 10 \end{bmatrix}$$

Since $X = (I - A)^{-1}D$, the product matrix is

$$X = \begin{bmatrix} 15 & 10 \\ 10 & 10 \end{bmatrix} \begin{bmatrix} 2 \\ 3 \end{bmatrix} = \begin{bmatrix} 60 \\ 50 \end{bmatrix}.$$

3. $A = \begin{bmatrix} 0.1 & 0.03 \\ 0.07 & 0.6 \end{bmatrix}, D = \begin{bmatrix} 5 \\ 10 \end{bmatrix}$

First, calculate $I - A$.

$$I - A = \begin{bmatrix} 0.9 & -0.03 \\ -0.07 & 0.4 \end{bmatrix}$$

Use row operations to find the inverse of $I - A$, which is

$$(I - A)^{-1} \approx \begin{bmatrix} 1.118 & 0.084 \\ 0.196 & 2.515 \end{bmatrix}.$$

Since $X = (I - A)^{-1}D$, the production matrix is

$$X = \begin{bmatrix} 1.118 & 0.084 \\ 0.196 & 2.515 \end{bmatrix} \begin{bmatrix} 5 \\ 10 \end{bmatrix} = \begin{bmatrix} 6.43 \\ 26.12 \end{bmatrix}.$$

5. $A = \begin{bmatrix} 0.8 & 0 & 0.1 \\ 0.1 & 0.5 & 0.2 \\ 0 & 0 & 0.7 \end{bmatrix}, D = \begin{bmatrix} 1 \\ 6 \\ 3 \end{bmatrix}$

To find the production matrix, first calculate $I - A$.

$$I - A = \begin{bmatrix} 1 & 0 & 0 \\ 0 & 1 & 0 \\ 0 & 0 & 1 \end{bmatrix} - \begin{bmatrix} 0.8 & 0 & 0.1 \\ 0.1 & 0.5 & 0.2 \\ 0 & 0 & 0.7 \end{bmatrix}$$

$$= \begin{bmatrix} 0.2 & 0 & -0.1 \\ -0.1 & 0.5 & -0.2 \\ 0 & 0 & 0.3 \end{bmatrix}$$

Using row operations, find the inverse of $I - A$.

$$[I - A \mid I] = \begin{bmatrix} 0.2 & 0 & -0.1 & 1 & 0 & 0 \\ -0.1 & 0.5 & -0.2 & 0 & 1 & 0 \\ 0 & 0 & 0.3 & 0 & 0 & 1 \end{bmatrix}$$

$$\begin{array}{c} 10R_1 \rightarrow R_1 \\ 10R_2 \rightarrow R_2 \\ 10R_3 \rightarrow R_3 \end{array} \begin{bmatrix} 2 & 0 & -1 & 10 & 0 & 0 \\ -1 & 5 & -2 & 0 & 10 & 0 \\ 0 & 0 & 3 & 0 & 0 & 10 \end{bmatrix}$$

$$\begin{array}{c} \\ R_1 + 2R_2 \rightarrow R_2 \\ \\ \end{array} \begin{bmatrix} 2 & 0 & -1 & 10 & 0 & 0 \\ 0 & 10 & -5 & 10 & 20 & 0 \\ 0 & 0 & 3 & 0 & 0 & 10 \end{bmatrix}$$

$$\begin{array}{c} R_3 + 3R_1 \rightarrow R_1 \\ 5R_3 + 3R_2 \rightarrow R_2 \\ \\ \end{array} \begin{bmatrix} 6 & 0 & 0 & 30 & 0 & 10 \\ 0 & 30 & 0 & 30 & 60 & 50 \\ 0 & 0 & 3 & 0 & 0 & 10 \end{bmatrix}$$

$$\begin{array}{c} \frac{1}{6}R_1 \rightarrow R_1 \\ \frac{1}{30}R_2 \rightarrow R_2 \\ \frac{1}{3}R_3 \rightarrow R_3 \end{array} \begin{bmatrix} 1 & 0 & 0 & 5 & 0 & \frac{5}{3} \\ 0 & 1 & 0 & 1 & 2 & \frac{5}{3} \\ 0 & 0 & 1 & 0 & 0 & \frac{10}{3} \end{bmatrix}$$

$$(I - A)^{-1} = \begin{bmatrix} 5 & 0 & \frac{5}{3} \\ 1 & 2 & \frac{5}{3} \\ 0 & 0 & \frac{10}{3} \end{bmatrix}$$

Since $X = (I - A)^{-1}D$, the product matrix is

$$X = \begin{bmatrix} 5 & 0 & \frac{5}{3} \\ 1 & 2 & \frac{5}{3} \\ 0 & 0 & \frac{10}{3} \end{bmatrix} \begin{bmatrix} 1 \\ 6 \\ 3 \end{bmatrix} = \begin{bmatrix} 10 \\ 18 \\ 10 \end{bmatrix}.$$

7.

$$\begin{array}{c} \\ A \\ B \\ C \end{array} \begin{array}{ccc} A & B & C \end{array}$$

$$\begin{array}{c} A \\ B \\ C \end{array} \begin{bmatrix} 0.3 & 0.1 & 0.8 \\ 0.5 & 0.6 & 0.1 \\ 0.2 & 0.3 & 0.1 \end{bmatrix} = A$$

$$I - A = \begin{bmatrix} 0.7 & -0.1 & -0.8 \\ -0.5 & 0.4 & -0.1 \\ -0.2 & -0.3 & 0.9 \end{bmatrix}$$

Set $(I - A)X = O$ to obtain the following.

$$\begin{bmatrix} 0.7 & -0.1 & -0.8 \\ -0.5 & 0.4 & -0.1 \\ -0.2 & -0.3 & 0.9 \end{bmatrix} \begin{bmatrix} x_1 \\ x_2 \\ x_3 \end{bmatrix} = \begin{bmatrix} 0 \\ 0 \\ 0 \end{bmatrix}$$

$$\begin{bmatrix} 0.7x_1 - 0.1x_2 - 0.8x_3 \\ -0.5x_1 + 0.4x_2 - 0.1x_3 \\ -0.2x_1 - 0.3x_2 + 0.9x_3 \end{bmatrix} = \begin{bmatrix} 0 \\ 0 \\ 0 \end{bmatrix}$$

Rewrite this matrix equation as a system of equations.

$$0.7x_1 - 0.1x_2 - 0.8x_3 = 0$$
$$-0.5x_1 + 0.4x_2 - 0.1x_3 = 0$$
$$-0.2x_1 - 0.3x_2 + 0.9x_3 = 0$$

Rewrite the equations without decimals.

$$7x_1 - x_2 - 8x_3 = 0 \quad (1)$$
$$-5x_1 + 4x_2 - x_3 = 0 \quad (2)$$
$$-2x_1 - 3x_2 + 9x_3 = 0 \quad (3)$$

Use row operations to solve this system of equations. Begin by eliminating x_1 in equations (2) and (3)

$$\begin{array}{c} \\ 5R_1 + 7R_2 \rightarrow R_1 \\ 2R_1 + 7R_3 \rightarrow R_3 \end{array} \begin{array}{l} 7x_1 - x_2 - 8x_3 = 0 \quad (1) \\ 23x_2 - 47x_3 = 0 \quad (4) \\ -23x_2 + 47x_3 = 0 \quad (5) \end{array}$$

Eliminate x_2 in equations (1) and (5).

$$\begin{array}{c} 23R_1 + R_2 \rightarrow R_1 \\ \\ R_2 + R_3 \rightarrow R_3 \end{array} \begin{array}{l} 161x_1 - 231x_3 = 0 \quad (6) \\ 23x_2 - 47x_3 = 0 \quad (4) \\ 0 = 0 \quad (7) \end{array}$$

The true statement in equation (7) indicates that the equations are dependent. Solve equation (6) for x_1 and equation (4) for x_2, each in terms of x_3.

$$x_1 = \frac{231}{161}x_3 = \frac{33}{23}x_3$$

$$x_2 = \frac{47}{23}x_3$$

The solution of the system is

$$\left(\frac{33}{23}x_3, \frac{47}{23}x_3, x_3\right).$$

If $x_3 = 23$, then $x_1 = 33$ and $x_2 = 47$, so the production of the three commodities should be in the ratio 33:47:23.

9. Use a graphing calculator or a computer to find the production matrix $X = (I - A)^{-1}D$. The answer is

$$X = \begin{bmatrix} 7697 \\ 4205 \\ 6345 \\ 4106 \end{bmatrix}.$$

Values have been rounded.

11. In Example 4, it was found that

$$(I - A)^{-1} \approx \begin{bmatrix} 1.3882 & 0.1248 \\ 0.5147 & 1.1699 \end{bmatrix}.$$

Since $X = (I - A)^{-1}D$, the production matrix is

$$X = \begin{bmatrix} 1.3882 & 0.1248 \\ 0.5147 & 1.1699 \end{bmatrix}\begin{bmatrix} 925 \\ 1250 \end{bmatrix} = \begin{bmatrix} 1440.085 \\ 1938.473 \end{bmatrix}.$$

Thus, about 1440. metric tons of wheat and 1938 metric tons of oil should be produced.

13. In Example 3, it was found that

$$(I - A)^{-1} \approx \begin{bmatrix} 1.3953 & 0.4961 & 0.5891 \\ 0.8372 & 1.3643 & 0.6202 \\ 0.5581 & 0.4651 & 1.3023 \end{bmatrix}.$$

Since $X = (I - A)^{-1}D$, the production matrix is

$$X = \begin{bmatrix} 1.3953 & 0.4961 & 0.5891 \\ 0.8372 & 1.3643 & 0.6202 \\ 0.5581 & 0.4651 & 1.3023 \end{bmatrix}\begin{bmatrix} 607 \\ 607 \\ 607 \end{bmatrix} = \begin{bmatrix} 1505.66 \\ 1712.77 \\ 1411.58 \end{bmatrix}.$$

Thus, about 1506 units of agriculture, 1713 units of manufacturing, and 1412 units of transportation should be produced.

15. From the given data, we get the input-output matrix

$$A = \begin{bmatrix} 0 & \frac{1}{2} & \frac{1}{4} \\ \frac{1}{4} & 0 & \frac{1}{4} \\ \frac{1}{2} & \frac{1}{4} & 0 \end{bmatrix}.$$

$$I - A = \begin{bmatrix} 1 & -\frac{1}{2} & -\frac{1}{4} \\ -\frac{1}{4} & 1 & -\frac{1}{4} \\ -\frac{1}{2} & -\frac{1}{4} & 1 \end{bmatrix}$$

Use row operations to find the inverse of $I - A$, which is

$$(I - A)^{-1} \approx \begin{bmatrix} 1.538 & 0.923 & 0.615 \\ 0.615 & 1.436 & 0.513 \\ 0.923 & 0.821 & 1.436 \end{bmatrix}.$$

Since $X = (I - A)^{-1}D$, the production matrix is

$$X = \begin{bmatrix} 1.538 & 0.923 & 0.615 \\ 0.615 & 1.436 & 0.513 \\ 0.923 & 0.821 & 1.436 \end{bmatrix}\begin{bmatrix} 1000 \\ 1000 \\ 1000 \end{bmatrix} \approx \begin{bmatrix} 3077 \\ 2564 \\ 3179 \end{bmatrix}.$$

Thus, the production should be about 3077 units of agriculture, 2564 units of manufacturing, and 3179 units of transportation.

17. From the given data, we get the input-output matrix

$$A = \begin{bmatrix} \frac{1}{4} & \frac{1}{6} \\ \frac{1}{2} & 0 \end{bmatrix}.$$

$$I - A = \begin{bmatrix} \frac{3}{4} & -\frac{1}{6} \\ -\frac{1}{2} & 1 \end{bmatrix}$$

Use row operations to find the inverse of $I - A$, which is

$$(I - A)^{-1} = \begin{bmatrix} \frac{3}{2} & \frac{1}{4} \\ \frac{3}{4} & \frac{9}{8} \end{bmatrix}.$$

(a) The production matrix is

$$X = (I - A)^{-1}D = \begin{bmatrix} \frac{3}{2} & \frac{1}{4} \\ \frac{3}{4} & \frac{9}{8} \end{bmatrix}\begin{bmatrix} 1 \\ 1 \end{bmatrix} = \begin{bmatrix} \frac{7}{4} \\ \frac{15}{8} \end{bmatrix}.$$

Thus, $\frac{7}{4}$ bushels of yams and $\frac{15}{8} \approx 2$ pigs should be produced.

(b) The production matrix is

$$X = (I - A)^{-1}D = \begin{bmatrix} \frac{3}{2} & \frac{1}{4} \\ \frac{3}{4} & \frac{9}{8} \end{bmatrix} \begin{bmatrix} 100 \\ 70 \end{bmatrix} = \begin{bmatrix} 167.5 \\ 153.75 \end{bmatrix}.$$

Thus, 167.5 bushels of yams and $153.75 \approx 154$ pigs should be produced.

19. Use a graphing calculator or a computer to find the production matrix $X = (I - A)^{-1}D$. The answer is

$$\begin{bmatrix} 848 \\ 516 \\ 2970 \end{bmatrix}.$$

Values have been rounded.

Produce 848 units of agriculture, 516 units of manufacturing, and 2970 units of households.

21. Use a graphing calculator or a computer to find the production matrix $X = (I - A)^{-1}D$. The answer is

$$\begin{bmatrix} 195,492 \\ 25,933 \\ 13,580 \end{bmatrix}.$$

Values have been rounded. Change from thousands of pounds to millions of pounds.

Produce about 195 million Israeli pounds of agriculture, 26 million Israeli pounds of manufacturing, and 13.6 million Israeli pounds of energy.

23. Use a graphing calculator or a computer to find the production matrix $X = (I - A)^{-1}D$. The answer is

$$\begin{bmatrix} 532 \\ 481 \\ 805 \\ 1185 \end{bmatrix}.$$

Values have been rounded.

Produce about 532 units of natural resources, 481 manufacturing units, 805 trade and service units, and 1185 personal consumption units. Units are millions of dollars.

25. **(a)** Use a graphing calculator or a computer to find the matrix $(I - A)^{-1}$. The answer is

$$\begin{bmatrix} 1.67 & 0.56 & 0.56 \\ 0.19 & 1.17 & 0.06 \\ 3.15 & 3.27 & 4.38 \end{bmatrix}.$$

Values have been rounded.

(b) These multipliers imply that if the demand for one community's output increases by \$1 then the output in the other community will increase by the amount in the row and column of this matrix. For example, if the demand for Hermitage's output increases by \$1, then output from Sharon will increase \$0.56, from Farrell by \$0.06, and from Hermitage by \$4.38.

27. Calculate $I - A$, and then set $(I - A)X = O$ to find X.

$$(I - A)X = \left(\begin{bmatrix} 1 & 0 \\ 0 & 1 \end{bmatrix} - \begin{bmatrix} \frac{3}{4} & \frac{1}{3} \\ \frac{1}{4} & \frac{2}{3} \end{bmatrix} \right) \begin{bmatrix} x_1 \\ x_2 \end{bmatrix}$$

$$= \begin{bmatrix} \frac{1}{4} & -\frac{1}{3} \\ -\frac{1}{4} & \frac{1}{3} \end{bmatrix} \begin{bmatrix} x_1 \\ x_2 \end{bmatrix}$$

$$= \begin{bmatrix} \frac{1}{4}x_1 - \frac{1}{3}x_2 \\ -\frac{1}{4}x_1 + \frac{1}{3}x_2 \end{bmatrix} \begin{bmatrix} 0 \\ 0 \end{bmatrix}$$

Thus,

$$\frac{1}{4}x_1 - \frac{1}{3}x_2 = 0$$

$$\frac{1}{4}x_1 = \frac{1}{3}x_2$$

$$x_1 = \frac{4}{3}x_2$$

If $x_2 = 3$, $x_1 = 4$. Therefore, produce 4 units of steel for every 3 units of coal.

29. For this economy,

$$A = \begin{bmatrix} \frac{1}{5} & \frac{3}{5} & 0 \\ \frac{2}{5} & \frac{1}{5} & \frac{4}{5} \\ \frac{2}{5} & \frac{1}{5} & \frac{1}{5} \end{bmatrix}.$$

Find the value of $I - A$, then set $(I - A)X = O$.

$$(I - A)X = \left(\begin{bmatrix} 1 & 0 & 0 \\ 0 & 1 & 0 \\ 0 & 0 & 1 \end{bmatrix} - \begin{bmatrix} \frac{1}{5} & \frac{3}{5} & 0 \\ \frac{2}{5} & \frac{1}{5} & \frac{4}{5} \\ \frac{2}{5} & \frac{1}{5} & \frac{1}{5} \end{bmatrix} \right) \begin{bmatrix} x_1 \\ x_2 \\ x_3 \end{bmatrix}$$

$$= \begin{bmatrix} \frac{4}{5} & -\frac{3}{5} & 0 \\ -\frac{2}{5} & \frac{4}{5} & -\frac{4}{5} \\ -\frac{2}{5} & -\frac{1}{5} & \frac{4}{5} \end{bmatrix} \begin{bmatrix} x_1 \\ x_2 \\ x_3 \end{bmatrix}$$

$$= \begin{bmatrix} \dfrac{4}{5}x_1 - \dfrac{3}{5}x_2 \\[2mm] -\dfrac{2}{5}x_1 + \dfrac{4}{5}x_2 - \dfrac{4}{5}x_3 \\[2mm] -\dfrac{2}{5}x_1 - \dfrac{1}{5}x_2 + \dfrac{4}{5}x_3 \end{bmatrix} = \begin{bmatrix} 0 \\ 0 \\ 0 \end{bmatrix}$$

The system to be solved is

$$\dfrac{4}{5}x_1 - \dfrac{3}{5}x_2 \qquad\qquad = 0$$

$$-\dfrac{2}{5}x_1 + \dfrac{4}{5}x_2 - \dfrac{4}{5}x_3 = 0$$

$$-\dfrac{2}{5}x_1 - \dfrac{1}{5}x_2 + \dfrac{4}{5}x_3 = 0.$$

Write the augmented matrix of the system.

$$\begin{bmatrix} \dfrac{4}{5} & -\dfrac{3}{5} & 0 & \Big| & 0 \\[2mm] -\dfrac{2}{5} & \dfrac{4}{5} & -\dfrac{4}{5} & \Big| & 0 \\[2mm] -\dfrac{2}{5} & -\dfrac{1}{5} & \dfrac{4}{5} & \Big| & 0 \end{bmatrix}$$

$$\begin{matrix} \dfrac{5}{4}R_1 \to R_1 \\[2mm] 5R_2 \to R_2 \\[2mm] 5R_3 \to R_3 \end{matrix} \begin{bmatrix} 1 & -\dfrac{3}{4} & 0 & \Big| & 0 \\[2mm] -2 & 4 & -4 & \Big| & 0 \\[2mm] -2 & -1 & 4 & \Big| & 0 \end{bmatrix}$$

$$\begin{matrix} \\ 2R_1 + R_2 \to R_2 \\[2mm] 2R_1 + R_3 \to R_3 \end{matrix} \begin{bmatrix} 1 & -\dfrac{3}{4} & 0 & \Big| & 0 \\[2mm] 0 & \dfrac{5}{2} & -4 & \Big| & 0 \\[2mm] 0 & -\dfrac{5}{2} & 4 & \Big| & 0 \end{bmatrix}$$

$$\dfrac{2}{5}R_2 \to R_2 \begin{bmatrix} 1 & -\dfrac{3}{4} & 0 & \Big| & 0 \\[2mm] 0 & 1 & -\dfrac{8}{5} & \Big| & 0 \\[2mm] 0 & -\dfrac{5}{2} & 4 & \Big| & 0 \end{bmatrix}$$

$$\begin{matrix} \dfrac{3}{4}R_2 + R_1 \to R_1 \\[2mm] \\ \dfrac{5}{2}R_2 + R_3 \to R_3 \end{matrix} \begin{bmatrix} 1 & 0 & -\dfrac{6}{5} & \Big| & 0 \\[2mm] 0 & 1 & -\dfrac{8}{5} & \Big| & 0 \\[2mm] 0 & 0 & 0 & \Big| & 0 \end{bmatrix}$$

Use x_3 as the parameter. Therefore, $x_1 = \dfrac{6}{5}x_3$ and

$x_2 = \dfrac{8}{5}x_3$, and the solution is $\left(\dfrac{6}{5}x_3, \dfrac{8}{5}x_3, x_3\right)$. If

$x_3 = 5$, then $x_1 = 6$ and $x_2 = 8$.

Produce 6 units of mining for every 8 units of manufacturing and 5 units of communication.

Chapter 2 Review Exercises

1. False; a system with three equations and four unknowns has an infinite number of solutions.

2. False; matrix A is a 2×2 matrix and matrix B is a 3×2 matrix. Only matrices having the same dimension can be added.

3. True

4. True

5. False; only matrices having the same dimension can be added.

6. True

7. False; in general, matrix multiplication is not commutative.

8. False; only square matrices can have inverses.

9. True; for example, $0 \cdot A = A \cdot 0 = 0$

10. False; only row operations can be used.

11. False; any $n \times n$ zero matrix does not have an inverse, and the matrix $\begin{bmatrix} 2 & -4 \\ 1 & -2 \end{bmatrix}$ is an example of another square matrix that doesn't have an inverse.

12. False; if $AB = C$ and A has an inverse, then
    $B = A^{-1}C.$

13. True

14. True

15. False; $AB = CB$ implies $A = C$ only if B is the identity matrix or B has an inverse.

16. True

17. For a system of m linear equations in n unknowns and $m = n$, there could be one, none, or an infinite number of solutions. If $m < n$, there are an infinite number of solutions. If $m > n$, there could be one, none, or an infinite number of solutions.

19. $2x - 3y = 14$ (*1*)
    $3x + 2y = -5$ (*2*)

Eliminate x in equation (2).

$$\begin{matrix} & 2x - 3y & = & 14 & (1) \\ -3R_1 + 2R_2 \to R_2 & 13y & = & -52 & (3) \end{matrix}$$

Make each leading coefficient equal 1.

$$\begin{matrix} \dfrac{1}{2}R_1 \to R_1 & x - \dfrac{3}{2}y = & 7 & (4) \\[2mm] \dfrac{1}{13}R_2 \to R_2 & y = -4 & (5) \end{matrix}$$

Substitute -4 for y in equation (4) to get $x = 1$.

The solution is $(1, -4)$.

21. $2x - 3y + z = -5$ (*1*)
    $5x + 5y + 3z = 14$ (*2*)

Eliminate x in equation (2).

$$2x - 3y + z = -5$$
$$5R_1 + (-2)R_2 \to R_2 \quad -25y - z = -53$$

Let z be the parameter. Solve for y and for x in terms of z.

$$-25y - z = -53$$
$$-25y = -53 + z$$
$$y = \frac{53 - z}{25}$$

$$2x - 3\left(\frac{53 - z}{25}\right) + z = -5$$
$$2x - \frac{159}{25} + \frac{3z}{25} + z = -5$$
$$2x + \frac{28}{25}z = \frac{34}{25}$$
$$2x = \frac{34}{25} - \frac{28}{25}z$$
$$x = \frac{34 - 28z}{50}$$

The solutions are $\left(\frac{34 - 28z}{50}, \frac{53 - z}{25}, z\right)$, where z is any real number.

23. $2x + 4y = -6$
$-3x - 5y = 12$

Write the augmented matrix and use row operations.

$$\begin{bmatrix} 2 & 4 & | & -6 \\ -3 & -5 & | & 12 \end{bmatrix}$$

$$3R_1 + 2R_2 \to R_2 \quad \begin{bmatrix} 2 & 4 & | & -6 \\ 0 & 2 & | & 6 \end{bmatrix}$$

$$-2R_2 + R_1 \to R_1 \quad \begin{bmatrix} 2 & 0 & | & -18 \\ 0 & 2 & | & 6 \end{bmatrix}$$

$$\frac{1}{2}R_1 \to R_1 \quad \begin{bmatrix} 1 & 0 & | & -9 \\ 0 & 1 & | & 3 \end{bmatrix}$$
$$\frac{1}{2}R_2 \to R_2$$

The solution is $(-9, 3)$.

25. $x - y + 3z = 13$
$4x + y + 2z = 17$
$3x + 2y + 2z = 1$

Write the augmented matrix and use row operations.

$$\begin{bmatrix} 1 & -1 & 3 & | & 13 \\ 4 & 1 & 2 & | & 17 \\ 3 & 2 & 2 & | & 1 \end{bmatrix}$$

$$-4R_1 + R_2 \to R_2 \quad \begin{bmatrix} 1 & -1 & 3 & | & 13 \\ 0 & 5 & -10 & | & -35 \\ 0 & 5 & -7 & | & -38 \end{bmatrix}$$
$$-3R_1 + R_3 \to R_3$$

$$R_2 + 5R_1 \to R_1 \quad \begin{bmatrix} 5 & 0 & 5 & | & 30 \\ 0 & 5 & -10 & | & -35 \\ 0 & 0 & 3 & | & -3 \end{bmatrix}$$
$$-1R_2 + R_3 \to R_3$$

$$5R_3 + (-3R_1) \to R_1 \quad \begin{bmatrix} -15 & 0 & 0 & | & -105 \\ 0 & 15 & 0 & | & -135 \\ 0 & 0 & 3 & | & -3 \end{bmatrix}$$
$$10R_3 + 3R_2 \to R_2$$

$$-\frac{1}{15}R_1 \to R_1 \quad \begin{bmatrix} 1 & 0 & 0 & | & 7 \\ 0 & 1 & 0 & | & -9 \\ 0 & 0 & 1 & | & -1 \end{bmatrix}$$
$$\frac{1}{15}R_2 \to R_2$$
$$\frac{1}{3}R_3 \to R_3$$

The solution is $(7, -9, -1)$.

27. $3x - 6y + 9z = 12$
$-x + 2y - 3z = -4$
$x + y + 2z = 7$

Write the augmented matrix and use row operations.

$$\begin{bmatrix} 3 & -6 & 9 & | & 12 \\ -1 & 2 & -3 & | & -4 \\ 1 & 1 & 2 & | & 7 \end{bmatrix}$$

$$R_1 + 3R_2 \to R_2 \quad \begin{bmatrix} 3 & -6 & 9 & | & 12 \\ 0 & 0 & 0 & | & 0 \\ 0 & 9 & -3 & | & 9 \end{bmatrix}$$
$$-1R_1 + 3R_3 \to R_3$$

The zero in row 2, column 2 is an obstacle. To proceed, interchange the second and third rows.

$$\begin{bmatrix} 3 & -6 & 9 & | & 12 \\ 0 & 9 & -3 & | & 9 \\ 0 & 0 & 0 & | & 0 \end{bmatrix}$$

$$3R_1 + 2R_2 \to R_1 \quad \begin{bmatrix} 9 & 0 & 21 & | & 54 \\ 0 & 9 & -3 & | & 9 \\ 0 & 0 & 0 & | & 0 \end{bmatrix}$$

$\frac{1}{9}R_1 \to R_1 \begin{bmatrix} 1 & 0 & \frac{7}{3} & 6 \\ 0 & 1 & -\frac{1}{3} & 1 \\ \frac{1}{9}R_2 \to R_2 & & & \\ 0 & 0 & 0 & 0 \end{bmatrix}$

The row of zeros indicates dependent equations.

Solve the first two equations respectively for x and y in terms of z to obtain

$$x = 6 - \frac{7}{3}z \text{ and } y = 1 + \frac{1}{3}z$$

The solution of the system is

$$\left(6 - \frac{7}{3}z, \ 1 + \frac{1}{3}z, \ z \right),$$

where z is any real number.

In Exercises 29–32, corresponding elements must be equal.

29. $\begin{bmatrix} 2 & 3 \\ 5 & q \end{bmatrix} = \begin{bmatrix} a & b \\ c & 9 \end{bmatrix}$

Size: 2×2; $a = 2, b = 3, c = 5, q = 9$; square matrix

31. $\begin{bmatrix} 2m & 4 & 3z & -12 \end{bmatrix}$
$= \begin{bmatrix} 12 & k+1 & -9 & r-3 \end{bmatrix}$

Size: 1×4; $m = 6, k = 3, z = -3, r = -9$; row matrix

33. $A + C = \begin{bmatrix} 4 & 10 \\ -2 & -3 \\ 6 & 9 \end{bmatrix} + \begin{bmatrix} 5 & 0 \\ -1 & 3 \\ 4 & 7 \end{bmatrix}$

$= \begin{bmatrix} 9 & 10 \\ -3 & 0 \\ 10 & 16 \end{bmatrix}$

35. $3C + 2A = 3\begin{bmatrix} 5 & 0 \\ -1 & 3 \\ 4 & 7 \end{bmatrix} + 2\begin{bmatrix} 4 & 10 \\ -2 & -3 \\ 6 & 9 \end{bmatrix}$

$= \begin{bmatrix} 15 & 0 \\ -3 & 9 \\ 12 & 21 \end{bmatrix} + \begin{bmatrix} 8 & 20 \\ -4 & -6 \\ 12 & 18 \end{bmatrix}$

$= \begin{bmatrix} 23 & 20 \\ -7 & 3 \\ 24 & 39 \end{bmatrix}$

37. $2A - 5C = 2\begin{bmatrix} 4 & 10 \\ -2 & -3 \\ 6 & 9 \end{bmatrix} - 5\begin{bmatrix} 5 & 0 \\ -1 & 3 \\ 4 & 7 \end{bmatrix}$

$= \begin{bmatrix} 8 & 20 \\ -4 & -6 \\ 12 & 18 \end{bmatrix} - \begin{bmatrix} 25 & 0 \\ -5 & 15 \\ 20 & 35 \end{bmatrix}$

$= \begin{bmatrix} -17 & 20 \\ 1 & -21 \\ -8 & -17 \end{bmatrix}$

39. A is 3×2 and C is 3×2, so finding the product AC is not possible.

$$\begin{array}{cc} A & C \\ 3 \times 2 & 3 \times 2 \end{array}$$

(The inner two numbers must match.)

41. $ED = \begin{bmatrix} 1 & 3 & -4 \end{bmatrix}\begin{bmatrix} 6 \\ 1 \\ 0 \end{bmatrix}$

$= [1·6 + 3·1 + (-4)0] = [9]$

43. $EC = \begin{bmatrix} 1 & 3 & -4 \end{bmatrix}\begin{bmatrix} 5 & 0 \\ -1 & 3 \\ 4 & 7 \end{bmatrix}$

$= [1·5 + 3(-1) + (-4)·4 \quad 1·0$
$+ 3·3 + (-4)·7]$

$= \begin{bmatrix} -14 & -19 \end{bmatrix}$

45. Find the inverse of $B = \begin{bmatrix} 2 & 3 & -2 \\ 2 & 4 & 0 \\ 0 & 1 & 2 \end{bmatrix}$, if it exists.

Write the augmented matrix to obtain

$$[B|I] = \begin{bmatrix} 2 & 3 & -2 & 1 & 0 & 0 \\ 2 & 4 & 0 & 0 & 1 & 0 \\ 0 & 1 & 2 & 0 & 0 & 1 \end{bmatrix}.$$

$-1R_1 + R_2 \to R_2 \begin{bmatrix} 2 & 3 & -2 & 1 & 0 & 0 \\ 0 & 1 & 2 & -1 & 1 & 0 \\ 0 & 1 & 2 & 0 & 0 & 1 \end{bmatrix}$

$-3R_2 + R_1 \to R_1 \begin{bmatrix} 2 & 0 & -8 & 4 & -3 & 0 \\ 0 & 1 & 2 & -1 & 1 & 0 \\ 0 & 0 & 0 & 1 & -1 & 1 \end{bmatrix}$
$-1R_2 + R_3 \to R_3$

No inverse exists, since the third row is all zeros to the left of the vertical bar.

47. Find the inverse of $A = \begin{bmatrix} 1 & 3 \\ 2 & 7 \end{bmatrix}$, if it exists.

Write the augmented matrix $\begin{bmatrix} A|I \end{bmatrix}$.

$$[A \mid I] = \begin{bmatrix} 1 & 3 & | & 1 & 0 \\ 2 & 7 & | & 0 & 1 \end{bmatrix}$$

Perform row operations on $\begin{bmatrix} A|I \end{bmatrix}$ to get a matrix of the form $\begin{bmatrix} I|B \end{bmatrix}$.

$$-2R_1 + R_2 \rightarrow R_2 \quad \begin{bmatrix} 1 & 3 & | & 1 & 0 \\ 0 & 1 & | & -2 & 1 \end{bmatrix}$$

$$-3R_2 + R_1 \rightarrow R_1 \quad \begin{bmatrix} 1 & 0 & | & 7 & -3 \\ 0 & 1 & | & -2 & 1 \end{bmatrix}$$

The last augmented matrix is of the form $\begin{bmatrix} I|B \end{bmatrix}$, so the desired inverse is

$$A^{-1} = \begin{bmatrix} 7 & -3 \\ -2 & 1 \end{bmatrix}.$$

49. Find the inverse of $A = \begin{bmatrix} 3 & -6 \\ -4 & 8 \end{bmatrix}$, if it exists.

Write the augmented matrix $\begin{bmatrix} A|I \end{bmatrix}$.

$$[A|I] = \begin{bmatrix} 3 & -6 & | & 1 & 0 \\ -4 & 8 & | & 0 & 1 \end{bmatrix}$$

Perform row operations on $\begin{bmatrix} A|I \end{bmatrix}$ to get a matrix of the form $\begin{bmatrix} I|B \end{bmatrix}$.

$$4R_1 + 3R_2 \rightarrow R_2 \quad \begin{bmatrix} 3 & -6 & | & 1 & 0 \\ 0 & 0 & | & 4 & 3 \end{bmatrix}$$

Since the entries left of the vertical bar in the second row are zeros, no inverse exists.

51. Find the inverse of $A = \begin{bmatrix} 2 & -1 & 0 \\ 1 & 0 & 1 \\ 1 & -2 & 0 \end{bmatrix}$, if it exists.

The augmented matrix is

$$[A|I] = \begin{bmatrix} 2 & -1 & 0 & | & 1 & 0 & 0 \\ 1 & 0 & 1 & | & 0 & 1 & 0 \\ 1 & -2 & 0 & | & 0 & 0 & 1 \end{bmatrix}.$$

$$\begin{array}{c} R_1 + (-2)R_2 \rightarrow R_2 \\ R_1 + (-2)R_3 \rightarrow R_3 \end{array} \begin{bmatrix} 2 & -1 & 0 & | & 1 & 0 & 0 \\ 0 & -1 & -2 & | & 1 & -2 & 0 \\ 0 & 3 & 0 & | & 1 & 0 & -2 \end{bmatrix}$$

$$\begin{array}{c} -1R_2 + R_1 \rightarrow R_1 \\ \\ 3R_2 + R_3 \rightarrow R_3 \end{array} \begin{bmatrix} 2 & 0 & 2 & | & 0 & 2 & 0 \\ 0 & -1 & -2 & | & 1 & -2 & 0 \\ 0 & 0 & -6 & | & 4 & -6 & -2 \end{bmatrix}$$

$$\begin{array}{c} R_3 + 3R_1 \rightarrow R_1 \\ R_3 + (-3)R_2 \rightarrow R_2 \end{array} \begin{bmatrix} 6 & 0 & 0 & | & 4 & 0 & -2 \\ 0 & 3 & 0 & | & 1 & 0 & -2 \\ 0 & 0 & -6 & | & 4 & -6 & -2 \end{bmatrix}$$

$$\begin{array}{c} \frac{1}{6}R_1 \rightarrow R_1 \\ \frac{1}{3}R_2 \rightarrow R_2 \\ -\frac{1}{6}R_3 \rightarrow R_3 \end{array} \begin{bmatrix} 1 & 0 & 0 & | & \frac{2}{3} & 0 & -\frac{1}{3} \\ 0 & 1 & 0 & | & \frac{1}{3} & 0 & -\frac{2}{3} \\ 0 & 0 & 1 & | & -\frac{2}{3} & 1 & \frac{1}{3} \end{bmatrix}$$

$$A^{-1} = \begin{bmatrix} \frac{2}{3} & 0 & -\frac{1}{3} \\ \frac{1}{3} & 0 & -\frac{2}{3} \\ -\frac{2}{3} & 1 & \frac{1}{3} \end{bmatrix}$$

53. Find the inverse of $A = \begin{bmatrix} 1 & 3 & 6 \\ 4 & 0 & 9 \\ 5 & 15 & 30 \end{bmatrix}$, if it exists

$$[A|I] = \begin{bmatrix} 1 & 3 & 6 & | & 1 & 0 & 0 \\ 4 & 0 & 9 & | & 0 & 1 & 0 \\ 5 & 15 & 30 & | & 0 & 0 & 1 \end{bmatrix}$$

$$\begin{array}{c} -4R_1 + R_2 \rightarrow R_2 \\ -5R_1 + R_3 \rightarrow R_3 \end{array} \begin{bmatrix} 1 & 3 & 6 & | & 1 & 0 & 0 \\ 0 & -12 & -15 & | & -4 & 1 & 0 \\ 0 & 0 & 0 & | & -5 & 0 & 1 \end{bmatrix}$$

The last row is all zeros to the left of the bar, so no inverse exists.

55. $A = \begin{bmatrix} 5 & 1 \\ -1 & -2 \end{bmatrix}, B = \begin{bmatrix} -8 \\ 24 \end{bmatrix}$

The matrix equation to be solved is $AX = B$, or

$$\begin{bmatrix} 5 & 1 \\ -1 & -2 \end{bmatrix}\begin{bmatrix} x \\ y \end{bmatrix} = \begin{bmatrix} -8 \\ 24 \end{bmatrix}.$$

Calculate the inverse of the coefficient matrix A to obtain

$$\begin{bmatrix} 5 & 1 \\ -1 & -2 \end{bmatrix}^{-1} = \begin{bmatrix} \frac{1}{4} & \frac{1}{8} \\ -\frac{1}{4} & -\frac{5}{8} \end{bmatrix}.$$

Now $X = A^{-1}B$, so

$$\begin{bmatrix} x \\ y \end{bmatrix} = \begin{bmatrix} \frac{1}{4} & \frac{1}{8} \\ -\frac{1}{4} & -\frac{5}{8} \end{bmatrix}\begin{bmatrix} -8 \\ 24 \end{bmatrix} = \begin{bmatrix} 1 \\ -13 \end{bmatrix}.$$

57. $A = \begin{bmatrix} 1 & 0 & 2 \\ -1 & 1 & 0 \\ 3 & 0 & 4 \end{bmatrix}, \; B = \begin{bmatrix} 8 \\ 4 \\ -6 \end{bmatrix}$

By the usual method, we find that the inverse of the coefficient matrix is

$$A^{-1} = \begin{bmatrix} -2 & 0 & 1 \\ -2 & 1 & 1 \\ \frac{3}{2} & 0 & -\frac{1}{2} \end{bmatrix}.$$

Since $X = A^{-1}B$,

$$X = \begin{bmatrix} -2 & 0 & 1 \\ -2 & 1 & 1 \\ \frac{3}{2} & 0 & -\frac{1}{2} \end{bmatrix} \begin{bmatrix} 8 \\ 4 \\ -6 \end{bmatrix} = \begin{bmatrix} -22 \\ -18 \\ 15 \end{bmatrix}.$$

59. $x + 2y = 4$
$2x - 3y = 1$

The coefficient matrix is

$$A = \begin{bmatrix} 1 & 2 \\ 2 & -3 \end{bmatrix}.$$

Calculate the inverse of A.

$$A^{-1} = \begin{bmatrix} \frac{3}{7} & \frac{2}{7} \\ \frac{2}{7} & -\frac{1}{7} \end{bmatrix}$$

Use $X = A^{-1}B$ to solve.

$$\begin{bmatrix} x \\ y \end{bmatrix} = \begin{bmatrix} \frac{3}{7} & \frac{2}{7} \\ \frac{2}{7} & -\frac{1}{7} \end{bmatrix} \begin{bmatrix} 4 \\ 1 \end{bmatrix} = \begin{bmatrix} 2 \\ 1 \end{bmatrix}$$

The solution is $(2, 1)$.

61. $x + y + z = 1$
$2x + y \quad\;\; = -2$
$\quad\;\; 3y + z = 2$

The coefficient matrix is

$$A = \begin{bmatrix} 1 & 1 & 1 \\ 2 & 1 & 0 \\ 0 & 3 & 1 \end{bmatrix}.$$

Find that the inverse of A is

$$A^{-1} = \begin{bmatrix} \frac{1}{5} & \frac{2}{5} & -\frac{1}{5} \\ -\frac{2}{5} & \frac{1}{5} & \frac{2}{5} \\ \frac{6}{5} & -\frac{3}{5} & -\frac{1}{5} \end{bmatrix}.$$

Now $X = A^{-1}B$, so

$$\begin{bmatrix} x \\ y \\ z \end{bmatrix} = \begin{bmatrix} \frac{1}{5} & \frac{2}{5} & -\frac{1}{5} \\ -\frac{2}{5} & \frac{1}{5} & \frac{2}{5} \\ \frac{6}{5} & -\frac{3}{5} & -\frac{1}{5} \end{bmatrix} \begin{bmatrix} 1 \\ -2 \\ 2 \end{bmatrix} = \begin{bmatrix} -1 \\ 0 \\ 2 \end{bmatrix}.$$

The solution is $(-1, 0, 2)$.

63. $A = \begin{bmatrix} 0.01 & 0.05 \\ 0.04 & 0.03 \end{bmatrix}, \; D = \begin{bmatrix} 200 \\ 300 \end{bmatrix}$

$$X = (I - A)^{-1}D$$

$$I - A = \begin{bmatrix} 1 & 0 \\ 0 & 1 \end{bmatrix} - \begin{bmatrix} 0.01 & 0.05 \\ 0.04 & 0.03 \end{bmatrix}$$

$$= \begin{bmatrix} 0.99 & -0.05 \\ -0.04 & 0.97 \end{bmatrix}$$

Use row operations to find the inverse of $I - A$, which is

$$(I - A)^{-1} = \begin{bmatrix} 1.0122 & 0.0522 \\ 0.0417 & 1.0331 \end{bmatrix}.$$

Since $X = (I - A)^{-1}D$, the production matrix is

$$X = \begin{bmatrix} 1.0122 & 0.0522 \\ 0.0417 & 1.0331 \end{bmatrix} \begin{bmatrix} 200 \\ 300 \end{bmatrix}$$

$$= \begin{bmatrix} 218.1 \\ 318.3 \end{bmatrix}.$$

65. $x + 2y + z = 7$ (1)
$2x - y - z = 2$ (2)
$3x - 3y + 2z = -5$ (3)

(a) To solve the system by the echelon method, begin by eliminating x in equations (2) and (3).

$$\begin{array}{rrcll} & x + 2y + z & = & 7 & (1) \\ -2R_1 + R_2 \to R_2 & -5y - 3z & = & -12 & (4) \\ -3R_1 + R_3 \to R_3 & -9y - z & = & -26 & (5) \end{array}$$

Eliminate y in equation (5).

$$\begin{array}{rrcll} & x + 2y + z & = & 7 & (1) \\ & -5y - 3z & = & -12 & (4) \\ -9R_2 + 5R_3 \to R_3 & 22z & = & -22 & (6) \end{array}$$

Make each leading coefficient equal 1.

$$x + 2y + z = 7 \qquad (1)$$

$$-\frac{1}{5}R_2 \to R_2 \qquad y + \frac{3}{5}z = \frac{12}{5} \qquad (7)$$

$$\frac{1}{22}R_3 \to R_3 \qquad z = -1 \qquad (8)$$

Substitute -1 for z in equation (7) to get $y = 3$. Substitute -1 for z and 3 for y in equation (1) to get $x = 2$.

The solution is $(2, 3, -1)$.

(b) The same system is to be solved using the Gauss-Jordan method. Write the augmented matrix and use row operations.

$$\begin{bmatrix} 1 & 2 & 1 & | & 7 \\ 2 & -1 & -1 & | & 2 \\ 3 & -3 & 2 & | & -5 \end{bmatrix}$$

$$\begin{matrix} -2R_1 + R_2 \to R_2 \\ -3R_1 + R_3 \to R_3 \end{matrix} \begin{bmatrix} 1 & 2 & 1 & | & 7 \\ 0 & -5 & -3 & | & -12 \\ 0 & -9 & -1 & | & -26 \end{bmatrix}$$

$$2R_2 + 5R_1 \to R_1 \begin{bmatrix} 5 & 0 & -1 & | & 11 \\ 0 & -5 & -3 & | & -12 \\ -9R_2 + 5R_3 \to R_3 \end{bmatrix}$$
$$\begin{bmatrix} 5 & 0 & -1 & | & 11 \\ 0 & -5 & -3 & | & -12 \\ 0 & 0 & 22 & | & -22 \end{bmatrix}$$

$$\begin{matrix} R_3 + 22R_1 \to R_1 \\ 3R_3 + 22R_2 \to R_2 \end{matrix} \begin{bmatrix} 110 & 0 & 0 & | & 220 \\ 0 & -110 & 0 & | & -330 \\ 0 & 0 & 22 & | & -22 \end{bmatrix}$$

$$\begin{matrix} \frac{1}{110}R_1 + R_1 \\ -\frac{1}{110}R_2 \to R_2 \\ \frac{1}{22}R_3 \to R_3 \end{matrix} \begin{bmatrix} 1 & 0 & 0 & | & 2 \\ 0 & 1 & 0 & | & 3 \\ 0 & 0 & 1 & | & -1 \end{bmatrix}$$

The corresponding system is $\begin{matrix} x = 2 \\ y = 3 \\ z = -1. \end{matrix}$

The solution is $(2, 3, -1)$

(c) The system can be written as a matrix equation $AX = B$ by writing

$$\begin{bmatrix} 1 & 2 & 1 \\ 2 & -1 & -1 \\ 3 & -3 & 2 \end{bmatrix} \begin{bmatrix} x \\ y \\ z \end{bmatrix} = \begin{bmatrix} 7 \\ 2 \\ -5 \end{bmatrix}.$$

(d) The inverse of the coefficient matrix A can be found by using row operations.

$$[A|I] = \begin{bmatrix} 1 & 2 & 1 & | & 1 & 0 & 0 \\ 2 & -1 & -1 & | & 0 & 1 & 0 \\ 3 & -3 & 2 & | & 0 & 0 & 1 \end{bmatrix}$$

$$\begin{matrix} -2R_1 + R_2 \to R_2 \\ -3R_1 + R_3 \to R_3 \end{matrix} \begin{bmatrix} 1 & 2 & 1 & | & 1 & 0 & 0 \\ 0 & -5 & -3 & | & -2 & 1 & 0 \\ 0 & -9 & -1 & | & -3 & 0 & 1 \end{bmatrix}$$

$$2R_2 + 5R_1 \to R_1 \begin{bmatrix} 5 & 0 & -1 & | & 1 & 2 & 0 \\ 0 & -5 & -3 & | & -2 & 1 & 0 \\ -9R_2 + 5R_3 \to R_3 \end{bmatrix}$$
$$\begin{bmatrix} 5 & 0 & -1 & | & 1 & 2 & 0 \\ 0 & -5 & -3 & | & -2 & 1 & 0 \\ 0 & 0 & 22 & | & 3 & -9 & 5 \end{bmatrix}$$

$$\begin{matrix} R_3 + 22R_1 \to R_1 \\ 3R_3 + 22R_2 \to R_2 \end{matrix} \begin{bmatrix} 110 & 0 & 0 & | & 25 & 35 & 5 \\ 0 & -110 & 0 & | & -35 & -5 & 15 \\ 0 & 0 & 22 & | & 3 & -9 & 5 \end{bmatrix}$$

$$\begin{matrix} \frac{1}{110}R_1 + R_1 \\ -\frac{1}{110}R_2 \to R_2 \\ \frac{1}{22}R_3 \to R_3 \end{matrix} \begin{bmatrix} 1 & 0 & 0 & | & \frac{5}{22} & \frac{7}{22} & \frac{1}{22} \\ 0 & 1 & 0 & | & \frac{7}{22} & \frac{1}{22} & -\frac{3}{22} \\ 0 & 0 & 1 & | & \frac{3}{22} & -\frac{9}{22} & \frac{5}{22} \end{bmatrix}$$

The inverse of matrix A is

$$A^{-1} = \begin{bmatrix} \frac{5}{22} & \frac{7}{22} & \frac{1}{22} \\ \frac{7}{22} & \frac{1}{22} & -\frac{3}{22} \\ \frac{3}{22} & -\frac{9}{22} & \frac{5}{22} \end{bmatrix} \approx \begin{bmatrix} 0.23 & 0.32 & 0.05 \\ 0.32 & 0.05 & -0.14 \\ 0.14 & -0.41 & 0.23 \end{bmatrix}.$$

(e) Since $X = A^{-1}B$,

$$\begin{bmatrix} x \\ y \\ z \end{bmatrix} = \begin{bmatrix} \frac{5}{22} & \frac{7}{22} & \frac{1}{22} \\ \frac{7}{22} & \frac{1}{22} & -\frac{3}{22} \\ \frac{3}{22} & -\frac{9}{22} & \frac{5}{22} \end{bmatrix} \begin{bmatrix} 7 \\ 2 \\ -5 \end{bmatrix} = \begin{bmatrix} 2 \\ 3 \\ -1 \end{bmatrix}.$$

Once again, the solution is $(2, 3, -1)$.

67. Let $x_1 =$ the number of blankets,

$x_2 =$ the number of rugs, and

$x_3 =$ the number of skirts.

The given information leads to the system

$$24x_1 + 30x_2 + 12x_3 = 306 \qquad (1)$$
$$4x_1 + 5x_2 + 3x_3 = 59 \qquad (2)$$
$$15x_1 + 18x_2 + 9x_3 = 201. \qquad (3)$$

Simplify equations (1) and (3).

$\frac{1}{6}R_1 \rightarrow R_1$ $4x_1 + 5x_2 + 2x_3 = 51$ (4)

$4x_1 + 5x_2 + 3x_3 = 59$ (2)

$\frac{1}{3}R_3 \rightarrow R_3$ $5x_1 + 6x_2 + 3x_3 = 67$ (5)

Solve this system by the Gauss-Jordan method. Write the augmented matrix and use row operations.

$$\begin{bmatrix} 4 & 5 & 2 & | & 51 \\ 4 & 5 & 3 & | & 59 \\ 5 & 6 & 3 & | & 67 \end{bmatrix}$$

$-1R_1 + R_2 \rightarrow R_2$
$-4R_3 + 5R_1 \rightarrow R_3$
$$\begin{bmatrix} 4 & 5 & 2 & | & 51 \\ 0 & 0 & 1 & | & 8 \\ 0 & 1 & -2 & | & -13 \end{bmatrix}$$

Interchange the second and third rows.

$$\begin{bmatrix} 4 & 5 & 2 & | & 51 \\ 0 & 1 & -2 & | & -13 \\ 0 & 0 & 1 & | & 8 \end{bmatrix}$$

$-5R_2 + R_1 \rightarrow R_1 \begin{bmatrix} 4 & 0 & 12 & | & 116 \\ 0 & 1 & -2 & | & -13 \\ 0 & 0 & 1 & | & 8 \end{bmatrix}$

$-12R_3 + R_1 \rightarrow R_1 \begin{bmatrix} 4 & 0 & 0 & | & 20 \\ 2R_3 + R_2 \rightarrow R_2 & 0 & 1 & 0 & | & 3 \\ & 0 & 0 & 1 & | & 8 \end{bmatrix}$

$\frac{1}{4}R_1 \rightarrow R_1 \begin{bmatrix} 1 & 0 & 0 & | & 5 \\ 0 & 1 & 0 & | & 3 \\ 0 & 0 & 1 & | & 8 \end{bmatrix}$

The solution of the system is $x = 5, y = 3$, $z = 8$. So, 5 blankets, 3 rugs, and 8 skirts can be made.

69. The 4×5 matrix of stock reports is

	div	ratio	sales	price	change
AT & T	1.33	17.6	152,000	26.75	1.88
GE	1.00	20	238,200	32.36	−1.50
SaraLee	0.79	25.4	39,110	16.51	−0.89
Disney	0.27	21.2	122,500	28.60	0.75

71. (a) The input-output matrix is

$$A = \begin{bmatrix} 0 & \frac{1}{2} \\ \frac{2}{3} & 0 \end{bmatrix}.$$

(b) $I - A = \begin{bmatrix} 1 & -\frac{1}{2} \\ -\frac{2}{3} & 1 \end{bmatrix}$, $D = \begin{bmatrix} 400 \\ 800 \end{bmatrix}$

Use row operations to find the inverse of $I - A$, which is

$$(I - A)^{-1} = \begin{bmatrix} \frac{3}{2} & \frac{3}{4} \\ 1 & \frac{3}{2} \end{bmatrix}.$$

Since $X = (I - A)^{-1}D$,

$$X = \begin{bmatrix} \frac{3}{2} & \frac{3}{4} \\ 1 & \frac{3}{2} \end{bmatrix}\begin{bmatrix} 400 \\ 800 \end{bmatrix} = \begin{bmatrix} 1200 \\ 1600 \end{bmatrix}.$$

The production required is 1200 units of cheese and 1600 units of goats.

73. The given information can be written as the following 4×3 matrix.

$$\begin{bmatrix} 8 & 8 & 8 \\ 10 & 5 & 9 \\ 7 & 10 & 7 \\ 8 & 9 & 7 \end{bmatrix}$$

75. (a) $a + b = 0.60$ (1)

$c + d = 0.75$ (2)

$a + c = 0.65$ (3)

$b + d = 0.70$ (4)

The augmented matrix of the system is

$$\begin{bmatrix} 1 & 1 & 0 & 0 & | & 0.60 \\ 0 & 0 & 1 & 1 & | & 0.75 \\ 1 & 0 & 1 & 0 & | & 0.65 \\ 0 & 1 & 0 & 1 & | & 0.70 \end{bmatrix}.$$

$-1R_1 + R_3 \rightarrow R_3 \begin{bmatrix} 1 & 1 & 0 & 0 & | & 0.60 \\ 0 & 0 & 1 & 1 & | & 0.75 \\ 0 & -1 & 1 & 0 & | & 0.05 \\ 0 & 1 & 0 & 1 & | & 0.70 \end{bmatrix}$

Interchange rows 2 and 4.

$$\begin{bmatrix} 1 & 1 & 0 & 0 & | & 0.60 \\ 0 & 1 & 0 & 1 & | & 0.70 \\ 0 & -1 & 1 & 0 & | & 0.05 \\ 0 & 0 & 1 & 1 & | & 0.75 \end{bmatrix}$$

$-1R_2 + R_1 \rightarrow R_1 \begin{bmatrix} 1 & 0 & 0 & -1 & | & -0.10 \\ & 0 & 1 & 0 & 1 & | & 0.70 \\ R_2 + R_3 \rightarrow R_3 & 0 & 0 & 1 & 1 & | & 0.75 \\ & 0 & 0 & 1 & 1 & | & 0.75 \end{bmatrix}$

Since R_3 and R_4 are identical, there will be infinitely many solutions. We do not have enough information to determine the values of a, b, c, and d.

(b) i. If $d = 0.33$, the system of equations in part (a) becomes

$$a + b = 0.60 \quad (1)$$
$$c + 0.33 = 0.75 \quad (2)$$
$$a + c = 0.65 \quad (3)$$
$$b + 0.33 = 0.70. \quad (4)$$

Equation (2) gives $c = 0.42$, and equation (4) gives $b = 0.37$. Substituting $c = 0.42$ into equation (3) gives $a = 0.23$. Therefore, $a = 0.23$, $b = 0.37$, $c = 0.42$, and $d = 0.33$.

Thus, A is healthy, B and D are tumorous, and C is bone.

ii. If $d = 0.43$, the system of equations in part (a) becomes

$$a + b = 0.60 \quad (1)$$
$$c + 0.43 = 0.75 \quad (2)$$
$$a + c = 0.65 \quad (3)$$
$$b + 0.43 = 0.70. \quad (4)$$

Equation (2) gives $c = 0.32$, and equation (4) gives $b = 0.27$. Substituting $c = 0.32$ into equation (3) gives $a = 0.33$. Therefore, $a = 0.33$, $b = 0.27$, $c = 0.32$, and $d = 0.43$.

Thus, A and C are tumorous, B could be healthy or tumorous, and D is bone.

(c) The original system now has two additional equations.

$$a + b = 0.60 \quad (1)$$
$$c + d = 0.75 \quad (2)$$
$$a + c = 0.65 \quad (3)$$
$$b + d = 0.70 \quad (4)$$
$$b + c = 0.85 \quad (5)$$
$$a + d = 0.50 \quad (6)$$

The augmented matrix of this system is

$$\begin{bmatrix} 1 & 1 & 0 & 0 & 0.60 \\ 0 & 0 & 1 & 1 & 0.75 \\ 1 & 0 & 1 & 0 & 0.65 \\ 0 & 1 & 0 & 1 & 0.70 \\ 0 & 1 & 1 & 0 & 0.85 \\ 1 & 0 & 0 & 1 & 0.50 \end{bmatrix}.$$

Using the Gauss-Jordan method we obtain

$$\begin{bmatrix} 1 & 0 & 0 & 0 & 0.20 \\ 0 & 1 & 0 & 0 & 0.40 \\ 0 & 0 & 1 & 0 & 0.45 \\ 0 & 0 & 0 & 1 & 0.30 \\ 0 & 0 & 0 & 0 & 0 \\ 0 & 0 & 0 & 0 & 0 \end{bmatrix}.$$

Therefore, $a = 0.20$, $b = 0.40$, $c = 0.45$, and $d = 0.30$. Thus, A is healthy, B and C are bone, and D is tumorous.

(d) As we saw in part (c), the six equations reduced to four independent equations. We need only four beams, correctly chosen, to obtain a solution. The four beams must pass through all four cells and must lead to independent equations. One such choice would be beams 1, 2, 3, and 6. Another choice would be beams 1, 2, 4, and 5.

77. $\dfrac{\sqrt{3}}{2}(W_1 + W_2) = 100 \quad (1)$

$$W_1 - W_2 = 0 \quad (2)$$

Equation (2) gives $W_1 = W_2$. Substitute W_1 for W_2 in equation (1).

$$\frac{\sqrt{3}}{2}(W_1 + W_1) = 100$$

$$\frac{\sqrt{3}}{2}(2W_1) = 100$$

$$\sqrt{3}W_1 = 100$$

$$W_1 = \frac{100}{\sqrt{3}} = \frac{100\sqrt{3}}{3} \approx 58$$

Therefore, $W_1 = W_2 \approx 58$ lb.

79. $C = at^2 + by + c$

Use the values for C from the table.

(a) For 1960, $t = 0$ and $C = 317$.

$$317 = a(0)^2 + b(0) + c$$
$$317 = c$$

For 1980, $t = 20$ and $C = 339$.

$$339 = a(20)^2 + b(20) + 317$$
$$22 = 400a + 20b$$

For 2004, $t = 44$ and $C = 377$.

$$377 = a(44)^2 + b(44) + 317$$
$$60 = 1936a + 44b$$

Thus, we need to solve the system

$$\begin{bmatrix} 400 & 20 \\ 1936 & 44 \end{bmatrix} = \begin{bmatrix} 22 \\ 60 \end{bmatrix}$$

$$\begin{bmatrix} a \\ b \end{bmatrix} = \begin{bmatrix} 400 & 20 \\ 1936 & 44 \end{bmatrix} \begin{bmatrix} 22 \\ 60 \end{bmatrix}$$

$$= \begin{bmatrix} -\dfrac{1}{480} & \dfrac{1}{1056} \\ \dfrac{11}{120} & -\dfrac{5}{264} \end{bmatrix} \begin{bmatrix} 22 \\ 60 \end{bmatrix}$$

$$= \begin{bmatrix} \dfrac{29}{2640} \\ \dfrac{581}{660} \end{bmatrix} \approx \begin{bmatrix} 0.010985 \\ 0.8830 \end{bmatrix}$$

Therefore,
$$C = 0.010985t^2 + 0.8803t + 317.$$

(b) In 1960, $C = 317$. So, double that level would be $C = 634$.

$$634 = 0.010985t^2 + 0.8803t + 317$$
$$0 = 0.010985t^2 + 0.8803t - 317$$

Multiply this equation by 2640 to clear the decimal values.

$$0 = 29t^2 + 2324t - 836,880$$

Use the quadratic formula with $a = 29$, $b = 2324$, and $c = -836,880$.

$$t = \frac{-2324 \pm \sqrt{2324^2 - 4(29)(-836,880)}}{2(29)}$$

$$= \frac{-2324 \pm \sqrt{5,400,976 - (-97,078,080)}}{2(29)}$$

$$\approx \frac{-2324 \pm 10,123}{58}$$

$$\approx 134.5 \text{ or} - 214.6$$

Ignore the negative value. If $t = 134.5$, then $1960 + 134.5 = 2094.5$. The 1960 CO_2 level will double in the year 2095.

81. Let x = the number of boys
and y = the number of girls.

$$0.2x + 0.3y = 500 \quad (1)$$
$$0.6x + 0.9y = 1500 \quad (2)$$

The augmented matrix is

$$\begin{bmatrix} 0.2 & 0.3 & | & 500 \\ 0.6 & 0.9 & | & 1500 \end{bmatrix}.$$

$$5R_1 \rightarrow R_1 \quad \begin{bmatrix} 1 & 1.5 & | & 2500 \\ 0.6 & 0.9 & | & 1500 \end{bmatrix}$$

$$-0.6R_1 + R_2 \rightarrow R_2 \quad \begin{bmatrix} 1 & 1.5 & | & 2500 \\ 0 & 0 & | & 0 \end{bmatrix}$$

Thus,
$$x + 1.5y = 2500$$
$$x = 2500 - 1.5y.$$

There are y girls and $2500 - 1.5y$ boys, where y is any even integer between 0 and 1666 since $y \geq 0$ and

$$2500 - 1.5y \geq 0$$
$$-1.5y \geq -2500$$
$$y \leq 1666.\overline{6}.$$

83. Let x = the weight of a single chocolate wafer and
y = the weight of a single layer of vanilla creme.

A serving of regular Oreo cookies is three cookies so that $3(2x + y) = 34$.

A serving of Double Stuf is two cookies so that $2(2x + 2y) = 29$.

Write the equations in proper form, obtain the augmented matrix, and use row operations to solve.

$$\begin{bmatrix} 6 & 3 & | & 34 \\ 4 & 4 & | & 29 \end{bmatrix}$$

$$-2R_1 + 3R_2 \rightarrow R_2 \quad \begin{bmatrix} 6 & 3 & | & 34 \\ 0 & 6 & | & 19 \end{bmatrix}$$

$$-1R_2 + 2R_1 \rightarrow R_1 \quad \begin{bmatrix} 12 & 0 & | & 49 \\ 0 & 6 & | & 19 \end{bmatrix}$$

$$\frac{1}{12}R_1 \rightarrow R_1 \quad \begin{bmatrix} 1 & 0 & | & \dfrac{49}{12} \\ 0 & 1 & | & \dfrac{19}{6} \end{bmatrix}$$
$$\frac{1}{6}R_2 \rightarrow R_2$$

The solution is $\left(\dfrac{49}{12}, \dfrac{19}{6} \right)$, or about $(4.08, 3.17)$.

A chocolate wafer weighs 4.08 g and a single layer of vanilla creme weighs 3.17g.

Chapter 3

LINEAR PROGRAMMING: THE GRAPHICAL METHOD

3.1 Graphing Linear Inequalities

Your Turn 1

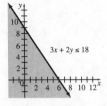

Your Turn 2

3.1 Exercises

1. $x + y \leq 2$

First graph the boundary line $x + y = 2$ using the points $(2, 0)$ and $(0, 2)$. Since the points on this line satisfy $x + y \leq 2$, draw a solid line. To find the correct region to shade, choose any point not on the line. If $(0, 0)$ is used as the test point, we have

$$x + y \leq 2$$
$$0 + 0 \leq 2$$
$$0 \leq 2,$$

which is a true statement. Shade the half-plane containing $(0, 0)$, or all points below the line.

3. $x \geq 2 - y$

First graph the boundary line $x = 2 - y$ using the points $(0, 2)$ and $(2, 0)$. This will be a solid line. Choose $(0, 0)$ as a test point.

$$x \geq 2 - y$$
$$0 \geq 2 - 0$$
$$0 \geq 2,$$

which is a false statement. Shade the half-plane that does not contain $(0, 0)$, or all points below the line.

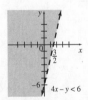

5. $4x - y < 6$

Graph $4x - y = 6$ as a dashed line, since the points on the line are not part of the solution; the line passes through the points $(0, -6)$ and $\left(\frac{3}{2}, 0\right)$.

Using the test point $(0, 0)$, we have $0 - 0 < 6$ or $0 < 6$, a true statement. Shade the half-plane containing $(0, 0)$, or all points above the line.

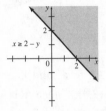

7. $4x + y < 8$

Graph $4x + y = 8$ as a dashed line through $(2, 0)$ and $(0, 8)$. Using the test point $(0, 0)$, we get $4 \cdot (0) + 0 < 8$ or $0 < 8, 0$ a true statement. Shade the half-plane containing $(0, 0)$, or all points below the line.

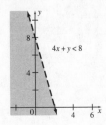

9. $x + 3y \geq -2$

The graph includes the line $x + 3y = -2$, whose intercepts are the points $\left(0, -\frac{2}{3}\right)$ and $(-2, 0)$. Graph $x + 3y = -2$ as a solid line and use the origin as a test point. Since $0 + 3(0) \geq -2$ is true, shade the half-plane containing $(0, 0)$, or all points above the line.

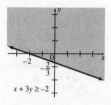

11. $x \leq 3y$

Graph $x = 3y$ as a solid line through the points $(0, 0)$ and $(3, 1)$. Since this line contains the origin, some point other than $(0, 0)$ must be used as a test point. If we use the point $(1, 2)$, we obtain $1 \leq 3(2)$ or $1 \leq 6$, a true statement. Shade the half-plane containing $(1, 2)$, or all points above the line.

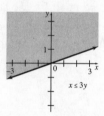

13. $x + y \leq 0$

Graph $x + y = 0$ as a solid line through the points $(0, 0)$ and $(1, -1)$. This line contains $(0, 0)$. If we use $(-1, 0)$ as a test point, we obtain $-1 + 0 \leq 0$ or $-1 \leq 0$, a true statement. Shade the half-plane containing $(-1, 0)$, or all points below the line.

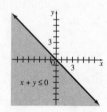

15. $y < x$

Graph $y = x$ as a dashed line through the points $(0, 0)$ and $(1, 1)$. Since this line contains the origin, choose a point other than $(0, 0)$ as a test point. If we use $(2, 3)$, we obtain $3 < 2$, which is false. Shade the half-plane that does not contain $(2, 3)$, or all points below the line.

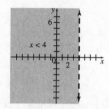

17. $x < 4$

Graph $x = 4$ as a dashed line. This is the vertical line crossing the x-axis at the point $(4, 0)$. Using $(0, 0)$ as a test point, we obtain $0 < 4$, which is true. Shade the half-plane containing $(0, 0)$, or all points to the left of the line.

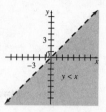

19. $y \leq -2$

Graph $y = -2$ as a solid horizontal line through the point $(0, -2)$. Using the origin as a test point, we obtain $0 \leq -2$, which is false. Shade the half-plane that does not contain $(0, 0)$, or all points below the line.

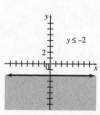

21. $x + y \leq 1$
    $x - y \geq 2$

Graph the solid lines

$$x + y = 1 \quad \text{and}$$
$$x - y = 2.$$

$0 + 0 \leq 1$ is true, and $0 - 0 \geq 2$ is false. In each case, the graph is the region below the line. Shade the overlapping part of these two half-planes, which is the region below both lines. The shaded region is the feasible region for this system.

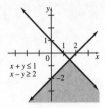

Unbounded

23. $x + 3y \le 6$

$2x + 4y \ge 7$

Graph the solid lines $x + 3y = 6$ and $2x + 4y = 7$. Use $(0, 0)$ as a test point. $0 + 0 \le 6$ is true, and $0 + 0 \ge 7$ is false. Shade all points below $x + 3y = 6$ and above $2x + 4y = 7$. The feasible region is the overlap of the two half-planes.

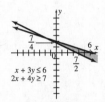

Unbounded

25. $x + y \le 7$

$x - y \le -4$

$4x + y \ge 0$

The graph of $x + y \le 7$ consists of the solid line $x + y = 7$ and all the points below it. The graph of $x - y \le -4$ consists of the solid line $x - y = -4$ and all the points above it. The graph of $4x + y \ge 0$ consists of the solid line $4x + y = 0$ and all the points above it. The feasible region is the overlapping part of these three half-planes.

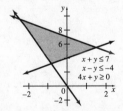

Bounded

27. $-2 < x < 3$

$-1 \le y \le 5$

$2x + y < 6$

The graph of $-2 < x < 3$ is the region between the vertical lines $x = -2$ and $x = 3$, but not including the lines themselves (so the two vertical

boundaries are drawn as dashed lines). The graph of $-1 \le y \le 5$ is the region between the horizontal lines $y = -1$ and $y = 5$, including the lines (so the two horizontal boundaries are drawn as solid lines). The graph of $2x + y < 6$ is the region below the line $2x + y = 6$ (so the boundary is drawn as a dashed line). Shade the region common to all three graphs to show the feasible region.

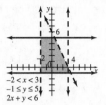

Bounded

29. $y - 2x \le 4$

$y \ge 2 - x$

$x \ge 0$

$y \ge 0$

The graph of $y - 2x \le 4$ consists of the boundary line $y - 2x = 4$ and the region below it. The graph of $y \ge 2 - x$ consists of the boundary line $y = 2 - x$ and the region above it. The inequalities $x \ge 0$ and $y \ge 0$ restrict the feasible region to the first quadrant. Shade the region in the first quadrant where the first two graphs overlap to show the feasible region.

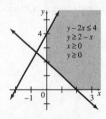

Unbounded

31. $3x + 4y > 12$

$2x - 3y < 6$

$0 \le y \le 2$

$x \ge 0$

$3x + 4y > 12$ is the set of points above the dashed line $3x + 4y = 12$; $2x - 3y < 6$ is the set of points above the dashed line $2x - 3y = 6$; $0 \le y \le 2$ is the set of points lying on or between the horizontal lines $y = 0$ and $y = 2$; and $x \ge 0$ consists of all the points on or to the right of the y-axis. Shade the feasible region, which is the triangular region satisfying all of the inequalities.

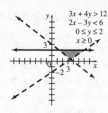

Bounded

33. $2x - 6y > 12$

Use a graphing calculator. The boundary line is the graph of $2x - 6y = 12$. Solve this equation for y.

$$-6y = -2x + 12$$

$$y = \frac{-2}{-6}x + \frac{12}{-6}$$

$$y = \frac{1}{3}x - 2$$

Enter $y_1 = \frac{1}{3}x - 2$ and graph it. Using the origin as a test point, we obtain $0 > 12$, which is false. Shade the region that does not contain the origin.

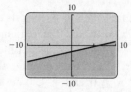

35. $3x - 4y < 6$

$2x + 5y > 15$

Use a graphing calculator. One boundary line is the graph of $3x - 4y = 6$. Solve this equation for y.

$$-4y = -3x + 6$$

$$y = \frac{-3}{-4}x + \frac{6}{-4}$$

$$y = \frac{3}{4}x - \frac{3}{2}$$

Enter $y_1 = \frac{3}{4}x - \frac{3}{2}$ and graph it. Using the origin as a test point, we obtain $0 < 6$, which is true. Shade the region that contains the origin.

The other boundary line is the graph of $2x + 5y = 15$. Solve this equation for y.

$$5y = -2x + 15$$

$$y = -\frac{2}{5}x + 3$$

Enter $y_2 = -\frac{2}{5}x + 3$ and graph it. Using the origin as a test point, we obtain $0 > 15$, which is false. Shade the region that does not contain the origin. The overlap of the two graphs is the feasible region.

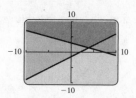

37. The region B is described by the inequalities

$$x + 3y \leq 6$$
$$x + y \leq 3$$
$$x - 2y \leq 2$$
$$x \geq 0$$
$$y \geq 0.$$

The region C is described by the inequalities

$$x + 3y \geq 6$$
$$x + y \geq 3$$
$$x - 2y \leq 2$$
$$x \geq 0$$
$$y \geq 0.$$

The region D is described by the inequalities

$$x + 3y \leq 6$$
$$x + y \geq 3$$
$$x - 2y \leq 2$$
$$x \geq 0$$
$$y \geq 0.$$

The region E is described by the inequalities

$$x + 3y \leq 6$$
$$x + y \leq 3$$
$$x - 2y \geq 2$$
$$x \geq 0$$
$$y \geq 0.$$

The region F is described by the inequalities

$$x + 3y \leq 6$$
$$x + y \geq 3$$
$$x - 2y \geq 2$$
$$x \geq 0$$
$$y \geq 0.$$

The region G is described by the inequalities

$$x + 3y \geq 6$$
$$x + y \geq 3$$
$$x - 2y \geq 2$$
$$x \geq 0$$
$$y \geq 0.$$

39. (a)

	Shawls	Afghans	Total
Number Made	x	y	
Spinning Time	1	2	≤ 8
Dyeing Time	1	1	≤ 6
Weaving Time	1	4	≤ 14

(b) $x + 2y \leq 8$ *Spinning inequality*

$\quad x + y \leq 6$ *Dyeing inequality*

$\quad x + 4y \leq 14$ *Weaving inequality*

$\quad\quad x \geq 0$ *Ensures a nonnegative*

$\quad\quad y \geq 0$ *number of each*

Graph the solid lines $x + 2y = 8$,

$x + y = 6$, $x + 4y = 14$, $x = 0$ and $y = 0$, and shade the appropriate half-planes to get the feasible region.

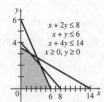

(c) Yes, 3 shawls and 2 afghans can be made because this corresponds to the point $(3, 2)$, which is in the feasible region.

No, 4 shawls and 3 afghans cannot be made because this corresponds to the point $(4, 3)$, which is not in the feasible region.

41. (a) The first sentence of the problem tells us that a total of $30 million or $x + y \leq 30$ has been set aside for loans. The second sentence of the problem gives $x \geq 4y$. The third and fourth sentences give

$\quad 0.06x + 0.08y \geq 1.6$

Also, $x \geq 0$ and $y \geq 0$ ensure nonnegative numbers. Thus,

$$x + y \leq 30$$
$$x \geq 4y$$
$$0.06x + 0.08y \geq 1.6$$
$$x \geq 0$$
$$y \geq 0$$

(b) Using the above system, graph solid lines and shade appropriate half-planes to get the feasible region.

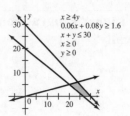

43. (a) The second sentence of the problem tells us that the number of M3 Power™ razors is never more than half the number of Fusion Power™ razors or $x \leq \frac{1}{2}y$. The third sentence tells us that a total of at most 800 razors can be produced per week or $x + y \leq 800$. The inequalities $x \geq 0$ and $y \geq 0$ ensure nonnegative numbers. Thus,

$$x \leq \frac{1}{2}y$$
$$x + y \leq 800$$
$$x \geq 0$$
$$y \geq 0.$$

(b)

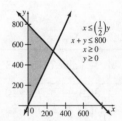

45. (a) The problem tells us that each ounce of fruit supplies 1 unit of protein and each ounce of nuts supplies 1 unit of protein. Thus, $1x + 1y$ is the number of units of protein per package. Since each package must provide at least 7 units of protein, we get the inequality $x + y \geq 7$. Similarly, we get the inequalities $2x + y \geq 10$ for carbohydrates and $x + y \leq 9$ for fat. The inequalities $x \geq 0$ and $y \geq 0$ ensure nonnegative numbers. Thus,

$$x + y \geq 7$$
$$2x + y \geq 10$$
$$x + y \leq 9$$
$$x \geq 0$$
$$y \geq 0.$$

(b)

3.2   Solving Linear Programming Problems Graphically

Your Turn 1

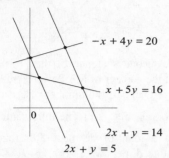

Corner Point	Value of $z = 3x + 4y$
$(0, 5)$	$3(0) + 4(5) = 20$
$(1, 3)$	$3(1) + 4(3) = 15$ Minimum
$(4, 6)$	$3(4) + 4(6) = 36$ Maximum
$(6, 2)$	$3(6) + 4(2) = 26$

The maximum value of 36 occurs at $(4, 6)$.
The minimum value of 15 occurs at $(1, 3)$.

3.2   Exercises

1. (a)

Corner Point	Value of $z = 3x + 2y$
$(0, 5)$	$3(0) + 2(5) = 10$ Minimum
$(3, 8)$	$3(3) + 2(8) = 25$
$(7, 4)$	$3(7) + 2(4) = 29$ Maximum
$(4, 1)$	$3(4) + 2(1) = 14$

The maximum value of 29 occurs at $(7, 4)$.
The minimum value of 10 occurs at $(0, 5)$.

(b)

Corner Point	Value of $z = x + 4y$
$(0, 5)$	$0 + 4(5) = 20$
$(3, 8)$	$3 + 4(8) = 35$ Maximum
$(7, 4)$	$7 + 4(4) = 23$
$(4, 1)$	$4 + 4(1) = 8$ Minimum

The maximum value of 35 occurs at $(3, 8)$.
The minimum value of 8 occurs at $(4, 1)$.

3. (a)

Corner Point	Value of $z = 0.40x + 0.75y$
$(0, 0)$	$0.40(0) + 0.75(0) = 0$ Minimum
$(0, 12)$	$0.40(0) + 0.75(12) = 9$ Maximum
$(4, 8)$	$0.40(4) + 0.75(8) = 7.6$
$(7, 3)$	$0.40(7) + 0.75(3) = 5.05$
$(8, 0)$	$0.40(8) + 0.75(0) = 3.2$

The maximum value of 9 occurs at $(0, 12)$.
The minimum value of 0 occurs at $(0, 0)$.

(b)

Corner Point	Value of $z = 1.50x + 0.25y$
$(0, 0)$	$1.50.(0) + 0.25(0) = 0$ Minimum
$(0, 12)$	$1.50(0) + 0.25(12) = 3$
$(4, 8)$	$1.50(4) + 0.25(8) = 8$
$(7, 3)$	$1.50(7) + 0.25(3) = 11.25$
$(8, 0)$	$1.50(8) + 0.25(0) = 12$ Maximum

The maximum value of 12 occurs at $(8, 0)$.
The minimum value of 0 occurs at $(0, 0)$.

5. (a)

Corner Point	Value of $z = 4x + 2y$
$(0, 8)$	$4(0) + 2(8) = 16$ Minimum
$(3, 4)$	$4(3) + 2(4) = 20$
$\left(\frac{13}{2}, 2\right)$	$4\left(\frac{13}{2}\right) + 2(2) = 30$
$(12, 0)$	$4(12) + 2(0) = 48$

The minimum value is 16 at $(0, 8)$. Since the feasible region is unbounded, there is no maximum value.

(b)

Corner Point	Value of $z = 2x + 3y$
$(0, 8)$	$2(0) + 3(8) = 24$
$(3, 4)$	$2(3) + 3(4) = 18$ Minimum
$\left(\frac{13}{2}, 2\right)$	$2\left(\frac{13}{2}\right) + 3(2) = 19$
$(12, 0)$	$2(12) + 3(0) = 24$

The minimum value is 18 at $(3, 4)$; there is no maximum value since the feasible region is unbounded.

(c)

Corner Point	Value of $z = 2x + 4y$
$(0, 8)$	$2(0) + 4(8) = 32$
$(3, 4)$	$2(3) + 4(4) = 22$
$\left(\frac{13}{2}, 2\right)$	$2\left(\frac{13}{2}\right) + 4(2) = 21$ Minimum
$(12, 0)$	$2(12) + 4(0) = 24$

The minimum value is 21 at $\left(\frac{13}{2}, 2\right)$; there is no maximum value since the feasible region is unbounded.

(d)

Corner Point	Value of $z = x + 4y$
$(0, 8)$	$0 + 4(8) = 32$
$(3, 4)$	$3 + 4(4) = 19$
$\left(\frac{13}{2}, 2\right)$	$\frac{13}{2} + 4(2) = \frac{29}{2}$
$(12, 0)$	$12 + 4(0) = 12$ Minimum

The minimum value is 12 at $(12, 0)$; there is no maximum value since the feasible region is unbounded.

7. Minimize $z = 4x + 7y$

subject to: $x - y \geq 1$

$3x + 2y \geq 18$

$x \geq 0$

$y \geq 0.$

Sketch the feasible region.

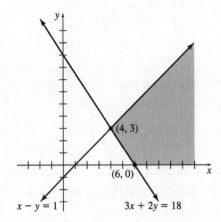

The sketch shows that the feasible region is unbounded. The corner points are $(4, 3)$ and $(6, 0)$. The corner point $(4, 3)$ can be found by solving the system

$$x - y = 1$$
$$3x + 2y = 18.$$

Use the corner points to find the minimum value of the objective function.

Corner Point	Value of $z = 4x + 7y$
$(4, 3)$	$4(4) + 7(3) = 37$
$(6, 0)$	$4(6) + 7(0) = 24$ Minimum

The minimum value is 24 when $x = 6$ and $y = 0$.

9. Maximize $z = 5x + 2y$

subject to: $4x - y \leq 16$

$2x + y \geq 11$

$x \geq 3$

$y \leq 8.$

Sketch the feasible region.

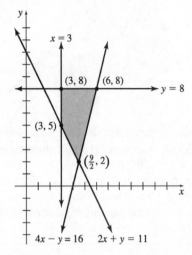

The sketch shows that the feasible region is bounded. The corner points are: $(3, 5)$, the intersection of $2x + y = 11$ and $x = 3$; $(3, 8)$, the intersection of $x = 3$ and $y = 8$; $(6, 8)$, the intersection of $y = 8$ and $4x - y = 16$; and $(9/2, 2)$, the intersection of $4x - y = 16$ and $2x + y = 11$. Use the corner points to find the maximum value of the objective function.

Corner Point	Value of $z = 5x + 2y$
$(3, 5)$	$5(3) + 2(5) = 25$
$(3, 8)$	$5(3) + 2(8) = 31$
$(6, 8)$	$5(6) + 2(8) = 46$ Maximum
$\left(\frac{9}{2}, 2\right)$	$5\left(\frac{9}{2}\right) + 2(2) = 26.5$

The maximum value is 46 when $x = 6$ and $y = 8$.

11. Maximize $z = 10x + 10y$

subject to: $5x + 8y \geq 200$

$25x - 10y \geq 250$

$x + y \leq 150$

$x \geq 0$

$y \geq 0.$

Sketch the feasible region.

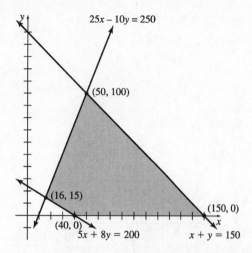

The sketch shows that the feasible region is bounded. The corner points are: $(16, 15)$, the intersection of $5x + 8y = 200$ and $25x - 10y = 250$; $(50, 100)$, the intersection of $25x - 10y = 250$ and $x + y = 150$; $(150, 0)$; and $(40, 0)$. Use the corner points to find the maximum value of the objective function.

Corner Point	Value of $z = 10x + 10y$	
$(16, 15)$	$10(16) + 10(15) = 310$	
$(50, 100)$	$10(50) + 10(100) = 1500$	Maximum
$(150, 0)$	$10(150) + 10(0) = 1500$	Maximum
$(40, 0)$	$10(40) + 10(0) = 400$	

The maximum value is 1500 when $x = 50$ and $y = 100$, as well as when $x = 150$ and $y = 0$ and all points on the line between.

13. Maximize $z = 3x + 6y$

    subject to: $2x - 3y \leq 12$
    $$x + y \geq 5$$
    $$3x + 4y \geq 24$$
    $$x \geq 0$$
    $$y \geq 0.$$

Sketch the feasible region.

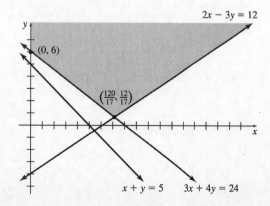

The graph shows that the feasible region is unbounded. Therefore, there is no maximum value of the objective function on the feasible region, hence, no solution.

15. Maximize $z = 10x + 12y$ subject to the following sets of constraints, with $x \geq 0$ and $y \geq 0$.

    (a) $x + y \leq 20$
    $$x + 3y \leq 24$$

    Sketch the feasible region in the first quadrant, and identify the corner points at $(0, 0)$, $(0, 8)$, $(18, 2)$, which is the intersection of $x + y = 20$ and $x + 3y = 24$, and $(20, 0)$.

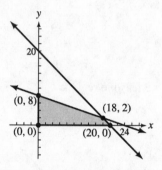

Corner Point	Value of $z = 10x + 12y$	
$(0, 0)$	$10(0) + 12(0) = 0$	
$(0, 8)$	$10(0) + 12(8) = 96$	
$(18, 2)$	$10(18) + 12(2) = 204$	Maximum
$(20, 0)$	$10(20) + 12(0) = 200$	

The maximum value of 204 occurs when $x = 18$ and $y = 2$.

    (b) $3x + y \leq 15$
    $$x + 2y \leq 18$$

    Sketch the feasible region in the first quadrant, and identify the corner points. The corner point $\left(\frac{12}{5}, \frac{39}{5}\right)$ can be found by solving the system
    $$3x + y = 15$$
    $$x + 2y = 18.$$

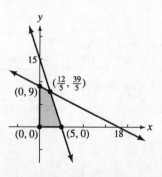

Corner Point	Value of $z = 10x + 12y$
$(0, 0)$	$10(0) + 12(0) = 0$
$(0, 9)$	$10(0) + 12(9) = 108$
$\left(\frac{12}{5}, \frac{39}{5}\right)$	$10\left(\frac{12}{5}\right) + 12\left(\frac{39}{5}\right) = \frac{588}{5} = 117\frac{3}{5}$ Maximum
$(5, 0)$	$10(5) + 12(0) = 50$

The maximum value of $\frac{588}{5}$ occurs when

$x = \frac{12}{5}$ and $y = \frac{39}{5}$.

(c) $2x + 5y \geq 22$

$4x + 3y \leq 28$

$2x + 2y \leq 17$

Sketch the feasible region in the first quadrant, and identify the corner points. The corner point $\left(\frac{5}{2}, 6\right)$ can be found by solving the system

$$4x + 3y = 28$$
$$2x + 2y = 17,$$

and the corner point $\left(\frac{37}{7}, \frac{16}{7}\right)$ can be found by solving the system

$$2x + 5y = 22$$
$$4x + 3y = 28.$$

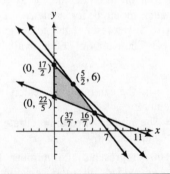

Corner Point	Value of $z = 10x + 12y$
$\left(0, \frac{22}{5}\right)$	$10(0) + 12\left(\frac{22}{5}\right) = \frac{264}{5} = 52.8$
$\left(0, \frac{17}{2}\right)$	$10(0) + 12\left(\frac{17}{2}\right) = 102$ Maximum
$\left(\frac{5}{2}, 6\right)$	$10\left(\frac{5}{2}\right) + 12(6) = 97$
$\left(\frac{37}{7}, \frac{16}{7}\right)$	$10\left(\frac{37}{7}\right) + 12\left(\frac{16}{7}\right) = \frac{562}{7} \approx 80.3$

The maximum value of 102 occurs when $x = 0$ and $y = \frac{17}{2}$.

17. Maximize $z = c_1 x_1 + c_2 x_2$

subject to: $2x_1 + x_2 \leq 11$

$\quad\quad\quad -x_1 + 2x_2 \leq 2$

$\quad\quad\quad x_1 \geq 0, \quad x_2 \geq 0.$

Sketch the feasible region.

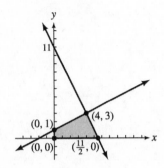

The region is bounded, with corner points $(0, 0)$,

$(0, 1)$, , $(4, 3)$, and $\left(\frac{11}{2}, 0\right)$.

Corner Point	Value of $z = c_1 x_1 + c_2 x_2$
$(0, 0)$	$c_1(0) + c_2(0) = 0$
$(0, 1)$	$c_1(0) + c_2(1) = c_2$
$(4, 3)$	$c_1(4) + c_2(3) = 4c_1 + 3c_2$
$\left(\frac{11}{2}, 0\right)$	$c_1\left(\frac{11}{2}\right) + c_2(0) = \frac{11}{2}c_1$

If we are to have $(x_1, x_2) = (4, 3)$ as an optimal solution, then it must be true that both $4c_1 + 3c_2 \geq c_2$ and $4c_1 + 3c_2 \geq \frac{11}{2}c_1$, because the value of z at $(4, 3)$ cannot be smaller than the other values of z in the table. Manipulate the symbols in these two inequalities in order to isolate $\frac{c_1}{c_2}$ in each; keep in mind the given information that $c_2 > 0$ when performing division by c_2. First,

$$4c_1 + 3c_2 \geq c_2$$
$$4c_1 \geq -2c_2$$
$$\frac{4c_1}{4c_2} \geq \frac{-2c_2}{4c_2}$$
$$\frac{c_1}{c_2} \geq -\frac{1}{2}.$$

Then,

$$4c_1 + 3c_2 \geq \frac{11}{2}c_1$$

$$-\frac{3}{2}c_1 + 3c_2 \geq 0$$

$$3c_1 - 6c_2 \leq 0$$

$$3c_1 \leq 6c_2$$

$$\frac{3c_1}{3c_2} \leq \frac{6c_2}{3c_2}$$

$$\frac{c_1}{c_2} \leq 2.$$

Since $\frac{c_1}{c_2} \geq -\frac{1}{2}$ and $\frac{c_1}{c_2} \leq 2$, the desired range for

$\frac{c_1}{c_2}$ is $\left[-\frac{1}{2}, 2\right]$, which corresponds to choice (b).

Your Turn 3

The new resource constraint is $25x + 75y \leq 1050$, or $x + 3y \leq 42$.

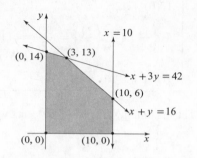

Corner Point	Value of $z = 12x + 40y$
$(0,0)$	$12(0) + 40(0) = 0$
$(0,14)$	$12(0) + 40(14) = 560$
$(3,13)$	$12(3) + 40(13) = 556$
$(10,6)$	$12(10) + 40(6) = 360$
$(10,0)$	$12(10) + 40(0) = 120$

14 pigs and 0 goats produces a maximum profit of $560.

3.3 Applications of Linear Programming

Your Turn 1

Corner Point	Value of $z = 25x + 35y$
$(0,0)$	$25(0) + 35(0) = 0$
$(65,0)$	$25(65) + 35(0) = 1625$
$(0,60)$	$25(0) + 35(60) = 2100$
$(25,40)$	$25(0) + 35(40) = 2025$

He should buy 60 kayaks and no canoes for a maximum revenue of $2100 a day.

Your Turn 2

Corner Point	Value of $z = 20x + 30y$
$(0,0)$	$20(0) + 30(0) = 0$
$(0,6)$	$20(0) + 30(6) = 180$
$(6,0)$	$20(6) + 30(0) = 120$
$(4,4)$	$20(4) + 30(4) = 200$

The company should make 4 batches of each for a maximum profit of $200.

3.3 Exercises

1. Let x represent the number of product A made and y represent the number of product B. Each item of A uses 3 hr on the machine, so $3x$ represents the total hours required for x items of product A. Similarly, $5y$ represents the total hours used for product B. There are only 60 hr available, so

$$3x + 5y \leq 60.$$

3. Let x = the amount of calcium carbonate supplement

and y = the amount of calcium citrate supplement.

Then $600x$ represents the number of units of calcium provided by the calcium carbonate supplement and $250y$ represents the number of units provided by the calcium citrate supplement. Since at least 1500 units are needed per day,

$$600x + 250y \geq 1500.$$

5. Let x represent the number of pounds of $8 coffee and y represent the number of pounds of $10 coffee. Since the mixture must weigh at least 40 lb,

$$x + y \geq 40.$$

(Notice that the price per pound is not used in setting up this inequality.)

7. Let $x =$ the number of engines to ship to plant I and $y =$ the number of engines to ship to plant II.

Minimize $z = 30x + 40y$

subject to:
$$x \geq 45$$
$$y \geq 32$$
$$x + y \leq 120$$
$$20x + 15y \geq 1500$$
$$x \geq 0.$$
$$y \geq 0.$$

Sketch the feasible region in quadrant I, and identify the corner points.

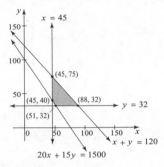

The corner points are:

$(45, 75)$, the intersection of $x = 45$ and $x + y = 120$,

$(45, 40)$, the intersection of $x = 45$ and $20x + 15y = 1500$,

$(88, 32)$, the intersection of $y = 32$ and $x + y = 120$, and

$(51, 32)$, the intersection of $y = 32$ and $20x + 15y = 1500$.

Use the corner points to find the minimum value of the objective function.

Corner Point	Value of $z = 30x + 40y$
$(45, 75)$	$30(45) + 40(75) = 4350$
$(45, 40)$	$30(45) + 40(40) = 2950$
$(88, 32)$	$30(88) + 40(32) = 3920$
$(51, 32)$	$30(51) + 40(32) = 2810$

The minimum value is \$2810, which occurs when 51 engines are shipped to plant I and 32 engines are shipped to plant II.

9. (a) Let $x =$ the number of units of policy A and $y =$ the number of units of policy B.

Minimize $z = 50x + 40y$

subject to:
$$10{,}000x + 15{,}000y \geq 300{,}000$$
$$180{,}000x + 120{,}000y \geq 3{,}000{,}000$$
$$x \geq 0$$
$$y \geq 0.$$

Sketch the feasible region in quadrant I, and identify the corner points. The corner point $(6, 16)$ can be found by solving the system

$$10{,}000x + 15{,}000y = 300{,}000$$
$$180{,}000x + 120{,}000y = 3{,}000{,}000,$$

which can be simplified as

$$2x + 3y = 60$$
$$3x + 2y = 50.$$

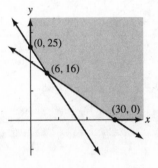

Corner Point	Value of $z = 50x + 40y$
$(0, 25)$	$50(0) + 40(25) = 1000$
$(6, 16)$	$50(6) + 40(16) = 940$ Minimum
$(30, 0)$	$50(30) + 40(0) = 1500$

The minimum cost is \$940, which occurs when 6 units of policy A and 16 units of policy B are purchased.

(b) The objective function changes to $z = 25x + 40y$, but the constraints remain the same. Use the same corner points as in part (a).

Corner Point	Value of $z = 25x + 40y$
$(0, 25)$	$25(0) + 40(25) = 1000$
$(6, 16)$	$25(6) + 40(16) = 790$
$(30, 0)$	$25(30) + 40(0) = 750$ Minimum

The minimum cost is \$750, which occurs when 30 units of policy A and no units of policy B are purchased.

11. (a) Let $x =$ the number of type I bolts
and $y =$ the number of type II bolts.

Maximize $z = 0.15x + 0.20y$

subject to: $0.2x + 0.2y \leq 300$

$0.6x + 0.2y \leq 720$

$0.04x + 0.08y \leq 100$

$x \geq 0$

$y \geq 0.$

Graph the feasible region and identify the corner points.

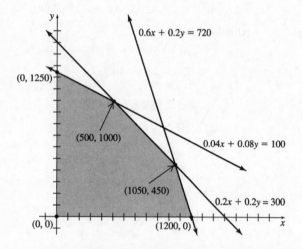

The corner points are $(0, 0)$; $(0, 1250)$; $(500, 1000)$, the intersection of $0.04x + 0.08y = 100$ and $0.02x + 0.2y = 300$; $(1050, 450)$, the intersection of $0.02x + 0.2y = 300$ and $0.6x + 0.2y = 720$; and $(1200, 0)$. Use the corner points to find the maximum value of the objective function.

Corner Point	Value of $z = 0.15x + 0.90y$
$(0, 0)$	$0.15(0) + 0.20(0) = 0$
$(0, 1250)$	$0.15(0) + 0.20(1250) = 250$
$(500, 1000)$	$0.15(500) + 0.20(1000) = 275$
	Maximum
$(1050, 450)$	$0.15(1050) + 0.20(450) = 247.5$
$(1200, 0)$	$0.15(1200) + 0.20(0) = 180$

The shop should manufacture 500 type I bolts and 1000 type II bolts to maximize revenue.

(b) The maximum revenue is $275.

(c) When the slope of the line $z = px + 0.20y$ for constant z matches the slope of the line joining the corner points $(500, 1000)$ and $(1050, 450)$, all points in the feasible region on this line produce the same maximum value of the objective function. This happens when

$$\frac{-p}{0.20} = \frac{1000 - 450}{500 - 1050} = -1,$$

or $p = 0.20$. As we increase the price of type I bolts beyond 20¢, the slope of $z = px + 0.20y$ becomes more negative and the corner point $(1050, 450)$ takes over and produces the maximum profit. So the answer is a price of 20¢ per bolt for type I bolts.

13. (a) Let $x =$ the number of kg of the half-and-half mixture
and $y =$ the number of kg of the second mixture.

Maximize $z = 7x + 9.5y$

subject to: $\dfrac{1}{2}x + \dfrac{3}{4}y \leq 150$

$\dfrac{1}{2}x + \dfrac{1}{4}y \leq 90$

$x \geq 0$

$y \geq 0.$

Sketch the feasible region and identify the corner points.

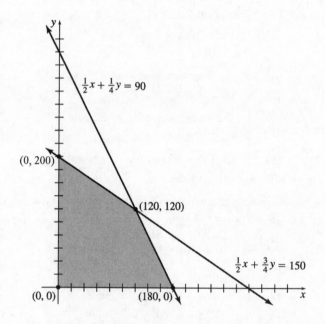

The corner points are $(0, 0)$; $(0, 200)$; $(120, 120)$, the intersection of $\frac{1}{2}x + \frac{3}{4}y = 150$ and $\frac{1}{2}x + \frac{1}{4}y = 90$; and $(180, 0)$. Use the corner points to find the maximum value of the objective function.

Corner Point	Value of $z = 7x + 9.5y$
$(0, 0)$	$7(0) + 9.5(0) = 0$
$(0, 200)$	$7(0) + 9.5(200) = 1900$
$(120, 120)$	$7(120) + 9.5(120) = 1980$ Maximum
$(180, 0)$	$7(180) + 9.5(0) = 1260$

The candy company should prepare 120 kg of the half-and-half mixture and 120 kg of the second mixture for a maximum revenue of $1980.

(b) The objective function to be maximized is now $z = 7x + 11y$. The corner points remain the same.

Corner Point	Value of $z = 7x + 11y$
$(0, 0)$	$7(0) + 11(0) = 0$
$(0, 200)$	$7(0) + 11(200) = 2200$
$(120, 120)$	$7(120) + 11(120) = 2160$ Maximum
$(180, 0)$	$7(180) + 11(0) = 1260$

In order to maximize the revenue under the altered conditions, the candy company should prepare 0 kg of the half-and-half mixture and 200 kg of the second mixture for a maximum revenue of $2200.

15. (a) Let $x =$ the number of gallons from dairy I and $y =$ the number of gallons from dairy II.

Maximize $z = 0.037x + 0.032y$

subject to: $0.60x + 0.20y \le 36$
$$x \le 50$$
$$y \le 80$$
$$x + y \le 100$$
$$x \ge 0$$
$$y \ge 0.$$

Sketch the feasible region, and identify the corner points.

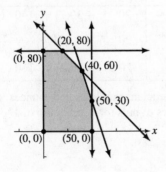

Corner Point	Value of $z = 0.037x + 0.032y$
$(0, 0)$	0
$(0, 80)$	2.56
$(20, 80)$	3.30
$(40, 60)$	3.40 Maximum
$(50, 30)$	2.81
$(50, 0)$	1.85

The maximum amount of butterfat is 3.4 gal, which occurs when 40 gal are purchased from dairy I and 60 gal are purchased from dairy II.

(b) In the solution to part (a), Mostpure uses 40 gallons from dairy I with a capacity for 50 gallons. Therefore, there is an excess capacity of 10 gallons from dairy I. Similarly, there is an excess capacity of $80 - 60$ or 20 gallons from dairy II. No, the excess capacity cannot be used.

17. Let $x =$ the amount (in millions) invested in U.S. Treasury bonds

and $y =$ the amount (in millions) invested in mutual funds.

Maximize $z = 0.04x + 0.08y$

subject to:
$$x + y \le 30$$
$$x \ge 5$$
$$y \ge 10$$
$$100x + 200y \ge 5000$$
$$x \ge 0$$
$$y \ge 0.$$

Sketch the feasible region, and identify the corner points.

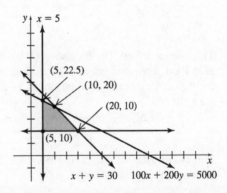

The corner points are $(5, 10)$; $(5, 22.5)$, the intersection of $x = 5$ and $100x + 200y = 5000$; $(10, 20)$, the intersection of $100x + 200y = 5000$ and $x + y = 30$; and $(20, 10)$. Use the corner points to find the maximum value of the objective function.

Corner Point	Value of $z = 0.04x + 0.08y$	
(5, 10)	$0.04(5) + 0.08(10) = 1$	
(5, 22.5)	$0.04(5) + 0.08(22.5) = 2$	Maximum
(10, 20)	$0.04(10) + 0.08(20) = 2$	Maximum
(20, 10)	$0.04(20) + 0.08(10) = 1.6$	

The maximum annual interest of $2 million can be achieved by investing $5 million in U.S. Treasury bonds and $22.5 million in mutual funds, or $10 million in bonds and $20 million in mutual funds (or in any solution on the line between those two points).

19. Beta is limited to 400 units per day, so Beta $\leq$ 400. The correct answer is choice (a).

21. (a) Let $x =$. the number of pill 1 and $y =$ the number of pill 2.

Minimize $z = 0.15x + 0.30y$

subject to: $240x + 60y \geq 480$

$$x + y \geq 5$$
$$2x + 7y \geq 20$$
$$x \geq 0$$
$$y \geq 0.$$

Sketch the feasible region in quadrant I.

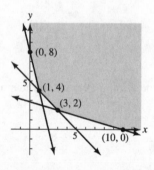

The corner points (0, 8) and (10, 0) can be identified from the graph. The coordinates of the corner point (1, 4) can be found by solving the system

$$240x + 60y = 480$$
$$x + \quad y = 5.$$

The coordinates of the corner point (3, 2) can be found by solving the system

$$2x + 7y = 20$$
$$x + \quad y = 5.$$

Corner Point	Value of $z = 0.15x + 0.30y$	
(1, 4)	1.35	
(3, 2)	1.05	Minimum
(0, 8)	2.40	
(10, 0)	1.50	

A minimum daily cost of $1.05 is incurred by taking three of pill 1 and two of pill 2.

(b) In the solution to part (a), Mark receives $240(3) + 60(2)$ or 840 units of vitamin A. This is a surplus of $840 - 480$ or 360 units of vitamin A. No, there is no way for him to avoid receiving a surplus.

23. Let $x =$ the number of ounces of fruit and $y =$ the number of ounces of nuts.

Minimize $z = 20x + 30y$

subject to:
$$3y \geq 6$$
$$2x + y \geq 10$$
$$x + 2y \leq 9$$
$$x \geq 0$$
$$y \geq 0.$$

Sketch the feasible region, and identify the corner points.

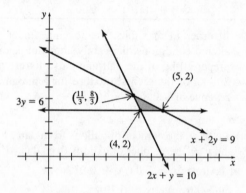

The corner points are: (4, 2), the intersection of $y = 2$ and $2x + y = 10$; $\left(\frac{11}{3}, \frac{8}{3}\right)$, the intersection of $2x + y = 10$ and $x + 2y = 9$; and (5, 2), the intersection of $x + 2y = 9$ and $y = 2$. Use the corner points to find the minimum value of the objective function.

Corner Point	Value of $z = 20x + 30y$	
(4, 2)	$20(4) + 30(2) = 140$	Minimum
$\left(\frac{11}{3}, \frac{8}{3}\right)$	$20\left(\frac{11}{3}\right) + 30\left(\frac{8}{3}\right) = \frac{460}{3} \approx 153$	
(5, 2)	$20(5) + 30(2) = 160$	

The dietician should use 4 ounces of fruit and 2 ounces of nuts for a minimum of 140 calories.

25. Let $x =$ the number of units of plants and $y =$ the number of animals.

Minimize $z = 30x + 15y$

subject to: $30x + 20y \geq 360$

$\qquad\qquad 10x + 25y \geq 300$

$\qquad\qquad\qquad y \geq 8$

$\qquad\qquad 0 \leq x \leq 25$

$\qquad\qquad 0 \leq y \leq 25.$

Sketch the feasible region in quadrant I.

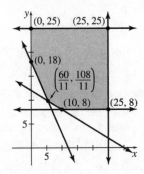

The corner points $(0, 18)$, $(0, 25)$, $(25, 25)$, and $(25, 8)$ can be determined from the graph. The corner point $\left(\frac{60}{11}, \frac{108}{11}\right)$ can be found by solving the system

$$30x + 20y = 360$$
$$10x + 25y = 300.$$

The corner point $(10, 8)$ can be found by solving the system

$$10x + 25y = 300$$
$$y = 8.$$

Corner Point	Value of $z = 30x + 15y$	
$(0, 18)$	270	Minimum
$(0, 25)$	375	
$(25, 25)$	1125	
$(25, 8)$	870	
$\left(\frac{60}{11}, \frac{108}{11}\right)$	$\frac{3420}{11} \approx 310.91$	
$(10, 8)$	420	

The minimum labor is 270 hours and is achieved when 0 units of plants and 18 animals are collected.

Chapter 3 Review Exercises

1. False; the graphical method is impossible with four or more variables.

2. True

3. False; $x \leq 2y$ means that the number of acres of wheat will be at most twice the number of acres of corn planted.

4. False; the total amount of time cannot exceed 60 hours, so the constraint is represented by $2x + y \leq 60$.

5. False; the objective function to maximize profit is $10x + 14y$.

6. False; a corner point of the feasible region is $\left(\frac{30}{13}, \frac{75}{26}\right)$, or $(2.308, 2.885)$, not $(2, 3)$.

7. True

8. False; the optimal solution occurs at a corner point, and $(2, 3)$ is not a corner point.

9. False; the half-planes of both sides of a linear inequality cannot contain the same points.

10. True

11. True

12. True

13. True

15. $y \geq 2x + 3$

Graph $y = 2x + 3$ as a solid line, using the intercepts $(0, 3)$ and $\left(-\frac{3}{2}, 0\right)$. Using the origin as a test point, we get $0 \geq 2(0) + 3$ or $0 \geq 3$, which is false. Shade the region that does not contain the origin, that is, the half-plane above the line.

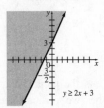

17. $2x + 6y \leq 8$

Graph $2x + 6y = 8$ as a solid line, using the intercepts $\left(0, \frac{4}{3}\right)$ and $(4, 0)$. Using the origin as a test point, we get $0 \leq 8$, which is true. Shade the region that contains the origin, that is, the half-plane below the line.

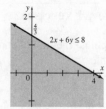

19. $y \geq x$

Graph $y = x$ as a solid line. Since this line contains the origin, choose a point other than $(0, 0)$ as a test point. If we use $(1, 4)$, we get $4 \geq 1$, which is true. Shade the region that contains the test point, that is, the half-plane above the line.

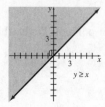

21. $x + y \leq 6$
   $2x - y \geq 3$

$x + y \leq 6$ is the half-plane on or below the line $x + y = 6$; $2x - y \geq 3$ is the half-plane on or below the line $2x - y = 3$. Shade the overlapping part of these two half-planes, which is the region below both lines. The only corner point is the intersection of the two boundary lines, the point $(3, 3)$.

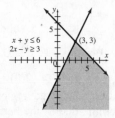

Unbounded

23. $-4 \leq x \leq 2$
    $-1 \leq y \leq 3$
     $x + y \leq 4$

$-4 \leq x \leq 2$ is the rectangular region lying on or between the two vertical lines, $x = -4$ and $x = 2$; $-1 \leq y \leq 3$ is the rectangular region lying on or between the two horizontal lines, $y = -1$ and $y = 3$; $x + y \leq 4$ is the half-plane lying on or below the line $x + y = 4$. Shade the overlapping part of these three regions. The corner points are $(-4, -1)$, $(-4, 3)$, $(1, 3)$, $(2, 2)$, and $(2, -1)$.

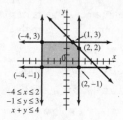

Bounded

25. $x + 2y \leq 4$
   $5x - 6y \leq 12$
     $x \geq 0$
     $y \geq 0$

$x + 2y \leq 4$ is the half-plane on or below the line $x + 2y = 4$; $5x - 6y \leq 12$ is the half-plane on or above the line $5x - 6y = 12$; $x \geq 0$ and $y \geq 0$ together restrict the graph to the first quadrant. Shade the portion of the first quadrant where the half-planes overlap. The corner points are $(0, 0)$, $(0, 2)$, $\left(\frac{12}{5}, 0\right)$, and $\left(3, \frac{1}{2}\right)$, which can be found by solving the system

$$x + 2y = 4$$
$$5x - 6y = 12.$$

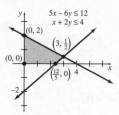

Bounded

27. Evaluate the objective function $z = 2x + 4y$ at each corner point.

Corner Point	Value of $z = 2x + 4y$	
$(0, 0)$	$2(0) + 4(0) = 0$	Minimum
$(0, 4)$	$2(0) + 4(4) = 16$	
$(3, 4)$	$2(3) + 4(4) = 22$	Maximum
$(6, 2)$	$2(6) + 4(2) = 20$	
$(4, 0)$	$2(4) + 4(0) = 8$	

The maximum value 22 occurs at $(3, 4)$, and the minimum value of 0 occurs at $(0, 0)$.

29. Maximize $z = 2x + 4y$

subject to: $3x + 2y \leq 12$

$5x + y \geq 5$

$x \geq 0$

$y \geq 0.$

Sketch the feasible region in quadrant I.

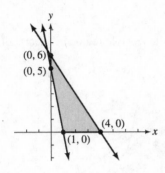

The corner points are $(0, 5)$, $(0, 6)$, $(4, 0)$, and $(1, 0)$.

Corner Point	Value of $z = 2x + 4y$
$(0, 5)$	20
$(0, 6)$	24 Maximum
$(4, 0)$	8
$(1, 0)$	2

The maximum value is 24 at $(0, 6)$.

31. Minimize $z = 4x + 2y$

subject to: $x + y \leq 50$

$2x + y \geq 20$

$x + 2y \geq 30$

$x \geq 0$

$y \geq 0.$

Sketch the feasible region.

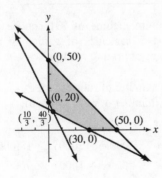

The corner points are $(0, 20)$, $\left(\frac{10}{3}, \frac{40}{3}\right)$, $(30, 0)$, $(50, 0)$, and $(0, 50)$. The corner point $\left(\frac{10}{3}, \frac{40}{3}\right)$ can be found by solving the system

$$2x + y = 20$$
$$x + 2y = 30.$$

Corner Point	Value of $z = 4x + 2y$	
$(0, 20)$	40	Minimum
$\left(\frac{10}{3}, \frac{40}{3}\right)$	40	Minimum
$(30, 0)$	120	
$(50, 0)$	200	
$(0, 50)$	100	

Thus, the minimum value is 40 and occurs at every point on the line segment joining $(0, 20)$ and $\left(\frac{10}{3}, \frac{40}{3}\right)$.

35. Maximize $z = 3x + 4y$

subject to: $2x + y \leq 4$

$-x + 2y \leq 4$

$x \geq 0$

$y \geq 0.$

(a) Sketch the feasible region. All corner points except one can be read from the graph. Solving the system

$$2x + y = 4$$
$$-x + 2y = 4$$

gives the final corner point, $\left(\frac{4}{5}, \frac{12}{5}\right)$.

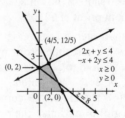

(b)

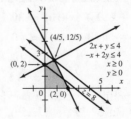

37. Let x = the number of batches of cakes and y = the number of batches of cookies.

Then we have the following inequalities.

$$2x + \frac{3}{2}y \le 15 \quad \text{(oven time)}$$

$$3x + \frac{2}{3}y \le 13 \quad \text{(decorating)}$$

$$x \ge 0$$

$$y \ge 0$$

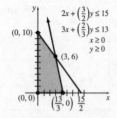

39. **(a)** From the graph for Exercise 37, the corner points are $(0, 10)$, $(3, 6)$, $\left(\frac{13}{3}, 0\right)$, and $(0, 0)$.

Since x was the number of batches of cakes and y the number of batches of cookies, the revenue function is

$$z = 30x + 20y.$$

Evaluate this objective function at each corner point.

Corner Point	Value of $z = 30x + 20y$
$(0, 10)$	200
$(3, 6)$	210 Maximum
$\left(\frac{13}{3}, 0\right)$	130
$(0, 0)$	0

Therefore, 3 batches of cakes and 6 batches of cookies should be made to produce a maximum profit of $210.

(b) Note that each $1 increase in the price of cookies increases the profit by $10 at $(0, 10)$ and by $6 at $(3, 6)$. (The increase has no effect at the other corner points since, for those, $y = 0$.) Therefore, the corner point $(0, 10)$ will begin to maximize the profit when x, the number of one-dollar increases, is larger than the solution to the following equation.

$$30(0) + (20 + 1x)(10) = 30(3) + (20 + 1x)(6)$$

$$(20 + x)(4) = 90$$

$$4x = 10$$

$$x = 2.5$$

If the profit per batch of cookies increases by more than $2.50 (to $22.50), then it will be more profitable to make 10 batches of cookies and no batches of cake.

41. Let x = number of packages of gardening mixture and y = number of packages of potting mixture.

Maximize $z = 3x + 5y$

subject to: $2x + y \le 16$

$$x + 2y \le 11$$

$$x + 3y \le 15$$

$$x \ge 0$$

$$y \ge 0.$$

Sketch the feasible region in quadrant I.

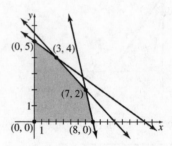

The corner points $(0, 0)$, $(0, 5)$, and $(8, 0)$ can be identified from the graph. The corner point $(3, 4)$ can be found by solving the system

$$x + 2y = 11$$

$$x + 3y = 15.$$

The corner point $(7, 2)$ can be found by solving the system

$$2x + y = 16$$

$$x + 2y = 11.$$

Corner Point	Value of $z = 3x + 5y$
$(0, 0)$	0
$(0, 5)$	25
$(3, 4)$	29
$(7, 2)$	31 Maximum
$(8, 0)$	24

A maximum income of $31 can be achieved by preparing 7 packages of gardening mixture and 2 packages of potting mixture.

43. Let x = number of runs of type I and y = number of runs of type II.

Minimize $z = 15,000x + 6000y$

subject to: $3000x + 3000y \ge 18,000$

$$2000x + 1000y \ge 7000$$

$$2000x + 3000y \ge 14,000$$

$$x \ge 0$$

$$y \ge 0.$$

Sketch the feasible region in quadrant I.

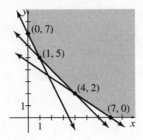

The corner points (0, 7) and (7, 0) can be identified from the graph. The corner point (1, 5) can be found by solving the system

$$3000x + 3000y = 18,000$$
$$2000x + 1000y = 7000.$$

The corner point (4, 2) can be found by solving the system

$$3000x + 3000y = 18,000$$
$$2000x + 3000y = 14,000.$$

Corner Point	Value of $z = 15,000x + 6000y$
(0, 7)	42,000 Minimum
(1, 5)	45,000
(4, 2)	72,000
(7, 0)	105,000

The company should produce 0 runs of type I and 7 runs of type II for a minimum cost of $42,000.

45. Let $x =$ the number of acres devoted to millet and $y =$ the number of acres devoted to wheat.

Maximize $z = 400x + 800y$

subject to: $36x + 8y \leq 48$
$$x + y \leq 2$$
$$x \geq 0$$
$$y \geq 0.$$

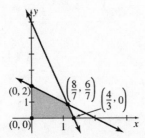

The corner points (0, 0) and (0, 2) can be identified from the graph. The corner point $\left(\frac{4}{3}, 0\right)$ can be found by solving the system

$$36x + 8y = 48$$
$$y = 0.$$

The corner point $\left(\frac{8}{7}, \frac{6}{7}\right)$ can be found by solving the system

$$36x + 8y = 48$$
$$x + y = 2.$$

Corner Point	Value of $z = 400x + 800y$
(0, 0)	0
(0, 2)	1600 Maximum
$\left(\frac{8}{7}, \frac{6}{7}\right)$	$\frac{8000}{7}$
$\left(\frac{4}{3}, 0\right)$	$\frac{1600}{3}$

The maximum amount of grain is 1600 pounds and can be obtained by planting 2 acres of wheat and no millet.

LINEAR PROGRAMMING: THE SIMPLEX METHOD

4.1 Slack Variables and the Pivot

Your Turn 1

The two new equations are:

$$300x_1 + 60x_2 + 180x_3 \leq 20{,}000$$
$$5x_1 + 10x_2 + 15x_3 \leq 900$$

The new answer tableau is:

$$
\begin{array}{ccccccc}
x_1 & x_2 & x_3 & s_1 & s_2 & s_3 & z \\
\end{array}
$$

$$
\left[
\begin{array}{ccccccc|c}
1 & 1 & 1 & 1 & 0 & 0 & 0 & 100 \\
15 & 3 & 9 & 0 & 1 & 0 & 0 & 1000 \\
1 & 2 & 3 & 0 & 0 & 1 & 0 & 180 \\
\hline
-120 & -40 & -60 & 0 & 0 & 0 & 1 & 0 \\
\end{array}
\right]
$$

Your Turn 2

Pivot around the indicated 6.

$$
\begin{array}{ccccccc}
x_1 & x_2 & x_3 & s_1 & s_2 & s_3 & z \\
\end{array}
$$

$$
\left[
\begin{array}{ccccccc|c}
3 & \boxed{6} & 2 & 1 & 0 & 0 & 0 & 60 \\
8 & 5 & 4 & 0 & 1 & 0 & 0 & 80 \\
3 & 6 & 7 & 0 & 0 & 1 & 0 & 120 \\
\hline
-30 & -50 & -15 & 0 & 0 & 0 & 1 & 0 \\
\end{array}
\right]
$$

The result is:

$$
\begin{array}{ccccccc}
x_1 & x_2 & x_3 & s_1 & s_2 & s_3 & z \\
\end{array}
$$

$$
\begin{array}{r}
 \\
-5R_1 + 6R_2 \to R_2 \\
-R_1 + R_3 \to R_3 \\
25R_1 + 3R_4 \to R_4
\end{array}
\left[
\begin{array}{ccccccc|c}
3 & \boxed{6} & 2 & 1 & 0 & 0 & 0 & 60 \\
33 & 0 & 14 & -5 & 6 & 0 & 0 & 180 \\
0 & 0 & 5 & -1 & 0 & 1 & 0 & 60 \\
\hline
-15 & 0 & 5 & 25 & 0 & 0 & 3 & 1500 \\
\end{array}
\right]
$$

$$
\begin{array}{ccccccc}
x_1 & x_2 & x_3 & s_1 & s_2 & s_3 & z \\
\end{array}
$$

$$
\begin{array}{r}
\frac{1}{6}R_1 \to R_1 \\
\frac{1}{6}R_2 \to R_2 \\
 \\
\frac{1}{3}R_4 \to R_4
\end{array}
\left[
\begin{array}{ccccccc|c}
1/2 & 1 & 1/3 & 1/6 & 0 & 0 & 0 & 10 \\
11 & 0 & 7/3 & -5/6 & 1 & 0 & 0 & 30 \\
0 & 0 & 5 & -1 & 0 & 1 & 0 & 60 \\
\hline
-5 & 0 & 5/3 & 25/3 & 0 & 0 & 1 & 500 \\
\end{array}
\right]
$$

The solution given by this tableau is:

$$x_1 = 0, x_2 = 10, x_3 = 0, s_1 = 0,$$
$$s_2 = 30, s_3 = 60, z = 500$$

4.1 Exercises

1. $x_1 + 2x_2 \leq 6$

    Add s_1 to the given inequality to obtain

    $$x_1 + 2x_2 + s_1 = 6.$$

3. $2.3x_1 + 5.7x_2 + 1.8x_3 \leq 17$

    Add s_1 to the given inequality to obtain

    $$2.3x_1 + 5.7x_2 + 1.8x_3 + s_1 = 17.$$

5. **(a)** Since there are three constraints to be converted into equations we need three slack variables.

    (b) We use s_1, s_2, and s_3 for the slack variables.

    (c) The equations are

    $$
    \begin{aligned}
    2x_1 + 3x_2 + s_1 \quad\;\;\;\;\;\;\;\; &= 15 \\
    4x_1 + 5x_2 \quad\;\; + s_2 \quad\;\; &= 35 \\
    x_1 + 6x_2 \quad\;\;\;\;\;\;\;\; + s_3 &= 20.
    \end{aligned}
    $$

7. **(a)** There are two constraints to be converted into equations, so we must introduce two slack variables.

    (b) Call the slack variables s_1 and s_2.

    (c) The equations are

    $$
    \begin{aligned}
    7x_1 + 6x_2 + 8x_3 + s_1 \quad\;\;\;\; &= 118 \\
    4x_1 + 5x_2 + 10x_3 \quad\;\; + s_2 &= 220.
    \end{aligned}
    $$

9. Find $x_1 \geq 0$ and $x_2 \geq 0$ such that

    $$
    \begin{aligned}
    4x_1 + 2x_2 &\leq 5 \\
    x_1 + 2x_2 &\leq 4
    \end{aligned}
    $$

    and $z = 7x_1 + x_2$ is maximized.

    We need two slack variables, s_1 and s_2. Then the problem can be restated as:

    Find $x_1 \geq 0, x_2 \geq 0, s_1 \geq 0$, and $s_2 \geq 0$ such that

    $$
    \begin{aligned}
    4x_1 + 2x_2 + s_1 \quad\;\; &= 5 \\
    x_1 + 2x_2 \quad\;\; + s_2 &= 4.
    \end{aligned}
    $$

    and $z = 7x_1 + x_2$ is maximized.

Rewrite the objective function as

$$-7x_1 - x_2 + z = 0.$$

The initial simplex tableau is

$$\begin{array}{ccccc} x_1 & x_2 & s_1 & s_2 & z \\ \left[\begin{array}{ccccc|c} 4 & 2 & 1 & 0 & 0 & 5 \\ 1 & 2 & 0 & 1 & 0 & 4 \\ \hline -7 & -1 & 0 & 0 & 1 & 0 \end{array}\right]. \end{array}$$

11. Find $x_1 \geq 0$ and $x_2 \geq 0$ such that

$$x_1 + x_2 \leq 10$$
$$5x_1 + 2x_2 \leq 20$$
$$x_1 + 2x_2 \leq 36$$

and $z = x_1 + 3x_2$ is maximized.

Using slack variables $s_1, s_2,$ and s_3, the problem can be restated as:

Find $x_1 \geq 0, x_2 \geq 0, s_1 \geq 0, \quad s_2 \geq 0, \quad$ and $s_3 \geq 0$ such that

$$x_1 + x_2 + s_1 \qquad\qquad = 10$$
$$5x_1 + 2x_2 \qquad + s_2 \qquad = 20$$
$$x_1 + 2x_2 \qquad\qquad + s_3 = 36$$

and $z = x_1 + 3x_2$ is maximized.

Rewrite the objective function as

$$-x_1 - 3x_2 + z = 0.$$

The initial simplex tableau is

$$\begin{array}{cccccc} x_1 & x_2 & s_1 & s_2 & s_3 & z \\ \left[\begin{array}{cccccc|c} 1 & 1 & 1 & 0 & 0 & 0 & 10 \\ 5 & 2 & 0 & 1 & 0 & 0 & 20 \\ 1 & 2 & 0 & 0 & 1 & 0 & 36 \\ \hline -1 & -3 & 0 & 0 & 0 & 1 & 0 \end{array}\right]. \end{array}$$

13. Find $x_1 \geq 0$ and $x_2 \geq 0$ such that

$$3x_1 + x_2 \leq 12$$
$$x_1 + x_2 \leq 15$$

and $z = 2x_1 + x_2$ is maximized.

Using slack variables s_1 and s_2, the problem can be restated as:

Find $x_1 \geq 0, x_2 \geq 0, s_1 \geq 0,$ and $s_2 \geq 0$ such that

$$3x_1 + x_2 + s_1 \qquad = 12$$
$$x_1 + x_2 \qquad + s_2 = 15$$

and $z = 2x_1 + x_2$ is maximized.

Rewrite the objective function as

$$-2x_1 - x_2 + z = 0.$$

The initial simplex tableau is

$$\begin{array}{ccccc} x_1 & x_2 & s_1 & s_2 & z \\ \left[\begin{array}{ccccc|c} 3 & 1 & 1 & 0 & 0 & 12 \\ 1 & 1 & 0 & 1 & 0 & 15 \\ \hline -2 & -1 & 0 & 0 & 1 & 0 \end{array}\right]. \end{array}$$

15.

$$\begin{array}{cccccc} x_1 & x_2 & x_3 & s_1 & s_2 & z \\ \left[\begin{array}{cccccc|c} 1 & 0 & 4 & 5 & 1 & 0 & 8 \\ 3 & 1 & 1 & 2 & 0 & 0 & 4 \\ \hline -2 & 0 & 2 & 3 & 0 & 1 & 28 \end{array}\right] \end{array}$$

The variables x_2 and s_2 are basic variables, because the columns for these variables have all zeros except for one nonzero entry. If the remaining variables x_1, x_3, and s_1 are zero, then $x_2 = 4$ and $s_2 = 8$. From the bottom row, $z = 28$. The basic feasible solution is $x_1 = 0$, $x_2 = 4$, $x_3 = 0$, $s_1 = 0$, $s_2 = 8$, and $z = 28$.

17.

$$\begin{array}{ccccccc} x_1 & x_2 & x_3 & s_1 & s_2 & s_3 & z \\ \left[\begin{array}{ccccccc|c} 6 & 2 & 2 & 3 & 0 & 0 & 0 & 16 \\ 2 & 2 & 0 & 1 & 0 & 5 & 0 & 35 \\ 2 & 1 & 0 & 3 & 1 & 0 & 0 & 6 \\ \hline -3 & -2 & 0 & 2 & 0 & 0 & 3 & 36 \end{array}\right] \end{array}$$

The basic variables are x_3, s_2, and s_3. If x_1, x_2, and s_1 are zero, then $2x_3 = 16$, so $x_3 = 8$. Similarly, $s_2 = 6$ and $5s_3 = 35$, so $s_3 = 7$. From the bottom row, $3z = 36$, so $z = 12$. The basic feasible solution is $x_1 = 0, x_2 = 0, x_3 = 8, s_1 = 0,$ $s_2 = 6, s_3 = 7,$ and $z = 12$.

19.

$$\begin{array}{cccccc} x_1 & x_2 & x_3 & s_1 & s_2 & z \\ \left[\begin{array}{cccccc|c} 1 & 2 & 4 & 1 & 0 & 0 & 56 \\ 2 & \boxed{2} & 1 & 0 & 1 & 0 & 40 \\ \hline -1 & -3 & -2 & 0 & 0 & 1 & 0 \end{array}\right] \end{array}$$

Clear the x_2 column.

$$\begin{array}{cccccc} & x_1 & x_2 & x_3 & s_1 & s_2 & z \\ -R_2 + R_1 \rightarrow R_1 & \left[\begin{array}{cccccc|c} -1 & 0 & 3 & 1 & -1 & 0 & 16 \\ 2 & \boxed{2} & 1 & 0 & 1 & 0 & 40 \\ \hline -1 & -3 & -2 & 0 & 0 & 1 & 0 \end{array}\right] \end{array}$$

$$\begin{array}{c} \\ \\ 3R_2 + 2R_3 \rightarrow R_3 \end{array} \begin{array}{cccccc} x_1 & x_2 & x_3 & s_1 & s_2 & z \\ \left[\begin{array}{cccccc|c} -1 & 0 & 3 & 1 & -1 & 0 & 16 \\ 2 & 2 & 1 & 0 & 1 & 0 & 40 \\ 4 & 0 & -1 & 0 & 3 & 2 & 120 \end{array}\right] \end{array}$$

x_2 and s_1 are now basic. The solution is $x_1 = 0$, $x_2 = 20$, $x_3 = 0$, $s_1 = 16$, $s_2 = 0$, and $z = 60$.

21.

$$\begin{array}{ccccccc} x_1 & x_2 & x_3 & s_1 & s_2 & s_3 & z \\ \left[\begin{array}{ccccccc|c} 2 & 2 & \boxed{1} & 1 & 0 & 0 & 0 & 12 \\ 1 & 2 & 3 & 0 & 1 & 0 & 0 & 45 \\ 3 & 1 & 1 & 0 & 0 & 1 & 0 & 20 \\ \hline -2 & -1 & -3 & 0 & 0 & 0 & 1 & 0 \end{array}\right] \end{array}$$

Clear the x_3 column.

$$\begin{array}{c} \\ \\ -3R_1 + R_2 \rightarrow R_2 \\ -R_1 + R_3 \rightarrow R_3 \\ 3R_1 + R_4 \rightarrow R_4 \end{array} \begin{array}{ccccccc} x_1 & x_2 & x_3 & s_1 & s_2 & s_3 & z \\ \left[\begin{array}{ccccccc|c} 2 & 2 & 1 & 1 & 0 & 0 & 0 & 12 \\ -5 & -4 & 0 & -3 & 1 & 0 & 0 & 9 \\ 1 & -1 & 0 & -1 & 0 & 1 & 0 & 8 \\ 4 & 5 & 0 & 3 & 0 & 0 & 1 & 36 \end{array}\right] \end{array}$$

x_3, s_2, and s_3 are now basic. The solution is $x_1 = 0$, $x_2 = 0$, $x_3 = 12$, $s_1 = 0$, $s_2 = 9$, $s_3 = 8$, and $z = 36$.

23.

$$\begin{array}{ccccccc} x_1 & x_2 & x_3 & s_1 & s_2 & s_3 & z \\ \left[\begin{array}{ccccccc|c} 2 & \boxed{2} & 3 & 1 & 0 & 0 & 0 & 500 \\ 4 & 1 & 1 & 0 & 1 & 0 & 0 & 300 \\ 7 & 2 & 4 & 0 & 0 & 1 & 0 & 700 \\ \hline -3 & -4 & -2 & 0 & 0 & 0 & 1 & 0 \end{array}\right] \end{array}$$

Clear the x_2 column.

$$\begin{array}{c} \\ \\ -R_1 + 2R_2 \rightarrow R_2 \\ -R_1 + R_3 \rightarrow R_3 \\ 2R_1 + R_4 \rightarrow R_4 \end{array} \begin{array}{ccccccc} x_1 & x_2 & x_3 & s_1 & s_2 & s_3 & z \\ \left[\begin{array}{ccccccc|c} 2 & \boxed{2} & 3 & 1 & 0 & 0 & 0 & 500 \\ 6 & 0 & -1 & -1 & 2 & 0 & 0 & 100 \\ 5 & 0 & 1 & -1 & 0 & 1 & 0 & 200 \\ 1 & 0 & 4 & 2 & 0 & 0 & 1 & 1000 \end{array}\right] \end{array}$$

x_2, s_2, and s_3 are now basic. Thus, the solution is $x_1 = 0$, $x_2 = 250$, $x_3 = 0$, $s_1 = 0$, $s_2 = 50$, $s_3 = 200$, and $z = 1000$.

25. A slack variable (a nonnegative quantity), converts a linear inequality into a linear equation by adding the amount needed in an expression to be equal to a specific value.

27. Let x_1 represent the number of simple figures, x_2 the number of figures with additions, and x_3 the number of computer-drawn sketches. Organize the information in a table.

	Simple Figures	Figures with Additions	Computer-Drawn Sketches	Maximum Allowed
Cost	20	35	60	2200
Royalties	95	200	325	

The cost constraint is

$$20x_1 + 35x_2 + 60x_3 \leq 2200.$$

The limit of 400 figures leads to the constraint

$$x_1 + x_2 + x_3 \leq 400.$$

The other stated constraints are

$$x_3 \leq x_1 + x_2 \text{ and } x_1 \geq 2x_2,$$

and these can be rewritten in standard form as

$$-x_1 - x_2 + x_3 \leq 0 \text{ and } -x_1 + 2x_2 \leq 0$$

respectively. The problem may be stated as:

Find $x_1 \geq 0$, $x_2 \geq 0$, and $x_3 \geq 0$ such that

$$\begin{aligned} 20x_1 + 35x_2 + 60x_3 &\leq 2200 \\ x_1 + x_2 + x_3 &\leq 400 \\ -x_1 - x_2 + x_3 &\leq 0 \\ -x_1 + 2x_2 &\leq 0 \end{aligned}$$

and $z = 95x_1 + 200x_2 + 325x_3$ is maximized.

Introduce slack variables s_1, s_2, s_3, and s_4, and the problem can be restated as:

Find $x_1 \geq 0$, $x_2 \geq 0$, $x_3 \geq 0$, $s_1 \geq 0$, $s_2 \geq 0$, $s_3 \geq 0$, and $s_4 \geq 0$ such that

$$\begin{aligned} 20x_1 + 35x_2 + 60x_3 + s_1 &= 2200 \\ x_1 + x_2 + x_3 + s_2 &= 400 \\ -x_1 - x_2 + x_3 + s_3 &= 0 \\ -x_1 + 2x_2 + s_4 &= 0 \end{aligned}$$

and $z = 95x_1 + 200x_2 + 325x_3$ is maximized.

Rewrite the objective function as

$$-95x_1 - 200x_2 - 325x_3 + z = 0.$$

The initial simlpex tableau is

$$\begin{array}{cccccccc} x_1 & x_2 & x_3 & s_1 & s_2 & s_3 & s_4 & z \\ \end{array}$$

$$\begin{bmatrix} 20 & 35 & 60 & 1 & 0 & 0 & 0 & 0 & 2200 \\ 1 & 1 & 1 & 0 & 1 & 0 & 0 & 0 & 400 \\ -1 & -1 & 1 & 0 & 0 & 1 & 0 & 0 & 0 \\ -1 & 2 & 0 & 0 & 0 & 0 & 1 & 0 & 0 \\ -95 & -200 & -325 & 0 & 0 & 0 & 0 & 1 & 0 \end{bmatrix}.$$

29. Let x_1 represent the number of redwood tables, x_2 the number of stained Douglas fir tables, and x_3 the number of stained white spruce tables. Organize the information in a table.

	Redwood	Douglas Fir	White Spruce	Maximum Available
Assembly Time	8	7	8	90 8-hr days = 720 hr
Staining Time	0	2	2	60 8-hr days = 480 hr
Cost	$159	$138.85	$129.35	$15,000

The limit of 720 hr for carpenters leads to the constraint

$$8x_1 + 7x_2 + 8x_3 \le 720.$$

The limit of 480 hr for staining leads to the constraint

$$2x_2 + 2x_3 \le 480.$$

The cost constraint is

$$159x_1 + 138.85x_2 + 129.35x_3 \le 15,000.$$

The problem may be stated as:

Find $x_1 \ge 0$, $x_2 \ge 0$, and $x_3 \ge 0$ such that

$$\begin{aligned} 8x_1 + \quad 7x_2 + \quad 8x_3 &\le \quad 720 \\ 2x_2 + \quad 2x_3 &\le \quad 480 \\ 159x_1 + 138.85x_2 + 129.35x_3 &\le 15,000 \end{aligned}$$

and $z = x_1 + x_2 + x_3$ is maximized.

Introduce slack variables s_1, s_2, and s_3, and the problem can be restated as:

Find $x_1 \ge 0$, $x_2 \ge 0$, $x_3 \ge 0$, $s_1 \ge 0$, $s_2 \ge 0$, and $s_3 \ge 0$ such that

$$\begin{aligned} 8x_1 + \quad 7x_2 + \quad 8x_3 + s_1 \qquad\qquad &= \quad 720 \\ 2x_2 + \quad 2x_3 \qquad + s_2 \qquad &= \quad 480 \\ 159x_1 + 138.85x_2 + 129.35x_3 \qquad\qquad + s_3 &= 15,000 \end{aligned}$$

and $z = x_1 + x_2 + x_3$ is maximized.

Rewrite the objective function as

$$-x_1 - x_2 - x_3 + z = 0.$$

The initial simplex tableau is

$$\begin{array}{ccccccc} x_1 & x_2 & x_3 & s_1 & s_2 & s_3 & z \\ \end{array}$$

$$\begin{bmatrix} 8 & 7 & 8 & 1 & 0 & 0 & 0 & 720 \\ 0 & 2 & 2 & 0 & 1 & 0 & 0 & 480 \\ 159 & 138.85 & 129.35 & 0 & 0 & 1 & 0 & 15,000 \\ -1 & -1 & -1 & 0 & 0 & 0 & 1 & 0 \end{bmatrix}.$$

31. Let $x_1 =$ the number of newspaper ads,

$x_2 =$ the number of internet banners,

and $x_3 =$ the number of TV ads.

Organize the information in a table.

	Newspaper Ads	Internet Banners	TV Ads
Cost per Ad	400	20	2000
Maximum Number	30	60	10
Women Seeing Ad	4000	3000	10,000

The cost constraint is

$$400x_1 + 20x_2 + 2000x_3 \le 8000$$

The constraints on the numbers of ads is

$$\begin{aligned} x_1 &\le 30 \\ x_2 &\le 60 \\ x_3 &\le 10. \end{aligned}$$

The problem may be stated as:

Find $x_1 \ge 0$, $x_2 \ge 0$, and $x_3 \ge 0$ such that

$$\begin{aligned} 400x_1 + 20x_2 + 2000x_3 &\le 8000 \\ x_1 \qquad\qquad &\le 30 \\ x_2 \qquad &\le 60 \\ x_3 &\le 10 \end{aligned}$$

and $z = 4000x_1 + 3000x_2 + 10,000x_3$ is maximized.

Introduce slack variables s_1, s_2, s_3 and s_4, and the problem can be restated as:

Find $x_1 \ge 0$, $x_2 \ge 0$, $x_3 \ge 0$, $s_1 \ge 0$, $s_2 \ge 0$, $s_3 \ge 0$, and $s_4 \ge 0$ such that

$$\begin{aligned} 400x_1 + 20x_2 + 2000x_3 + s_1 \qquad\qquad\qquad &= 8000 \\ x_1 \qquad\qquad + s_2 \qquad\qquad &= 30 \\ x_2 \qquad\qquad + s_3 \qquad &= 60 \\ x_3 \qquad\qquad + s_4 &= 10 \end{aligned}$$

and $z = 4000x_1 + 3000x_2 + 10{,}000x_3$ is maximized.

Rewrite the objective function as

$$-4000x_1 - 3000x_2 - 10{,}000x_3 + z = 0.$$

The initial simplex tableau is

$$
\begin{array}{cccccccc|c}
x_1 & x_2 & x_3 & s_1 & s_2 & s_3 & s_4 & z & \\
\hline
400 & 20 & 2000 & 1 & 0 & 0 & 0 & 0 & 8000 \\
1 & 0 & 0 & 0 & 1 & 0 & 0 & 0 & 30 \\
0 & 1 & 0 & 0 & 0 & 1 & 0 & 0 & 60 \\
0 & 0 & 1 & 0 & 0 & 0 & 1 & 0 & 10 \\
\hline
-4000 & -3000 & -10{,}000 & 0 & 0 & 0 & 0 & 1 & 0
\end{array}
$$

4.2 Maximization Problems

Your Turn 1

Example 2 of Section 3.3 yields the following linear programming problem, where we have renamed x and y as x_1 and x_2.

Maximize $\quad z = 12x_1 + 40x_2$

subject to: $\quad x_1 + x_2 \le 16$

$$x_1 + 3x_2 \le 36$$
$$x_1 \le 10$$
$$x_1 \ge 0$$
$$x_2 \ge 0$$

Add a slack variable to each of the first three constraints:

$$x_1 + x_2 + s_1 \le 16$$
$$x_1 + 3x_2 + s_2 \le 36$$
$$x_1 + s_3 \le 10$$

with $x_1 \ge 0,\ x_2 \ge 0,\ s_1 \ge 0,\ s_2 \ge 0,\ s_3 \ge 0$

The corresponding initial tableau is

$$
\begin{array}{cccccc|c}
x_1 & x_2 & s_1 & s_2 & s_3 & z & \\
\hline
1 & 1 & 1 & 0 & 0 & 0 & 16 \\
1 & \boxed{3} & 0 & 1 & 0 & 0 & 36 \\
1 & 0 & 0 & 0 & 1 & 0 & 10 \\
\hline
-12 & -40 & 0 & 0 & 0 & 1 & 0
\end{array}
$$

Since the most negative indicator is -40 and the quotient $36/3$ is smaller than $16/1$, we pivot on the 3 in column 2:

$$
\begin{array}{l}
-R_2 + 3R_1 \to R_1 \\
\\
\\
40R_2 + 3R_4 \to R_4
\end{array}
\quad
\begin{array}{cccccc|c}
x_1 & x_2 & s_1 & s_2 & s_3 & z & \\
\hline
2 & 0 & 3 & -1 & 0 & 0 & 12 \\
1 & 3 & 0 & 1 & 0 & 0 & 36 \\
1 & 0 & 0 & 0 & 1 & 0 & 10 \\
\hline
4 & 0 & 0 & 40 & 0 & 3 & 1440
\end{array}
$$

There are now no negative indicators, so we can read the solution:

$$x_1 = 0,\ x_2 = \frac{36}{3} = 12,\ z = \frac{1440}{3} = 480$$

Your Turn 2

Pivot on the 4 in column 2 of the following tableau.

$$
\begin{array}{cccccc|c}
x_1 & x_2 & s_1 & s_2 & s_3 & z & \\
\hline
1 & -2 & 1 & 0 & 0 & 0 & 100 \\
3 & \boxed{4} & 0 & 1 & 0 & 0 & 200 \\
5 & 0 & 0 & 0 & 1 & 0 & 150 \\
\hline
-10 & -25 & 0 & 0 & 0 & 1 & 0
\end{array}
$$

$$
\begin{array}{l}
R_2 + 2R_1 \to R_1 \\
\\
\\
25R_2 + 4R_4 \to R_4
\end{array}
\quad
\begin{array}{cccccc|c}
x_1 & x_2 & s_1 & s_2 & s_3 & z & \\
\hline
5 & 0 & 2 & 1 & 0 & 0 & 400 \\
3 & 4 & 0 & 1 & 0 & 0 & 200 \\
5 & 0 & 0 & 0 & 1 & 0 & 150 \\
\hline
35 & 0 & 0 & 25 & 0 & 4 & 5000
\end{array}
$$

There are no negative indicators, so the optimal solution is:

$$x_1 = 0,\ x_2 = \frac{200}{4} = 50,\ s_1 = 200,$$

$$s_2 = 0,\ s_3 = 150,\ z = \frac{5000}{4} = 1250$$

4.2 Exercises

1.
$$
\begin{array}{cccccc|c}
x_1 & x_2 & x_3 & s_1 & s_2 & z & \\
\hline
1 & 4 & 4 & 1 & 0 & 0 & 16 \\
2 & 1 & 5 & 0 & 1 & 0 & 20 \\
\hline
-3 & -1 & -2 & 0 & 0 & 1 & 0
\end{array}
$$

The most negative indicator is -3, in the first column. Find the quotients $\frac{16}{1} = 16$ and $\frac{20}{2} = 10$; since 10 is the smaller quotient, 2 in row 2, column 1 is the pivot.

$$
\begin{array}{l}
\frac{16}{1} = 16 \\
\frac{20}{2} = 10 \\
\\
\end{array}
\quad
\begin{array}{cccccc|c}
x_1 & x_2 & x_3 & s_1 & s_2 & z & \\
\hline
1 & 4 & 4 & 1 & 0 & 0 & 16 \\
\boxed{2} & 1 & 5 & 0 & 1 & 0 & 20 \\
\hline
-3 & -1 & -2 & 0 & 0 & 1 & 0
\end{array}
$$

Performing row transformations, we get the following tableau.

$$\begin{array}{c} \\ -R_2 + 2R_1 \to R_1 \\ \\ 3R_2 + 2R_3 \to R_3 \end{array} \begin{array}{c} x_1 \;\; x_2 \;\; x_3 \;\; s_1 \;\; s_2 \;\; z \\ \left[\begin{array}{cccccc|c} 0 & 7 & 3 & 2 & -1 & 0 & 12 \\ 2 & 1 & 5 & 0 & 1 & 0 & 20 \\ 0 & 1 & 11 & 0 & 3 & 2 & 60 \end{array}\right] \end{array}$$

All of the numbers in the last row are nonnegative, so we are finished pivoting. Create a 1 in the columns corresponding to x_1, s_1 and z.

$$\begin{array}{c} \frac{1}{2}R_1 \to R_1 \\ \\ \frac{1}{2}R_2 \to R_2 \\ \\ \frac{1}{2}R_3 \to R_3 \end{array} \begin{array}{c} x_1 \;\;\; x_2 \;\;\; x_3 \;\; s_1 \;\; s_2 \;\; z \\ \left[\begin{array}{cccccc|c} 0 & \frac{7}{2} & \frac{3}{2} & 1 & -\frac{1}{2} & 0 & 6 \\ 1 & \frac{1}{2} & \frac{5}{2} & 0 & \frac{1}{2} & 0 & 10 \\ 0 & \frac{1}{2} & \frac{11}{2} & 0 & \frac{3}{2} & 1 & 30 \end{array}\right] \end{array}$$

The maximum value is 30 and occurs when $x_1 = 10$, $x_2 = 0$, $x_3 = 0$, $s_1 = 6$, and $s_2 = 0$.

3.
$$\begin{array}{c} x_1 \;\;\; x_2 \;\; s_1 \;\; s_2 \;\; s_3 \;\; z \\ \left[\begin{array}{cccccc|c} 1 & 3 & 1 & 0 & 0 & 0 & 12 \\ 2 & 1 & 0 & 1 & 0 & 0 & 10 \\ 1 & 1 & 0 & 0 & 1 & 0 & 4 \\ \hline -2 & -1 & 0 & 0 & 0 & 1 & 0 \end{array}\right] \end{array}$$

The most negative indicator is -2, in the first column. Find the quotients $\frac{12}{1} = 12$, $\frac{10}{2} = 5$, and $\frac{4}{1} = 4$; since 4 is the smallest quotient, 1 in row 3, column 1 is the pivot.

$$\begin{array}{c} x_1 \;\;\; x_2 \;\; s_1 \;\; s_2 \;\; s_3 \;\; z \\ \left[\begin{array}{cccccc|c} 1 & 3 & 1 & 0 & 0 & 0 & 12 \\ 2 & 1 & 0 & 1 & 0 & 0 & 10 \\ \boxed{1} & 1 & 0 & 0 & 1 & 0 & 4 \\ \hline -2 & -1 & 0 & 0 & 0 & 1 & 0 \end{array}\right] \end{array}$$

$$\begin{array}{c} \\ -R_3 + R_1 \to R_1 \\ -2R_3 + R_2 \to R_2 \\ \\ 2R_3 + R_4 \to R_4 \end{array} \begin{array}{c} x_1 \;\;\; x_2 \;\; s_1 \;\; s_2 \;\; s_3 \;\; z \\ \left[\begin{array}{cccccc|c} 0 & 2 & 1 & 0 & -1 & 0 & 8 \\ 0 & -1 & 0 & 1 & -2 & 0 & 2 \\ 1 & 1 & 0 & 0 & 1 & 0 & 4 \\ \hline 0 & 1 & 0 & 0 & 2 & 1 & 8 \end{array}\right] \end{array}$$

This is a final tableau since all of the numbers in the last row are nonnegative. The maximum value is 8 when $x_1 = 4$, $x_2 = 0$, $s_1 = 8$, $s_2 = 2$, and $s_3 = 0$.

5.
$$\begin{array}{c} x_1 \;\;\; x_2 \;\;\; x_3 \;\; s_1 \;\; s_2 \;\; s_3 \;\; z \\ \left[\begin{array}{ccccccc|c} 2 & 2 & 8 & 1 & 0 & 0 & 0 & 40 \\ 4 & -5 & 6 & 0 & 1 & 0 & 0 & 60 \\ 2 & -2 & 6 & 0 & 0 & 1 & 0 & 24 \\ \hline -14 & -10 & -12 & 0 & 0 & 0 & 1 & 0 \end{array}\right] \end{array}$$

The most negative indicator is -14, in the first column. Find the quotients $\frac{40}{2} = 20$, $\frac{60}{4} = 15$, and $\frac{24}{2} = 12$; since 12 is the smallest quotient, 2 in row 3, column 1 is the pivot.

$$\begin{array}{c} x_1 \;\;\; x_2 \;\;\; x_3 \;\; s_1 \;\; s_2 \;\; s_3 \;\; z \\ \left[\begin{array}{ccccccc|c} 2 & 2 & 8 & 1 & 0 & 0 & 0 & 40 \\ 4 & -5 & 6 & 0 & 1 & 0 & 0 & 60 \\ \boxed{2} & -2 & 6 & 0 & 0 & 1 & 0 & 24 \\ \hline -14 & -10 & -12 & 0 & 0 & 0 & 1 & 0 \end{array}\right] \end{array}$$

Performing row transformations, we get the following tableau.

$$\begin{array}{c} -R_3 + R_1 \to R_1 \\ -2R_3 + R_2 \to R_2 \\ \\ 7R_3 + R_4 \to R_4 \end{array} \begin{array}{c} x_1 \;\;\; x_2 \;\;\; x_3 \;\; s_1 \;\; s_2 \;\;\; s_3 \;\; z \\ \left[\begin{array}{ccccccc|c} 0 & \boxed{4} & 2 & 1 & 0 & -1 & 0 & 16 \\ 0 & -1 & -6 & 0 & 1 & -2 & 0 & 12 \\ 2 & -2 & 6 & 0 & 0 & 1 & 0 & 24 \\ \hline 0 & -24 & 30 & 0 & 0 & 7 & 1 & 168 \end{array}\right] \end{array}$$

Since there is still a negative indicator, we must repeat the process. The second pivot is the 4 in column 2, since $\frac{16}{4}$ is the only nonnegative quotient in the only column with a negative indicator. Performing row transformations again, we get the following tableau.

$$\begin{array}{c} \\ R_1 + 4R_2 \to R_2 \\ R_1 + 2R_3 \to R_3 \\ 6R_1 + R_4 \to R_4 \end{array} \begin{array}{c} x_1 \; x_2 \;\;\; x_3 \;\; s_1 \;\; s_2 \;\;\; s_3 \;\; z \\ \left[\begin{array}{ccccccc|c} 0 & 4 & 2 & 1 & 0 & -1 & 0 & 16 \\ 0 & 0 & -22 & 1 & 4 & -9 & 0 & 64 \\ 4 & 0 & 14 & 1 & 0 & 1 & 0 & 64 \\ \hline 0 & 0 & 42 & 6 & 0 & 1 & 1 & 264 \end{array}\right] \end{array}$$

All of the numbers in the last row are nonnegative, so we are finished pivoting. Create a 1 in the columns corresponding to x_1, x_2, and s_2.

$$\begin{array}{c} \frac{1}{4}R_1 \to R_1 \\ \\ \frac{1}{4}R_2 \to R_2 \\ \\ \frac{1}{4}R_3 \to R_3 \\ \\ \end{array} \begin{array}{c} x_1 \; x_2 \;\;\; x_3 \;\;\; s_1 \;\; s_2 \;\;\; s_3 \;\; z \\ \left[\begin{array}{ccccccc|c} 0 & 1 & \frac{1}{2} & \frac{1}{4} & 0 & -\frac{1}{4} & 0 & 4 \\ 0 & 0 & -\frac{11}{2} & \frac{1}{4} & 1 & -\frac{9}{4} & 0 & 16 \\ 1 & 0 & \frac{7}{2} & \frac{1}{4} & 0 & \frac{1}{4} & 0 & 16 \\ \hline 0 & 0 & 42 & 6 & 0 & 1 & 1 & 264 \end{array}\right] \end{array}$$

The maximum value is 264 and occurs when $x_1 = 16$, $x_2 = 4$, $x_3 = 0$, $s_1 = 0$, $s_2 = 16$, and $s_3 = 0$.

7. Maximize $z = 3x_1 + 5x_2$

subject to: $4x_1 + x_2 \le 25$

$2x_1 + 3x_2 \le 15$

with $x_1 \ge 0, x_2 \ge 0$.

Two slack variables, s_1 and s_2, need to be introduced. The problem can be restated as:

Maximize $z = 3x_1 + 5x_2$

subject to: $4x_1 + x_2 + s_1 = 25$

$2x_1 + 3x_2 + s_2 = 15$

with $x_1 \ge 0, x_2 \ge 0, s_1 \ge 0, s_2 \ge 0$.

Rewrite the objective function as

$$-3x_1 - 5x_2 + z = 0.$$

The initial simplex tableau follows.

$$\begin{array}{ccccc} x_1 & x_2 & s_1 & s_2 & z \\ \end{array}$$
$$\left[\begin{array}{cccc|c} 4 & 1 & 1 & 0 & 0 \\ 2 & 3 & 0 & 1 & 0 \\ \hline -3 & -5 & 0 & 0 & 1 \end{array}\begin{array}{c} 25 \\ 15 \\ 0 \end{array}\right]$$

The most negative indicator is -5, in the second column. To select the pivot from column 2, find the quotients $\frac{25}{1} = 25$ and $\frac{15}{3} = 5$. The smaller quotient is 5, so 3 is the pivot.

$$\begin{array}{ccccc} x_1 & x_2 & s_1 & s_2 & z \\ \end{array}$$
$$\left[\begin{array}{cccc|c} 4 & 1 & 1 & 0 & 0 \\ 2 & \boxed{3} & 0 & 1 & 0 \\ \hline -3 & -5 & 0 & 0 & 1 \end{array}\begin{array}{c} 25 \\ 15 \\ 0 \end{array}\right]$$

$$\begin{array}{ccccc} x_1 & x_2 & s_1 & s_2 & z \\ \end{array}$$
$$\begin{array}{c} -R_2 + 3R_1 \to R_1 \\ \\ 5R_2 + 3R_3 \to R_3 \end{array}\left[\begin{array}{ccccc|c} 10 & 0 & 3 & -1 & 0 & 60 \\ 2 & 3 & 0 & 1 & 0 & 15 \\ \hline 1 & 0 & 0 & 5 & 3 & 75 \end{array}\right]$$

All of the indicators are nonnegative. Create a 1 in the columns corresponding to x_2, s_1, and z.

$$\begin{array}{ccccc} x_1 & x_2 & s_1 & s_2 & z \\ \end{array}$$
$$\begin{array}{c} \frac{1}{3}R_1 \to R_1 \\ \frac{1}{3}R_2 \to R_2 \\ \frac{1}{3}R_3 \to R_3 \end{array}\left[\begin{array}{ccccc|c} \frac{10}{3} & 0 & 1 & -\frac{1}{3} & 0 & 20 \\ \frac{2}{3} & 1 & 0 & \frac{1}{3} & 0 & 5 \\ \frac{1}{3} & 0 & 0 & \frac{5}{3} & 1 & 25 \end{array}\right]$$

The maximum value is 25 when $x_1 = 0$, $x_2 = 5$, $s_1 = 20$, and $s_2 = 0$.

9. Maximize $z = 10x_1 + 12x_2$

subject to: $4x_1 + 2x_2 \le 20$

$5x_1 + x_2 \le 50$

$2x_1 + 2x_2 \le 24$

with $x_1 \ge 0, x_2 \ge 0$.

Three slack variables, s_1, s_2, and s_3, need to be introduced. The initial tableau is as follows.

$$\begin{array}{cccccc} x_1 & x_2 & s_1 & s_2 & s_3 & z \\ \end{array}$$
$$\left[\begin{array}{cccccc|c} 4 & 2 & 1 & 0 & 0 & 0 & 20 \\ 5 & 1 & 0 & 1 & 0 & 0 & 50 \\ 2 & 2 & 0 & 0 & 1 & 0 & 24 \\ \hline -10 & -12 & 0 & 0 & 0 & 1 & 0 \end{array}\right]$$

The most negative indicator is -12, in column 2.

The quotients are $\frac{20}{2} = 10, \frac{50}{1} = 50$, and $\frac{24}{2} = 12$; the smallest is 10, so 2 in row 1, column 2 is the pivot.

$$\begin{array}{cccccc} x_1 & x_2 & s_1 & s_2 & s_3 & z \\ \end{array}$$
$$\left[\begin{array}{cccccc|c} 4 & \boxed{2} & 1 & 0 & 0 & 0 & 20 \\ 5 & 1 & 0 & 1 & 0 & 0 & 50 \\ 2 & 2 & 0 & 0 & 1 & 0 & 24 \\ \hline -10 & -12 & 0 & 0 & 0 & 1 & 0 \end{array}\right]$$

$$\begin{array}{cccccc} x_1 & x_2 & s_1 & s_2 & s_3 & z \\ \end{array}$$
$$\begin{array}{c} \\ -R_1 + 2R_2 \to R_2 \\ -R_1 + R_3 \to R_3 \\ 6R_1 + R_4 \to R_4 \end{array}\left[\begin{array}{cccccc|c} 4 & 2 & 1 & 0 & 0 & 0 & 20 \\ 6 & 0 & -1 & 2 & 0 & 0 & 80 \\ -2 & 0 & -1 & 0 & 1 & 0 & 4 \\ \hline 14 & 0 & 6 & 0 & 0 & 1 & 120 \end{array}\right]$$

All of the indicators are nonnegative, so we are finished pivoting. Create a 1 in the columns corresponding to x_2 and s_2.

$$\begin{array}{cccccc} x_1 & x_2 & s_1 & s_2 & s_3 & z \\ \end{array}$$
$$\begin{array}{c} \\ \frac{1}{2}R_1 \to R_1 \\ \frac{1}{2}R_2 \to R_2 \\ \end{array}\left[\begin{array}{cccccc|c} 2 & 1 & \frac{1}{2} & 0 & 0 & 0 & 10 \\ 3 & 0 & -\frac{1}{2} & 1 & 0 & 0 & 40 \\ -2 & 0 & -1 & 0 & 1 & 0 & 4 \\ \hline 14 & 0 & 6 & 0 & 0 & 1 & 120 \end{array}\right]$$

The maximum value is 120 when $x_1 = 0$, $x_2 = 10$, $s_1 = 0$, $s_2 = 40$, and $s_3 = 4$.

11. Maximize $z = 8x_1 + 3x_2 + x_3$

subject to: $x_1 + 6x_2 + 8x_3 \le 118$

$x_1 + 5x_2 + 10x_3 \le 220$

with $x_1 \ge 0, x_2 \ge 0, x_3 \ge 0.$

Two slack variables, s_1 and s_2, need to be introduced. The initial simplex tableau is as follows.

$$
\begin{array}{cccccc|c}
x_1 & x_2 & x_3 & s_1 & s_2 & z & \\
\hline
\boxed{1} & 6 & 8 & 1 & 0 & 0 & 118 \\
1 & 5 & 10 & 0 & 1 & 0 & 220 \\
\hline
-8 & -3 & -1 & 0 & 0 & 1 & 0
\end{array}
$$

The most negative indicator is -8, in the first column. The quotients are $\frac{118}{1} = 118$ and $\frac{220}{1} = 220$; since 118 is the smaller, 1 in row 1, column 1 is the pivot. Performing row transformations, we get the following tableau.

$$
\begin{array}{c}
 \\
-R_1 + R_2 \rightarrow R_2 \\
8R_1 + R_3 \rightarrow R_3
\end{array}
\begin{array}{cccccc|c}
x_1 & x_2 & x_3 & s_1 & s_2 & z & \\
\hline
1 & 6 & 8 & 1 & 0 & 0 & 118 \\
0 & -1 & 2 & -1 & 1 & 0 & 102 \\
0 & 45 & 63 & 8 & 0 & 1 & 944
\end{array}
$$

All of the indicators are nonnegative, so we are finished pivoting. The maximum value is 944 when $x_1 = 118$, $x_2 = 0$, $x_3 = 0$, $s_1 = 0$, and $s_2 = 102$.

13. Maximize $z = 10x_1 + 15x_2 + 10x_3 + 5x_4$

subject to: $x_1 + x_2 + x_3 + x_4 \le 300$

$x_1 + 2x_2 + 3x_3 + x_4 \le 360$

with $x_1 \ge 0, x_2 \ge 0, x_3 \ge 0, x_4 \ge 0.$

The initial tableau is as follows.

$$
\begin{array}{ccccccc|c}
x_1 & x_2 & x_3 & x_4 & s_1 & s_2 & z & \\
\hline
1 & 1 & 1 & 1 & 1 & 0 & 0 & 300 \\
1 & \boxed{2} & 3 & 1 & 0 & 1 & 0 & 360 \\
\hline
-10 & -15 & -10 & -5 & 0 & 0 & 1 & 0
\end{array}
$$

In the column with the most negative indicator, -15, the quotients are $\frac{300}{1} = 300$ and $\frac{360}{2} = 180$. The smaller quotient is 180, so the 2 in row 2, column 2, is the pivot.

$$
\begin{array}{c}
-R_2 + 2R_1 \rightarrow R_1 \\
 \\
15R_2 + 2R_3 \rightarrow R_3
\end{array}
\begin{array}{ccccccc|c}
x_1 & x_2 & x_3 & x_4 & s_1 & s_2 & z & \\
\hline
\boxed{1} & 0 & -1 & 1 & 2 & -1 & 0 & 240 \\
1 & 2 & 3 & 1 & 0 & 1 & 0 & 360 \\
-5 & 0 & 25 & 5 & 0 & 15 & 2 & 5400
\end{array}
$$

Pivot on the 1 in row 1, column 1.

$$
\begin{array}{c}
 \\
-R_1 + R_2 \rightarrow R_2 \\
5R_1 + R_3 \rightarrow R_3
\end{array}
\begin{array}{ccccccc|c}
x_1 & x_2 & x_3 & x_4 & s_1 & s_2 & z & \\
\hline
1 & 0 & -1 & 1 & 2 & -1 & 0 & 240 \\
0 & 2 & 4 & 0 & -2 & 2 & 0 & 120 \\
0 & 0 & 20 & 10 & 10 & 10 & 2 & 6600
\end{array}
$$

Create a 1 in the columns corresponding to x_2 and z.

$$
\begin{array}{c}
 \\
\frac{1}{2}R_2 \rightarrow R_2 \\
\frac{1}{2}R_3 \rightarrow R_3
\end{array}
\begin{array}{ccccccc|c}
x_1 & x_2 & x_3 & x_4 & s_1 & s_2 & z & \\
\hline
1 & 0 & -1 & 1 & 2 & -1 & 0 & 240 \\
0 & 1 & 2 & 0 & -1 & 1 & 0 & 60 \\
0 & 0 & 10 & 5 & 5 & 5 & 1 & 3300
\end{array}
$$

The maximum value is 3300 when $x_1 = 240$, $x_2 = 60$, $x_3 = 0$, $x_4 = 0$, $s_1 = 0$, and $s_2 = 0$.

15. Maximize $z = 4x_1 + 6x_2$

subject to: $x_1 - 5x_2 \le 25$

$4x_1 - 3x_2 \le 12$

with $x_1 \ge 0, x_2 \ge 0.$

$$
\begin{array}{ccccc|c}
x_1 & x_2 & s_1 & s_2 & z & \\
\hline
1 & -5 & 1 & 0 & 0 & 25 \\
4 & -3 & 0 & 1 & 0 & 12 \\
\hline
-4 & -6 & 0 & 0 & 1 & 0
\end{array}
$$

The most negative indicator is -6. The negative quotients $25/(-5)$ and $12/(-3)$ indicate an unbounded feasible region, so there is no unique optimum solution.

17. Maximize

$$z = 37x_1 + 34x_2 + 36x_3 + 30x_4 + 35x_5$$

subject to:

$$16x_1 + 19x_2 + 23x_3 + 15x_4 + 21x_5 \leq 42{,}000$$
$$15x_1 + 10x_2 + 19x_3 + 23x_4 + 10x_5 \leq 25{,}000$$
$$9x_1 + 16x_2 + 14x_3 + 12x_4 + 11x_5 \leq 23{,}000$$
$$18x_1 + 20x_2 + 15x_3 + 17x_4 + 19x_5 \leq 36{,}000$$

with $x_1 \geq 0, x_2 \geq 0, x_3 \geq 0, x_4 \geq 0, x_5 \geq 0.$

Four slack variables, s_1, s_2, s_3, and s_4, need to be introduced. The initial simplex tableau follows.

$$
\begin{array}{ccccccccccc|c}
x_1 & x_2 & x_3 & x_4 & x_5 & s_1 & s_2 & s_3 & s_4 & z & \\
16 & 19 & 23 & 15 & 21 & 1 & 0 & 0 & 0 & 0 & 42{,}000 \\
15 & 10 & 19 & 23 & 10 & 0 & 1 & 0 & 0 & 0 & 25{,}000 \\
9 & 16 & 14 & 12 & 11 & 0 & 0 & 1 & 0 & 0 & 23{,}000 \\
18 & 20 & 15 & 17 & 19 & 0 & 0 & 0 & 1 & 0 & 36{,}000 \\
\hline
-37 & -34 & -36 & -30 & -35 & 0 & 0 & 0 & 0 & 1 & 0
\end{array}
$$

Using a graphing calculator or computer program, the maximum value is found to be 70,818.18 when $x_1 = 181.82$, $x_2 = 0$, $x_3 = 454.55$, $x_4 = 0$, $x_5 = 1363.64$, $s_1 = 0$, $s_2 = 0$, $s_3 = 0$, and $s_4 = 0$.

23. Organize the information in a table.

	Church Group	Labor Union	Maximum Time Available
Letter Writing	2	2	16
Follow-up	1	3	12
Money Raised	$100	$200	

Let x_1 and x_2 be the number of church groups and labor unions contacted respectively. We need two slack variables, s_1 and s_2.

Maximize $z = 100x_1 + 200x_2$

subject to: $2x_1 + 2x_2 + s_1 \quad\quad = 16$
$\quad\quad\quad\quad x_1 + 3x_2 \quad\quad + s_2 = 12$

with $x_1 \geq 0, x_2 \geq 0, s_1 \geq 0, s_2 \geq 0.$

The initial simplex tableau is as follows.

$$
\begin{array}{cccc|c}
x_1 & x_2 & s_1 & s_2 & z & \\
2 & 2 & 1 & 0 & 0 & 16 \\
1 & \boxed{3} & 0 & 1 & 0 & 12 \\
\hline
-100 & -200 & 0 & 0 & 1 & 0
\end{array}
$$

Pivot on the 3 in row 2, column 2.

$$
\begin{array}{l}
-2R_2 + 3R_1 \to R_1 \\
\\
200R_2 + 3R_3 \to R_3
\end{array}
\begin{array}{ccccc|c}
x_1 & x_2 & s_1 & s_2 & z & \\
\boxed{4} & 0 & 3 & -2 & 0 & 24 \\
1 & 3 & 0 & 1 & 0 & 12 \\
\hline
-100 & 0 & 0 & 200 & 3 & 2400
\end{array}
$$

Pivot on the 4 in row 1, column 1.

$$
\begin{array}{l}
\\
-R_1 + 4R_2 \to R_2 \\
25R_1 + R_3 \to R_3
\end{array}
\begin{array}{ccccc|c}
x_1 & x_2 & s_1 & s_2 & z & \\
4 & 0 & 3 & -2 & 0 & 24 \\
0 & 12 & -3 & 6 & 0 & 24 \\
0 & 0 & 75 & 150 & 3 & 3000
\end{array}
$$

This is a final tableau, since all of the indicators are nonnegative. Create a 1 in the columns corresponding to $x_1, x_2,$ and z.

$$
\begin{array}{l}
\frac{1}{4}R_1 \to R_1 \\
\\
\frac{1}{12}R_2 \to R_2 \\
\\
\frac{1}{3}R_3 \to R_3
\end{array}
\begin{array}{ccccc|c}
x_1 & x_2 & s_1 & s_2 & z & \\
1 & 0 & \frac{3}{4} & -\frac{1}{2} & 0 & 6 \\
0 & 1 & -\frac{1}{4} & \frac{1}{2} & 0 & 2 \\
0 & 0 & 25 & 50 & 1 & 1000
\end{array}
$$

The maximum amount of money raised is $1000/mo when $x_1 = 6$ and $x_2 = 2$, that is, when 6 churches and 2 labor unions are contacted.

25. (a) Let x_1 be the number of Royal Flush poker sets, x_2 be the number of Deluxe Diamond sets, and x_3 be the number of Full House sets. The problem can be stated as follows.

Maximize $z = 38x_1 + 22x_2 + 12x_3$

subject to:

$$1000x_1 + 600x_2 + 300x_3 \leq 2{,}800{,}000$$
$$4x_1 + 2x_2 + 2x_3 \leq 10{,}000$$
$$10x_1 + 5x_2 + 5x_3 \leq 25{,}000$$
$$2x_1 + x_2 + x_3 \leq 6000$$

with $x_1 \geq 0, x_2 \geq 0, x_3 \geq 0.$

Since there are four constraints, introduce slack variables, s_1, s_2, s_3, and s_4 and set up the initial simplex tableau.

$$
\begin{array}{cccccccc}
x_1 & x_2 & x_3 & s_1 & s_2 & s_3 & s_4 & z \\
\end{array}
$$

$$
\left[\begin{array}{cccccccc|c}
1000 & 600 & 300 & 1 & 0 & 0 & 0 & 0 & 2{,}800{,}000 \\
4 & 2 & 2 & 0 & 1 & 0 & 0 & 0 & 10{,}000 \\
10 & 5 & 5 & 0 & 0 & 1 & 0 & 0 & 25{,}000 \\
2 & 1 & 1 & 0 & 0 & 0 & 1 & 0 & 6000 \\
\hline
-38 & -22 & -12 & 0 & 0 & 0 & 0 & 1 & 0
\end{array}\right]
$$

Using a graphing calculator or computer program, the maximum profit is $104,000 and is obtained when 1000 Royal Flush poker sets, 3000 Deluxe Diamond poker sets, and no Full House poker sets are assembled.

(b) According to the poker chip constraint:

$$1000(1000) + 600(3000) + 300(0) + s_2$$
$$= 2{,}800{,}000 \quad s_1 = 0.$$

So all of the poker chips are used. Checking the card constraint:

$$4(1000) + 2(3000) + 2(0) + s_2 = 10{,}000$$
$$s_2 = 0.$$

So all of the cards are used. Checking the dice constraint:

$$10(1000) + 5(3000) + 5(0) + s_3 = 25{,}000$$
$$s_3 = 0.$$

So all of the dice are used. Finally, checking the dealer button constraint:

$$2(1000) + 3000 + 0 + s_4 = 6000$$
$$s_4 = 1000.$$

This means there are 1000 unused dealer buttons.

27. (a) Let x_1 represent the number of racing bicycles, x_2 the number of touring bicycles, and x_3 the number of mountain bicycles.

From Exercise 28 in Section 4.1, the initial simplex tableau is as follows.

$$
\begin{array}{cccccc}
x_1 & x_2 & x_3 & s_1 & s_2 & z \\
\end{array}
$$

$$
\left[\begin{array}{cccccc|c}
17 & 27 & \boxed{34} & 1 & 0 & 0 & 91{,}800 \\
12 & 21 & 15 & 0 & 1 & 0 & 42{,}000 \\
\hline
-8 & -12 & -22 & 0 & 0 & 1 & 0
\end{array}\right]
$$

Pivot on the 34 in row 1, column 3.

$$
\begin{array}{cccccc}
x_1 & x_2 & x_3 & s_1 & s_2 & z \\
\end{array}
$$

$$
\begin{array}{c}
\\
-15R_1 + 34R_2 \to R_2 \\
11R_1 + 17R_3 \to R_3
\end{array}
\left[\begin{array}{cccccc|c}
17 & 27 & 34 & 1 & 0 & 0 & 91{,}800 \\
153 & 309 & 0 & -15 & 34 & 0 & 51{,}000 \\
51 & 93 & 0 & 11 & 0 & 17 & 1{,}009{,}800
\end{array}\right]
$$

This is a final tableau, since all of the indicators are nonnegative. Create a 1 in the columns corresponding to x_3, s_2, and z.

$$
\begin{array}{cccccc}
x_1 & x_2 & x_3 & s_1 & s_2 & z \\
\end{array}
$$

$$
\begin{array}{c}
\frac{1}{34}R_1 \to R_1 \\
\frac{1}{34}R_2 \to R_2 \\
\frac{1}{17}R_3 \to R_3
\end{array}
\left[\begin{array}{cccccc|c}
\frac{1}{2} & \frac{27}{34} & 1 & \frac{1}{34} & 0 & 0 & 2700 \\
\frac{9}{2} & \frac{309}{34} & 0 & -\frac{15}{34} & 1 & 0 & 1500 \\
3 & \frac{93}{17} & 0 & \frac{11}{17} & 0 & 1 & 59{,}400
\end{array}\right]
$$

From the tableau, $x_1 = 0$, $x_2 = 0$, and $x_3 = 2700$. The company should make no racing or touring bicycles and 2700 mountain bicycles.

(b) From the third row of the final tableau, the maximum profit is $59,400.

(c) When $x_1 = 0$, $x_2 = 0$, and $x_3 = 2700$, the number of units of steel used is

$$17(0) + 27(0) + 34(2700) = 91{,}800$$

which is all the steel available. The number of units of aluminum used is

$$12(0) + 21(0) + 15(2700) = 40{,}500$$

which leaves $42{,}000 - 40{,}500 = 1500$ units of aluminum unused.

Checking the second constraint:

$$12x_1 + 21x_2 + 15x_3 + s_2 = 42{,}000$$
$$12(0) + 21(0) + 15(2700) + s_2 = 42{,}000$$
$$s_2 = 1500.$$

29. **(a)** Let x_1 be the number of newspaper ads, x_2 be the number of Internet banner ads, and x_3 be the number of TV ads. Here is the initial tableau:

$$
\begin{array}{ccccccccc}
x_1 & x_2 & x_3 & s_1 & s_2 & s_3 & s_4 & z & \\
\end{array}
$$

$$
\left[
\begin{array}{cccccccc|c}
400 & 20 & \boxed{2000} & 1 & 0 & 0 & 0 & 0 & 8000 \\
1 & 0 & 0 & 0 & 1 & 0 & 0 & 0 & 30 \\
0 & 1 & 0 & 0 & 0 & 1 & 0 & 0 & 60 \\
0 & 0 & 1 & 0 & 0 & 0 & 1 & 0 & 10 \\
\hline
-4000 & -3000 & -10{,}000 & 0 & 0 & 0 & 0 & 1 & 0 \\
\end{array}
\right]
$$

Pivot on the 2000 in row 1, column 3.

$$
\begin{array}{ccccccccc}
x_1 & x_2 & x_3 & s_1 & s_2 & s_3 & s_4 & z & \\
\end{array}
$$

$$
\begin{array}{r}
\\ \\ \\ \\
-R_1 + 2000R_4 \to R_4 \\
5R_1 + R_5 \to R_5
\end{array}
\left[
\begin{array}{cccccccc|c}
400 & 20 & 2000 & 1 & 0 & 0 & 0 & 0 & 8000 \\
1 & 0 & 0 & 0 & 1 & 0 & 0 & 0 & 30 \\
0 & \boxed{1} & 0 & 0 & 0 & 1 & 0 & 0 & 60 \\
-400 & -20 & 0 & -1 & 0 & 0 & 2000 & 0 & 12{,}000 \\
-2000 & -2900 & 0 & 5 & 0 & 0 & 0 & 1 & 40{,}000 \\
\end{array}
\right]
$$

Pivot on the 1 in row 3, column 2.

$$
\begin{array}{ccccccccc}
x_1 & x_2 & x_3 & s_1 & s_2 & s_3 & s_4 & z & \\
\end{array}
$$

$$
\begin{array}{r}
-20R_3 + R_1 \to R_1 \\
\\ \\
20R_3 + R_4 \to R_4 \\
2900R_3 + R_5 \to R_5
\end{array}
\left[
\begin{array}{cccccccc|c}
\boxed{400} & 20 & 2000 & 1 & 0 & -20 & 0 & 0 & 6800 \\
1 & 0 & 0 & 0 & 1 & 0 & 0 & 0 & 30 \\
0 & 1 & 0 & 0 & 0 & 1 & 0 & 0 & 60 \\
-400 & 0 & 0 & -1 & 0 & 20 & 2000 & 0 & 13{,}200 \\
-2000 & 0 & 0 & 5 & 0 & 2900 & 0 & 1 & 214{,}000 \\
\end{array}
\right]
$$

Pivot on the 400 in row 1, column 1.

$$
\begin{array}{cccccccc}
x_1 & x_2 & x_3 & s_1 & s_2 & s_3 & s_4 & z \\
\end{array}
$$

$$
\begin{array}{r}
\\
-R_1 + 400R_2 \to R_2 \\
\\
R_1 + R_4 \to R_4 \\
5R_1 + R_5 \to R_5
\end{array}
\left[
\begin{array}{cccccccc|c}
400 & 0 & 2000 & 1 & 0 & -20 & 0 & 0 & 6800 \\
0 & 0 & -2000 & -1 & 400 & 20 & 0 & 0 & 5200 \\
0 & 1 & 0 & 0 & 0 & 1 & 0 & 0 & 60 \\
0 & 0 & 2000 & 0 & 0 & 0 & 2000 & 0 & 20{,}000 \\
0 & 0 & 10{,}000 & 10 & 0 & 2800 & 0 & 1 & 248{,}000 \\
\end{array}
\right]
$$

Create a 1 in the columns corresponding to $x_1, s_2,$ and s_4.

$$
\begin{array}{cccccccc}
x_1 & x_2 & x_3 & s_1 & s_2 & s_3 & s_4 & z \\
\end{array}
$$

$$
\begin{array}{r}
\frac{1}{400}R_1 \to R_1 \\
\frac{1}{400}R_2 \to R_2 \\
\\
\frac{1}{2000}R_4 \to R_4
\end{array}
\left[
\begin{array}{cccccccc|c}
1 & 0 & 5 & \frac{1}{400} & 0 & -\frac{1}{20} & 0 & 0 & 17 \\
0 & 0 & -5 & -\frac{1}{400} & 1 & \frac{1}{20} & 0 & 0 & 13 \\
0 & 1 & 0 & 0 & 0 & 1 & 0 & 0 & 60 \\
0 & 0 & 1 & 0 & 0 & 0 & 1 & 0 & 10 \\
\hline
0 & 0 & 10{,}000 & 10 & 0 & 2800 & 0 & 1 & 248{,}000 \\
\end{array}
\right]
$$

This is the final tableau. The maximum exposure is 248,000 women when 17 newspaper ads, 60 Internet banner ads, and no TV ads are used.

31. (a) The coefficients of the objective function are the profit coefficients from the table: 5, 4, and 3; choice (3) is correct.

(b) The constraints are the available man-hours for the 2 departments, 400 and 600; choice (4) is correct.

(c) $2X_1 + 3X_2 + 1X_3 \leq 400$ is the constraint on department 1; choice (3) is correct.

33. Maximize $z = 100x + 200y$

     subject to: $2x + 2y \leq 16$

                  $x + 3y \leq 12$

     with $x \geq 0, y \geq 0.$

Using Excel, we enter the variables x and y in cells Al and Bl, respectively. Enter the x- and y-coordinates of the initial corner point of the feasible region, (0, 0), in cells A2 and B2, respectively, and NAME these cells x and y, respectively. In cells C2, C4, C5, C6, and C7, enter the formula for the function to maximize and each of the constraints: $100x + 200y$, $2x + 2y$, $x + 3y, x$, and y. Since x and y have been set to 0, all the cells containing formulas should also show the value 0, as below.

	A	B	C
1	x	y	
2	0	0	0
3			
4			0
5			0
6			0
7			0

Using the SOLVER, ask Excel to maximize the value in cell C2 subject to the constraints $C4 \leq 16$, $C5 \leq 12$, $C6 \geq 0$, $C7 \geq 0$. Make sure you have checked off the box *Assume Linear Model* in SOLVER OPTIONS.

Excel returns the following values and allows you to choose a report.

	A	B	C
1	x	y	
2	6	2	1000
3			
4			16
5			12
6			6
7			2

Select the sensitivity report. The report will appear on a new sheet of the spread sheet.

Adjustable Cells

Cell	Name	Final Value	Reduced Cost	Objective Coefficient	Allowable Increase	Allowable Decrease
A2	x	6	0	100	100	33.33333333
B2	y	2	0	200	100	100

Constraints

Cell	Name	Final Value	Shadow Price	Constraint R.H. Side	Allowable Increase	Allowable Decrease
C4		16	25	16	8	8
C5		12	50	12	12	4
C6		6	0	0	6	1E + 30
C7		2	0	0	2	1E + 30

The church group's allowable increase is $100 and the allowable decrease is $33.33. So their contribution can be as high as $100 + $100 = 200 or as low as $100 - $33.33 = 66.67 and the original solution is still optimal. The unions' allowable increase is $100 and the allowable decrease is $100. So their contribution can be as high as $200 + $100 = 300 or as low as $200 - $100 = 100 and the original solution is still optimal.

35. Let x_1 = number of hours running, x_2 be the number of hours biking, and x_3 be the number hours walking. The problem can be stated as follows.

Maximize $z = 531x_1 + 472x_2 + 354x_3$

subject to:
$$x_1 + x_2 + x_3 \le 15$$
$$x_1 \le 3$$
$$2x_2 - x_3 \le 0$$

with $x_1 \ge 0, x_2 \ge 0, x_3 \ge 0.$

We need three slack variables, s_1, s_2, and s_3. The initial simplex tableau as follows.

$$
\begin{array}{ccccccc|c}
x_1 & x_2 & x_3 & s_1 & s_2 & s_3 & z & \\
\hline
1 & 1 & 1 & 1 & 0 & 0 & 0 & 15 \\
\boxed{1} & 0 & 0 & 0 & 1 & 0 & 0 & 3 \\
0 & 2 & -1 & 0 & 0 & 1 & 0 & 0 \\
\hline
-531 & -472 & -354 & 0 & 0 & 0 & 1 & 0
\end{array}
$$

Pivot on the 1 in row 2, column 1.

$$
\begin{array}{r}
-R_2 + R_1 \to R_1 \\
\\
\\
531R_2 + R_4 \to R_4
\end{array}
\begin{array}{ccccccc|c}
x_1 & x_2 & x_3 & s_1 & s_2 & s_3 & z & \\
\hline
0 & 1 & 1 & 1 & -1 & 0 & 0 & 12 \\
1 & 0 & 0 & 0 & 1 & 0 & 0 & 3 \\
0 & \boxed{2} & -1 & 0 & 0 & 1 & 0 & 0 \\
0 & -472 & -354 & 0 & 351 & 0 & 1 & 1593
\end{array}
$$

Pivot on the 2 in row 3, column 2.

$$
\begin{array}{r}
-R_3 + 2R_1 \to R_1 \\
\\
\\
236R_3 + R_4 \to R_4
\end{array}
\begin{array}{ccccccc|c}
x_1 & x_2 & x_3 & s_1 & s_2 & s_3 & z & \\
\hline
0 & 0 & \boxed{3} & 2 & -2 & -1 & 0 & 24 \\
1 & 0 & 0 & 0 & 1 & 0 & 0 & 3 \\
0 & 2 & -1 & 0 & 0 & 1 & 0 & 0 \\
0 & 0 & -590 & 0 & 531 & 236 & 1 & 1593
\end{array}
$$

Finally pivot on the 3 in row 1, column 3.

$$
\begin{array}{r}
\\
\\
R_1 + 3R_3 \to R_3 \\
590R_1 + 3R_4 \to R_4
\end{array}
\begin{array}{ccccccc|c}
x_1 & x_2 & x_3 & s_1 & s_2 & s_3 & z & \\
\hline
0 & 1 & 3 & 2 & -2 & -1 & 0 & 24 \\
1 & 0 & 0 & 0 & 1 & 0 & 0 & 3 \\
0 & 6 & 0 & 2 & -2 & 2 & 0 & 24 \\
0 & 0 & 0 & 1180 & 413 & 118 & 3 & 18,939
\end{array}
$$

Create a 1 in the columns corresponding to x_2, x_3, and z.

$$\begin{array}{c} \\ \frac{1}{3}R_1 \to R_1 \\ \\ \\ \frac{1}{6}R_3 \to R_3 \\ \\ \frac{1}{3}R_4 \to R_4 \end{array} \begin{array}{cccccccc} x_1 & x_2 & x_3 & s_1 & s_2 & s_3 & z & \\ \left[\begin{array}{ccccccc|c} 0 & 0 & 1 & \frac{2}{3} & -\frac{2}{3} & -\frac{1}{3} & 0 & 8 \\ 1 & 0 & 0 & 0 & 1 & 0 & 0 & 3 \\ 0 & 1 & 0 & \frac{1}{3} & -\frac{1}{3} & \frac{1}{3} & 0 & 4 \\ \hline 0 & 0 & 0 & \frac{1180}{3} & \frac{413}{3} & \frac{118}{3} & 1 & 6313 \end{array}\right] \end{array}$$

Rachel should run 3 hours, bike 4 hours, and walk 8 hours for a maximum calorie expenditure of 6313 calories.

37. **(a)** Let x_1 represent the number of species A, x_2 represent the number of species B, and x_3 represent the number of species C.

Maximize $\quad z = 1.62x_1 + 2.14x_2 + 3.01x_3$

subject to: $\quad 1.32x_1 + 2.1x_2 + 0.86x_3 \le 490$
$\qquad\qquad 2.9x_1 + 0.95x_2 + 1.52x_3 \le 897$
$\qquad\qquad 1.75x_1 + 0.6x_2 + 2.01x_3 \le 653$

with $\qquad x_1 \ge 0, x_2 \ge 0, x_3 \ge 0.$

Use a graphing calculator or computer to solve this problem and find that the answer is to stock none of species A, 114 of species B, and 291 of species C for a maximum combined weight of 1119.72 kg.

(b) When $x_1 = 0$, $x_2 = 114$, and $x_3 = 291$, the number of units used are as follows.

$$\text{Food I: } 1.32(0) + 2.1 + (114) + 0.86(291) = 489.66$$

or 490 units, which is the total amount available of Food I.

$$\text{Food II: } 2.9(0) + 0.95(114) + 1.52(291) = 550.62$$

or 551 units, which leaves $897 - 551$, or 346 units of Food II available.

$$\text{Food III: } 1.75(0) + 0.6(114) + 2.01(291) = 653.31$$

or 653 units, which is the total amount available of Food III.

(c) Many answers are possible. The idea is to choose average weights for species B and C that are considerably smaller than the average weight chosen for species A, so that species A dominates the objective function.

(d) Many answers are possible. The idea is to choose average weights for species A and B that are considerably smaller than the average weight chosen for species C.

39. Let x_1 represent the number of minutes for the senator, x_2 the number of minutes for the congresswoman, and x_3 the number of minutes for the governor.

Of the half-hour show's time, at most only $30 - 3 = 27$ min are available to be allotted to the politicians. The given information leads to the inequality

$$x_1 + x_2 + x_3 \le 27$$

and the inequalities

$$x_1 \ge 2x_3 \quad \text{and} \quad x_1 + x_3 \ge 2x_2,$$

and we are to maximize the objective function

$$z = 35x_1 + 40x_2 + 45x_3.$$

Rewrite the equation as

$$x_3 \le 27 - x_1 - x_2$$

and the inequalities as

$$-x_1 + 2x_3 \leq 0 \quad \text{and} \quad -x_1 + 2x_2 - x_3 \leq 0.$$

Substitute $27 - x_1 - x_2$ for x_3 in the objective function and the inequalities, and the problem is as follows.

Maximize $\quad z = 35x_1 + 40x_2 + 45x_3$

subject to:
$$\begin{aligned} -x_1 \qquad\quad + 2x_3 &\leq 0 \\ -x_1 + 2x_2 - \ x_3 &\leq 0 \\ x_1 + \ x_2 + \ x_3 &\leq 27 \end{aligned}$$

with $x_1 \geq 0,\ x_2 \geq 0,\ x_3 \geq 0$.

We need three slack variables. The initial simplex tableau is as follows.

$$
\begin{array}{ccccccc|c}
x_1 & x_2 & x_3 & s_1 & s_2 & s_3 & z & \\
\hline
-1 & 0 & 2 & 1 & 0 & 0 & 0 & 0 \\
-1 & \boxed{2} & -1 & 0 & 1 & 0 & 0 & 0 \\
1 & 1 & 1 & 0 & 0 & 1 & 0 & 27 \\
\hline
-35 & -40 & -45 & 0 & 0 & 0 & 1 & 0
\end{array}
$$

Pivot on the 2 in row 2, column 2, and then pivot on the 9 in row 3, column 1.

$$
\begin{array}{c}
 \\
 \\
-R_2 + 2R_3 \to R_3 \\
20R_2 + R_4 \to R_4
\end{array}
\begin{array}{ccccccc|c}
x_1 & x_2 & x_3 & s_1 & s_2 & s_3 & z & \\
\hline
-1 & 0 & \boxed{2} & 1 & 0 & 0 & 0 & 0 \\
-1 & 2 & -1 & 0 & 1 & 0 & 0 & 0 \\
3 & 0 & 3 & 0 & -1 & 2 & 0 & 54 \\
-55 & 0 & -65 & 0 & 20 & 0 & 1 & 0
\end{array}
$$

$$
\begin{array}{c}
 \\
R_1 + 2R_2 \to R_2 \\
-3R_1 + 2R_3 \to R_3 \\
65R_1 + 2R_4 \to R_4
\end{array}
\begin{array}{ccccccc|c}
x_1 & x_2 & x_3 & s_1 & s_2 & s_3 & z & \\
\hline
-1 & 0 & 2 & 1 & 0 & 0 & 0 & 0 \\
-3 & 4 & 0 & 1 & 2 & 0 & 0 & 0 \\
\boxed{9} & 0 & 0 & -3 & -2 & 4 & 0 & 108 \\
-175 & 0 & 0 & 65 & 40 & 0 & 2 & 0
\end{array}
$$

Pivot on the 9 in row 3, column 1.

$$
\begin{array}{c}
R_3 + 9R_1 \to R_1 \\
R_3 + 3R_2 \to R_2 \\
 \\
175R_3 + 9R_4 \to R_4
\end{array}
\begin{array}{ccccccc|c}
x_1 & x_2 & x_3 & s_1 & s_2 & s_3 & z & \\
\hline
0 & 0 & 18 & 6 & -2 & 4 & 0 & 108 \\
0 & 12 & 0 & 0 & 4 & 4 & 0 & 108 \\
9 & 0 & 0 & -3 & -2 & 4 & 0 & 108 \\
0 & 0 & 0 & 60 & 10 & 700 & 18 & 18{,}900
\end{array}
$$

Create a 1 in the columns corresponding to x_1, x_2, x_3, and z.

$$
\begin{array}{c}
\frac{1}{18}R_1 \to R_1 \\[4pt]
\frac{1}{12}R_2 \to R_2 \\[4pt]
\frac{1}{9}R_3 \to R_3 \\[4pt]
\frac{1}{18}R_4 \to R_4
\end{array}
\begin{array}{ccccccc|c}
x_1 & x_2 & x_3 & s_1 & s_2 & s_3 & z & \\
\hline
0 & 0 & 1 & \frac{1}{3} & -\frac{1}{9} & \frac{2}{9} & 0 & 6 \\[4pt]
0 & 1 & 0 & 0 & \frac{1}{3} & \frac{1}{3} & 0 & 9 \\[4pt]
1 & 0 & 0 & -\frac{1}{3} & -\frac{2}{9} & \frac{4}{9} & 0 & 12 \\[4pt]
0 & 0 & 0 & \frac{10}{3} & \frac{5}{9} & \frac{350}{9} & 1 & 1050
\end{array}
$$

The maximum value of z is 1050 when $x_1 = 12$, $x_2 = 9$, and $x_3 = 6$. That is, for a maximum of 1,050,000 viewers, the time allotments should be 12 minutes for the senator, 9 minutes for the congresswoman, and 6 minutes for the governor.

4.3 Minimization Problems; Duality

Your Turn 1

Write the augmented matrix.

$$\begin{bmatrix} 3 & 3 & 4 & 24 \\ 5 & 1 & 3 & 27 \\ 25 & 12 & 27 & 0 \end{bmatrix}$$

Transpose to get the matrix for the dual problem.

$$\begin{bmatrix} 3 & 5 & 25 \\ 3 & 1 & 12 \\ 4 & 3 & 27 \\ 24 & 27 & 0 \end{bmatrix}$$

Write the dual problem.

Maximize $z = 24x_1 + 27x_2$

subject to: $3x_1 + 5x_2 \le 25$

$3x_1 + x_2 \le 12$

$4x_1 + 3x_2 \le 27$

with $x_1 \ge 0,\ x_2 \ge 0$

Your Turn 2

Write the augmented matrix.

$$\begin{bmatrix} 3 & 5 & 20 \\ 3 & 1 & 18 \\ 15 & 12 & 0 \end{bmatrix}$$

Transpose to get the matrix for the dual problem.

$$\begin{bmatrix} 3 & 3 & 15 \\ 5 & 1 & 12 \\ 20 & 18 & 0 \end{bmatrix}$$

Write the dual problem.

Maximize $z = 20x_1 + 18x_2$

subject to: $3x_1 + 3x_2 \le 15$

$5x_1 + x_2 \le 12$

with $x_1 \ge 0,\ x_2 \ge 0.$

The initial tableau for this problem is

$$\begin{array}{c} \begin{array}{ccccc} x_1 & x_2 & s_1 & s_2 & z \end{array} \\ \left[\begin{array}{ccccc|c} 3 & 3 & 1 & 0 & 0 & 15 \\ \boxed{5} & 1 & 0 & 1 & 0 & 12 \\ \hline -20 & -18 & 0 & 0 & 1 & 0 \end{array}\right] \end{array}$$

Pivot around the indicated 5.

$$\begin{array}{c} \begin{array}{ccccc} x_1 & x_2 & s_1 & s_2 & z \end{array} \\ \begin{array}{r} -3R_2 + 5R_1 \to R_1 \\ \\ 4R_2 + R_3 \to R_3 \end{array} \left[\begin{array}{ccccc|c} 0 & \boxed{12} & 5 & -3 & 0 & 39 \\ 5 & 1 & 0 & 1 & 0 & 12 \\ 0 & -14 & 0 & 4 & 1 & 48 \end{array}\right] \end{array}$$

Now pivot around the indicated 12:

$$\begin{array}{c} \begin{array}{ccccc} x_1 & x_2 & s_1 & s_2 & z \end{array} \\ \begin{array}{r} \\ -R_1 + 12R_2 \to R_2 \\ 7R_1 + 6R_3 \to R_3 \end{array} \left[\begin{array}{ccccc|c} 0 & 12 & 5 & -3 & 0 & 39 \\ 60 & 0 & -5 & 15 & 0 & 105 \\ 0 & 0 & 35 & 3 & 6 & 561 \end{array}\right] \end{array}$$

Finally divide the last row by 6 to produce a 1 in the z column:

$$\begin{array}{c} \begin{array}{ccccc} x_1 & x_2 & s_1 & s_2 & z \end{array} \\ \begin{array}{r} \\ \\ R_3/6 \to R_3 \end{array} \left[\begin{array}{ccccc|c} 0 & 12 & 5 & -3 & 0 & 39 \\ 60 & 0 & -5 & 15 & 0 & 105 \\ 0 & 0 & \frac{35}{6} & \frac{1}{2} & 1 & \frac{187}{2} \end{array}\right] \end{array}$$

In the original problem, w has a minimum of $\frac{187}{2}$ when $y_1 = \frac{35}{6}$ and $y_2 = \frac{1}{2}.$

4.3   Exercises

1. To form the transpose of a matrix, the rows of the original matrix are written as the columns of the transpose. The transpose of

$$\begin{bmatrix} 1 & 2 & 3 \\ 3 & 2 & 1 \\ 1 & 10 & 0 \end{bmatrix}$$

is

$$\begin{bmatrix} 1 & 3 & 1 \\ 2 & 2 & 10 \\ 3 & 1 & 0 \end{bmatrix}.$$

3. The transpose of

$$\begin{bmatrix} 4 & 5 & -3 & 15 \\ 7 & 14 & 20 & -8 \\ 5 & 0 & -2 & 23 \end{bmatrix}$$

is

$$\begin{bmatrix} 4 & 7 & 5 \\ 5 & 14 & 0 \\ -3 & 20 & -2 \\ 15 & -8 & 23 \end{bmatrix}.$$

5. Maximize $z = 4x_1 + 3x_2 + 2x_3$

subject to: $\quad x_1 + x_2 + x_3 \leq 5$

$\qquad\qquad x_1 + x_2 \qquad\; \leq\; 4$

$\qquad\quad 2x_1 + x_2 + 3x_3 \leq 15$

with $\qquad x_1 \geq 0, x_2 \geq 0, x_3 \geq 0.$

To form the dual, first write the augmented matrix for the given problem.

$$\begin{bmatrix} 1 & 1 & 1 & 5 \\ 1 & 1 & 0 & 4 \\ 2 & 1 & 3 & 15 \\ \hline 4 & 3 & 2 & 0 \end{bmatrix}$$

Then form the transpose of this matrix.

$$\begin{bmatrix} 1 & 1 & 2 & 4 \\ 1 & 1 & 1 & 3 \\ 1 & 0 & 3 & 2 \\ \hline 5 & 4 & 15 & 0 \end{bmatrix}$$

The dual problem is stated from this second matrix (using y instead of x).

Minimize $\quad w = 5y_1 + 4y_2 + 15y_3$

subject to: $\quad y_1 + y_2 + 2y_3 \geq 4$

$\qquad\qquad y_1 + y_2 + \; y_3 \geq 3$

$\qquad\qquad y_1 \qquad\; + 3y_3 \geq 2$

with $\qquad y_1 \geq 0, y_2 \geq 0, y_3 \geq 0.$

7. Minimize $\quad w = 3y_1 + 6y_2 + 4y_3 + y_4$

subject to: $\quad y_1 + \; y_2 + \; y_3 + \; y_4 \geq 150$

$\qquad\qquad 2y_1 + 2y_2 + 3y_3 + 4y_4 \geq 275$

with $\qquad y_1 \geq 0, y_2 \geq 0, y_3 \geq 0, y_4 \geq 0.$

To find the dual problem, first write the augmented matrix for the problem.

$$\begin{bmatrix} 1 & 1 & 1 & 1 & 150 \\ 2 & 2 & 3 & 4 & 275 \\ \hline 3 & 6 & 4 & 1 & 0 \end{bmatrix}$$

Then form the transpose of this matrix.

$$\begin{bmatrix} 1 & 2 & 3 \\ 1 & 2 & 6 \\ 1 & 3 & 4 \\ 1 & 4 & 1 \\ \hline 150 & 275 & 0 \end{bmatrix}$$

The dual problem is

Maximize $\quad z = 150x_1 + 275x_2$

subject to: $\quad x_1 + 2x_2 \leq 3$

$\qquad\qquad x_1 + 2x_2 \leq 6$

$\qquad\qquad x_1 + 3x_2 \leq 4$

$\qquad\qquad x_1 + 4x_2 \leq 1$

with $\qquad x_1 \geq 0, \; x_2 \geq 0.$

9. Find $y_1 \geq 0$ and $y_2 \geq 0$ such that

$$2y_1 + 3y_2 \geq 6$$
$$2y_1 + \; y_2 \geq 7$$

and $w = 5y_1 + 2y_2$ is minimized.

Write the augmented matrix for this problem.

$$\begin{bmatrix} 2 & 3 & 6 \\ 2 & 1 & 7 \\ \hline 5 & 2 & 0 \end{bmatrix}$$

Form the transpose of this matrix.

$$\begin{bmatrix} 2 & 2 & 5 \\ 3 & 1 & 2 \\ \hline 6 & 7 & 0 \end{bmatrix}$$

Use this matrix to write the dual problem.

Find $x_1 \geq 0$ and $x_2 \geq 0$ such that

$$2x_1 + 2x_2 \leq 5$$
$$3x_1 + \; x_2 \leq 2$$

and $z = 6x_1 + 7x_2$ is maximized.

Introduce slack variables s_1 and s_2. The initial tableau is as follows.

$$\begin{array}{ccccc} x_1 & x_2 & s_1 & s_2 & z \\ \end{array}$$
$$\begin{bmatrix} 2 & 2 & 1 & 0 & 0 & 5 \\ 3 & \boxed{1} & 0 & 1 & 0 & 2 \\ \hline -6 & -7 & 0 & 0 & 1 & 0 \end{bmatrix}$$

Pivot on the 1 in row 2, column 2, since that column has the most negative indicator and that row has the smallest nonnegative quotient.

$$\begin{array}{ccccc} x_1 & x_2 & s_1 & s_2 & z \\ \end{array}$$
$$\begin{array}{r} -2R_2 + R_1 \rightarrow R_1 \\ \\ 7R_2 + R_3 \rightarrow R_3 \end{array} \begin{bmatrix} -4 & 0 & 1 & -2 & 0 & 1 \\ 3 & 1 & 0 & 1 & 0 & 2 \\ \hline 15 & 0 & 0 & 7 & 1 & 14 \end{bmatrix}$$

The minimum value of w is the same as the maximum value of z. The minimum value of w is 14 when $y_1 = 0$ and $y_2 = 7$. (Note that the values of y_1 and y_2 are given by the entries in the bottom row of the columns corresponding to the slack variables in the final tableau.)

11. Find $y_1 \geq 0$ and $y_2 \geq 0$ such that

$$10y_1 + 5y_2 \geq 100$$
$$20y_1 + 10y_2 \geq 150$$

and $w = 4y_1 + 5y_2$ is minimized.

Write the augmented matrix for this problem.

$$\begin{bmatrix} 10 & 5 & | & 100 \\ 20 & 10 & | & 150 \\ \hline 4 & 5 & | & 0 \end{bmatrix}$$

Form the transpose of this matrix.

$$\begin{bmatrix} 10 & 20 & | & 4 \\ 5 & 10 & | & 5 \\ \hline 100 & 150 & | & 0 \end{bmatrix}$$

Write the dual problem from this matrix.

Find $x_1 \geq 0$ and $x_2 \geq 0$ such that

$$10x_1 + 20x_2 \leq 4$$
$$5x_1 + 10x_2 \leq 5$$

and $z = 100x_1 + 150x_2$ is maximized.

The initial simplex tableau is as follows.

$$\begin{array}{ccccc} x_1 & x_2 & s_1 & s_2 & z \\ \left[\begin{array}{ccccc|c} 10 & \boxed{20} & 1 & 0 & 0 & 4 \\ 5 & 10 & 0 & 1 & 0 & 5 \\ \hline -100 & -150 & 0 & 0 & 1 & 0 \end{array}\right] \end{array}$$

Pivot on the 20 in row 1, column 2.

$$\begin{array}{l} \\ -R_1 + 2R_1 \to R_2 \\ 15R_1 + 2R_3 \to R_3 \end{array} \begin{array}{ccccc} x_1 & x_2 & s_1 & s_2 & z \\ \left[\begin{array}{ccccc|c} \boxed{10} & 20 & 1 & 0 & 0 & 4 \\ 0 & 0 & -1 & 2 & 0 & 6 \\ -50 & 0 & 15 & 0 & 2 & 60 \end{array}\right] \end{array}$$

Pivot on the 10 in row 1, column 1.

$$5R_1 + R_3 \to R_3 \begin{array}{ccccc} x_1 & x_2 & s_1 & s_2 & z \\ \left[\begin{array}{ccccc|c} 10 & 20 & 1 & 0 & 0 & 4 \\ 0 & 0 & -1 & 2 & 0 & 6 \\ 0 & 100 & 20 & 0 & 2 & 80 \end{array}\right] \end{array}$$

Create a 1 in the columns corresponding to x_1, s_2, and z.

$$\begin{array}{l} \dfrac{1}{10}R_1 \to R_1 \\[4pt] \dfrac{1}{2}R_2 \to R_2 \\[4pt] \dfrac{1}{2}R_3 \to R_3 \end{array} \begin{array}{ccccc} x_1 & x_2 & s_1 & s_2 & z \\ \left[\begin{array}{ccccc|c} 1 & 2 & \dfrac{1}{10} & 0 & 0 & \dfrac{2}{5} \\[4pt] 0 & 0 & -\dfrac{1}{2} & 1 & 0 & 3 \\[4pt] 0 & 50 & 10 & 0 & 1 & 40 \end{array}\right] \end{array}$$

The minimum value of w is 40 when $y_1 = 10$ and $y_2 = 0$. (These values of y_1 and y_2 are read from the last row of the columns corresponding to s_1 and s_2 in the final tableau.)

13. Minimize $w = 6y_1 + 10y_2$

subject to: $3y_1 + 5y_2 \geq 15$
               $4y_1 + 7y_2 \geq 20$

with $y_1 \geq 0, \ y_2 \geq 0.$

Write the augmented matrix.

$$\begin{bmatrix} 3 & 5 & | & 15 \\ 4 & 7 & | & 20 \\ \hline 6 & 10 & | & 0 \end{bmatrix}$$

Transpose to get the matrix for the dual problem.

$$\begin{bmatrix} 3 & 4 & | & 6 \\ 5 & 7 & | & 10 \\ \hline 15 & 20 & | & 0 \end{bmatrix}$$

Write the dual problem.

Maximize $z = 15x_1 + 20x_2$

subject to: $3x_1 + 4x_2 \leq 6$
               $5x_1 + 7x_2 \leq 10$

with $x_1 \geq 0, \ x_2 \geq 0.$

Write the initial tableau for this problem.

$$\begin{array}{ccccc} x_1 & x_2 & s_1 & s_2 & z \\ \end{array}$$

$$\left[\begin{array}{ccccc|c} \boxed{3} & 4 & 1 & 0 & 0 & 6 \\ 5 & 7 & 0 & 1 & 0 & 10 \\ \hline -15 & -20 & 0 & 0 & 1 & 0 \end{array}\right]$$

Pivot around the indicated 3 to obtain this final tableau:

$$\begin{array}{ccccc} x_1 & x_2 & s_1 & s_2 & z \\ \end{array}$$

$$\begin{array}{c} \\ -5R_1 + 3R_2 \to R_2 \\ 5R_1 + R_3 \to R_3 \end{array} \left[\begin{array}{ccccc|c} 3 & 4 & 1 & 0 & 0 & 6 \\ 0 & 1 & -5 & 3 & 0 & 0 \\ \hline 0 & 0 & 5 & 0 & 1 & 30 \end{array}\right]$$

Instead we could pivot around the 5 in the first row, second column. This produces the following final tableau:

$$\begin{array}{ccccc} x_1 & x_2 & s_1 & s_2 & z \\ \end{array}$$

$$\begin{array}{c} -3R_2 + 5R_1 \to R_1 \\ \\ 3R_2 + R_3 \to R_3 \end{array} \left[\begin{array}{ccccc|c} 0 & -1 & 5 & -3 & 0 & 6 \\ 5 & 7 & 0 & 1 & 0 & 10 \\ \hline 0 & 1 & 0 & 3 & 1 & 30 \end{array}\right]$$

In the original problem, w has a minimum of 30 when $y_1 = 5$ and $y_2 = 0$ (reading from the first final tableau) or when $y_1 = 0$ and $y_2 = 3$ (reading from the second final tableau). Any point on the line segment between $(5, 0)$ and $(0, 3)$ also gives the minimum of 30.

15. Minimize $w = 2y_1 + y_2 + 3y_3$

subject to: $y_1 + y_2 + y_3 \geq 100$
$$2y_1 + y_2 \qquad \geq 50$$

with $y_1 \geq 0,\ y_2 \geq 0,\ y_3 \geq 0.$

Write the augmented matrix.

$$\left[\begin{array}{ccc|c} 1 & 1 & 1 & 100 \\ 2 & 1 & 0 & 50 \\ 2 & 1 & 3 & 0 \end{array}\right]$$

Form the transpose of this matrix.

$$\left[\begin{array}{cc|c} 1 & 2 & 2 \\ 1 & 1 & 1 \\ 1 & 0 & 3 \\ \hline 100 & 50 & 0 \end{array}\right]$$

The dual problem is as follows.

Maximize $z = 100x_1 + 50x_2$

subject to: $x_1 + 2x_2 \leq 2$
$$x_1 + x_2 \leq 1$$
$$x_1 \qquad \leq 3$$

with $x_1 \geq 0,\ x_2 \geq 0.$

The initial simplex tableau is as follows.

$$\begin{array}{cccccc} x_1 & x_2 & s_1 & s_2 & s_3 & z \\ \end{array}$$

$$\left[\begin{array}{cccccc|c} 1 & 2 & 1 & 0 & 0 & 0 & 2 \\ \boxed{1} & 1 & 0 & 1 & 0 & 0 & 1 \\ 1 & 0 & 0 & 0 & 1 & 0 & 3 \\ \hline -100 & -50 & 0 & 0 & 0 & 1 & 0 \end{array}\right]$$

Pivot on the 1 in row 2, column 1.

$$\begin{array}{cccccc} x_1 & x_2 & s_1 & s_2 & s_3 & z \\ \end{array}$$

$$\begin{array}{c} -R_2 + R_1 \to R_1 \\ \\ -R_2 + R_3 \to R_3 \\ 100R_2 + R_4 \to R_4 \end{array} \left[\begin{array}{cccccc|c} 0 & 1 & 1 & -1 & 0 & 0 & 1 \\ 1 & 1 & 0 & 1 & 0 & 0 & 1 \\ 0 & -1 & 0 & -1 & 1 & 0 & 2 \\ \hline 0 & 50 & 0 & 100 & 0 & 1 & 100 \end{array}\right]$$

The minimum value of w is 100 when $y_1 = 0$, $y_2 = 100$, and $y_3 = 0$.

17. Minimize $z = x_1 + 2x_2$

subject to: $-2x_1 + x_2 \geq 1$
$$x_1 - 2x_2 \geq 1$$

with $x_1 \geq 0,\ x_2 \geq 0.$

A quick sketch of the constraints $-2x_1 + x_2 \geq 1$ and $x_1 - 2x_2 \geq 1$ will verify that the two corresponding half planes do not overlap in the first quadrant of the $x_1 x_2$-plane. Therefore, this problem (P) has no feasible solution. The dual of the given problem is as follows:

Maximize $w = y_1 + y_2$

subject to: $-2y_1 + y_2 \leq 1$
$$y_1 - 2y_2 \leq 2$$

with $y_1 \geq 0,\ y_2 \geq 0.$

A quick sketch here will verify that there is a feasible region in the $y_1 y_2$-plane, and it is unbounded. Therefore, there is no maximum value of w in this problem (D).

(P) has no feasible solution and the objective function of (D) is unbounded; this is choice (a).

19. (a) Let $y_1 = $ the number of units of regular beer

and $y_2 = $ the number of units of light beer.

Minimize $w = 32{,}000y_1 + 50{,}000y_2$

subject to:

$$y_1 \geq 10$$
$$y_2 \geq 15$$
$$y_1 + y_2 \geq 45$$
$$120{,}000y_1 + 300{,}000y_2 \geq 9{,}000{,}000$$
$$y_1 + y_2 \geq 20$$

with $y_1 \geq 0, y_2 \geq 0.$

Write the augmented matrix for this problem, and form the transpose to give the matrix for the dual problem.

$$\left[\begin{array}{cc|c} 1 & 0 & 10 \\ 0 & 1 & 15 \\ 1 & 1 & 45 \\ 120{,}000 & 300{,}000 & 9{,}000{,}000 \\ 1 & 1 & 20 \\ \hline 32{,}000 & 50{,}000 & 0 \end{array}\right]$$

$$\left[\begin{array}{ccccc|c} 1 & 0 & 1 & 120{,}000 & 1 & 32{,}000 \\ 0 & 1 & 1 & 300{,}000 & 1 & 50{,}000 \\ \hline 10 & 15 & 45 & 9{,}000{,}000 & 20 & 0 \end{array}\right]$$

The dual problem is

Maximize $z = 10x_1 + 15x_2 + 45x_3 + 9{,}000{,}000x_4 + 20x_5$

subject to: $x_1 + x_3 + 120{,}000x_4 + x_5 \leq 32{,}000$
$x_2 + x_3 + 300{,}000x_4 + x_5 \leq 50{,}000$

with $x_1 \geq 0, \ x_2 \geq 0, \ x_3 \geq 0, \ x_4 \geq 0, \ x_5 \geq 0.$

Write the initial simplex tableau.

$$\begin{array}{cccccccc} x_1 & x_2 & x_3 & x_4 & x_5 & s_1 & s_2 & z \end{array}$$
$$\left[\begin{array}{cccccccc|c} 1 & 0 & 1 & 120{,}000 & 1 & 1 & 0 & 0 & 32{,}000 \\ 0 & 1 & 1 & \boxed{300{,}000} & 1 & 0 & 1 & 0 & 50{,}000 \\ \hline -10 & -15 & -45 & -9{,}000{,}000 & -20 & 0 & 0 & 1 & 0 \end{array}\right]$$

Pivot on the 300,000 in row 2, column 3.

$$\begin{array}{ccccccccc} & x_1 & x_2 & x_3 & x_4 & x_5 & s_1 & s_2 & z \end{array}$$
$$\begin{array}{c} -2R_2 + 5R_1 \to R_1 \\ \\ 30R_2 + R_3 \to R_3 \end{array}\left[\begin{array}{cccccccc|c} 5 & -2 & \boxed{3} & 0 & 3 & 5 & -2 & 0 & 60{,}000 \\ 0 & 1 & 1 & 300{,}000 & 1 & 0 & 1 & 0 & 50{,}000 \\ \hline -10 & 15 & -15 & 0 & 10 & 0 & 30 & 1 & 1{,}500{,}000 \end{array}\right]$$

Pivot on the 3 in row 1, column 3.

$$
\begin{array}{c}
\\
-R_1+3R_2 \to R_2 \\
5R_1+R_3 \to R_3
\end{array}
\begin{array}{cccccccc}
x_1 & x_2 & x_3 & x_4 & x_5 & s_1 & s_2 & z \\
\end{array}
\left[
\begin{array}{cccccccc|c}
5 & -2 & 3 & 0 & 3 & 5 & -2 & 0 & 60{,}000 \\
-5 & 5 & 0 & 900{,}000 & 0 & -5 & 5 & 0 & 90{,}000 \\
15 & 5 & 0 & 0 & 25 & 25 & 20 & 1 & 1{,}800{,}000
\end{array}
\right]
$$

Create a 1 in the columns corresponding to x_3 and x_4.

$$
\begin{array}{c}
\frac{1}{3}R_1 \to R_1 \\[2mm]
\frac{1}{900{,}000}R_3 \to R_3
\end{array}
\begin{array}{ccccccc}
x_1 & x_2 & x_3 & x_4 & x_5 & s_1 & s_2 & z
\end{array}
\left[
\begin{array}{ccccccc|c}
\frac{5}{3} & -\frac{2}{3} & 1 & 0 & 1 & \frac{5}{3} & -\frac{2}{3} & 0 & 20{,}000 \\[2mm]
-\frac{1}{180{,}000} & \frac{1}{180{,}000} & 0 & 1 & 0 & -\frac{1}{180{,}000} & \frac{1}{180{,}000} & 0 & \frac{1}{10} \\[2mm]
15 & 5 & 0 & 0 & 25 & 25 & 20 & 1 & 1{,}800{,}000
\end{array}
\right]
$$

The minimum value of w is 1,800,000 when $y_1 = 25$ and $y_2 = 20$.

Therefore, 25 units of regular beer and 20 units of light beer should be made for a minimum cost of $1,800,000.

(b) The shadow cost for revenue is $\frac{1}{10}$ dollar or $0.10. An increase in $500,000 in revenue will increase costs to

$$\$1{,}800{,}000 + \$0.10(500{,}000) = \$1{,}850{,}000.$$

21. (a) The initial matrix for the original problem is

$$
\left[
\begin{array}{ccc|c}
1 & 1 & 1 & 100 \\
400 & 160 & 280 & 20{,}000 \\
\hline
120 & 40 & 60 & 0
\end{array}
\right].
$$

The transposed matrix, for the dual problem, is

$$
\left[
\begin{array}{cc|c}
1 & 400 & 120 \\
1 & 160 & 40 \\
1 & 280 & 60 \\
\hline
100 & 20{,}000 & 0
\end{array}
\right].
$$

Minimize $w = 100y_1 + 20{,}000y_2$

subject to: $y_1 + 400y_2 \ge 120$

$\qquad\qquad y_1 + 160y_2 \ge 40$

$\qquad\qquad y_1 + 280y_2 \ge 60$

with $\qquad y_1 \ge 0, y_2 \ge 0.$

(b) We apply the simplex algorithm to the original maximization problem. The initial tableau is

$$
\begin{array}{cccccc}
x_1 & x_2 & x_3 & s_1 & s_2 & z
\end{array}
\left[
\begin{array}{cccccc|c}
1 & 1 & 1 & 1 & 0 & 0 & 100 \\
\boxed{400} & 160 & 280 & 0 & 1 & 0 & 20{,}000 \\
\hline
-120 & -40 & -60 & 0 & 0 & 1 & 0
\end{array}
\right]
$$

Pivot on the 400 in row 2, column 1.

$$
\begin{array}{c}
-R_2 + 400\,R_1 \rightarrow R_1 \\[4pt]
\\[4pt]
\frac{3}{10}R_2 + R_3 \rightarrow R_3
\end{array}
\begin{array}{c}
\begin{array}{cccccc}
x_1 & x_2 & x_3 & s_1 & s_2 & z
\end{array} \\
\left[
\begin{array}{cccccc|c}
0 & 240 & 120 & 400 & -1 & 0 & 20{,}000 \\
400 & 160 & 280 & 0 & 1 & 0 & 20{,}000 \\
0 & 8 & 24 & 0 & 0.3 & 1 & 6000
\end{array}
\right]
\end{array}
$$

Create a 1 in the columns corresponding to x_1 and s_1.

$$
\begin{array}{c}
\frac{1}{400}R_1 \rightarrow R_1 \\[10pt]
\frac{1}{400}R_2 \rightarrow R_2 \\[10pt]
\\
\end{array}
\begin{array}{c}
\begin{array}{cccccc}
x_1 & x_2 & x_3 & s_1 & s_2 & z
\end{array} \\
\left[
\begin{array}{cccccc|c}
0 & 0.6 & 0.3 & 1 & -\dfrac{1}{400} & 0 & 50 \\[6pt]
1 & 0.4 & 0.7 & 0 & \dfrac{1}{400} & 0 & 50 \\[6pt]
0 & 8 & 24 & 0 & 0.3 & 1 & 6000
\end{array}
\right]
\end{array}
$$

This solution is optimal. A maximum profit of \$6000 is achieved by planting 50 acres of potatoes, 0 acres of corn, and 0 acres of cabbage.

From the dual solution, the shadow cost of acreage is 0 and of capital is $\frac{3}{10}$.

$$
\text{New profit} = 6000 + 0(-10) + \left(\frac{3}{10}\right)1000
$$

$$
= \$6300
$$

Now calculate the number of acres of each:

$$
\begin{aligned}
\text{Profit} &= 120\,P + 40\,C + 60\,B \\
6300 &= 120\,P + 40(0) + 60(0) \\
P &= 52.5.
\end{aligned}
$$

The farmer will make a profit of \$6300 by planting 52.5 acres of potatoes and no corn or cabbage.

(c) New profit $= 6000 + 0(10) + \left(\dfrac{3}{10}\right)(-1000)$

$$
= \$5700
$$

Calculate the number of acres of each:

$$
\begin{aligned}
\text{Profit} &= 120\,P + 40\,C + 60\,B \\
5700 &= 120\,P + 40(0) + 60(0) \\
P &= 47.5.
\end{aligned}
$$

The farmer will make a profit of \$5700 by planting 47.5 acres of potatoes and no corn or cabbage.

23. Let $y_1 =$ the number of political interviews
conducted

and $y_2 =$ the number of market interviews
conducted.

The problem is:

Minimize $w = 45y_1 + 55y_2$

subject to: $y_1 + y_2 \geq 8$
$8y_1 + 10y_2 \geq 60$
$6y_1 + 5y_2 \geq 40$

with $y_1 \geq 0, y_2 \geq 0.$

Write the augmented matrix.

$$\begin{bmatrix} 1 & 1 & 8 \\ 8 & 10 & 60 \\ 6 & 5 & 40 \\ \hline 45 & 55 & 0 \end{bmatrix}$$

Transpose to get the matrix for the dual problem.

$$\begin{bmatrix} 1 & 8 & 6 & 45 \\ 1 & 10 & 5 & 55 \\ \hline 8 & 60 & 40 & 0 \end{bmatrix}$$

Write the dual problem:

Maximize $z = 8x_1 + 60x_2 + 40x_3$

subject to: $x_1 + 8x_2 + 6x_3 \leq 45$
$x_1 + 10x_2 + 5x_3 \leq 55$

with $x_1 \geq 0, x_2 \geq 0, x_3 \geq 0.$

Write the initial tableau.

$$\begin{array}{ccccccc} x_1 & x_2 & x_3 & s_1 & s_2 & z & \\ \end{array}$$
$$\begin{bmatrix} 1 & \boxed{8} & 6 & 1 & 0 & 0 & 45 \\ 1 & \boxed{10} & 5 & 0 & 1 & 0 & 55 \\ -8 & -60 & -40 & 0 & 0 & 1 & 0 \end{bmatrix}$$

Pivot on the 10 in row 2, column 2.

$$\begin{array}{c} \\ -4R_2 + 5R_1 \to R_1 \\ \\ 6R_2 + R_3 \to R_3 \end{array} \begin{array}{cccccc} x_1 & x_2 & x_3 & s_1 & s_2 & z \\ \end{array}$$
$$\begin{bmatrix} 1 & 0 & \boxed{10} & 5 & -4 & 0 & 5 \\ 1 & 10 & 5 & 0 & 1 & 0 & 55 \\ -2 & 0 & -10 & 0 & 6 & 1 & 330 \end{bmatrix}$$

Pivot on the 10 in row 1, column 3.

$$\begin{array}{cccccc} x_1 & x_2 & x_3 & s_1 & s_2 & z \\ \end{array}$$
$$\begin{array}{c} \\ -R_1 + 2R_2 \to R_2 \\ R_1 + R_3 \to R_3 \end{array} \begin{bmatrix} \boxed{1} & 0 & 10 & 5 & -4 & 0 & 5 \\ 1 & 20 & 0 & -5 & 6 & 0 & 105 \\ -1 & 0 & 0 & 5 & 2 & 1 & 335 \end{bmatrix}$$

Pivot on the 1 in row 1, column 1.

$$\begin{array}{cccccc} x_1 & x_2 & x_3 & s_1 & s_2 & z \\ \end{array}$$
$$\begin{array}{c} \\ -R_1 + R_2 \to R_2 \\ R_1 + R_3 \to R_3 \end{array} \begin{bmatrix} 1 & 0 & 10 & 5 & -4 & 0 & 5 \\ 0 & 20 & -10 & -10 & \boxed{10} & 0 & 100 \\ 0 & 0 & 10 & 10 & -2 & 1 & 340 \end{bmatrix}$$

Pivot on the 10 in row 2, column 5.

$$\begin{array}{cccccc} x_1 & x_2 & x_3 & s_1 & s_2 & z \\ \end{array}$$
$$\begin{array}{c} 2R_2 + 5R_1 \to R_1 \\ \\ R_2 + 5R_3 \to R_3 \end{array} \begin{bmatrix} 5 & 40 & 30 & 5 & 0 & 0 & 225 \\ 0 & 20 & -10 & -10 & 10 & 0 & 100 \\ 0 & 20 & 40 & 40 & 0 & 5 & 1800 \end{bmatrix}$$

Create a 1 in the columns corresponding to x_1, s_2, and z.

$$\begin{array}{cccccc} x_1 & x_2 & x_3 & s_1 & s_2 & z \\ \end{array}$$
$$\begin{array}{c} \frac{1}{5}R_1 \to R_1 \\ \frac{1}{10}R_2 \to R_2 \\ \frac{1}{5}R_3 \to R_3 \end{array} \begin{bmatrix} 1 & 8 & 6 & 1 & 0 & 0 & 45 \\ 0 & 2 & -1 & -1 & 1 & 0 & 10 \\ 0 & 4 & 8 & 8 & 0 & 1 & 360 \end{bmatrix}$$

The minimum time spent is 360 min when $y_1 = 8$ and $y_2 = 0$, that is, when 8 political interviews and no market interviews are done.

25. Organize the information in a table.

	Units of Nutrient A (per bag)	Units of Nutrient B (per bag)	Cost (per bag)
Feed 1	1	2	$3
Feed 2	3	1	$2
Minimum	7	4	

Let $y_1 =$ the number of bags of feed 1

and $y_2 =$ the number of bags of feed 2.

(a) We want the cost to equal $7 for 7 units of A and 4 units of B exactly. Therefore, use a system of equations rather than a system of inequalities.

$$3y_1 + 2y_2 = 7$$
$$y_1 + 3y_2 = 7$$
$$2y_1 + y_2 = 4$$

Use Gauss-Jordan elimination to solve this system of equations.

$$\begin{bmatrix} 3 & 2 & 7 \\ 1 & 3 & 7 \\ 2 & 1 & 4 \end{bmatrix}$$

$$\begin{matrix} \\ -R_1 + 3R_2 \to R_2 \\ -2R_1 + 3R_3 \to R_3 \end{matrix} \begin{bmatrix} 3 & 2 & 7 \\ 0 & 7 & 14 \\ 0 & -1 & -2 \end{bmatrix}$$

$$\begin{matrix} -2R_2 + 7R_1 \to R_1 \\ \\ R_2 + 7R_3 \to R_3 \end{matrix} \begin{bmatrix} 21 & 0 & 21 \\ 0 & 7 & 14 \\ 0 & 0 & 0 \end{bmatrix}$$

$$\begin{matrix} \frac{1}{21}R_1 \to R_1 \\ \frac{1}{7}R_2 \to R_2 \end{matrix} \begin{bmatrix} 1 & 0 & 1 \\ 0 & 1 & 2 \\ 0 & 0 & 0 \end{bmatrix}$$

Thus, $y_1 = 1$ and $y_2 = 2$, so use 1 bag of feed 1 and 2 bags of feed 2. The cost will be $3(1) + 2(2) = \$7$ as desired. The number of units of A is $1(1) + 3(2) = 7$, and the number of units of B is $2(1) + 1(2) = 4$.

(b)

	Units of Nutrient A (per bag)	Units of Nutrient B (per bag)	Cost (per bag)
Feed 1	1	2	\$3
Feed 2	3	1	\$2
Minimum	5	4	

The problem is:

Minimize $w = 3y_1 + 2y_2$

subject to: $y_1 + 3y_2 \ge 5$
$$2y_1 + y_2 \ge 4$$

with $y_1 \ge 0, y_2 \ge 0$.

The dual problem is as follows.

Maximize $z = 5x_1 + 4x_2$

subject to: $x_1 + 2x_2 \le 3$
$$3x_1 + x_2 \le 2$$

with $x_1 \ge 0,\ x_2 \ge 0.$

The initial tableau is as follows.

$$\begin{array}{ccccc|c} x_1 & x_2 & s_1 & s_2 & z & \\ \hline 1 & 2 & 1 & 0 & 0 & 3 \\ \boxed{3} & 1 & 0 & 1 & 0 & 2 \\ \hline -5 & -4 & 0 & 0 & 1 & 0 \end{array}$$

Pivot as indicated.

$$\begin{matrix} -R_2 + 3R_1 \to R_1 \\ \\ 5R_2 + 3R_3 \to R_3 \end{matrix} \begin{array}{ccccc|c} x_1 & x_2 & s_1 & s_2 & z & \\ 0 & \boxed{5} & 3 & -1 & 0 & 7 \\ 3 & 1 & 0 & 1 & 0 & 2 \\ 0 & -7 & 0 & 5 & 3 & 10 \end{array}$$

$$\begin{matrix} -R_1 + 5R_2 \to R_2 \\ 7R_1 + 5R_3 \to R_3 \end{matrix} \begin{array}{ccccc|c} x_1 & x_2 & s_1 & s_2 & z & \\ 0 & 5 & 3 & -1 & 0 & 7 \\ 15 & 0 & -3 & 6 & 0 & 3 \\ 0 & 0 & 21 & 18 & 15 & 99 \end{array}$$

Create a 1 in the columns corresponding to $x_1, x_2,$ and z.

$$\begin{matrix} \frac{1}{5}R_1 \to R_1 \\ \frac{1}{15}R_2 \to R_2 \\ \frac{1}{15}R_3 \to R_3 \end{matrix} \begin{array}{ccccc|c} x_1 & x_2 & s_1 & s_2 & z & \\ 0 & 1 & \frac{3}{5} & -\frac{1}{5} & 0 & \frac{7}{5} \\ 1 & 0 & -\frac{1}{5} & \frac{2}{5} & 0 & \frac{1}{5} \\ 0 & 0 & \frac{7}{5} & \frac{6}{5} & 1 & \frac{33}{5} \end{array}$$

Reading from the final column of the final tableau, $x_2 = \$1.40$ is the cost of nutrient B and $x_1 = \$0.20$ is the cost of nutrient A. With 5 units of A and 4 units of B, this gives a minimum cost of

$$5(\$0.20) + 4(\$1.40) = \$6.60$$

as given in the lower right corner. $1.4 \left(\text{or } \frac{7}{5}\right)$ bags of feed 1 and $1.2 \left(\text{or } \frac{6}{5}\right)$ bags of feed 2 should be used.

27. Let $y_1 =$ the number of minutes spent walking,

$y_2 =$ the number of minutes spent cycling,

and $y_3 =$ the number of minutes spent swimming.

Minimize $w = y_1 + y_2 + y_3$

subject to: $3.5y_1 + 4y_2 + 8y_3 \ge 1500$
$$y_1 + y_2 \ge 3y_3$$
$$y_1 \ge 30$$

with $\qquad y_1 \geq 0, y_2 \geq 0, y_3 \geq 0.$

The second constraint can be written as

$$y_1 + y_2 - 3y_3 \geq 0.$$

Write the augmented matrix for this problem.

$$\begin{bmatrix} 3.5 & 4 & 8 & | & 1500 \\ 1 & 1 & -3 & | & 0 \\ 1 & 0 & 0 & | & 30 \\ \hline 1 & 1 & 1 & | & 0 \end{bmatrix}$$

Transpose to get the matrix for the dual problem.

$$\begin{bmatrix} 3.5 & 1 & 1 & | & 1 \\ 4 & 1 & 0 & | & 1 \\ 8 & -3 & 0 & | & 1 \\ \hline 1500 & 0 & 30 & | & 0 \end{bmatrix}$$

Write the dual problem.

Maximize $\quad z = 1500x_1 + 30x_3$

subject to: $\quad 3.5x_1 + x_2 + x_3 \leq 1$

$$4x_1 + x_2 \qquad \leq 1$$
$$8x_1 - 3x_2 \qquad \leq 1$$

with $\qquad x_1 \geq 0, s_2 \geq 0, x_3 \geq 0.$

Write the initial simplex tableau.

$$\begin{array}{ccccccc} x_1 & x_2 & x_3 & x_4 & s_1 & s_2 & z \end{array}$$
$$\begin{bmatrix} 3.5 & 1 & 1 & 1 & 0 & 0 & 0 & | & 1 \\ 4 & 1 & 0 & 0 & 1 & 0 & 0 & | & 1 \\ 8 & -3 & 0 & 0 & 0 & 1 & 0 & | & 1 \\ \hline -1500 & 0 & -30 & 0 & 0 & 0 & 1 & | & 0 \end{bmatrix}$$

Using a graphing calculator or computer program, such as Solver in Microsoft Excel, we obtain the optimal answer: 30 minutes walking, 197.25 minutes cycling, and 75.75 minutes swimming for a total minimum time of 303 minutes per week.

29. Let $y_1 =$ the number of units of ingredient I;

$\quad y_2 =$ the number of units of ingredient II;

and $y_3 =$ the number of units of ingredient III.

The problem is:

Minimize $\quad w = 4y_1 + 7y_2 + 5y_3$

subject to: $\quad 4y_1 + y_2 + 10y_3 \geq 10$

$$3y_1 + 2y_2 + y_3 \geq 12$$
$$4y_2 + 5y_3 \geq 20$$

with $\qquad y_1 \geq 0, y_2 \geq 0, y_3 \geq 0.$

The dual problem is as follows.

Maximize $\quad z = 10x_1 + 12x_2 + 20x_3$

subject to: $\quad 4x_1 + 3x_2 \qquad \leq 4$

$$x_1 + 2x_2 + 4x_3 \leq 7$$
$$10x_1 + x_2 + 5x_3 \leq 5$$

with $\qquad x_1 \geq 0, x_2 \geq 0, x_3 \geq 0.$

The initial tableau is as follows.

$$\begin{array}{ccccccc} x_1 & x_2 & x_3 & s_1 & s_2 & s_3 & z \end{array}$$
$$\begin{bmatrix} 4 & 3 & 0 & 1 & 0 & 0 & 0 & | & 4 \\ 1 & 2 & 4 & 0 & 1 & 0 & 0 & | & 7 \\ 10 & 1 & \boxed{5} & 0 & 0 & 1 & 0 & | & 5 \\ \hline -10 & -12 & -20 & 0 & 0 & 0 & 1 & | & 0 \end{bmatrix}$$

Pivot as indicated.

$$\begin{array}{ccccccc} x_1 & x_2 & x_3 & s_1 & s_2 & s_3 & z \end{array}$$
$$\begin{array}{r} \\ -4R_3 + 5R_2 \to R_2 \\ \\ 4R_3 + R_4 \to R_4 \end{array} \begin{bmatrix} 4 & \boxed{3} & 0 & 1 & 0 & 0 & 0 & | & 4 \\ -35 & 6 & 0 & 0 & 5 & -4 & 0 & | & 15 \\ 10 & 1 & 5 & 0 & 0 & 1 & 0 & | & 5 \\ 30 & -8 & 0 & 0 & 0 & 4 & 1 & | & 20 \end{bmatrix}$$

$$\begin{array}{ccccccc} x_1 & x_2 & x_3 & s_1 & s_2 & s_3 & z \end{array}$$
$$\begin{array}{r} \\ -2R_1 + R_2 \to R_2 \\ -R_1 + 3R_3 \to R_3 \\ 8R_1 + 3R_4 \to R_4 \end{array} \begin{bmatrix} 4 & 3 & 0 & 1 & 0 & 0 & 0 & | & 4 \\ -43 & 0 & 0 & -2 & 5 & -4 & 0 & | & 7 \\ 26 & 0 & 15 & -1 & 0 & 3 & 0 & | & 11 \\ 122 & 0 & 0 & 8 & 0 & 12 & 3 & | & 92 \end{bmatrix}$$

Create a 1 in the columns corresponding to x_2, x_3, and z.

$$\begin{array}{ccccccc} x_1 & x_2 & x_3 & s_1 & s_2 & s_3 & z \end{array}$$
$$\begin{array}{r} \frac{1}{3}R_1 \to R_1 \\ \\ \frac{1}{15}R_3 \to R_3 \\ \frac{1}{3}R_4 \to R_4 \end{array} \begin{bmatrix} \frac{4}{3} & 1 & 0 & \frac{1}{3} & 0 & 0 & 0 & | & \frac{4}{3} \\ -43 & 0 & 0 & -2 & 5 & -4 & 0 & | & 7 \\ \frac{26}{15} & 0 & 1 & -\frac{1}{15} & 0 & \frac{1}{5} & 0 & | & \frac{11}{5} \\ \frac{122}{3} & 0 & 0 & \frac{8}{3} & 0 & 4 & 1 & | & \frac{92}{3} \end{bmatrix}$$

From the last row, the minimum value is $\frac{92}{3}$ when

$y_1 = \frac{8}{3}$, $y_2 = 0$, and $y_3 = 4$. The biologist can meet his needs at a minimum cost of \$30.67 by using $\frac{8}{3}$ units of ingredient I and 4 units of ingredient III. (Ingredient II should not be used at all.)

4.4 Nonstandard Problems

Your Turn 1

Minimize $w = 6y_1 + 4y_2$

subject to: $3y_1 + 4y_2 \geq 10$

$9y_1 + 7y_2 \leq 18$

with $y_1 \geq 0, \; y_2 \geq 0.$

Instead we maximize $z = -w = -6y_1 - 4y_2$ subject to the same constraints. Inserting slack and surplus variables produces the following initial tableau.

$$\begin{array}{ccccc|c} y_1 & y_2 & s_1 & s_2 & z & \\ 3 & 4 & -1 & 0 & 0 & 10 \\ \boxed{9} & 7 & 0 & 1 & 0 & 18 \\ \hline 6 & 4 & 0 & 0 & 1 & 0 \end{array}$$

Because s_1 is negative, we choose the positive entry farthest to the left in row 1, which is the 3 in column 1. The entry 9 in this column gives the smallest quotient so we choose 9 as the pivot.

$$\begin{array}{c} \\ -R_2 + 3R_1 \to R_1 \\ \\ -2R_2 + 3R_3 \to R_3 \end{array} \begin{array}{ccccc|c} y_1 & y_2 & s_1 & s_2 & z & \\ 0 & \boxed{5} & -3 & -1 & 0 & 12 \\ 9 & 7 & 0 & 1 & 0 & 18 \\ \hline 0 & -2 & 0 & -2 & 3 & -36 \end{array}$$

s_2 is still negative, so we pivot on the 5 in column 2.

$$\begin{array}{c} \\ -7R_1 + 5R_2 \to R_2 \\ 2R_1 + 5R_3 \to R_3 \end{array} \begin{array}{ccccc|c} y_1 & y_2 & s_1 & s_2 & z & \\ 0 & 5 & -3 & -1 & 0 & 12 \\ 45 & 0 & 21 & \boxed{12} & 0 & 6 \\ \hline 0 & 0 & -6 & -12 & 15 & -156 \end{array}$$

Now we work on the largest negative indicator and pivot on the 12 in column 4.

$$\begin{array}{c} R_2 + 12R_1 \to R_1 \\ \\ R_2 + R_3 \to R_3 \end{array} \begin{array}{ccccc|c} y_1 & y_2 & s_1 & s_2 & z & \\ 45 & 60 & -15 & 0 & 0 & 150 \\ 45 & 0 & 21 & 12 & 0 & 6 \\ \hline 45 & 0 & 15 & 0 & 15 & -150 \end{array}$$

From this we can read the solution: The minimum is $-\left(\frac{-150}{15}\right) = 10$ when $y_1 = 0$ and $y_2 = \frac{150}{60} = \frac{5}{2}.$

Your Turn 2

We start with this tableau.

$$\begin{array}{ccccccccc|c} y_1 & y_2 & y_3 & y_4 & s_1 & s_2 & s_3 & s_4 & z & \\ 0 & 1 & 0 & 1 & 0 & 0 & 0 & -1 & 0 & 16 \\ 0 & 0 & 0 & 0 & 1 & 1 & 1 & 1 & 0 & 0 \\ 1 & 0 & 0 & -1 & 1 & 0 & 0 & 1 & 0 & 12 \\ 0 & 0 & 1 & \boxed{1} & -1 & 0 & -1 & -1 & 0 & 8 \\ \hline 0 & 0 & 0 & -300 & 180 & 0 & 400 & 480 & 1 & -10,640 \end{array}$$

We pivot on the 1 in row 4 of column 4.

$$\begin{array}{c} -R_4 + R_1 \to R_1 \\ \\ R_4 + R_3 \to R_3 \\ \\ 300R_4 + R_5 \to R_5 \end{array} \begin{array}{ccccccccc|c} y_1 & y_2 & y_3 & y_4 & s_1 & s_2 & s_3 & s_4 & z & \\ 0 & 1 & -1 & 0 & 1 & 0 & 1 & 0 & 0 & 8 \\ 0 & 0 & 0 & 0 & \boxed{1} & 1 & 1 & 1 & 0 & 0 \\ 1 & 0 & 1 & 0 & 0 & 0 & -1 & 0 & 0 & 20 \\ 0 & 0 & 1 & 1 & -1 & 0 & -1 & -1 & 0 & 8 \\ \hline 0 & 0 & 300 & 0 & -120 & 0 & 100 & 180 & 1 & -8240 \end{array}$$

Finally we pivot on the 1 in row 2 of column 5.

$$\begin{array}{c} -R_2 + R_1 \to R_1 \\ \\ \\ R_2 + R_4 \to R_4 \\ 120R_2 + R_5 \to R_5 \end{array} \begin{array}{ccccccccc|c} y_1 & y_2 & y_3 & y_4 & s_1 & s_2 & s_3 & s_4 & z & \\ 0 & 1 & -1 & 0 & 0 & -1 & 0 & -1 & 0 & 8 \\ 0 & 0 & 0 & 0 & 1 & 1 & 1 & 1 & 0 & 0 \\ 1 & 0 & 1 & 0 & 0 & 0 & -1 & 0 & 0 & 20 \\ 0 & 0 & 1 & 1 & 0 & 1 & 0 & 0 & 0 & 8 \\ \hline 0 & 0 & 300 & 0 & 0 & 120 & 220 & 300 & 1 & -8240 \end{array}$$

Since there are now no negative indicators this tableau gives the solution:
$y_1 = 20, \; y_2 = 8, \; y_3 = 0, \; y_4 = 8,$ with a minimum cost of $8240.

4.4 Exercises

1. $2x_1 + 3x_2 \leq 8$

$x_1 + 4x_2 \geq 7$

Introduce the slack variable s_1 and the surplus variable s_2 to obtain the following equations:

$$2x_1 + 3x_2 + s_1 \qquad = 8$$
$$x_1 + 4x_2 \qquad - s_2 = 7.$$

3. $2x_1 + x_2 + 2x_3 \leq 50$

$x_1 + 3x_2 + x_3 \geq 35$

$x_1 + 2x_2 \qquad \geq 15$

Introduce the slack variable s_1 and the surplus variables s_2 and s_3 to obtain the following equations:

$$2x_1 + x_2 + 2x_3 + s_1 \qquad\qquad = 50$$
$$x_1 + 3x_2 + x_3 \qquad - s_2 \qquad = 35$$
$$x_1 + 2x_2 \qquad\qquad - s_3 = 15.$$

5. Minimize $w = 3y_1 + 4y_2 + 5y_3$

subject to: $y_1 + 2y_2 + 3y_3 \geq 9$

$y_2 + 2y_3 \geq 8$

$2y_1 + y_2 + 2y_3 \geq 6$

with $y_1 \geq 0, y_2 \geq 0, y_3 \geq 0$.

Change this to a maximization problem by letting $z = -w$. The problem can now be stated equivalently as follows:

Maximize $z = -3y_1 - 4y_2 - 5y_3$

subject to: $y_1 + 2y_2 + 3y_3 \geq 9$

$y_2 + 2y_3 \geq 8$

$2y_1 + y_2 + 2y_3 \geq 6$

with $y_1 \geq 0, y_2 \geq 0, y_3 \geq 0$.

7. Minimize $w = y_1 + 2y_2 + y_3 + 5y_4$

subject to: $y_1 + y_2 + y_3 + y_4 \geq 50$

$3y_1 + y_2 + 2y_3 + y_4 \geq 100$

with $y_1 \geq 0, y_2 \geq 0, y_3 \geq 0, y_4 \geq 0$.

Change this to a maximization problem by letting $z = -w$. The problem can now be stated equivalently as follows:

Maximize $z = -y_1 - 2y_2 - y_3 - 5y_4$

subject to: $y_1 + y_2 + y_3 + y_4 \geq 50$

$3y_1 + y_2 + 2y_3 + y_4 \geq 100$

with $y_1 \geq 0, y_2 \geq 0, y_3 \geq 0, y_4 \geq 0$.

9. Find $x_1 \geq 0$ and $x_2 \geq 0$ such that

$x_1 + 2x_2 \geq 24$

$x_1 + x_2 \leq 40$

and $z = 12x_1 + 10x_2$ is maximized.

Subtracting the surplus variable s_1 and adding the slack variable s_2 leads to the equations

$x_1 + 2x_2 - s_1 \qquad = 24$

$x_1 + x_2 \qquad + s_2 = 40$.

The initial simplex tableau is as follows.

$$
\begin{array}{ccccc|c}
x_1 & x_2 & s_1 & s_2 & z & \\
\hline
\boxed{1} & 2 & -1 & 0 & 0 & 24 \\
1 & 1 & 0 & 1 & 0 & 40 \\
\hline
-12 & -10 & 0 & 0 & 1 & 0
\end{array}
$$

The initial basic solution is not feasible since $s_1 = -24$ is negative, so row transformations must be used. Pivot on the 1 in row 1, column 1, since it is the positive entry that is farthest to the left in the first row (the row containing the -1) and since, in the first column, $\frac{24}{1} = 24$ is a smaller quotient than $\frac{40}{1} = 40$. After row transformations, we obtain the following tableau.

$$
\begin{array}{c}
\\
-R_1 + R_2 \rightarrow R_2 \\
12R_1 + R_3 \rightarrow R_3
\end{array}
\begin{array}{ccccc|c}
x_1 & x_2 & s_1 & s_2 & z & \\
\hline
1 & 2 & -1 & 0 & 0 & 24 \\
0 & -1 & \boxed{1} & 1 & 0 & 16 \\
0 & 14 & -12 & 0 & 1 & 288
\end{array}
$$

The basic solution is now feasible, but the problem is not yet finished since there is a negative indicator. Continue in the usual way. The 1 in column 3 is the next pivot. After row transformations, we get the following tableau.

$$
\begin{array}{c}
R_1 + R_2 \rightarrow R_1 \\
\\
12R_2 + R_3 \rightarrow R_3
\end{array}
\begin{array}{ccccc|c}
x_1 & x_2 & s_1 & s_2 & z & \\
\hline
1 & 1 & 0 & 1 & 0 & 40 \\
0 & -1 & 1 & 1 & 0 & 16 \\
0 & 2 & 0 & 12 & 1 & 480
\end{array}
$$

This is a final tableau since the entries in the last row are all nonnegative. The maximum value is 480 when $x_1 = 40$ and $x_2 = 0$.

11. Find $x_1 \geq 0$, $x_2 \geq 0$, and $x_3 \geq 0$ such that

$x_1 + x_2 + x_3 \leq 150$

$x_1 + x_2 + x_3 \geq 100$

and $z = 2x_1 + 5x_2 + 3x_3$ is maximized.

The initial tableau is as follows.

$$
\begin{array}{cccccc|c}
x_1 & x_2 & x_3 & s_1 & s_2 & z & \\
\hline
1 & 1 & 1 & 1 & 0 & 0 & 150 \\
\boxed{1} & 1 & 1 & 0 & -1 & 0 & 100 \\
\hline
-2 & -5 & -3 & 0 & 0 & 1 & 0
\end{array}
$$

Note that s_1 is a slack variable, while s_2 is a surplus variable. The initial basic solution is not feasible, since $s_2 = -100$ is negative. Pivot on the 1 in row 2, column 1.

$$
\begin{array}{c}
-R_2 + R_1 \rightarrow R_1 \\
\\
2R_2 + R_3 \rightarrow R_3
\end{array}
\begin{array}{cccccc|c}
x_1 & x_2 & x_3 & s_1 & s_2 & z & \\
\hline
0 & 0 & 0 & 1 & 1 & 0 & 50 \\
1 & \boxed{1} & 1 & 0 & -1 & 0 & 100 \\
0 & -3 & -1 & 0 & -2 & 1 & 200
\end{array}
$$

Pivot on the 1 in row 2, column 2.

$$\begin{array}{c} \\ \\ \\ 3R_2 + R_3 \to R_3 \end{array} \begin{array}{cccccc} x_1 & x_2 & x_3 & s_1 & s_2 & z \\ \end{array} \left[\begin{array}{cccccc|c} 0 & 0 & 0 & 1 & \boxed{1} & 0 & 50 \\ 1 & 1 & 1 & 0 & -1 & 0 & 100 \\ 3 & 0 & 2 & 0 & -5 & 1 & 500 \end{array}\right]$$

Pivot on the 1 in row 1, column 5.

$$\begin{array}{c} \\ R_1 + R_2 \to R_2 \\ 5R_1 + R_3 \to R_3 \end{array} \begin{array}{cccccc} x_1 & x_2 & x_3 & s_1 & s_2 & z \\ \end{array} \left[\begin{array}{cccccc|c} 0 & 0 & 0 & 1 & 1 & 0 & 50 \\ 1 & 1 & 1 & 1 & 0 & 0 & 150 \\ 3 & 0 & 2 & 5 & 0 & 1 & 750 \end{array}\right]$$

This is a final tableau. The maximum value is 750 when $x_1 = 0$, $x_2 = 150$, and $x_3 = 0$.

13. Find $x_1 \geq 0$ and $x_2 \geq 0$ such that

$$\begin{aligned} x_1 + x_2 &\leq 100 \\ 2x_1 + 3x_2 &\leq 75 \\ x_1 + 4x_2 &\geq 50 \end{aligned}$$

and $z = 5x_1 - 3x_2$ is maximized.

The initial simplex tableau is

$$\begin{array}{cccccc} x_1 & x_2 & s_1 & s_2 & s_3 & z \\ \end{array} \left[\begin{array}{cccccc|c} 1 & 1 & 1 & 0 & 0 & 0 & 100 \\ \boxed{2} & 3 & 0 & 1 & 0 & 0 & 75 \\ 1 & 4 & 0 & 0 & -1 & 0 & 50 \\ \hline -5 & 3 & 0 & 0 & 0 & 1 & 0 \end{array}\right]$$

The initial basic solution is not feasible since $s_3 = -50$. Pivot on the 2 in row 2, column 1.

$$\begin{array}{c} -R_2 + 2R_1 \to R_1 \\ \\ -R_2 + 2R_3 \to R_3 \\ 5R_2 + 2R_4 \to R_4 \end{array} \begin{array}{cccccc} x_1 & x_2 & x_3 & s_1 & s_2 & z \\ \end{array} \left[\begin{array}{cccccc|c} 0 & -1 & 2 & -1 & 0 & 0 & 125 \\ 2 & 3 & 0 & 1 & 0 & 0 & 75 \\ 0 & \boxed{5} & 0 & -1 & -2 & 0 & 25 \\ 0 & 21 & 0 & 5 & 0 & 2 & 375 \end{array}\right]$$

This solution is still not feasible since $s_3 = -\dfrac{25}{2}$. Pivot on the 5 in row 3, column 2.

$$\begin{array}{c} R_3 + 5R_1 \to R_1 \\ -3R_3 + 5R_2 \to R_2 \\ \\ -21R_3 + 5R_4 \to R_4 \end{array} \begin{array}{cccccc} x_1 & x_2 & x_3 & s_1 & s_2 & z \\ \end{array} \left[\begin{array}{cccccc|c} 0 & 0 & 10 & -6 & -2 & 0 & 650 \\ 10 & 0 & 0 & 8 & 6 & 0 & 300 \\ 0 & \boxed{5} & 0 & -1 & -2 & 0 & 25 \\ \hline 0 & 0 & 0 & 46 & 42 & 10 & 1350 \end{array}\right]$$

Create a 1 in the columns corresponding to x_1, x_2, s_1, and z.

$$\begin{array}{c} \frac{1}{10}R_1 \to R_1 \\ \frac{1}{10}R_2 \to R_2 \\ \frac{1}{5}R_3 \to R_3 \\ \frac{1}{10}R_4 \to R_4 \end{array} \begin{array}{cccccc} x_1 & x_2 & x_3 & s_1 & s_2 & z \\ \end{array} \left[\begin{array}{cccccc|c} 0 & 0 & 1 & -\frac{3}{5} & -\frac{1}{5} & 0 & 65 \\ 1 & 0 & 0 & \frac{4}{5} & \frac{3}{5} & 0 & 30 \\ 0 & 1 & 0 & -\frac{1}{5} & -\frac{2}{5} & 0 & 5 \\ \hline 0 & 0 & 0 & \frac{23}{5} & \frac{21}{5} & 1 & 135 \end{array}\right]$$

This is a final tableau. The maximum is 135 when $x_1 = 30$, $x_2 = 5$.

15. Find $y_1 \geq 0$, $y_2 \geq 0$, and $y_3 \geq 0$ such that

$$\begin{aligned} 5y_1 + 3y_2 + 2y_3 &\leq 150 \\ 5y_1 + 10y_2 + 3y_3 &\geq 90 \end{aligned}$$

and $w = 10y_1 + 12y_2 + 10y_3$ is minimized.

Let $z = -w = -10y - 12y_2 - 10y_3$.
Maximize z.

The initial simplex tableau is

$$\begin{array}{cccccc} y_1 & y_2 & y_3 & s_1 & s_2 & z \\ \end{array} \left[\begin{array}{cccccc|c} 5 & 3 & 2 & 1 & 0 & 0 & 150 \\ \boxed{5} & 10 & 3 & 0 & -1 & 0 & 90 \\ \hline 10 & 12 & 10 & 0 & 0 & 1 & 0 \end{array}\right]$$

The initial basic solution is not feasible since $s_2 = -90$. Pivot on the 5 in row 2, column 1.

$$\begin{array}{c} -R_2 + R_1 \to R_1 \\ \\ -2R_2 + R_3 \to R_3 \end{array} \begin{array}{cccccc} y_1 & y_2 & y_3 & s_1 & s_2 & z \\ \end{array} \left[\begin{array}{cccccc|c} 0 & -7 & -1 & 1 & 1 & 0 & 60 \\ 5 & \boxed{10} & 3 & 0 & -1 & 0 & 90 \\ 0 & -8 & 4 & 0 & 2 & 1 & -180 \end{array}\right]$$

Pivot on the 10 in row 2, column 2.

$$\begin{array}{c} 7R_2 + 10R_1 \to R_1 \\ \\ 8R_2 + 10R_3 \to R_3 \end{array} \begin{array}{cccccc} y_1 & y_2 & y_3 & s_1 & s_2 & z \\ \end{array} \left[\begin{array}{cccccc|c} 35 & 0 & 11 & 10 & 3 & 0 & 1230 \\ 5 & 10 & 3 & 0 & -1 & 0 & 90 \\ 40 & 0 & 64 & 0 & 12 & 10 & -1080 \end{array}\right]$$

Create a 1 in the columns corresponding to y_2, s_1, and z.

$$
\begin{array}{c}
\quad\quad\quad\quad y_1 \;\; y_2 \;\; y_3 \;\; s_1 \quad\quad s_2 \;\; z \\
\begin{array}{l}
\frac{1}{10}R_1 \rightarrow R_1 \\[4pt]
\frac{1}{10}R_2 \rightarrow R_2 \\[4pt]
\frac{1}{10}R_3 \rightarrow R_3
\end{array}
\left[
\begin{array}{cccccc|c}
\frac{7}{2} & 0 & \frac{11}{10} & 1 & \frac{3}{10} & 0 & 123 \\[4pt]
\frac{1}{2} & 1 & \frac{3}{10} & 0 & -\frac{1}{10} & 0 & 9 \\[4pt]
4 & 0 & \frac{32}{5} & 0 & \frac{6}{5} & 1 & -108
\end{array}
\right]
\end{array}
$$

This is a final tableau. The minimum is 108 when $y_1 = 0$, $y_2 = 9$, and $y_3 = 0$.

17. Maximize $z = 3x_1 + 2x_1$

subject to:
$$x_1 + x_2 = 50$$
$$4x_1 + 2x_2 \geq 120$$
$$5x_1 + 2x_2 \leq 200$$

with $x_1 \geq 0, x_2 \geq 0$.

The artificial variable a_1 is used to rewrite $x_1 + x_2 = 50$ as $x_1 + x_2 + a_1 = 50$; note that a_1 must equal 0 for this equation to be a true statement. Also the surplus variable s_1 and the slack variable s_2 are needed. The initial tableau is as follows.

$$
\begin{array}{c}
x_1 \;\; x_2 \;\; a_1 \;\; s_1 \;\; s_2 \;\; z \\
\left[
\begin{array}{cccccc|c}
1 & 1 & 1 & 0 & 0 & 0 & 50 \\
\boxed{4} & 2 & 0 & -1 & 0 & 0 & 120 \\
5 & 2 & 0 & 0 & 1 & 0 & 200 \\
\hline
-3 & -2 & 0 & 0 & 0 & 1 & 0
\end{array}
\right]
\end{array}
$$

The initial basic solution is not feasible. Pivot on the 4 in row 2, column 1.

$$
\begin{array}{c}
\quad\quad\quad\quad\quad x_1 \;\; x_2 \;\; a_1 \;\; s_1 \;\; s_2 \;\; z \\
\begin{array}{l}
-R_2 + 4R_1 \rightarrow R_1 \\[2pt]
\\[2pt]
-5R_2 + 4R_3 \rightarrow R_3 \\[2pt]
3R_2 + 4R_4 \rightarrow R_4
\end{array}
\left[
\begin{array}{cccccc|c}
0 & 2 & 4 & 1 & 0 & 0 & 80 \\
4 & 2 & 0 & -1 & 0 & 0 & 120 \\
0 & -2 & 0 & \boxed{5} & 4 & 0 & 200 \\
0 & -2 & 0 & -3 & 0 & 4 & 360
\end{array}
\right]
\end{array}
$$

The basic solution is now feasible, but there are negative indicators. Pivot on the 5 in row 3, column 4 (which is the column with the most negative indicator and the row with the smallest nonnegative quotient).

$$
\begin{array}{c}
\quad\quad\quad\quad\quad x_1 \;\; x_2 \;\; a_1 \;\; s_1 \;\; s_2 \;\; z \\
\begin{array}{l}
-R_3 + 5R_1 \rightarrow R_1 \\[2pt]
R_3 + 5R_2 \rightarrow R_2 \\[2pt]
\\[2pt]
3R_3 + 5R_4 \rightarrow R_4
\end{array}
\left[
\begin{array}{cccccc|c}
0 & \boxed{12} & 20 & 0 & -4 & 0 & 200 \\
20 & 8 & 0 & 0 & 4 & 0 & 800 \\
0 & -2 & 0 & 5 & 4 & 0 & 200 \\
0 & -16 & 0 & 0 & 12 & 20 & 2400
\end{array}
\right]
\end{array}
$$

Pivot on the 12 in row 1, column 2.

$$
\begin{array}{c}
\quad\quad\quad\quad\quad x_1 \;\; x_2 \;\; a_1 \;\; s_1 \;\; s_2 \;\; z \\
\begin{array}{l}
\\[2pt]
-2R_1 + 3R_2 \rightarrow R_2 \\[2pt]
R_1 + 6R_3 \rightarrow R_3 \\[2pt]
4R_1 + 3R_4 \rightarrow R_4
\end{array}
\left[
\begin{array}{cccccc|c}
0 & 12 & 20 & 0 & -4 & 0 & 200 \\
60 & 0 & -40 & 0 & 20 & 0 & 2000 \\
0 & 0 & 20 & 30 & 20 & 0 & 1400 \\
0 & 0 & 80 & 0 & 20 & 60 & 8000
\end{array}
\right]
\end{array}
$$

We now have $a_1 = 0$, so drop the a_1 column.

$$
\begin{array}{c}
x_1 \;\; x_2 \;\; s_1 \quad s_2 \;\; z \\
\left[
\begin{array}{ccccc|c}
0 & 12 & 0 & -4 & 0 & 200 \\
60 & 0 & 0 & 20 & 0 & 2000 \\
0 & 0 & 30 & 20 & 0 & 1400 \\
\hline
0 & 0 & 0 & 20 & 60 & 8000
\end{array}
\right]
\end{array}
$$

We are finished pivoting. Create a 1 in the columns corresponding to x_1, x_2, s_1, and z.

$$
\begin{array}{c}
\quad\quad\quad\quad\quad x_1 \;\; x_2 \;\; s_1 \quad s_2 \;\; z \\
\begin{array}{l}
\frac{1}{12}R_1 \rightarrow R_1 \\[4pt]
\frac{1}{60}R_2 \rightarrow R_2 \\[4pt]
\frac{1}{30}R_3 \rightarrow R_3 \\[4pt]
\frac{1}{60}R_4 \rightarrow R_4
\end{array}
\left[
\begin{array}{ccccc|c}
0 & 1 & 0 & -\frac{1}{3} & 0 & \frac{50}{3} \\[4pt]
1 & 0 & 0 & \frac{1}{3} & 0 & \frac{100}{3} \\[4pt]
0 & 0 & 1 & \frac{2}{3} & 0 & \frac{140}{3} \\[4pt]
0 & 0 & 0 & \frac{1}{3} & 1 & \frac{400}{3}
\end{array}
\right]
\end{array}
$$

The maximum value is $\frac{400}{3}$ when $x_1 = \frac{100}{3}$ and $x_2 = \frac{50}{3}$.

19. Minimize $w = 32y_1 + 40y_2 + 48y_3$

subject to:
$$20y_1 + 10y_2 + 5y_3 = 200$$
$$25y_1 + 40y_2 + 50y_3 \leq 500$$
$$18y_1 + 24y_2 + 12y_3 \geq 300$$

with $y_1 \geq 0, y_2 \geq 0, y_3 \geq 0$

With artificial, slack, and surplus variables, this problem becomes

Maximize $z = -32y_1 - 40y_2 - 48y_3$

subject to:

$$20y_1 + 10y_2 + 5y_3 + a_1 \quad\quad\quad = 200$$
$$25y_1 + 40y_2 + 50y_3 + \quad s_1 \quad = 500$$
$$18y_1 + 24y_2 + 12y_3 \quad\quad\quad - s_2 = 300.$$

The initial tableau is as follows.

$$\begin{array}{ccccccc|c} y_1 & y_2 & y_3 & a_1 & s_1 & s_2 & z & \\ \boxed{20} & 10 & 5 & 1 & 0 & 0 & 0 & 200 \\ 25 & 40 & 50 & 0 & 1 & 0 & 0 & 500 \\ 18 & 24 & 12 & 0 & 0 & -1 & 0 & 300 \\ \hline 32 & 40 & 48 & 0 & 0 & 0 & 1 & 0 \end{array}$$

The initial basic tableau is not feasible. Pivot on the 20 in row 1, column 1.

$$\begin{array}{r} \\ \\ -5R_1 + 4R_2 \rightarrow R_2 \\ -9R_1 + 10R_3 \rightarrow R_3 \\ -8R_1 + 5R_4 \rightarrow R_4 \end{array} \begin{array}{ccccccc|c} y_1 & y_2 & y_3 & a_1 & s_1 & s_2 & z & \\ 20 & 10 & 5 & 1 & 0 & 0 & 0 & 200 \\ 0 & 110 & 175 & -5 & 4 & 0 & 0 & 1000 \\ 0 & 150 & 75 & -9 & 0 & -10 & 0 & 1200 \\ 0 & 120 & 200 & -8 & 0 & 0 & 5 & -1600 \end{array}$$

Eliminate the a_1 column.

$$\begin{array}{cccccc|c} y_1 & y_2 & y_3 & s_1 & s_2 & z & \\ 20 & 10 & 5 & 0 & 0 & 0 & 200 \\ 0 & 110 & 175 & 4 & 0 & 0 & 1000 \\ 0 & \boxed{150} & 75 & 0 & -10 & 0 & 1200 \\ \hline 0 & 120 & 200 & 0 & 0 & 5 & -1600 \end{array}$$

Pivot on the 150 in row 3, column 2.

$$\begin{array}{r} -R_3 + 15R_1 \rightarrow R_1 \\ -11R_3 + 15R_2 \rightarrow R_2 \\ \\ -4R_3 + 5R_4 \rightarrow R_4 \end{array} \begin{array}{cccccc|c} y_1 & y_2 & y_3 & s_1 & s_2 & z & \\ 300 & 0 & 0 & 0 & 10 & 0 & 1800 \\ 0 & 0 & 1800 & 60 & 110 & 0 & 1800 \\ 0 & 150 & 75 & 0 & -10 & 0 & 1200 \\ 0 & 0 & 700 & 0 & 40 & 25 & -12{,}800 \end{array}$$

Create ones in the columns corresponding to y_1, y_2, s_1, and z.

$$\begin{array}{r} \dfrac{1}{300}R_1 \rightarrow R_1 \\[2mm] \dfrac{1}{60}R_2 \rightarrow R_2 \\[2mm] \dfrac{1}{150}R_3 \rightarrow R_3 \\[2mm] \dfrac{1}{25}R_4 \rightarrow R_4 \end{array} \begin{array}{cccccc|c} y_1 & y_2 & y_3 & s_1 & s_2 & z & \\ 1 & 0 & 0 & 0 & \dfrac{1}{30} & 0 & 6 \\[2mm] 0 & 0 & 30 & 1 & \dfrac{11}{6} & 0 & 30 \\[2mm] 0 & 1 & \dfrac{1}{2} & 0 & -\dfrac{1}{15} & 0 & 8 \\[2mm] 0 & 0 & 28 & 0 & \dfrac{8}{5} & 1 & -512 \end{array}$$

This is a final tableau. The minimum value is 512 when $y_1 = 6$, $y_2 = 8$, and $y_3 = 0$.

23. **(a)** Let $y_1 = $ amount shipped from S_1 to D_1,

   $y_2 = $ amount shipped from S_1 to D_2,

   $y_3 = $ amount shipped from S_2 to D_1,

and $y_4 = $ amount shipped from S_2 to D_2.

Minimize $w = 30y_1 + 20y_2 + 25y_3 + 22y_4$

subject to:
$$\begin{aligned}
y_1 + y_3 &\geq 3000 \\
y_2 + y_4 &\geq 5000 \\
y_1 + y_2 &\leq 5000 \\
y_3 + y_4 &\leq 5000 \\
2y_1 + 6y_2 + 5y_3 + 4y_4 &\leq 40,000
\end{aligned}$$

with $y_1 \geq 0, y_2 \geq 0, y_3 \geq 0, y_4 \geq 0.$

Maximize $z = -w = -30y_1 - 20y_2 - 25y_3 - 22y_4.$

y_1	y_2	y_3	y_4	s_1	s_2	s_3	s_4	s_5	z	
⊡1	0	1	0	−1	0	0	0	0	0	3000
0	1	0	1	0	−1	0	0	0	0	5000
1	1	0	0	0	0	1	0	0	0	5000
0	0	1	1	0	0	0	1	0	0	5000
2	6	5	4	0	0	0	0	1	0	40,000
30	20	25	22	0	0	0	0	0	1	0

Pivot on the 1 in row 1, column 1 since the feasible solution has a negative value, $s_1 = -3000$.

	y_1	y_2	y_3	y_4	s_1	s_2	s_3	s_4	s_5	z	
	1	0	1	0	−1	0	0	0	0	0	3000
	0	1	0	1	0	−1	0	0	0	0	5000
$-R_1 + R_3 \to R_3$	0	⊡1	−1	0	1	0	1	0	0	0	2000
	0	0	1	1	0	0	0	1	0	0	5000
$-2R_1 + R_5 \to R_5$	0	6	3	4	2	0	0	0	1	0	34,000
$-30R_1 + R_6 \to R_6$	0	20	−5	22	30	0	0	0	0	1	−90,000

Since the feasible solution has a negative value ($s_2 = -5000$), pivot on the 1 in row 3, column 2.

	y_1	y_2	y_3	y_4	s_1	s_2	s_3	s_4	s_5	z	
	1	0	1	0	−1	0	0	0	0	0	3000
$-R_3 + R_2 \to R_2$	0	0	1	1	−1	−1	−1	0	0	0	3000
	0	1	−1	0	1	0	1	0	0	0	2000
	0	0	1	1	0	0	0	1	0	0	5000
$-6R_3 + R_5 \to R_5$	0	0	⊡9	4	−4	0	−6	0	1	0	22,000
$-20R_3 + R_6 \to R_6$	0	0	15	22	10	0	−20	0	0	1	−130,000

Since the feasible solution has a negative value ($s_2 = -3000$), pivot on the 9 in row 5, column 3.

$$
\begin{array}{r}
-R_5 + 9R_1 \to R_1 \\
-R_5 + 9R_2 \to R_2 \\
R_5 + 9R_3 \to R_3 \\
-R_5 + 9R_4 \to R_4 \\
\\
-5R_5 + 3R_6 \to R_6
\end{array}
$$

	y_1	y_2	y_3	y_4	s_1	s_2	s_3	s_4	s_5	z	
	9	0	0	-4	-5	0	6	0	-1	0	5000
	0	0	0	[5]	-5	-9	-3	0	-1	0	5000
	0	9	0	4	5	0	3	0	1	0	40,000
	0	0	0	5	4	0	6	9	-1	0	23,000
	0	0	9	4	-4	0	-6	0	1	0	22,000
	0	0	0	46	50	0	-30	0	-5	3	-500,000

Pivot on the 5 in row 2, column 4.

$$
\begin{array}{r}
4R_2 + 5R_1 \to R_1 \\
\\
-4R_2 + 5R_3 \to R_3 \\
-R_2 + R_4 \to R_4 \\
-4R_2 + 5R_5 \to R_5 \\
-46R_2 + 5R_6 \to R_6
\end{array}
$$

	y_1	y_2	y_3	y_4	s_1	s_2	s_3	s_4	s_5	z	
	45	0	0	0	-45	-36	18	0	-9	0	45,000
	0	0	0	5	-5	-9	-3	0	-1	0	5000
	0	45	0	0	45	36	27	0	9	0	180,000
	0	0	0	0	9	9	[9]	9	0	0	18,000
	0	0	45	0	0	36	-18	0	9	0	90,000
	0	0	0	0	480	414	-12	0	21	15	-2,730,000

Pivot on the 9 in row 4, column 7.

$$
\begin{array}{r}
-2R_4 + R_1 \to R_1 \\
R_4 + 3R_2 \to R_2 \\
-3R_4 + R_3 \to R_3 \\
\\
2R_4 + R_5 \to R_5 \\
4R_4 + 3R_6 \to R_6
\end{array}
$$

	y_1	y_2	y_3	y_4	s_1	s_2	s_3	s_4	s_5	z	
	45	0	0	0	-63	-54	0	-18	-9	0	9000
	0	0	0	15	-6	-18	0	9	-3	0	33,000
	0	45	0	0	18	9	0	-27	9	0	126,000
	0	0	0	0	9	9	9	9	0	0	18,000
	0	0	45	0	18	54	0	18	9	0	126,000
	0	0	0	0	1476	450	0	36	63	45	-8,118,000

Create a 1 in the columns corresponding to y_1, y_2, y_3, y_4, and z.

$$
\begin{array}{r}
\frac{1}{45}R_1 \to R_1 \\
\\
\frac{1}{15}R_2 \to R_2 \\
\\
\frac{1}{45}R_3 \to R_3 \\
\\
\\
\frac{1}{45}R_5 \to R_5 \\
\\
\frac{1}{45}R_6 \to R_6
\end{array}
$$

	y_1	y_2	y_3	y_4	s_1	s_2	s_3	s_4	s_5	z	
	1	0	0	0	$-\frac{7}{5}$	$-\frac{6}{5}$	0	$-\frac{2}{5}$	$-\frac{1}{5}$	0	200
	0	0	0	1	$-\frac{2}{5}$	$-\frac{6}{5}$	0	$\frac{3}{5}$	$-\frac{1}{5}$	0	2200
	0	1	0	0	$\frac{2}{5}$	$\frac{1}{5}$	0	$-\frac{3}{5}$	$\frac{1}{5}$	0	2800
	0	0	0	0	9	9	9	9	0	0	18,000
	0	0	1	0	$\frac{2}{5}$	$\frac{6}{5}$	0	$\frac{2}{5}$	$\frac{1}{5}$	0	2800
	0	0	0	0	$\frac{164}{5}$	10	0	$\frac{4}{5}$	$\frac{7}{5}$	1	-180,400

Here, $y_1 = 200$, $y_2 = 2800$, $y_3 = 2800$, $y_4 = 2200$, and $-z = w = 180,400$. So, ship 200 barrels of oil from supplier S_1 to distributor D_1. Ship 2800 barrels of oil from supplier S_1 to distributor D_2. Ship 2800 barrels of oil from supplier S_2 to distributor D_1. Ship 2200 barrels of oil from supplier S_2 to distributor D_2. The minimum cost is $180,400.

(b) From the final tableau, $9s_3 = 18,000$, so $s_3 = 2000$. Therefore, S_1 could furnish 2000 more barrels of oil.

25. Let $x_1 =$ the number of million dollars for home loans

and $x_2 =$ the number of million dollars for commercial loans.

Maximize $z = 0.12x_1 + 0.10x_2$

subject to: $x_1 \geq 4x_2$ or $x_1 - 4x_2 \geq 0$

$\qquad\quad x_1 + x_2 \geq 10$

$\qquad 3x_1 + 2x_2 \leq 72$

$\qquad\quad x_1 + x_2 \leq 25$

with $x_1 \geq 0, x_2 \geq 0.$

$$\begin{array}{c}
\begin{array}{ccccccc} x_1 & x_2 & s_1 & s_2 & s_3 & s_4 & z \end{array} \\
\left[\begin{array}{ccccccc|c}
1 & -4 & -1 & 0 & 0 & 0 & 0 & 0 \\
1 & 1 & 0 & -1 & 0 & 0 & 0 & 10 \\
3 & 2 & 0 & 0 & 1 & 0 & 0 & 72 \\
1 & 1 & 0 & 0 & 0 & 1 & 0 & 25 \\
\hline
-0.12 & -0.10 & 0 & 0 & 0 & 0 & 1 & 0
\end{array}\right]
\end{array}$$

Eliminate the decimals in the last row by multiplying by 100

$$\begin{array}{c}
\begin{array}{ccccccc} x_1 & x_2 & s_1 & s_2 & s_3 & s_4 & z \end{array} \\
\left[\begin{array}{ccccccc|c}
1 & -4 & -1 & 0 & 0 & 0 & 0 & 0 \\
\boxed{1} & 1 & 0 & -1 & 0 & 0 & 0 & 10 \\
3 & 2 & 0 & 0 & 1 & 0 & 0 & 72 \\
1 & 1 & 0 & 0 & 0 & 1 & 0 & 25 \\
\hline
-12 & -10 & 0 & 0 & 0 & 0 & 100 & 0
\end{array}\right]
\end{array}$$

Pivot on the 1 in row 2, column 1.

$$\begin{array}{c}
\hspace{3cm}\begin{array}{ccccccc} x_1 & x_2 & s_1 & s_2 & s_3 & s_4 & z \end{array} \\
\begin{array}{r}
-R_2 + R_1 \rightarrow R_1 \\
\\
-3R_2 + R_3 \rightarrow R_3 \\
-R_2 + R_4 \rightarrow R_4 \\
12R_2 + R_5 \rightarrow R_5
\end{array}
\left[\begin{array}{ccccccc|c}
0 & -5 & -1 & 1 & 0 & 0 & 0 & -10 \\
1 & 1 & 0 & -1 & 0 & 0 & 0 & 10 \\
0 & -1 & 0 & \boxed{3} & 1 & 0 & 0 & 42 \\
0 & 0 & 0 & 1 & 0 & 1 & 0 & 15 \\
0 & 2 & 0 & -12 & 0 & 0 & 100 & 120
\end{array}\right]
\end{array}$$

Pivot on the 3 in row 3, column 4.

$$\begin{array}{c}
\hspace{3cm}\begin{array}{ccccccc} x_1 & x_2 & s_1 & s_2 & s_3 & s_4 & z \end{array} \\
\begin{array}{r}
-R_3 + 3R_1 \rightarrow R_1 \\
R_3 + 3R_2 \rightarrow R_2 \\
\\
-R_3 + 3R_4 \rightarrow R_4 \\
4R_3 + R_5 \rightarrow R_5
\end{array}
\left[\begin{array}{ccccccc|c}
0 & -14 & -3 & 0 & -1 & 0 & 0 & -72 \\
3 & 2 & 0 & 0 & -2 & 0 & 0 & 72 \\
0 & -1 & 0 & 3 & 1 & 0 & 0 & 42 \\
0 & \boxed{1} & 0 & 0 & -1 & 3 & 0 & 3 \\
0 & -2 & 0 & 0 & 4 & 0 & 100 & 288
\end{array}\right]
\end{array}$$

Pivot on the 1 in row 4, column 2.

$$\begin{array}{c}
\hspace{3cm}\begin{array}{ccccccc} x_1 & x_2 & s_1 & s_2 & s_3 & s_4 & z \end{array} \\
\begin{array}{r}
14R_4 + R_1 \rightarrow R_1 \\
-2R_4 + R_2 \rightarrow R_2 \\
R_4 + R_3 \rightarrow R_3 \\
\\
2R_4 + R_5 \rightarrow R_5
\end{array}
\left[\begin{array}{ccccccc|c}
0 & 0 & -3 & 0 & -15 & 42 & 0 & -30 \\
3 & 0 & 0 & 0 & 0 & -6 & 0 & 66 \\
0 & 0 & 0 & 3 & 0 & 3 & 0 & 45 \\
0 & 1 & 0 & 0 & -1 & 3 & 0 & 3 \\
0 & 0 & 0 & 0 & 2 & 6 & 100 & 294
\end{array}\right]
\end{array}$$

Create a 1 in the columns corresponding to x_1 and z.

$$\begin{array}{c}
\hspace{3cm}\begin{array}{ccccccc} x_1 & x_2 & s_1 & s_2 & s_3 & s_4 & z \end{array} \\
\begin{array}{r}
\\
\frac{1}{3}R_2 \rightarrow R_2 \\
\\
\\
\frac{1}{100}R_5 \rightarrow R_5
\end{array}
\left[\begin{array}{ccccccc|c}
0 & 0 & -3 & 0 & -15 & 42 & 0 & -30 \\
1 & 0 & 0 & 0 & 0 & -2 & 0 & 22 \\
0 & 0 & 0 & 3 & 0 & 3 & 0 & 45 \\
0 & 1 & 0 & 0 & -1 & 3 & 0 & 3 \\
0 & 0 & 0 & 0 & 0.02 & 0.06 & 1 & 2.94
\end{array}\right]
\end{array}$$

Here, $x_1 = 22$, $x_2 = 3$, and $z = 2.94$. Make $22 million ($22,000,000) in home loans and $3 million ($3,000,000) in commercial loans for a maximum return of $2.94 million, or $2,940,000.

27. Let $x_1 =$ the number of pounds of bluegrass seed,

$x_2 =$ the number of pounds of rye seed,

and $x_3 =$ the number of pounds of Bermuda seed.

If each batch must contain at least 25% bluegrass seed, then

$$y_1 \geq 0.25(y_1 + y_2 + y_3)$$

$0.75y_1 - 0.25y_2 - 0.25y_3 \geq 0.$

And if the amount of Bermuda must be no more than $\frac{2}{3}$ the amount of rye, then

$$y_3 \leq \frac{2}{3}y_2$$

$-2y_2 + 3y_3 = 0.$

Using these forms for our constraints, we can now state the problem as follows.

Minimize $w = 16y_1 + 14y_2 + 12y_3$

subject to: $0.75y_1 - 0.25y_2 - 0.25y_3 \geq 0$

$\qquad\qquad - 2y_2 + 3y_3 \leq 0$

$\qquad y_1 + y_2 + y_3 \geq 6000$

with $y_1 \geq 0, y_2 \geq 0, y_3 \geq 0.$

The initial simplex tableau is

$$\begin{array}{ccccccc} y_1 & y_2 & y_3 & s_1 & s_2 & a & z \\ \end{array}$$

$$\left[\begin{array}{ccccccc|c} 0.75 & -0.25 & -0.25 & -1 & 0 & 0 & 0 & 0 \\ 0 & -2 & 3 & 0 & 1 & 0 & 0 & 0 \\ \boxed{1} & 1 & 1 & 0 & 0 & 1 & 0 & 6000 \\ \hline 16 & 14 & 12 & 0 & 0 & 0 & 1 & 0 \end{array}\right]$$

First eliminate the artificial variable a. Pivot on the 1 in row 3, column 1.

$$\begin{array}{ccccccc} & y_1 & y_2 & y_3\ s_1\ s_2 & a & z \end{array}$$

$$\begin{array}{r} 0.75R_3 - R_1 \rightarrow R_1 \\ \\ \\ -16R_3 + R_4 \rightarrow R_4 \end{array} \left[\begin{array}{ccccccc|c} 0 & 1 & 1 & 1 & 0 & 0.75 & 0 & 4500 \\ 0 & -2 & 3 & 0 & 1 & 0 & 0 & 0 \\ 1 & 1 & 1 & 0 & 0 & 1 & 0 & 6000 \\ 0 & -2 & -4 & 0 & 0 & -16 & 1 & -96{,}000 \end{array}\right]$$

Since $a = 0$, we can drop the a column.

$$\begin{array}{cccccc} y_1 & y_2 & y_3 & s_1 & s_2 & z \end{array}$$

$$\left[\begin{array}{cccccc|c} 0 & 1 & 1 & 1 & 0 & 0 & 4500 \\ 0 & -2 & \boxed{3} & 0 & 1 & 0 & 0 \\ 1 & 1 & 1 & 0 & 0 & 0 & 6000 \\ \hline 0 & -2 & -4 & 0 & 0 & 1 & -96{,}000 \end{array}\right]$$

Pivot on the 3 in row 2, column 3.

$$\begin{array}{cccccc} & y_1 & y_2 & y_3 & s_1 & s_2 & z \end{array}$$

$$\begin{array}{r} -R_2 + 3R_1 \rightarrow R_1 \\ \\ -R_2 + 3R_3 \rightarrow R_3 \\ 4R_2 + 3R_4 \rightarrow R_4 \end{array} \left[\begin{array}{cccccc|c} 0 & \boxed{5} & 0 & 3 & -1 & 0 & 13{,}500 \\ 0 & -2 & 3 & 0 & 1 & 0 & 0 \\ 3 & 5 & 0 & 0 & -1 & 0 & 18{,}000 \\ 0 & -14 & 0 & 0 & 4 & 3 & -288{,}000 \end{array}\right]$$

Pivot on the 5 in row 1, column 2.

$$\begin{array}{cccccc} & y_1 & y_2 & y_3 & s_1 & s_2 & z \end{array}$$

$$\begin{array}{r} \\ 2R_1 + 5R_2 \rightarrow R_2 \\ -R_1 + R_3 \rightarrow R_3 \\ 14R_1 + 5R_4 \rightarrow R_4 \end{array} \left[\begin{array}{cccccc|c} 0 & 5 & 0 & 3 & -1 & 0 & 13{,}500 \\ 0 & 0 & 15 & 6 & 3 & 0 & 27{,}000 \\ 3 & 0 & 0 & -3 & 0 & 0 & 4500 \\ 0 & 0 & 0 & 42 & 6 & 15 & -1{,}251{,}000 \end{array}\right]$$

Create a 1 in the columns corresponding to y_1, y_2, y_3, and z.

$$\begin{array}{cccccc} & y_1 & y_2 & y_3 & s_1 & s_2 & z \end{array}$$

$$\begin{array}{r} \frac{1}{5}R_1 \rightarrow R_1 \\ \frac{1}{15}R_2 \rightarrow R_2 \\ \frac{1}{3}R_3 \rightarrow R_3 \\ \frac{1}{15}R_4 \rightarrow R_4 \end{array} \left[\begin{array}{cccccc|c} 0 & 1 & 0 & 0.6 & -0.2 & 0 & 2700 \\ 0 & 0 & 1 & 0.4 & 0.2 & 0 & 1800 \\ 1 & 0 & 0 & -1 & 0 & 0 & 1500 \\ 0 & 0 & 0 & 2.8 & 0.4 & 1 & -83{,}400 \end{array}\right]$$

Here, $y_1 = 1500$, $y_2 = 2700$, $y_3 = 1800$, and $z = -w = 83{,}400$. Therefore, use 1500 lb of bluegrass, 2700 lb of rye, and 1800 lb of Bermuda for a minimum cost of \$834.

29. (a) Let $x_1 =$ the number of computers shipped from W_1 to D_1,
$x_2 =$ the number of computers shipped from W_1 to D_2,
$x_3 =$ the number of computers shipped from W_2 to D_1,
and $x_4 =$ the number of computers shipped from W_2 to D_2.

Minimize $\quad w = 14x_1 + 12x_2 + 12x_3 + 10x_4$

subject to:
$$\begin{aligned} x_1 + x_3 &\geq 32 \\ x_2 + x_4 &\geq 20 \\ x_1 + x_2 &\leq 25 \\ x_3 + x_4 &\leq 30 \end{aligned}$$

with $\quad x_1 \geq 0,\ x_2 \geq 0,\ x_3 \geq 0,\ x_4 \geq 0.$

Maximize
$$z = -w = -14x_1 - 12x_2 - 12x_3 - 10x_4.$$

The initial tableau looks like the following.

$$\begin{array}{ccccccccc} x_1 & x_2 & x_3 & x_4 & s_1 & s_2 & s_3 & s_4 & z \end{array}$$

$$\left[\begin{array}{ccccccccc|c} 1 & 0 & 1 & 0 & -1 & 0 & 0 & 0 & 0 & 32 \\ 0 & 1 & 0 & 1 & 0 & -1 & 0 & 0 & 0 & 20 \\ \boxed{1} & 1 & 0 & 0 & 0 & 0 & 1 & 0 & 0 & 25 \\ 0 & 0 & 1 & 1 & 0 & 0 & 0 & 1 & 0 & 30 \\ \hline 14 & 12 & 12 & 10 & 0 & 0 & 0 & 0 & 1 & 0 \end{array}\right]$$

The variable s_1 is negative; we pivot on the 1 in row 3 of column 1.

$$\begin{array}{ccccccccc} x_1 & x_2 & x_3 & x_4 & s_1 & s_2 & s_3 & s_4 & z \end{array}$$

$$\begin{array}{r} -R_3 + R_1 \rightarrow R_1 \\ \\ \\ \\ -14R_3 + R_5 \rightarrow R_5 \end{array} \left[\begin{array}{ccccccccc|c} 0 & -1 & \boxed{1} & 0 & -1 & 0 & -1 & 0 & 0 & 7 \\ 0 & 1 & 0 & 1 & 0 & -1 & 0 & 0 & 0 & 20 \\ 1 & 1 & 0 & 0 & 0 & 0 & 1 & 0 & 0 & 25 \\ 0 & 0 & 1 & 1 & 0 & 0 & 0 & 1 & 0 & 30 \\ \hline 0 & -2 & 12 & 10 & 0 & 0 & -14 & 0 & 1 & -350 \end{array}\right]$$

The variable s_1 is still negative; we pivot on the 1 in row 1 of column 3.

$$\begin{array}{ccccccccc} x_1 & x_2 & x_3 & x_4 & s_1 & s_2 & s_3 & s_4 & z \end{array}$$

$$\begin{array}{r} \\ \\ \\ -R_1 + R_4 \rightarrow R_4 \\ -12R_1 + R_5 \rightarrow R_5 \end{array} \left[\begin{array}{ccccccccc|c} 0 & -1 & 1 & 0 & -1 & 0 & -1 & 0 & 0 & 7 \\ 0 & \boxed{1} & 0 & 1 & 0 & -1 & 0 & 0 & 0 & 20 \\ 1 & 1 & 0 & 0 & 0 & 0 & 1 & 0 & 0 & 25 \\ 0 & 1 & 0 & 1 & 1 & 0 & 1 & 1 & 0 & 23 \\ \hline 0 & 10 & 0 & 10 & 12 & 0 & -2 & 0 & 1 & -434 \end{array}\right]$$

The variable s_2 is still negative; we pivot on the 1 in row 2 of column 2.

$$\begin{array}{c} \\ R_2 + R_1 \to R_1 \\ \\ \\ -R_2 + R_3 \to R_3 \\ \\ -R_2 + R_4 \to R_4 \\ -10R_2 + R_5 \to R_5 \end{array} \begin{array}{cccccccccc} x_1 & x_2 & x_3 & x_4 & s_1 & s_2 & s_3 & s_4 & z & \\ 0 & 0 & 1 & 1 & -1 & -1 & -1 & 0 & 0 & 27 \\ 0 & 1 & 0 & 1 & 0 & -1 & 0 & 0 & 0 & 20 \\ 1 & 0 & 0 & -1 & 0 & 1 & 1 & 0 & 0 & 5 \\ 0 & 0 & 0 & 0 & 1 & 1 & \boxed{1} & 1 & 0 & 3 \\ \hline 0 & 0 & 0 & 0 & 12 & 10 & -2 & 0 & 1 & -634 \end{array}$$

Now we eliminate the only negative indicator by pivoting on the 1 in row 4 of column 7.

$$\begin{array}{c} \\ R_4 + R_1 \to R_1 \\ \\ \\ \\ -R_4 + R_3 \to R_3 \\ \\ \\ 2R_4 + R_5 \to R_5 \end{array} \begin{array}{cccccccccc} x_1 & x_2 & x_3 & x_4 & s_1 & s_2 & s_3 & s_4 & z & \\ 0 & 0 & 1 & 1 & 0 & 0 & 0 & 1 & 0 & 30 \\ 0 & 1 & 0 & 1 & 0 & 0 & 0 & 0 & 0 & 20 \\ 1 & 0 & 0 & -1 & -1 & 0 & 0 & -1 & 0 & 2 \\ 0 & 0 & 0 & 0 & 1 & 1 & 1 & 1 & 0 & 3 \\ \hline 0 & 0 & 0 & 0 & 14 & 12 & 0 & 2 & 1 & -628 \end{array}$$

From this we can read the solution: Ship 2 computers from W_1 to D_1, ship 20 computers from W_1 to D_2, ship 30 computers from W_2 to D_1, and 0 computers from W_2 to D_2. The resulting minimum cost is $628.

(b) From the final tableau, $s_3 = 3$. Therefore, warehouse W_1 has three more computers that it could ship.

31. Let $x_1 = $ the amount of chemical I,

$x_2 = $ the amount of chemical II,

and $x_3 = $ the amount of chemical III.

Minimize $w = 1.09x_1 + 0.87x_2 + 0.65x_3$

subject to: $\qquad x_1 + x_2 + x_3 \geq 750$

$0.09x_1 + 0.04x_2 + 0.03x_3 \geq 30$

$3x_2 = 4x_3$

with $\qquad x_1 \geq 0, \ x_2 \geq 0, \ x_3 \geq 0.$

We follow the suggestion in the note in the text to reduce the number of variables by using the fact that $x_3 = 0.75x_2$ to express our constraints as follows:

Minimize $\quad w = 1.09x_1 + 1.3575x_2$

subject to $\qquad x_1 + 1.75x_2 \geq 750$

$0.09x_1 + 0.0625x_2 \geq 30$

We maximize $z = -w$ and after multiplying the second constraint through by 100, our initial tableau is the following:

$$\begin{array}{ccccc} x_1 & x_2 & s_1 & s_2 & z \\ \left[\begin{array}{ccccc} 1 & 1.75 & -1 & 0 & 0 \\ \boxed{9} & 6.25 & 0 & -100 & 0 \\ \hline 1.09 & 1.3575 & 0 & 0 & 1 \end{array} \right. & & & & \left. \begin{array}{c} 750 \\ 3000 \\ \hline 0 \end{array} \right] \end{array}$$

Since s_1 is negative, we look for a pivot in the first column, and choose 9 because it has the smallest ratio with the corresponding entry in the last column.

$$\begin{array}{c} \\ -R_2 + 9R_1 \to R_1 \\ \\ -1.09R_2 + 9R_3 \to R_3 \end{array} \begin{array}{ccccc} x_1 & x_2 & s_1 & s_2 & z & \\ 0 & \boxed{9.5} & -9 & 100 & 0 & 3750 \\ 9 & 6.25 & 0 & -100 & 0 & 3000 \\ \hline 0 & 5.405 & 0 & 109 & 9 & -3270 \end{array}$$

s_1 is still negative so we pivot on the 9.5 in column 2.

$$\begin{array}{c} \\ -6.25R_1 + 9.5R_2 \to R_2 \\ -5.405R_1 + 9.5R_3 \to R_3 \end{array} \begin{array}{ccccc} x_1 & x_2 & s_1 & s_2 & z & \\ 0 & 9.5 & -9 & 100 & 0 & 3750 \\ 85.5 & 0 & 56.25 & -1575 & 0 & 5062.5 \\ \hline 0 & 0 & 48.645 & 495 & 85.5 & -51,333.75 \end{array}$$

This tableau yields the following solution.

$$x_1 = \frac{5062.5}{85.5} = 59.21, \ x_2 = \frac{3750}{9.5} = 394.74,$$

$$x_3 = \frac{3750}{9.5} \cdot \frac{3}{4} = 296.05$$

$$\text{Minimum} = -\left(\frac{-51,333.75}{85.5} \right) = 600.39$$

So use 59.21 kg of chemical I, 394.74 kg of chemical II, and 296.05 kg of chemical III, for a minimum cost of $600.39.

33. Let $y_1 = $ the number of ounces of ingredient I,

$y_2 = $ the number of ounces of ingredient II,

and $y_3 = $ the number of ounces of ingredient III.

Expressing the problem in cents, the problem is:

Minimize $w = 30y_1 + 9y_2 + 27y_3$

subject to $\quad y_1 + y_2 + y_3 \geq 10$

$y_1 + y_2 + y_3 \leq 15$

$y_1 \geq \frac{1}{4}y_2$

$y_3 \geq y_1$

with $\qquad y_1 \geq 0, \ y_2 \geq 0, \ y_3 \geq 0.$

Rewrite the last two inequalities so that the problem becomes:

Minimize $w = 30y_1 + 9y_2 + 27y_3$

subject to:
$$y_1 + y_2 + y_3 \geq 10$$
$$y_1 + y_2 + y_3 \leq 15$$
$$-4y_1 + y_2 \leq 0$$
$$y_1 - y_3 \leq 0$$

with $\quad y_1 \geq 0, y_2 \geq 0, y_3 \geq 0.$

We maximize $z = -w$ and have the following initial tableau.

$$
\begin{array}{ccccccccc}
y_1 & y_2 & y_3 & s_1 & s_2 & s_3 & s_4 & z & \\
\end{array}
$$
$$
\left[\begin{array}{cccccccc|c}
1 & 1 & 1 & -1 & 0 & 0 & 0 & 0 & 10 \\
1 & 1 & 1 & 0 & 1 & 0 & 0 & 0 & 15 \\
-4 & 1 & 0 & 0 & 0 & 1 & 0 & 0 & 0 \\
\boxed{1} & 0 & -1 & 0 & 0 & 0 & 1 & 0 & 0 \\
\hline
30 & 9 & 27 & 0 & 0 & 0 & 0 & 1 & 0
\end{array}\right]
$$

Because the solution is not feasible ($s_1 = -10$), pivot on the 1 in row 4, column 1.

$$
\begin{array}{ccccccccc}
 & y_1 & y_2 & y_3 & s_1 & s_2 & s_3 & s_4 & z \\
\end{array}
$$
$$
\begin{array}{l}
-R_4 + R_1 \to R_1 \\
-R_4 + R_2 \to R_2 \\
4R_4 + R_3 \to R_3 \\
\\
-30R_4 + R_5 \to R_5
\end{array}
\left[\begin{array}{cccccccc|c}
0 & 1 & 2 & -1 & 0 & 0 & -1 & 0 & 10 \\
0 & 1 & 2 & 0 & 1 & 0 & -1 & 0 & 15 \\
0 & \boxed{1} & -4 & 0 & 0 & 1 & 4 & 0 & 0 \\
1 & 0 & -1 & 0 & 0 & 0 & 1 & 0 & 0 \\
\hline
0 & 9 & 57 & 0 & 0 & 0 & -30 & 1 & 0
\end{array}\right]
$$

Because the solution is still not feasible ($s_1 = -10$), pivot on the 1 in row 3, column 2.

$$
\begin{array}{ccccccccc}
 & y_1 & y_2 & y_3 & s_1 & s_2 & s_3 & s_4 & z \\
\end{array}
$$
$$
\begin{array}{l}
-R_3 + R_1 \to R_1 \\
-R_3 + R_2 \to R_2 \\
\\
\\
-9R_3 + R_5 \to R_5
\end{array}
\left[\begin{array}{cccccccc|c}
0 & 1 & \boxed{6} & -1 & 0 & -1 & -5 & 0 & 10 \\
0 & 1 & 6 & 0 & 1 & -1 & -5 & 0 & 15 \\
0 & 1 & -4 & 0 & 0 & 1 & 4 & 0 & 0 \\
1 & 0 & -1 & 0 & 0 & 0 & 1 & 0 & 0 \\
\hline
0 & 0 & 93 & 0 & 0 & -9 & -66 & 1 & 0
\end{array}\right]
$$

Because the solution is still not feasible ($s_1 = -10$), pivot on the 6 in row 1, column 3.

$$
\begin{array}{ccccccccc}
 & y_1 & y_2 & y_3 & s_1 & s_2 & s_3 & s_4 & z \\
\end{array}
$$
$$
\begin{array}{l}
\\
-R_1 + R_2 \to R_2 \\
2R_1 + 3R_3 \to R_3 \\
R_1 + 6R_4 \to R_4 \\
-31R_3 + 2R_5 \to R_5
\end{array}
\left[\begin{array}{cccccccc|c}
0 & 0 & 6 & -1 & 0 & -1 & -5 & 0 & 10 \\
0 & 0 & 0 & 1 & 1 & 0 & 0 & 0 & 5 \\
0 & 3 & 0 & -2 & 0 & 1 & 2 & 0 & 20 \\
6 & 0 & 0 & -1 & 0 & -1 & 1 & 0 & 10 \\
\hline
0 & 0 & 0 & 31 & 0 & 13 & 23 & 2 & -310
\end{array}\right]
$$

Create a 1 in the columns corresponding to y_1, y_2, y_2, and z.

$$
\begin{array}{ccccccccc}
y_1 & y_2 & y_3 & s_1 & s_2 & s_3 & s_4 & z & \\
\end{array}
$$
$$
\begin{array}{l}
\\
\frac{1}{6}R_2 \to R_2 \\
\frac{1}{3}R_3 \to R_3 \\
\frac{1}{6}R_4 \to R_4 \\
\frac{1}{2}R_5 \to R_5
\end{array}
\left[\begin{array}{cccccccc|c}
0 & 0 & 1 & -\frac{1}{6} & 0 & -\frac{1}{6} & -\frac{5}{6} & 0 & \frac{5}{3} \\
0 & 0 & 0 & 1 & 1 & 0 & 0 & 0 & 5 \\
0 & 1 & 0 & -\frac{2}{3} & 0 & \frac{1}{3} & \frac{2}{3} & 0 & \frac{20}{3} \\
1 & 0 & 0 & -\frac{1}{6} & 0 & -\frac{1}{6} & \frac{1}{6} & 0 & \frac{5}{3} \\
\hline
0 & 0 & 0 & \frac{31}{2} & 0 & \frac{13}{2} & \frac{23}{2} & 1 & -155
\end{array}\right]
$$

Here $y_1 = \dfrac{5}{3}$, $y_2 = \dfrac{20}{3}$, $y_3 = \dfrac{5}{3}$, and $w = -z = 155$.

Therefore, the additive should consist of $\dfrac{5}{3}$ oz of ingredient I, $\dfrac{20}{3}$ oz of ingredient II, and $\dfrac{5}{3}$ oz of ingredient III, for a minimum cost of 155¢/gal, or $1.55/gal. The amount of additive that should be used per gallon of gasoline is $\dfrac{5}{3} + \dfrac{20}{3} + \dfrac{5}{3} = 10$ oz.

Chapter 4 Review Exercises

1. True

2. False

3. True

4. False

5. False

6. True

7. True

8. False

9. False

10. True

11. False

12. True

13. False

14. True

15. The simplex method should be used for problems with more than two variables or problems with two variables and many constants.

17. (a) Maximize $z = 2x_1 + 7x_2$

subject to: $4x_1 + 6x_2 \leq 60$
$3x_1 + x_2 \leq 18$
$2x_1 + 5x_2 \leq 20$
$x_1 + x_2 \leq 15$

with $x_1 \geq 0, x_2 \geq 0$.

Adding slack variables s_1, s_2, s_3, and s_4, we obtain the following equations.

$4x_1 + 6x_2 + s_1 \qquad\qquad = 60$
$3x_1 + x_2 \qquad + s_2 \qquad\quad = 18$
$2x_1 + 5x_2 \qquad\qquad + s_3 \qquad = 20$
$x_1 + x_2 \qquad\qquad\qquad + s_4 = 15.$

(b) The initial simplex tableau is as follows.

$$\begin{array}{ccccccc|c}
x_1 & x_2 & s_1 & s_2 & s_3 & s_4 & z & \\
\hline
4 & 6 & 1 & 0 & 0 & 0 & 0 & 60 \\
3 & 1 & 0 & 1 & 0 & 0 & 0 & 18 \\
2 & 5 & 0 & 0 & 1 & 0 & 0 & 20 \\
1 & 1 & 0 & 0 & 0 & 1 & 0 & 15 \\
\hline
-2 & -7 & 0 & 0 & 0 & 0 & 1 & 0
\end{array}$$

19. Maximize $z = 5x_1 + 8x_2 + 6x_3$

subject to: $x_1 + x_2 + x_3 \leq 90$
$2x_1 + 5x_2 + x_3 \leq 120$
$x_1 + 3x_2 \geq 80$

with $x_1 \geq 0, x_2 \geq 0, x_3 \geq 0$.

(a) Adding the slack variables s_1 and s_2 and subtracting the surplus variable s_3, we obtain the following equations:

$x_1 + x_2 + x_3 + s_1 \qquad\qquad = 90$
$2x_1 + 5x_2 + x_3 \qquad + s_2 \qquad = 120$
$x_1 + 3x_2 \qquad\qquad\qquad - s_3 = 80.$

(b) The initial tableau is

$$\begin{array}{ccccccc|c}
x_1 & x_2 & x_3 & s_1 & s_2 & s_3 & z & \\
\hline
1 & 1 & 1 & 1 & 0 & 0 & 0 & 90 \\
2 & 5 & 1 & 0 & 1 & 0 & 0 & 120 \\
1 & 3 & 0 & 0 & 0 & -1 & 0 & 80 \\
\hline
-5 & -8 & -6 & 0 & 0 & 0 & 1 & 0
\end{array}$$

21.

$$\begin{array}{cccccc|c}
x_1 & x_2 & x_3 & s_1 & s_2 & z & \\
\hline
4 & 5 & 2 & 1 & 0 & 0 & 18 \\
2 & 8 & \boxed{6} & 0 & 1 & 0 & 24 \\
\hline
-5 & -3 & -6 & 0 & 0 & 1 & 0
\end{array}.$$

The most negative entry in the last row is -6, and the smaller of the two quotients is $\frac{24}{6} = 4$. Hence, the 6 in row 2, column 3, is the first pivot. Performing row transformations leads to the following tableau.

$$\begin{array}{c}
-R_2 + 3R_1 \rightarrow R_1 \\
\\
R_2 + R_3 \rightarrow R_3
\end{array}
\begin{array}{cccccc|c}
x_1 & x_2 & x_3 & s_1 & s_2 & z & \\
\hline
\boxed{10} & 7 & 0 & 3 & -1 & 0 & 30 \\
2 & 8 & 6 & 0 & 1 & 0 & 24 \\
\hline
-3 & 5 & 0 & 0 & 1 & 1 & 24
\end{array}.$$

Pivot on the 10 in row 1, column 1.

$$\begin{array}{c}
\\
-R_1 + 5R_2 \rightarrow R_2 \\
3R_1 + 10R_3 \rightarrow R_3
\end{array}
\begin{array}{cccccc|c}
x_1 & x_2 & x_3 & s_1 & s_2 & z & \\
\hline
10 & 7 & 0 & 3 & -1 & 0 & 30 \\
0 & 33 & 30 & -3 & 6 & 0 & 90 \\
0 & 71 & 0 & 9 & 7 & 10 & 330
\end{array}.$$

Create a 1 in the columns corresponding to x_1, x_3, and z.

$$\begin{array}{c}
\frac{1}{10}R_1 \rightarrow R_1 \\
\frac{1}{30}R_2 \rightarrow R_2 \\
\frac{1}{10}R_3 \rightarrow R_3
\end{array}
\begin{array}{cccccc|c}
x_1 & x_2 & x_3 & s_1 & s_2 & z & \\
\hline
1 & \frac{7}{10} & 0 & \frac{3}{10} & -\frac{1}{10} & 0 & 3 \\
0 & \frac{11}{10} & 1 & -\frac{1}{10} & \frac{1}{5} & 0 & 3 \\
0 & \frac{71}{10} & 0 & \frac{9}{10} & \frac{7}{10} & 10 & 33
\end{array}.$$

The maximum value is 33 when $x_1 = 3$, $x_2 = 0$, $x_3 = 3$, $s_1 = 0$, and $s_2 = 0$.

23.

$$\begin{array}{ccccccc|c}
x_1 & x_2 & x_3 & s_1 & s_2 & s_3 & z & \\
\hline
1 & 2 & 2 & 1 & 0 & 0 & 0 & 50 \\
\boxed{3} & 1 & 0 & 0 & 1 & 0 & 0 & 20 \\
1 & 0 & 2 & 0 & 0 & -1 & 0 & 15 \\
\hline
-5 & -3 & -2 & 0 & 0 & 0 & 1 & 0
\end{array}$$

The initial basic solution is not feasible since $s_3 = -15$. In the third row where the negative coefficient appears, the nonnegative entry that appears farthest to the left is the 1 in the first column. In the first column, the smallest nonnegative quotient is $\frac{20}{3}$. Pivot on the 3 in row 2, column 1.

$$
\begin{array}{c}
\\
-R_2 + 3R_1 \rightarrow R_1 \\
\\
-R_2 + 3R_3 \rightarrow R_3 \\
5R_2 + 3R_4 \rightarrow R_4
\end{array}
\begin{array}{ccccccc|c}
x_1 & x_2 & x_3 & s_1 & s_2 & s_3 & z & \\
0 & 5 & 6 & 3 & -1 & 0 & 0 & 130 \\
3 & 1 & 0 & 0 & 1 & 0 & 0 & 20 \\
0 & -1 & \boxed{6} & 0 & -1 & -3 & 0 & 25 \\
0 & -4 & -6 & 0 & 5 & 0 & 3 & 100
\end{array}
$$

Continue by pivoting on each boxed entry.

$$
\begin{array}{c}
\\
-R_3 + R_2 \rightarrow R_1 \\
\\
\\
R_3 + R_4 \rightarrow R_4
\end{array}
\begin{array}{ccccccc|c}
x_1 & x_2 & x_3 & s_1 & s_2 & s_3 & z & \\
0 & \boxed{6} & 0 & 3 & 0 & 3 & 0 & 105 \\
3 & 1 & 0 & 0 & 1 & 0 & 0 & 20 \\
0 & -1 & 6 & 0 & -1 & -3 & 0 & 25 \\
0 & -5 & 0 & 0 & 4 & -3 & 3 & 125
\end{array}
$$

The basic solution is now feasible, but there are negative indicators.

Continue pivoting.

$$
\begin{array}{c}
\\
-R_1 + 6R_2 \rightarrow R_2 \\
R_1 + 6R_3 \rightarrow R_3 \\
5R_1 + 6R_4 \rightarrow R_4
\end{array}
\begin{array}{ccccccc|c}
x_1 & x_2 & x_3 & s_1 & s_2 & s_3 & z & \\
0 & 6 & 0 & 3 & 0 & \boxed{3} & 0 & 105 \\
18 & 0 & 0 & -3 & 6 & -3 & 0 & 15 \\
0 & 0 & 36 & 3 & 0 & -15 & 0 & 255 \\
0 & 0 & 0 & 15 & 24 & -3 & 18 & 1275
\end{array}
$$

$$
\begin{array}{c}
\\
R_1 + R_2 \rightarrow R_2 \\
5R_1 + R_3 \rightarrow R_3 \\
R_1 + R_4 \rightarrow R_4
\end{array}
\begin{array}{ccccccc|c}
x_1 & x_2 & x_3 & s_1 & s_2 & s_3 & z & \\
0 & 6 & 0 & 3 & 0 & 3 & 0 & 105 \\
18 & 6 & 0 & 0 & 6 & 0 & 0 & 120 \\
0 & 30 & 36 & 18 & 0 & 0 & 0 & 780 \\
0 & 6 & 0 & 18 & 24 & 0 & 18 & 1380
\end{array}
$$

Create a 1 in the columns corresponding to x_1, x_3, s_3, and z.

$$
\begin{array}{c}
\frac{1}{3}R_1 \rightarrow R_1 \\
\frac{1}{18}R_2 \rightarrow R_2 \\
\frac{1}{36}R_3 \rightarrow R_3 \\
\frac{1}{18}R_4 \rightarrow R_4
\end{array}
\begin{array}{ccccccc|c}
x_1 & x_2 & x_3 & s_1 & s_2 & s_3 & z & \\
0 & 2 & 0 & 1 & 0 & 1 & 0 & 35 \\
1 & .33 & 0 & 0 & .33 & 0 & 0 & 6.67 \\
0 & .83 & 1 & .5 & 0 & 0 & 0 & 21.67 \\
0 & .33 & 0 & 1 & 1.33 & 0 & 1 & 76.67
\end{array}
$$

The maximum value is about 76.67 when $x_1 \approx 6.67$, $x_2 = 0$, $x_3 \approx 21.67$, $s_1 = 0$, $s_2 = 0$, and $s_3 = 35$.

25. Minimize $w = 10y_1 + 15y_2$

subject to: $y_1 + y_2 \geq 17$
$$5y_1 + 8y_2 \geq 42$$

with $y_1 \geq 0, y_2 \geq 0.$

Using the dual method:

To form the dual, write the augmented matrix for the given problem.

$$
\begin{bmatrix}
1 & 1 & 17 \\
5 & 8 & 42 \\
\hline
10 & 15 & 0
\end{bmatrix}
$$

Form the transpose of this matrix.

$$
\begin{bmatrix}
1 & 5 & 10 \\
1 & 8 & 15 \\
\hline
17 & 42 & 0
\end{bmatrix}
$$

Write the dual problem.

Maximize $z = 17x_1 + 42x_2$

subject to: $x_1 + 5x_2 \leq 10$
$$x_1 + 8x_2 \leq 15$$

with $x_1 \geq 0, x_2 \geq 0.$

The initial simplex tableau is as follows.

$$
\begin{array}{ccccc|c}
x_1 & x_2 & s_1 & s_2 & z & \\
1 & 5 & 1 & 0 & 0 & 10 \\
1 & \boxed{8} & 0 & 1 & 0 & 15 \\
\hline
-17 & -42 & 0 & 0 & 1 & 0
\end{array}
$$

Pivot on the 8 in row 2 column 2.

$$
\begin{array}{c}
-5R_2 + 8R_1 \rightarrow R_1 \\
\\
21R_2 + 4R_3 \rightarrow R_3
\end{array}
\begin{array}{ccccc|c}
x_1 & x_2 & s_1 & s_2 & z & \\
\boxed{3} & 0 & 8 & -5 & 0 & 5 \\
1 & 8 & 0 & 1 & 0 & 15 \\
\hline
-47 & 0 & 0 & 21 & 4 & 315
\end{array}
$$

Pivot on the 3 in row 1, column 1.

$$
\begin{array}{c}
\\
-R_1 + 3R_2 \rightarrow R_2 \\
47R_1 + 3R_3 \rightarrow R_3
\end{array}
\begin{array}{ccccc|c}
x_1 & x_2 & s_1 & s_2 & z & \\
3 & 0 & 8 & -5 & 0 & 5 \\
0 & 24 & -8 & \boxed{8} & 0 & 40 \\
\hline
0 & 0 & 376 & -172 & 12 & 1180
\end{array}
$$

Pivot on the 8 in row 2, column 4.

$$
\begin{array}{c}
5R_2 + 8R_1 \rightarrow R_1 \\
\\
43R_2 + 2R_3 \rightarrow R_3
\end{array}
\begin{array}{ccccc|c}
x_1 & x_2 & s_1 & s_2 & z & \\
24 & 120 & 24 & 0 & 0 & 240 \\
0 & 24 & -8 & 8 & 0 & 40 \\
\hline
0 & 1032 & 408 & 0 & 24 & 4080
\end{array}
$$

Create a 1 in the columns corresponding to x_1, x_2, and z.

$$\frac{1}{24}R_1 \rightarrow R_1$$
$$\frac{1}{8}R_2 \rightarrow R_2$$
$$\frac{1}{24}R_3 \rightarrow R_3$$

$$\begin{array}{ccccc}x_1 & x_2 & s_1 & s_2 & z\end{array}$$
$$\begin{bmatrix} 1 & 5 & 1 & 0 & 0 & 10 \\ 0 & 3 & -1 & 1 & 0 & 5 \\ 0 & 43 & 17 & 0 & 1 & 170 \end{bmatrix}$$

The minimum value is 170 when $y_1 = 17$ and $y_2 = 0$.

Using the method of 4.4:

Change the objective function to

$$\text{Maximize } z = -w = -10y_1 - 15y_2.$$

The constraints are not changed.

The initial simplex tableau is as follows.

$$\begin{array}{ccccc}y_1 & y_2 & s_1 & s_2 & z\end{array}$$
$$\begin{bmatrix} 1 & 1 & -1 & 0 & 0 & 17 \\ \boxed{5} & 8 & 0 & -1 & 0 & 42 \\ \hline 10 & 15 & 0 & 0 & 1 & 0 \end{bmatrix}$$

The solution is not feasible since $s_1 = -17$ and $s_2 = -42$. Pivot on the 5 in row 2, column 1.

$$\begin{array}{ccccc}y_1 & y_2 & s_1 & s_2 & z\end{array}$$
$$-R_2 + 5R_1 \rightarrow R_1$$
$$-2R_2 + R_3 \rightarrow R_3$$
$$\begin{bmatrix} 0 & -3 & -5 & 1 & 0 & 43 \\ 5 & 8 & 0 & -1 & 0 & 42 \\ \hline 0 & -1 & 0 & 2 & 1 & -84 \end{bmatrix}$$

The solution is still not feasible since $s_1 = -\dfrac{43}{5}$.

But there are no positive entries to the left of the -5 in column 3 so it is not possible to choose a pivot element. The method of 4.4 fails to provide a solution in this case.

27. Minimize $w = 7y_1 + 2y_2 + 3y_3$

subject to: $y_1 + y_2 + 2y_3 \geq 48$
$$y_1 + y_2 \qquad \geq 12$$
$$y_3 \geq 10$$
$$3y_1 \qquad + y_3 \geq 30$$

with $y_1 \geq 0, y_2 \geq 0, y_3 \geq 0.$

Using the dual method:

To form the dual, write the augmented matrix for the given problem.

$$\begin{bmatrix} 1 & 1 & 2 & 48 \\ 1 & 1 & 0 & 12 \\ 0 & 0 & 1 & 10 \\ 3 & 0 & 1 & 30 \\ \hline 7 & 2 & 3 & 0 \end{bmatrix}$$

Form the transpose of this matrix.

$$\begin{bmatrix} 1 & 1 & 0 & 3 & 7 \\ 1 & 1 & 0 & 0 & 2 \\ 2 & 0 & 1 & 1 & 3 \\ \hline 48 & 12 & 10 & 30 & 0 \end{bmatrix}$$

Write the dual problem.

Maximize $z = 48x_1 + 12x_2 + 10x_3 + 30x_4$

subject to: $x_1 + x_2 \qquad + 3x_4 \leq 7$
$$x_1 + x_2 \qquad \leq 2$$
$$2x_1 + \quad x_3 + x_4 \leq 3$$

with $x_1 \geq 0, x_2 \geq 0, x_3 \geq 0, x_4 \geq 0.$

The initial simplex tableau is as follows.

$$\begin{array}{cccccccc}x_1 & x_2 & x_3 & x_4 & s_1 & s_2 & s_3 & z\end{array}$$
$$\begin{bmatrix} 1 & 1 & 0 & 3 & 1 & 0 & 0 & 0 & 7 \\ 1 & 1 & 0 & 0 & 0 & 1 & 0 & 0 & 2 \\ \boxed{2} & 0 & 1 & 1 & 0 & 0 & 1 & 0 & 3 \\ \hline -48 & -12 & -10 & -30 & 0 & 0 & 0 & 1 & 0 \end{bmatrix}$$

Pivot on the 2 in row 3, column 1.

$$\begin{array}{cccccccc}x_1 & x_2 & x_3 & x_4 & s_1 & s_2 & s_3 & z\end{array}$$
$$-R_3 + 2R_1 \rightarrow R_1$$
$$-R_3 + 2R_2 \rightarrow R_2$$
$$24R_3 + R_4 \rightarrow R_4$$
$$\begin{bmatrix} 0 & 2 & -1 & 5 & 2 & 0 & -1 & 0 & 11 \\ 0 & \boxed{2} & -1 & -1 & 0 & 2 & -1 & 0 & 1 \\ 2 & 0 & 1 & 1 & 0 & 0 & 1 & 0 & 3 \\ \hline 0 & -12 & 14 & -6 & 0 & 0 & 24 & 1 & 72 \end{bmatrix}$$

Pivot on the 2 in row 2, column 2.

$$\begin{array}{cccccccc}x_1 & x_2 & x_3 & x_4 & s_1 & s_2 & s_3 & z\end{array}$$
$$-R_2 + R_1 \rightarrow R_1$$
$$6R_2 + R_4 \rightarrow R_4$$
$$\begin{bmatrix} 0 & 0 & 0 & \boxed{6} & 2 & -2 & 0 & 0 & 10 \\ 0 & 2 & -1 & -1 & 0 & 2 & -1 & 0 & 1 \\ 2 & 0 & 1 & 1 & 0 & 0 & 1 & 0 & 3 \\ \hline 0 & 0 & 8 & -12 & 0 & 12 & 18 & 1 & 78 \end{bmatrix}$$

Pivot on the 6 in row 1, column 4.

$$\begin{array}{l} \\ R_1 + 6R_2 \to R_2 \\ -R_1 + 6R_3 \to R_3 \\ 2R_1 + R_4 \to R_4 \end{array}
\begin{array}{cccccccc|c} x_1 & x_2 & x_3 & x_4 & s_1 & s_2 & s_3 & z & \\ 0 & 0 & 0 & 6 & 2 & -2 & 0 & 0 & 10 \\ 0 & 12 & -6 & 0 & 2 & 10 & -6 & 0 & 16 \\ 12 & 0 & 6 & 0 & -2 & 2 & 6 & 0 & 8 \\ 0 & 0 & 8 & 0 & 4 & 8 & 18 & 1 & 98 \end{array}$$

Create a 1 in the columns corresponding to x_1, x_2, and x_4.

$$\begin{array}{l} \frac{1}{6}R_1 \to R_1 \\ \frac{1}{12}R_2 \to R_2 \\ \frac{1}{12}R_3 \to R_3 \\ \end{array}
\begin{array}{cccccccc|c} x_1 & x_2 & x_3 & x_4 & s_1 & s_2 & s_3 & z & \\ 0 & 0 & 0 & 1 & \frac{1}{3} & -\frac{1}{3} & 0 & 0 & \frac{5}{3} \\ 0 & 1 & -\frac{1}{2} & 0 & \frac{1}{6} & \frac{5}{6} & -\frac{1}{2} & 0 & \frac{4}{3} \\ 1 & 0 & \frac{1}{2} & 0 & -\frac{1}{6} & \frac{1}{6} & \frac{1}{2} & 0 & \frac{2}{3} \\ \hline 0 & 0 & 8 & 0 & 4 & 8 & 18 & 1 & 98 \end{array}$$

The minimum value is 98 when $y_1 = 4$, $y_2 = 8$, and $y_3 = 18$.

<u>Using the method of 4.4:</u>

Change the objective function to

$$\text{Maximize } z = -w = -7y_1 - 2y_2 - 3y_3.$$

The constraints are not changed.

The initial simplex tableau is as follows.

$$\begin{array}{cccccccc|c} y_1 & y_2 & y_3 & s_1 & s_2 & s_3 & s_4 & z & \\ 1 & 1 & 2 & -1 & 0 & 0 & 0 & 0 & 48 \\ 1 & 1 & 0 & 0 & -1 & 0 & 0 & 0 & 12 \\ 0 & 0 & 1 & 0 & 0 & -1 & 0 & 0 & 10 \\ \boxed{3} & 0 & 1 & 0 & 0 & 0 & -1 & 0 & 30 \\ \hline 7 & 2 & 3 & 0 & 0 & 0 & 0 & 1 & 0 \end{array}$$

The solution is not feasible since $s_1 = -48$, $s_2 = -12$, $s_3 = -10$, and $s_4 = -30$. Pivot on the 3 in row 4, column 1.

$$\begin{array}{l} -R_4 + 3R_1 \to R_1 \\ -R_4 + 3R_2 \to R_2 \\ \\ \\ -7R_4 + 3R_5 \to R_5 \end{array}
\begin{array}{cccccccc|c} y_1 & y_2 & y_3 & s_1 & s_2 & s_3 & s_4 & z & \\ 0 & \boxed{3} & 5 & -3 & 0 & 0 & 1 & 0 & 114 \\ 0 & 3 & -1 & 0 & -3 & 0 & 1 & 0 & 6 \\ 0 & 0 & 1 & 0 & 0 & -1 & 0 & 0 & 10 \\ 3 & 0 & 1 & 0 & 0 & 0 & -1 & 0 & 30 \\ \hline 0 & 6 & 2 & 0 & 0 & 0 & 7 & 3 & -210 \end{array}$$

The solution is still not feasible since $s_1 = -38$, $s_2 = -2$, and $s_3 = -10$. Pivot on the 3 in row 1, column 2.

$$\begin{array}{l} \\ R_1 - R_2 \to R_2 \\ \\ \\ -2R_1 + R_5 \to R_5 \end{array}
\begin{array}{cccccccc|c} y_1 & y_2 & y_3 & s_1 & s_2 & s_3 & s_4 & z & \\ 0 & 3 & 5 & -3 & 0 & 0 & 1 & 0 & 114 \\ 0 & 0 & 6 & -3 & 3 & 0 & 0 & 0 & 108 \\ 0 & 0 & \boxed{1} & 0 & 0 & -1 & 0 & 0 & 10 \\ 3 & 0 & 1 & 0 & 0 & 0 & -1 & 0 & 30 \\ \hline 0 & 0 & -8 & 6 & 0 & 0 & 5 & 3 & -438 \end{array}$$

Again, the solution is not feasible since $s_3 = -10$ and $s_4 = -30$. Pivot on the 1 in row 3, column 3.

$$\begin{array}{l} -5R_3 + R_1 \to R_1 \\ -6R_3 + R_2 \to R_2 \\ \\ -R_3 + R_4 \to R_4 \\ 8R_3 + R_5 \to R_5 \end{array}
\begin{array}{cccccccc|c} y_1 & y_2 & y_3 & s_1 & s_2 & s_3 & s_4 & z & \\ 0 & 3 & 0 & -3 & 0 & 5 & 1 & 0 & 64 \\ 0 & 0 & 0 & -3 & 3 & \boxed{6} & 0 & 0 & 48 \\ 0 & 0 & 1 & 0 & 0 & -1 & 0 & 0 & 10 \\ 3 & 0 & 0 & 0 & 0 & 1 & -1 & 0 & 20 \\ \hline 0 & 0 & 0 & 6 & 0 & -8 & 5 & 3 & -358 \end{array}$$

The solution is feasible because all variables are nonnegative. But it is still not optimal. Pivot on the 6 in row 2, column 6.

$$\begin{array}{l} -5R_2 + 6R_1 \to R_1 \\ \\ R_2 + 6R_3 \to R_3 \\ -R_2 + 6R_4 \to R_4 \\ 4R_2 + 3R_5 \to R_5 \end{array}
\begin{array}{cccccccc|c} y_1 & y_2 & y_3 & s_1 & s_2 & s_3 & s_4 & z & \\ 0 & 18 & 0 & -3 & -15 & 0 & 6 & 0 & 144 \\ 0 & 0 & 0 & -3 & 3 & 6 & 0 & 0 & 48 \\ 0 & 0 & 6 & -3 & 3 & 0 & 0 & 0 & 108 \\ 18 & 0 & 0 & 3 & -3 & 0 & -6 & 0 & 72 \\ \hline 0 & 0 & 0 & 6 & 12 & 0 & 15 & 9 & -882 \end{array}$$

Create a 1 in the columns corresponding to y_1, y_2, y_3, s_3 and z.

$$\begin{array}{l} \frac{1}{18}R_1 \to R_1 \\ \frac{1}{6}R_2 \to R_2 \\ \frac{1}{6}R_3 \to R_3 \\ \frac{1}{18}R_4 \to R_4 \\ \frac{1}{9}R_5 \to R_5 \end{array}
\begin{array}{cccccccc|c} y_1 & y_2 & y_3 & s_1 & s_2 & s_3 & s_4 & z & \\ 0 & 1 & 0 & -\frac{1}{6} & -\frac{5}{6} & 0 & \frac{1}{3} & 0 & 8 \\ 0 & 0 & 0 & -\frac{1}{2} & \frac{1}{2} & 1 & 0 & 0 & 8 \\ 0 & 0 & 1 & -\frac{1}{2} & \frac{1}{2} & 0 & 0 & 0 & 18 \\ 1 & 0 & 0 & \frac{1}{6} & -\frac{1}{6} & 0 & -\frac{1}{3} & 0 & 4 \\ \hline 0 & 0 & 0 & \frac{2}{3} & \frac{4}{3} & 0 & \frac{5}{3} & 1 & -98 \end{array}$$

Since $z = -w = -98$, the minimum value is 98 when $y_1 = 4$, $y_2 = 8$, and $y_3 = 18$.

29.

$$\begin{array}{cccc|c} x_1 & x_2 & s_1 & s_2 & z & \\ 5 & 10 & 1 & 0 & 0 & 120 \\ \boxed{10} & 15 & 0 & -1 & 0 & 200 \\ \hline -20 & -30 & 0 & 0 & 1 & 0 \end{array}$$

The initial tableau is not feasible. Pivot on the 10 in row 2, column 1.

$$\begin{array}{c} \\ -R_2 + 2R_1 \to R_1 \\ \\ 2R_2 + R_3 \to R_3 \end{array} \begin{array}{ccccc} x_1 & x_2 & s_1 & s_2 & z \\ \hline 0 & 5 & 2 & \boxed{1} & 0 & 40 \\ 10 & 15 & 0 & -1 & 0 & 200 \\ \hline 0 & 0 & 0 & -2 & 1 & 400 \end{array}$$

The basic solution is feasible, but there are negative indicators. Pivot on the 1 in row 1, column 4.

$$\begin{array}{c} \\ R_2 + 2R_1 \to R_2 \\ 2R_1 + R_3 \to R_3 \end{array} \begin{array}{ccccc} x_1 & x_2 & s_1 & s_2 & z \\ 0 & 5 & 2 & 1 & 0 & 40 \\ 10 & 20 & 2 & 0 & 0 & 240 \\ 0 & 10 & 4 & 0 & 1 & 480 \end{array}$$

Create a one in the column corresponding to x_1.

$$\begin{array}{c} \\ \frac{1}{10}R_2 \to R_2 \\ \\ \end{array} \begin{array}{ccccc} x_1 & x_2 & s_1 & s_2 & z \\ 0 & 5 & 2 & 1 & 0 & 40 \\ 1 & 2 & \frac{1}{5} & 0 & 0 & 24 \\ \hline 0 & 10 & 4 & 0 & 1 & 480 \end{array}$$

The maximum value is $z = 480$ when $x_1 = 24$ and $x_2 = 0$.

31. Maximize $z = 10x_1 + 12x_2$

subject to: $2x_1 + 2x_2 = 17$
$2x_1 + 5x_2 \geq 22$
$4x_1 + 3x_2 \leq 28$

with $x_1 \geq 0,\ x_2 \geq 0.$

Introduce artificial variable a, surplus variable s_1, and slack variable s_2. The initial simplex tableau as follows.

$$\begin{array}{cccccc} x_1 & x_2 & a & s_1 & s_2 & z \\ \boxed{2} & 2 & 1 & 0 & 0 & 0 & 17 \\ 2 & 5 & 0 & -1 & 0 & 0 & 22 \\ 4 & 3 & 0 & 0 & 1 & 0 & 28 \\ \hline -10 & -12 & 0 & 0 & 0 & 1 & 0 \end{array}$$

First, eliminate the artificial variable a. Pivot on the 2 in row 1, column 1.

$$\begin{array}{c} \\ -R_1 + R_2 \to R_2 \\ 2R_1 - R_3 \to R_3 \\ 5R_1 + R_4 \to R_4 \end{array} \begin{array}{cccccc} x_1 & x_2 & a & s_1 & s_2 & z \\ 2 & 2 & 1 & 0 & 0 & 0 & 17 \\ 0 & 3 & -1 & -1 & 0 & 0 & 5 \\ 0 & 1 & 2 & 0 & -1 & 0 & 6 \\ \hline 0 & -2 & 5 & 0 & 0 & 1 & 85 \end{array}$$

Now $a = 0$, so we can drop the a column.

$$\begin{array}{ccccc} x_1 & x_2 & s_1 & s_2 & z \\ 2 & 2 & 0 & 0 & 0 & 17 \\ 0 & \boxed{3} & -1 & 0 & 0 & 5 \\ 0 & 1 & 0 & -1 & 0 & 6 \\ \hline 0 & -2 & 0 & 0 & 1 & 85 \end{array}$$

Because $s_1 = -5$, we choose the 3 in row 2, column 2, as the next pivot.

$$\begin{array}{c} \\ -2R_2 + 3R_1 \to R_1 \\ \\ -R_2 + 3R_3 \to R_3 \\ 2R_2 + 3R_4 \to R_4 \end{array} \begin{array}{ccccc} x_1 & x_2 & s_1 & s_2 & z \\ 6 & 0 & 2 & 0 & 0 & 41 \\ 0 & 3 & -1 & 0 & 0 & 5 \\ 0 & 0 & \boxed{1} & -3 & 0 & 13 \\ 0 & 0 & -2 & 0 & 3 & 265 \end{array}$$

The solution is still not feasible since $s_2 = -\dfrac{13}{3}$. Pivot on the 1 in row 3, column 3.

$$\begin{array}{c} \\ -2R_3 + R_1 \to R_1 \\ R_3 + R_2 \to R_2 \\ \\ 2R_3 + R_4 \to R_4 \end{array} \begin{array}{ccccc} x_1 & x_2 & s_1 & s_2 & z \\ 6 & 0 & 0 & \boxed{6} & 0 & 15 \\ 0 & 3 & 0 & -3 & 0 & 18 \\ 0 & 0 & 1 & -3 & 0 & 13 \\ \hline 0 & 0 & 0 & -6 & 3 & 291 \end{array}$$

The solution is now feasible but is not yet optimal. Pivot on the 6 in row 1, column 4.

$$\begin{array}{c} \\ R_1 + 2R_2 \to R_2 \\ R_1 + 2R_3 \to R_3 \\ R_1 + R_4 \to R_4 \end{array} \begin{array}{ccccc} x_1 & x_2 & s_1 & s_2 & z \\ 6 & 0 & 0 & 6 & 0 & 15 \\ 6 & 6 & 0 & 0 & 0 & 51 \\ 6 & 0 & 2 & 0 & 0 & 41 \\ 6 & 0 & 0 & 0 & 3 & 306 \end{array}$$

Create a 1 in the columns corresponding to x_2, s_1, s_2, and z.

$$\begin{array}{c} \\ \frac{1}{6}R_1 \to R_1 \\ \frac{1}{6}R_2 \to R_2 \\ \frac{1}{2}R_3 \to R_3 \\ \frac{1}{3}R_4 \to R_4 \end{array} \begin{array}{ccccc} x_1 & x_2 & s_1 & s_2 & z \\ 1 & 0 & 0 & 1 & 0 & \frac{5}{2} \\ 1 & 1 & 0 & 0 & 0 & \frac{17}{2} \\ 3 & 0 & 1 & 0 & 0 & \frac{41}{2} \\ 2 & 0 & 0 & 0 & 1 & 102 \end{array}$$

The maximum is 102 when $x_1 = 0$ and $x_2 = \dfrac{17}{2}$.

33. Any maximizing or minimizing problems can be solved using slack, surplus, and artificial variables. Slack variables are used in problems involving "$\leq$" constraints. Surplus variables are used in problems involving "$\geq$" constraints. Artificial variables are used in problems involving "$=$" constraints.

35.
$$\left[\begin{array}{cccccc|c} 4 & 2 & 3 & 1 & 0 & 0 & 9 \\ 5 & 4 & 1 & 0 & 1 & 0 & 10 \\ \hline -6 & -7 & -5 & 0 & 0 & 1 & 0 \end{array}\right]$$

(a) The 1 in column 4 and the 1 in column 5 indicate that the constraints involve $\leq$. The problem being solved with this tableau is:

Maximize $\qquad z = 6x_1 + 7x_2 + 5x_3$

subject to: $\qquad 4x_1 + 2x_2 + 3x_3 \leq 9$

$\qquad\qquad\qquad 5x_1 + 4x_2 + x_3 \leq 10$

with $\quad x_1 \geq 0, x_2 \geq 0, x_3 \geq 0.$

(b) If the 1 in row 1, column 4 was -1 rather than 1, then the first constraint would have a surplus variable rather than a slack variable, which means the first constraint would be $4x_1 + 2x_2 + 3x_3 \geq 9$ instead of $4x_1 + 2x_2 + 3x_3 \leq 9.$

(c)
$$\begin{array}{cccccc} x_1 & x_2 & x_3 & s_1 & s_2 & z \end{array}$$
$$\left[\begin{array}{cccccc|c} 3 & 0 & 5 & 2 & -1 & 0 & 8 \\ 11 & 10 & 0 & -1 & 3 & 0 & 21 \\ \hline 47 & 0 & 0 & 13 & 11 & 10 & 227 \end{array}\right]$$

From this tableau, the solution is $x_1 = 0,$

$x_2 = \frac{21}{10} = 2.1, \; x_3 = \frac{8}{5} = 1.6,$ and

$z = \frac{227}{10} = 22.7.$

(d) The dual of the original problem is as follows:

Minimize $\qquad w = 9y_1 + 10y_2$

subject to: $\qquad 4y_1 + 5y_2 \geq 6$

$\qquad\qquad\qquad 2y_1 + 4y_2 \geq 7$

$\qquad\qquad\qquad 3y_1 + y_2 \geq 5$

with $\qquad y_1 \geq 0, \; y_2 \geq 0.$

(e) From the tableau in part (c), the solution of the dual in part (d) is $y_1 = \frac{13}{10} = 1.3,$

$y_2 = \frac{11}{10} = 1.1,$ and $w = \frac{227}{10} = 22.7.$

37. (a) Let $x_1 = $ the number of cake plates,

$x_2 = $ the number of bread plates,

and $x_3 = $ the number of dinner plates.

(b) The objective function to maximize is $z = 15x_1 + 12x_2 + 5x_3.$

(c) The constraints are

$$15x_1 + 10x_2 + 8x_3 \leq 1500$$
$$5x_1 + 4x_2 + 4x_3 \leq 2700$$
$$6x_1 + 5x_2 + 5x_3 \leq 1200.$$

39. (a) Let $x_1 = $ number of gallons of Fruity wine

and $x_2 = $ number of gallons of Crystal wine.

(b) The profit function is

$$z = 12x_1 + 15x_2.$$

(c) The ingredients available are the limitations; the constraints are

$$2x_1 + x_2 \leq 110$$
$$2x_1 + 3x_2 \leq 125$$
$$2x_1 + x_2 \leq 90.$$

41. Maximize $z = 15x_1 + 12x_2 + 5x_3$

subject to: $\quad 15x_1 + 10x_2 + 8x_3 \leq 1500$

$\qquad\qquad\qquad 5x_1 + 4x_2 + 4x_3 \leq 2700$

$\qquad\qquad\qquad 6x_1 + 5x_2 + 5x_3 \leq 1200$

with $\qquad x_1 \geq 0, \; x_2 \geq 0, \; x_3 \geq 0.$

The initial tableau is as follows.

$$\begin{array}{ccccccc} x_1 & x_2 & x_3 & s_1 & s_2 & s_3 & z \end{array}$$
$$\left[\begin{array}{ccccccc|c} \boxed{15} & 10 & 8 & 1 & 0 & 0 & 0 & 1500 \\ 5 & 4 & 4 & 0 & 1 & 0 & 0 & 2700 \\ 6 & 5 & 5 & 0 & 0 & 1 & 0 & 1200 \\ \hline -15 & -12 & -5 & 0 & 0 & 0 & 1 & 0 \end{array}\right].$$

Pivot on the 15 in row 1, column 1.

$$\begin{array}{ccccccc} x_1 & x_2 & x_3 & s_1 & s_2 & s_3 & z \end{array}$$
$$\begin{array}{c} \\ -R_1 + 3R_2 \to R_2 \\ -2R_1 + 5R_3 \to R_3 \\ R_1 + R_4 \to R_4 \end{array}\left[\begin{array}{ccccccc|c} 15 & \boxed{10} & 8 & 1 & 0 & 0 & 0 & 1500 \\ 0 & 2 & 4 & -1 & 3 & 0 & 0 & 6600 \\ 0 & 5 & 9 & -2 & 0 & 5 & 0 & 3000 \\ 0 & -2 & 3 & 1 & 0 & 0 & 1 & 1500 \end{array}\right]$$

Pivot on the 10 in row 1, column 2.

$$\begin{array}{ccccccc} x_1 & x_2 & x_3 & s_1 & s_2 & s_3 & z \end{array}$$
$$\begin{array}{c} \\ -R_1 + 5R_2 \to R_2 \\ -R_1 + 2R_3 \to R_3 \\ R_1 + 5R_4 \to R_4 \end{array}\left[\begin{array}{ccccccc|c} 15 & 10 & 8 & 1 & 0 & 0 & 0 & 1500 \\ -15 & 0 & 12 & -6 & 15 & 0 & 0 & 31{,}500 \\ -15 & 0 & 10 & -5 & 0 & 10 & 0 & 4500 \\ 15 & 0 & 23 & 6 & 0 & 0 & 5 & 9000 \end{array}\right]$$

Create a 1 in the columns corresponding to $x_2,$

$s_2, \; s_3,$ and $z.$

$$\begin{array}{c}
\\
\frac{1}{10}R_1 \rightarrow R_1 \\
\frac{1}{15}R_2 \rightarrow R_2 \\
\frac{1}{10}R_3 \rightarrow R_3 \\
\frac{1}{5}R_4 \rightarrow R_4
\end{array}
\begin{array}{ccccccc|c}
x_1 & x_2 & x_3 & s_1 & s_2 & s_3 & z & \\
\hline
\frac{3}{2} & 1 & \frac{4}{5} & \frac{1}{10} & 0 & 0 & 0 & 150 \\
-1 & 0 & \frac{4}{5} & -\frac{2}{5} & 1 & 0 & 0 & 2100 \\
-\frac{3}{2} & 0 & 1 & -\frac{1}{2} & 0 & 1 & 0 & 450 \\
\hline
3 & 0 & \frac{23}{5} & \frac{6}{5} & 0 & 0 & 1 & 1800
\end{array}$$

The maximum profit of $1800 occurs when no cake plates, 150 bread plates, and no dinner plates are produced.

43. Based on Exercise 39, the initial tableau is

$$\begin{array}{cccccc|c}
x_1 & x_2 & s_1 & s_2 & s_3 & z & \\
\hline
2 & 1 & 1 & 0 & 0 & 0 & 110 \\
2 & \boxed{3} & 0 & 1 & 0 & 0 & 125 \\
2 & 1 & 0 & 0 & 1 & 0 & 90 \\
\hline
-12 & -15 & 0 & 0 & 0 & 1 & 0
\end{array}$$

Locating the first pivot in the usual way, it is found to be the 3 in row 2, column 2. After row transformations, we get the next tableau.

$$\begin{array}{c}
-R_2 + 3R_1 \rightarrow R_1 \\
\\
-R_2 + 3R_3 \rightarrow R_3 \\
5R_2 + R_4 \rightarrow R_4
\end{array}
\begin{array}{cccccc|c}
x_1 & x_2 & s_1 & s_2 & s_3 & z & \\
\hline
4 & 0 & 3 & -1 & 0 & 0 & 205 \\
2 & 3 & 0 & 1 & 0 & 0 & 125 \\
\boxed{4} & 0 & 0 & -1 & 3 & 0 & 145 \\
\hline
-2 & 0 & 0 & 5 & 0 & 1 & 625
\end{array}$$

Pivot on the 4 in row 3, column 1.

$$\begin{array}{c}
-R_3 + R_1 \rightarrow R_1 \\
-R_3 + 2R_2 \rightarrow R_2 \\
\\
R_3 + 2R_4 \rightarrow R_4
\end{array}
\begin{array}{cccccc|c}
x_1 & x_2 & s_1 & s_2 & s_3 & z & \\
\hline
0 & 0 & 3 & 0 & -3 & 0 & 60 \\
0 & 6 & 0 & 3 & -3 & 0 & 105 \\
4 & 0 & 0 & -1 & 3 & 0 & 145 \\
\hline
0 & 0 & 0 & 9 & 3 & 2 & 1395
\end{array}$$

$$\begin{array}{c}
\frac{1}{3}R_1 \rightarrow R_1 \\
\frac{1}{6}R_2 \rightarrow R_2 \\
\frac{1}{4}R_3 \rightarrow R_3 \\
\frac{1}{2}R_4 \rightarrow R_4
\end{array}
\begin{array}{cccccc|c}
x_1 & x_2 & s_1 & s_2 & s_3 & z & \\
\hline
0 & 0 & 1 & 0 & -1 & 0 & 20 \\
0 & 1 & 0 & \frac{1}{2} & -\frac{1}{2} & 0 & \frac{35}{2} \\
1 & 0 & 0 & -\frac{1}{4} & \frac{3}{4} & 0 & \frac{145}{4} \\
\hline
0 & 0 & 0 & \frac{9}{2} & \frac{3}{2} & 1 & \frac{1395}{2}
\end{array}$$

The final tableau gives the solution $x_1 = \frac{145}{4}$, $x_2 = \frac{35}{2}$, and $z = \frac{1395}{2} = 697.5$. 36.25 gal of Fruity wine and 17.5 gal of Crystal wine should be produced for a maximum profit of $697.50.

45. (a) Let $y_1 =$ the number of cases of corn,

$y_2 =$ the number of cases of beans

and $y_3 =$ the number of cases of carrots.

Minimize $\qquad w = 10y_1 + 15y_2 + 25y_3$

subject to: $\qquad y_1 + y_2 + y_3 \geq 1000$

$\qquad\qquad\qquad y_1 \geq 2y_2$

$\qquad\qquad\qquad y_3 \geq 340$

with $\qquad\qquad y_1 \geq 0, y_2 \geq 0.$

The second constraint can be rewritten as $y_1 - 2y_2 \geq 0$. Change this to a maximization problem by letting $z = -w = -10y_1 - 15y_2 - 25y_3$. Now maximize $z = -10y_1 - 15y_2 - 25y_3$ subject to the constraints above. Begin by inserting surplus variables to set up the first tableau.

$$\begin{array}{ccccccc|c}
y_1 & y_2 & y_3 & s_1 & s_2 & s_3 & z & \\
\hline
\boxed{1} & 1 & 1 & -1 & 0 & 0 & 0 & 1000 \\
1 & -2 & 0 & 0 & -1 & 0 & 0 & 0 \\
0 & 0 & 1 & 0 & 0 & -1 & 0 & 340 \\
\hline
10 & 15 & 25 & 0 & 0 & 0 & 1 & 0
\end{array}$$

Multiply row 2 by -1 so that s_2 is positive.

$$-R_2 \rightarrow R_2 \quad
\begin{array}{ccccccc|c}
y_1 & y_2 & y_3 & s_1 & s_2 & s_3 & z & \\
\hline
\boxed{1} & 1 & 1 & -1 & 0 & 0 & 0 & 1000 \\
-1 & 2 & 0 & 0 & 1 & 0 & 0 & 0 \\
0 & 0 & 1 & 0 & 0 & -1 & 0 & 340 \\
\hline
10 & 15 & 25 & 0 & 0 & 0 & 1 & 0
\end{array}$$

Pivot on the 1 in row 1, column 1.

$$\begin{array}{c}
\\
R_1 + R_2 \rightarrow R_2 \\
\\
-10R_1 + R_4 \rightarrow R_4
\end{array}
\begin{array}{ccccccc|c}
y_1 & y_2 & y_3 & s_1 & s_2 & s_3 & z & \\
\hline
1 & 1 & 1 & -1 & 0 & 0 & 0 & 1000 \\
0 & 3 & 1 & -1 & 1 & 0 & 0 & 1000 \\
0 & 0 & \boxed{1} & 0 & 0 & -1 & 0 & 340 \\
\hline
0 & 5 & 15 & 10 & 0 & 0 & 1 & -10{,}000
\end{array}$$

Pivot on the 1 in row 3, column 3.

$$\begin{array}{c}
-R_3 + R_1 \rightarrow R_1 \\
-R_3 + R_2 \rightarrow R_2 \\
\\
-15R_3 + R_4 \rightarrow R_4
\end{array}
\begin{array}{ccccccc|c}
y_1 & y_2 & y_3 & s_1 & s_2 & s_3 & z & \\
\hline
1 & 1 & 0 & -1 & 0 & 1 & 0 & 660 \\
0 & 3 & 0 & -1 & 1 & 1 & 0 & 660 \\
0 & 0 & 1 & 0 & 0 & -1 & 0 & 340 \\
\hline
0 & 5 & 0 & 10 & 0 & 15 & 1 & -15{,}100
\end{array}$$

The maximum value of z is $-15{,}100$ when $y_1 = 660$, $y_2 = 0$, and $y_3 = 340$. Hence the minimum value of w is $15{,}100$ when $y_1 = 660$, $y_2 = 0$, and $y_3 = 340$.

Produce 660 cases of corn and 340 cases of carrots for a minimum cost of $15.000.

(b) The dual problem is as follows.

Maximize $z = 1000x_1 + 340x_3$

subject to: $x_1 + x_2 \le 10$

$x_1 - 2x_2 \le 15$

$x_1 + x_3 \le 25$

with $x_1 \ge 0, x_2 \ge 0, x_3 \ge 0$.

The initial simplex tableau is as follows.

$$\begin{array}{ccccccc|c}
x_1 & x_2 & x_3 & s_1 & s_2 & s_3 & z & \\
\hline
\boxed{1} & 1 & 0 & 1 & 0 & 0 & 0 & 10 \\
1 & -2 & 0 & 0 & 1 & 0 & 0 & 15 \\
1 & 0 & 1 & 0 & 0 & 1 & 0 & 25 \\
\hline
-1000 & 0 & -340 & 0 & 0 & 0 & 1 & 0
\end{array}$$

Pivot on the 1 in row 1, column 1.

$$\begin{array}{l}
\\
-R_1 + R_2 \to R_2 \\
-R_1 + R_3 \to R_3 \\
1000R_1 + R_4 \to R_4
\end{array}
\begin{array}{ccccccc|c}
x_1 & x_2 & x_3 & s_1 & s_2 & s_3 & z & \\
1 & 1 & 0 & 1 & 0 & 0 & 0 & 10 \\
0 & -3 & 0 & -1 & 1 & 0 & 0 & 5 \\
0 & -1 & \boxed{1} & -1 & 0 & 1 & 0 & 15 \\
0 & 1000 & -340 & 1000 & 0 & 0 & 1 & 10{,}000
\end{array}$$

Pivot on the 1 in row 3, column 3.

$$\begin{array}{l}
\\
\\
340R_3 + R_4 \to R_4
\end{array}
\begin{array}{ccccccc|c}
x_1 & x_2 & x_3 & s_1 & s_2 & s_3 & z & \\
1 & 1 & 0 & 1 & 0 & 0 & 0 & 10 \\
0 & -3 & 0 & -1 & 1 & 0 & 0 & 5 \\
0 & -1 & 1 & -1 & 0 & 1 & 0 & 15 \\
0 & 660 & 0 & 660 & 0 & 340 & 1 & 15{,}100
\end{array}$$

The minimum value of w is $15{,}100$ when $y_1 = 660$, $y_2 = 0$, and $y_3 = 340$, that is, 660 cases of corn, 0 cases of beans, and 340 cases of carrots should be produced to minimize costs, and the minimum cost is $15,100.

(c) The final tableau for the dual solution shows that the shadow cost of acreage (x_1) is $10 acre, so increasing the number of acres planted by 100 will increase the minimum cost by ($10)(100) or $1000, so the new minimum will be $15,100 + $1000 = $16,100.

47. **(a)** Let $x_1 =$ the number of hours doing tai chi,

$x_2 =$ the number of hours riding a unicycle,

and $x_3 =$ the number of hours fencing.

If Ginger wants the total time doing tai chi to be at least twice as long as she rides a unicycle, then

$$x_1 \ge 2x_2$$

or $-x_1 + 2x_2 \le 0$.

The problem can be stated as follows.

Maximize $z = 236x_1 + 295x_2 + 354x_3$

subject to: $x_1 + x_2 + x_3 \le 10$

$x_3 \le 2$

$-x_1 + 2x_2 \qquad \le 0$

with $x_1 \ge 0, x_2 \ge 0, x_3 \ge 0$.

The initial simplex tableau is as follows.

$$\begin{array}{ccccccc|c}
x_1 & x_2 & x_3 & s_1 & s_2 & s_3 & z & \\
\hline
1 & 1 & 1 & 1 & 0 & 0 & 0 & 10 \\
0 & 0 & \boxed{1} & 0 & 1 & 0 & 0 & 2 \\
-1 & 2 & 0 & 0 & 0 & 1 & 0 & 0 \\
\hline
-236 & -295 & -354 & 0 & 0 & 0 & 1 & 0
\end{array}$$

Pivot on the 1 in row 2, column 3.

$$\begin{array}{l}
-R_2 + R_1 \to R_1 \\
\\
\\
354R_2 + R_4 \to R_4
\end{array}
\begin{array}{ccccccc|c}
x_1 & x_2 & x_3 & s_1 & s_2 & s_3 & z & \\
1 & 1 & 0 & 1 & -1 & 0 & 0 & 8 \\
0 & 0 & 1 & 0 & 1 & 0 & 0 & 2 \\
-1 & \boxed{2} & 0 & 0 & 0 & 1 & 0 & 0 \\
-236 & -295 & 0 & 0 & 354 & 0 & 1 & 708
\end{array}$$

Pivot on the 2 in row 3, column 2.

$$\begin{array}{l}
-R_3 + 2R_1 \to R_1 \\
\\
\\
295R_3 + 2R_4 \to R_4
\end{array}
\begin{array}{ccccccc|c}
x_1 & x_2 & x_3 & s_1 & s_2 & s_3 & z & \\
\boxed{3} & 0 & 0 & 2 & -2 & -1 & 0 & 16 \\
0 & 0 & 1 & 0 & 1 & 0 & 0 & 2 \\
-1 & 2 & 0 & 0 & 0 & 1 & 0 & 0 \\
-767 & 0 & 0 & 0 & 708 & 295 & 2 & 1416
\end{array}$$

Pivot on the 3 in row 1, column 1.

$$\begin{array}{l}
\\
\\
R_1 + 3R_3 \to R_3 \\
767R_1 + 3R_4 \to R_4
\end{array}
\begin{array}{ccccccc|c}
x_1 & x_2 & x_3 & s_1 & s_2 & s_3 & z & \\
3 & 0 & 0 & 2 & -2 & -1 & 0 & 16 \\
0 & 0 & 1 & 0 & 1 & 0 & 0 & 2 \\
0 & 6 & 0 & 2 & -2 & 2 & 0 & 16 \\
0 & 0 & 0 & 1534 & 590 & 118 & 6 & 16{,}520
\end{array}$$

Create a 1 in the columns corresponding to x_1, x_2, and z.

$$
\begin{array}{c}
\\
\tfrac{1}{3}R_1 \to R_1 \\
\\
\\
\tfrac{1}{6}R_3 \to R_3 \\
\tfrac{1}{6}R_4 \to R_4
\end{array}
\begin{array}{c}
x_1 \;\; x_2 \;\; x_3 \;\;\; s_1 \;\;\;\; s_2 \;\;\;\; s_3 \;\; z \\
\left[\begin{array}{ccccccc|c}
1 & 0 & 0 & \tfrac{2}{3} & -\tfrac{2}{3} & -\tfrac{1}{3} & 0 & \tfrac{16}{3} \\
0 & 0 & 1 & 0 & 1 & 0 & 0 & 2 \\
0 & 1 & 0 & \tfrac{1}{3} & -\tfrac{1}{3} & \tfrac{1}{3} & 0 & \tfrac{8}{3} \\
\hline
0 & 0 & 0 & \tfrac{767}{3} & \tfrac{295}{3} & \tfrac{59}{3} & 1 & \tfrac{8260}{3}
\end{array}\right]
\end{array}
$$

Ginger will burn a maximum of $2753\tfrac{1}{3}$ calories if she does $\tfrac{16}{3}$ hours of tai chi, $\tfrac{8}{3}$ hours riding a unicycle, and 2 hours fencing.

(b) Since fencing burns the most calories, she should do as much fencing as possible, which is 2 hours. This leaves 8 hours to divide between tai chi and the unicycle. The unicycle burns more calories, so she wants as much of unicycle as possible subject to the tai chi getting at least twice as much time as the unicycle. This requires devoting $\tfrac{1}{3}$ of the remaining 8 hours to the unicycle and $\tfrac{2}{3}$ of the 8 hours to tai chi. So the times are: $\tfrac{16}{3}$ hours of tai chi, $\tfrac{8}{3}$ hours of unicycle, and 2 hours of fencing.

Chapter 5

MATHEMATICS OF FINANCE

5.1 Simple and Compound Interest

Your Turn 1

Use the formula for maturity value, with $P = 3000$, $r = 0.058$, and $t = \frac{100}{360}$. We assume a year of 360 days.

$$A = P(1 + rt)$$

$$A = 3000\left[1 + 0.058\left(\frac{100}{360}\right)\right]$$

$$A = 3048.333$$

The maturity value is $3048.33.

Your Turn 2

Use the formula $A = P(1 + rt)$ with $t = 0.75$ or three quarters of a year.

$$5243.75 = 5000\,[1 + r(0.75)]$$

Solve for r:

$$5243.75 = 5000 + 3750r$$

$$243.75 = 3750r$$

$$r = \frac{243.75}{3750} = 0.065$$

The interest rate is 6.5%.

Your Turn 3

For 7 years compounded quarterly there are $(7)(12) = 84$ periods. The interest rate per month is $\frac{0.042}{12} = 0.0035$. The interest earned on a principal of $1600 is

$$1600(1 + 0.0035)^{84} - 1600 = 545.75 \text{ or } \$545.75.$$

Your Turn 4

The number of compounding periods is $(8)(12) = 96$. We need to solve the equation

$$6500\left(1 + \frac{r}{12}\right)^{96} = 8665.69$$

$$\left(1 + \frac{r}{12}\right)^{96} = \frac{8665.69}{6500} = 1.33318 \quad \begin{array}{l}\text{Divide both} \\ \text{sides by 6500.}\end{array}$$

$$1 + \frac{r}{12} = 1.33318^{1/96} = 1.003 \quad \begin{array}{l}\text{Raise both} \\ \text{sides to the} \\ \text{1/96 power.}\end{array}$$

$$\frac{r}{12} = 0.003 \quad \begin{array}{l}\text{Subtract 1 from} \\ \text{both sides.}\end{array}$$

$$r = 0.036 \quad \begin{array}{l}\text{Multiply both} \\ \text{sides by 12.}\end{array}$$

The annual interest rate is 3.6%.

Your Turn 5

Use the formula $r_E = \left(1 + \frac{r}{m}\right)^m - 1$ with $r = 0.027$ and $m = 12$.

$$r_E = \left(1 + \frac{0.027}{12}\right)^{12} - 1$$

$$r_E = 1.0273 - 1 = 0.0273$$

The effective rate is 2.73%.

Your Turn 6

The interest rate per quarter is $\frac{0.0425}{4} = 0.010625$. Use the formula for present value with $i = 0.010625$ and $n = (7)(4) = 28$.

$$P = \frac{A}{(1 + i)^n}$$

$$P = \frac{10,000}{(1 + 0.010625)^{28}}$$

$$P = 7438.39$$

The present value of the investment is $7438.39.

Your Turn 7

The semiannual interest rate is $\frac{0.035}{2} = 0.0175$. Let n be the number of compounding periods. Then we want $7000 = 3800(1 + 0.0175)^n$.

Solving, we find

$$(1 + 0.0175)^n = \frac{7000}{3800} = 1.842$$

Using logarithms,

$$n\log(1.0175) = \log(1.842)$$

$$n = \frac{\log(1.842)}{\log(1.0175)} = 35.21$$

Since each period is half a year, this corresponds to $\frac{35.21}{2} = 17.605$ years. Rounding up to the next whole period we get an answer of 18 years.

Your Turn 8

Use the formula for continuous compounding with $P = 5000$, $r = 0.038$ and $t = 9$.

$$A = Pe^{rt}$$
$$A = 5000e^{(0.038)(9)}$$
$$A = 7038.80$$

Subtracting the initial investment we get $7038.80 - 5000 = 2038.80$, or $2038.80 interest earned.

5.1 Exercises

5. $25,000 at 3% for 9 mo

Use the formula for simple interest.

$$I = Prt$$
$$= 25,000(0.03)\left(\frac{9}{12}\right)$$
$$= 562.50$$

The simple interest is $562.50.

7. $1974 at 6.3% for 25 wk

Use the formula for simple interest.

$$I = Prt$$
$$= 1974(0.063)\left(\frac{25}{52}\right) \approx 59.79$$

The simple interest is $59.79.

9. $8192.17 at 3.1% for 72 days

Use the formula for simple interest.

$$I = Prt$$
$$= 8192.17(0.031)\left(\frac{72}{360}\right)$$
$$\approx 50.79$$

The simple interest is $50.79.

11. Use the formula for future value for simple interest.

$$A = P(1 + rt)$$
$$= 3125\left[1 + 0.0285\left(\frac{7}{12}\right)\right]$$
$$\approx 3176.95$$

The maturity value is $3176.95. The interest earned is $3176.95 - 3125 = $51.95.

13. Use the formula for simple interest.

$$I = Prt$$
$$56.25 = 1500r\left(\frac{6}{12}\right)$$
$$r = 0.075$$

The interest rate was 7.5%.

19. Use the formula for compound amount with $P = 1000$, $i = 0.06$, and $n = 8$.

$$A = P(1 + i)^n$$
$$= 1000(1 + 0.06)^8$$
$$\approx 1593.85$$

The compound amount is $1593.85. The interest earned is $1593.85 - 1000 = $593.85.

21. Use the formula for compound amount with $P = 470$, $i = \frac{0.054}{2} = 0.027$, and $n = 12(2) = 24$.

$$A = P(1 + i)^n$$
$$= 470(1 + 0.027)^{24}$$
$$\approx 890.82$$

The compound amount is $890.82. The interest earned is $890.82 - 470 = $420.82.

23. Use the formula for compound amount with $P = 8500$, $i = \frac{0.08}{4} = 0.02$, and $n = 5(4) = 20$.

$$A = P(1 + i)^n$$
$$= 8500(1 + 0.02)^{20}$$
$$\approx 12,630.55$$

The compound amount is $12,630.55. The interest earned is $12,630.55 - 8500 = $4130.55.

25. The number of compounding periods is $(4)(8) = 32$.

$$8000\left(1 + \frac{r}{4}\right)^{32} = 11672.12$$

$$\left(1 + \frac{r}{4}\right)^{32} = \frac{11672.12}{8000} = 1.45902$$

$$1 + \frac{r}{4} = 1.45902^{1/32} = 1.011875$$

$$\frac{r}{4} = 0.011875$$

$$r = 0.0475$$

The answer is 4.75%.

27. The number of compounding periods is $(12)(5) = 60$.

$$4500\left(1 + \frac{r}{12}\right)^{60} = 5994.79$$

$$\left(1 + \frac{r}{12}\right)^{60} = \frac{5994.79}{4500} = 1.332176$$

$$1 + \frac{r}{12} = 1.332176^{1/60} = 1.004792$$

$$\frac{r}{12} = 0.004792$$

$$r = 0.0575$$

The answer is 5.75%

29. 4% compounded quarterly.

Use the formula for effective rate with $r = 0.04$ and $m = 4$.

$$r_E = \left(1 + \frac{r}{m}\right)^m - 1$$

$$= \left(1 + \frac{0.04}{4}\right)^4 - 1$$

$$\approx 0.04060$$

The effective rate is about 4.06%.

31. 7.25% compounded semiannually.

Use the formula for effective rate with $r = 0.0725$ and $m = 2$.

$$r_E = \left(1 + \frac{r}{m}\right)^m - 1$$

$$= \left(1 + \frac{0.0725}{2}\right)^2 - 1$$

$$\approx 0.07381$$

The effective rate is about 7.381%, or rounding to two decimal places, 7.38%.

33. Use the formula for present value for compound interest with $A = 12{,}820.77$, $i = 0.048$, and $n = 6$.

$$P = \frac{A}{(1 + r)^n}$$

$$= \frac{12{,}820.77}{(1 + 0.048)^6}$$

$$\approx 9677.13$$

The present value is $9677.13.

35. Use the formula for present value for compound interest with $A = 2000$, $i = \frac{0.06}{2} = 0.03$, and $n = 8(2) = 16$.

$$P = \frac{A}{(1 + r)^n}$$

$$= \frac{2000}{(1 + 0.03)^{16}}$$

$$\approx 1246.33$$

The present value is $1246.33.

37. Use the formula for present value for compound interest with $A = 8800$, $i = \frac{0.05}{4} = 0.0125$, and $n = 5(4) = 20$.

$$P = \frac{A}{(1 + r)^n}$$

$$= \frac{8800}{(1 + 0.0125)^{20}}$$

$$\approx 6864.08$$

The present value is $6864.08.

41. The quarterly interest rate is $\frac{0.04}{4} = 0.01$. Let n be the number of compounding periods. Then we want $9000 = 5000(1 + 0.01)^n$

or

$$(1 + 0.01)^n = \frac{9000}{5000} = 1.8$$

Using logarithms, we have

$$n\log(1.01) = \log(1.8)$$

$$n = \frac{\log(1.8)}{\log(1.01)}$$

$$n = 59.07$$

Since each period is one quarter of a year, the number of years is $\frac{59.07}{4} = 14.768$. Rounding up to the next whole quarter gives us an answer of 15 years.

43. The monthly interest rate is $\frac{0.036}{12} = 0.003$. Let n be the number of compounding periods. Then we want $11000 = 4500(1 + 0.003)^n$

or

$$(1 + 0.003)^n = \frac{11000}{4500} = 2.444$$

Using logarithms, we have

$$n\log(1.003) = \log(2.444)$$
$$n = \frac{\log(2.444)}{\log(1.003)}$$
$$n = 298.325$$

Since each period is one twelfth of a year, the number of years is $\frac{298.325}{12} = 24.86$.

$$\frac{10}{12} = 0.833 \text{ and } \frac{11}{12} = 0.917$$

so rounding up to the next whole month gives us an answer of 24 years and 11 months.

45. (a) The doubling time for an inflation rate of 3.3% is the solution of $2 = (1.033)^n$. Taking logarithms on both sides we have

$$\log(2) = n\log(1.033)$$
$$n = \frac{\log(2)}{\log(1.033)}$$
$$n = 21.349$$

The doubling time is about 21.35 years.

(b) Since $0.001 < 0.033 < 0.05$, this is a small growth rate and we may use the rule of 70 which estimates the doubling time as

$$\frac{70}{(100)(0.033)} = 21.212$$

or about 21.21 years.

47. (a) The future value is $5500e^{(0.031)(9)} = 7269.94$, or \$7269.94.

(b) The effective rate is $e^{0.031} - 1 = 0.0315$ or 3.15%.

(c) To find the time to reach \$10,000, we solve

$$10,000 = 5500e^{0.031t}$$
$$e^{0.031t} = \frac{10,000}{5500}$$

Using logarithms with base e we have

$$0.031t = \ln\left(\frac{10,000}{5500}\right)$$
$$t = \frac{\ln\left(\frac{10,000}{5500}\right)}{0.031}$$
$$t = 19.285$$

The time to reach \$10,000 is 19.29 years.

49. Start by finding the total amount repaid. Use the formula for future value for simple interest, with $P = 2700$, $r = 0.062$, and $t = \frac{9}{12}$.

$$A = P(1 + rt)$$
$$= 7200\left[1 + 0.062\left(\frac{9}{12}\right)\right]$$
$$= 7534.80$$

Tanya repaid her father \$7534.80. To find the amount of this which was interest, subtract the original loan amount from the repayment amount.

$$7534.80 - 7200 = 334.80$$

Of the amount repaid, \$334.80 was interest.

51. The interest earned was
$$\$1521.25 - \$1500 = \$21.25$$

Use the formula for simple interest, with $I = 21.25$, $P = 1500$, and $t = \frac{75}{360}$.

$$I = Prt$$
$$21.25 = 1500r\left(\frac{75}{360}\right)$$
$$0.068 = r$$

The interest rate was 6.8%.

53. Start by finding the total interest earned.
$$I = (\$24 - \$22) + \$0.50 = \$2.50$$

Now use the formula for simple interest, with $I = 2.50$, $P = 22$, and $t = 1$.

$$I = Prt$$
$$2.50 = 22r(1)$$
$$0.11364 \approx r$$

The interest rate was about 11.36% or, rounding to one decimal place, 11.4%.

55. Use the formula for compound amount with
$P = 40,000$, $i = \frac{0.0654}{12}$, and $n = 6$.

$$A = P(1 + i)^n$$

$$= 40,000\left(1 + \frac{0.0654}{12}\right)^6$$

$$\approx 41,325.95$$

When Kelly begins paying off his loan, he will owe $41,325.95.

57. (a) Use the formula for compound amount to find the value of $1000 in 5 yr.

$$A = P(1 + i)^n$$

$$= 1000(1.06)^5$$

$$\approx 1338.23$$

In 5 yr, $1000 will be worth $1338.23. Since this is larger than the $1210 one would receive in 5 yr, it would be more profitable to take the $1000 now.

59. Let $P = 150,000$, $i = -2.4\% = -.024$, and $n = 4$.

$$A = P(1 + i)^n$$

$$= 150,000[1 + (-.024)]^4$$

$$= 150,000(.976)^4$$

$$\approx 136,110.16$$

After 4 yr, the amount on deposit will be $136,110.16.

61. Use the formula
$$A = P(1 + i)^n$$

with $P = \frac{2}{8}$ cent $= \$0.0025$ and $r = 0.04$
compounded quarterly for 2000 yr.

$$A = 0.0025\left(1 + \frac{0.04}{4}\right)^{4(2000)}$$

$$= 0.0025(1.01)^{8000}$$

$$\approx 9.31 \times 10^{31}$$

2000 years later, the money would be worth 9.31×10^{31}.

63. Use the formula
$$A = P(1 + i)^n$$

with $P = 10,000$ and $r = 0.05$ for 10 years777.

(a) If interest is compounding annually,

$$A = 10,000(1 + 0.05)^{10}$$

$$\approx 16,288.95.$$

The future value is $16,288.95.

(b) If interest is compounding quarterly,

$$A = 10,000\left(1 + \frac{0.05}{4}\right)^{40}$$

$$\approx 16,436.19.$$

The future value is $16,436.19.

(c) If interest is compounding monthly,

$$A = 10,000\left(1 + \frac{0.05}{12}\right)^{120}$$

$$\approx 16,470.09.$$

The future value is $16,470.09.

(d) If interest is compounding daily,

$$A = 10,000\left(1 + \frac{0.05}{365}\right)^{3650}$$

$$\approx 16,486.65.$$

The future value is $16,486.65.

(e) If the interest is compounded continuously for 10 years at 5%, the future value is
$10,000e^{(0.05)(10)} = 16,487.213$ or $16,487.21.

65. First consider the case of earning interest at a rate of k per annnm compounded quarterly for all 8 yr and earning $2203.76 on the $1000 investment.

$$2203.76 = 1000\left(1 + \frac{k}{4}\right)^{8(4)}$$

$$2.20376 = \left(1 + \frac{k}{4}\right)^{32}$$

Use a calculator to raise both sides to the power $\frac{1}{32}$.

$$1.025 = 1 + \frac{k}{4}$$

$$0.025 = \frac{k}{4}$$

$$0.1 = k$$

Next consider the actual investments. The $1000 was invested for the first 5 yr at a rate of j per annum compounded semiannually.

$$A = 1000\left(1 + \frac{j}{2}\right)^{5(2)}$$

$$A = 1000\left(1 + \frac{j}{2}\right)^{10}$$

This amount was then invested for the remaining 3 yr at $k = .1$ per annum compounded quarterly for a final compound amount of $1990.76.

$$1990.76 = A\left(1 + \frac{0.1}{4}\right)^{3(4)}$$

$$1990.76 = A(1.025)^{12}$$

$$1480.24 \approx A$$

Recall that $A = 1000\left(1 + \frac{j}{2}\right)^{10}$ and substitute this value into the above equation.

$$1480.24 = 1000\left(1 + \frac{j}{2}\right)^{10}$$

$$1.48024 = \left(1 + \frac{j}{2}\right)^{10}$$

Use a calculator to raise both sides to the power $\frac{1}{10}$.

$$1.04 \approx 1 + \frac{j}{2}$$

$$0.04 = \frac{j}{2}$$

$$0.08 = j$$

The ratio of k to j is

$$\frac{k}{j} = \frac{0.1}{0.08} = \frac{10}{8} = \frac{5}{4}.$$

67. For each quoted effective rate, find the corresponding nominal rate by using the formula for effective rate. Regardless of the CD's term, m always equals 4, since compounding is always quarterly.

For the 6-month CD, use $r_E = 0.025$.

$$r_E = \left(1 + \frac{r}{m}\right)^m - 1$$

$$0.025 = \left(1 + \frac{r}{4}\right)^4 - 1$$

$$(1 + 0.025)^{1/4} = 1 + \frac{r}{4}.$$

$$0.02477 \approx r$$

For the 6-month CD, the nominal rate is about 2.48%.

For the 9-month CD, use $r_E = 0.051$.

$$r_E = \left(1 + \frac{r}{m}\right)^m - 1$$

$$0.051 = \left(1 + \frac{r}{4}\right)^4 - 1$$

$$(1 + 0.051)^{1/4} = 1 + \frac{r}{4}.$$

$$0.050053 \approx r$$

For the 9-month CD, the nominal rate is about 5.01%.

For the 1-year CD, use $r_E = 0.0425$ and $m = 4$.

$$r_E = \left(1 + \frac{r}{m}\right)^m - 1$$

$$0.0425 = \left(1 + \frac{r}{4}\right)^4 - 1$$

$$(1 + 0.0425)^{1/4} = 1 + \frac{r}{4}.$$

$$0.04184 \approx r$$

For the 1-year CD, the nominal rate is about 4.18%.

For the 2-year CD, use $r_E = 0.045$.

$$r_E = \left(1 + \frac{r}{m}\right)^m - 1$$

$$0.045 = \left(1 + \frac{r}{4}\right)^4 - 1$$

$$(1 + 0.045)^{1/4} = 1 + \frac{r}{4}.$$

$$0.04426 \approx r$$

For the 2-year CD, the nominal rate is about 4.43%.

For the 3-year CD, use $r_E = 0.0525$.

$$r_E = \left(1 + \frac{r}{m}\right)^m - 1$$

$$0.0525 = \left(1 + \frac{r}{4}\right)^4 - 1$$

$$(1 + 0.0525)^{1/4} = 1 + \frac{r}{4}.$$

$$0.05150 \approx r$$

For the 3-year CD, the nominal rate is about 5.15%.

69. Start by finding the effective rate for the CD offered by Centennial Bank of Fountain Valley. Use the formula for effective rate, with $r = 0.055$ and $m = 12$.

$$r_E = \left(1 + \frac{r}{m}\right)^m - 1$$

$$= \left(1 + \frac{0.055}{12}\right)^{12} - 1$$

$$\approx 0.05641$$

The effective rate is about 5.64%.

Since the CD offered by First Source Bank of South Bend is compounded annually, the quoted rate of 5.63% is also the effective rate.

Centennial Bank of Fountain Valley pays a slightly higher effective rate.

71. Use the formula for present value for compound interest with $A = 30,000$, $i = \frac{0.055}{4} = 0.01375$, and $n = 5(4) = 20$.

$$P = \frac{A}{(1 + r)^n}$$

$$= \frac{30,000}{(1 + 0.01375)^{20}}$$

$$\approx 22,829.89$$

The present value is $22,829.89, or rounding up to the nearest cent (to make sure that the investment really grows to $30,000), $22,829.90. That is how much of the inherited $25,000 Phyllis should invest in order to have $30,000 for a down payment in 5 years.

73. To find the number of years it will take prices to double at 4% annual inflation, find n in the equation

$$2 = (1 + 0.04)^n,$$

which simplifies to

$$2 = (1.04)^n.$$

By trying various values of n, find that $n = 18$ is approximately correct, because

$$1.04^{18} \approx 2.0258 \approx 2.$$

Prices will double in about 18 yr.

75. To find the number of years it will be until the generating capacity will need to be doubled, find n in the equation

$$2 = (1 + 0.06)^n,$$

which simplifies to

$$2 = (1.06)^n.$$

By trying various values of n, find that $n = 12$ is approximately correct, because

$$1.06^{12} \approx 2.0122 \approx 2.$$

The generating capacity will need to be doubled in about 12 yr.

77. (a) To find this rate of return we must solve $14 = 1(1 + r)^{14}$.

$$14 = 1(1 + r)^{14}$$

$$1 + r = 14^{1/14} = 1.207$$

$$r = 0.207$$

The required rate of return is 20.7%.

(b) With an annual rate of return of 113%, in 14 years an initial investment of $1 million would be worth $1(1 + 1.13)^{14} = 39,565.299$ million dollars or about $39.6 billion.

5.2 Future Value of an Annuity

Your Turn 1

For the geometric series 4, 12, 36,... the common ratio r is $\frac{12}{4} = 3$. The first term is $a = 4$, and to find the sum of the first nine terms we set $n = 9$ and use the formula for the sum of the first n terms of a geometric series:

$$S_n = \frac{a(r^n - 1)}{r - 1}.$$

$$S_9 = \frac{4(3^9 - 1)}{3 - 1}$$

$$S_9 = 39,364$$

Your Turn 2

Use the formula for the future value of an ordinary annuity,

$$S = R\left[\frac{(1 + i)^n - 1}{i}\right],$$

with $R = 250$, $i = 0.033/12 = 0.00275$, and $n = (11)(12) = 132$.

$$S = 250\left[\frac{(1 + 0.00275)^{132} - 1}{0.00275}\right] = 39719.98$$

The accumulated amount after 11 years is $39,719.98.

Your Turn 3

Use the formula for a sinking fund payment,

$$R = \frac{Si}{(1+r)^n - 1},$$

with $S = 13{,}500$, $i = 0.0375/4 = 0.009375$, and $n = (4)(14) = 56$.

$$R = \frac{(13{,}500)(0.009375)}{(1 + 0.009375)^{56} - 1}$$

$$R = 184.41$$

The quarterly payment will be $184.41.

Your Turn 4

Use the formula for the future value of an annuity due,

$$S = R\left[\frac{(1+i)^{n+1} - 1}{i}\right] - R,$$

with $R = 325$, $i = 0.033/12 = 0.0025$, and $n = (12)(5) = 60$.

$$S = 325\left[\frac{(1 + 0.00275)^{61} - 1}{0.00275}\right] - 325 = 21{,}227.66$$

The future value of this annuity due is $21,227.66.

5.2   Exercises

1. $a = 3; r = 2$

The first five terms are

$$3, 3(2), 3(2)^2, 3(2)^3, 3(2)^4$$

or

$$3, 6, 12, 24, 48.$$

The fifth term is 48.

Or, use the formula $a_n = ar^{n-1}$ with $n = 5$.

$$a_5 = ar^{5-1} = 3(2)^4 = 3(16) = 48$$

3. $a = -8; r = 3; n = 5$

$$a_5 = ar^{5-1} = -8(3)^4 = -8(81) = -648$$

The fifth term is -648.

5. $a = 1; r = -3; n = 5$

$$a_5 = ar^{5-1} = 1(-3)^4 = 81$$

The fifth term is 81.

7. $a = 256; r = \frac{1}{4}; n = 5$

$$a_5 = ar^{5-1} = 256\left(\frac{1}{4}\right)^4 = 256\left(\frac{1}{256}\right) = 1$$

The fifth term is 1.

9. $a = 1; r = 2; n = 4$

To find the sum of the first 4 terms, S_4, use the formula for the sum of the first n terms of a geometric sequence.

$$S_n = \frac{a(r^n - 1)}{r - 1}$$

$$S_4 = \frac{1(2^4 - 1)}{2 - 1} = \frac{16 - 1}{1} = 15$$

11. $a = 5; r = \frac{1}{5}; n = 4$

$$S_n = \frac{a(r^n - 1)}{r - 1}$$

$$S_4 = \frac{5\left[\left(\frac{1}{5}\right)^4 - 1\right]}{\frac{1}{5} - 1} = \frac{5\left(-\frac{624}{625}\right)}{-\frac{4}{5}}$$

$$= \frac{-\frac{624}{125}}{-\frac{4}{5}} = \left(-\frac{624}{125}\right)\left(-\frac{5}{4}\right) = \frac{156}{25}$$

13. $a = 128; r = -\frac{3}{2}; n = 4$

$$S_n = \frac{a(r^n - 1)}{r - 1}$$

$$S_4 = \frac{128\left[\left(-\frac{3}{2}\right)^4 - 1\right]}{-\frac{3}{2} - 1} = \frac{128\left(\frac{65}{16}\right)}{-\frac{5}{2}}$$

$$= -208$$

17. $R = 100; i = 0.06; n = 4$

Use the formula for the future value of an ordinary annuity.

$$S = R\left[\frac{(1+i)^n - 1}{i}\right]$$

$$= 100\left[\frac{(1.06)^4 - 1}{0.06}\right]$$

$$= 100\left(\frac{1.262477 - 1}{0.06}\right)$$

$$\approx 437.46$$

The future value is $437.46.

19. $R = 25,000;\ i = 0.045;\ n = 36$

$$S = R\left[\frac{(1+i)^n - 1}{i}\right]$$

$$= 25,000\left[\frac{(1 + 0.045)^{36} - 1}{0.045}\right]$$

$$\approx 2,154,099.15$$

The future value is $2,154,099.15.

21. $R = 9200;$ 10% interest compounded semiannually for 7 yr

Interest of $\frac{10\%}{2} = 5\%$ is earned semiannually, so $i = 0.05.$ In 7 yr, there are $7(2) = 14$ semiannual periods, so $n = 14.$

$$S = R\left[\frac{(1+i)^n - 1}{i}\right]$$

$$= 9200\left[\frac{(1.05)^{14} - 1}{0.05}\right]$$

$$\approx 180,307.41$$

The future value is $180,307.41.

$9200 is contributed in each of 14 periods. The total contribution is

$$\$9200(14) = \$128,800.$$

The amount from interest is

$$\$180,307.41 - 128,800 = \$51,507.41$$

23. $R = 800;$ 6.51% interest compounded semiannually for 12 yr

Interest of $\frac{6.51\%}{2}$ is earned semiannually, so $i = \frac{0.0651}{2} = 0.03255.$ In 12 yr, there are $12(2) = 24$ semiannual periods, so $n = 24.$

$$S = R\left[\frac{(1+i)^n - 1}{i}\right]$$

$$= 800\left[\frac{(1 + 0.03255)^{24} - 1}{0.03255}\right]$$

$$\approx 28,438.21$$

The future value is $28,438.21.

$800 is contributed in each of 24 periods. The total contribution is

$$\$800(24) = \$19,200.$$

The amount from interest is

$$\$28,438.21 - 19,200 = \$9238.21.$$

25. $R = 12,000;\ i = \frac{0.048}{2} = 0.012;\ n = 16(4) = 64$

$$S = R\left[\frac{(1+i)^n - 1}{i}\right]$$

$$= 12,000\left[\frac{(1 + 0.012)^{64} - 1}{0.012}\right]$$

$$\approx 1,145,619.96$$

The future value is $1,145,619.96.

$12,000 is contributed in each of 64 periods. The total contribution is

$$\$12,000(64) = \$768,000.$$

The amount from interest is

$$\$1,145,619.96 - 768,000 = \$377,619.96.$$

29. Using the TMV Solver under the FINANCE menu on the TI-84 Plus calculator, set up the following input:

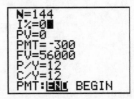

Put the cursor next to I% and press SOLVE to show the solution:

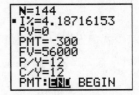

The required interest rate is 4.19%.

31. $S = \$10,000;$ interest is 5% compounded annually; payments are made at the end of each year for 12 yr.

This is a sinking fund. Use the formula for an ordinary annuity with $S = 10,000,\ i = 0.05,$ and $n = 12$ to find the value of R, the amount of each payment.

$$10,000 = Rs_{\overline{12}|0.05}$$

$$R = \frac{10,000}{s_{\overline{12}|0.05}}$$

$$= \frac{10,000}{\frac{(1+0.05)^{12}-1}{0.05}}$$

$$\approx 628.25$$

The required periodic payment is $628.25.

33. $S = 8500; \ i = 0.08; \ n = 7$

$$R = \frac{S}{s_{\overline{n}|i}}$$

$$= \frac{8500}{s_{\overline{7}|0.08}}$$

$$= \frac{8500(0.08)}{(1 + 0.08)^7 - 1}$$

$$\approx 952.62$$

The payment is $952.62.

35. $S = 75,000; \ i = \frac{0.06}{2} = 0.03; \ n = 4\frac{1}{2}(2) = 9$

$$R = \frac{S}{s_{\overline{n}|i}}$$

$$= \frac{75,000}{s_{\overline{9}|0.03}}$$

$$= \frac{75,000(0.03)}{(1 + 0.03)^9 - 1}$$

$$\approx 7382.54$$

The payment is $7382.54.

37. $65,000; money earns 7.5% compounded quarterly for $2\frac{1}{2}$ years

Thus, $i = \frac{0.075}{4} = 0.01875$ and $n = \left(2\frac{1}{2}\right)4 = 10.$

$$R = \frac{65,000}{s_{\overline{10}|0.01875}}$$

$$= \frac{65,000}{\frac{(1+0.01875)^{10}-1}{0.01875}}$$

$$\approx 5970.23$$

The amount of each payment is $5970.23.

39. $R = 600; \ i = 0.06; \ n = 8$

To find the future value of an annuity due, use the formula for the future value of an ordinary annuity, but include one additional time period and subtract the amount of one payment.

$$S = R\left[\frac{(1 + i)^{n+1} - 1}{i}\right] - R$$

$$= 600\left[\frac{(1 + 0.06)^9 - 1}{0.06}\right] - 600$$

$$\approx 6294.79$$

The future value is $6294.79.

41. $R = 16,000; \ i = 0.05; \ n = 7$

$$S = R\left[\frac{(1 + i)^{n+1} - 1}{i}\right] - R$$

$$= 16,000\left[\frac{(1 + 0.05)^8 - 1}{0.05}\right] - 16,000$$

$$\approx 136,785.74$$

The future value is $136,785.74.

43. $R = 1000; \ i = \frac{0.0815}{2} = 0.04075; \ n = 9(2) = 18$

$$S = R\left[\frac{(1 + i)^{n+1} - 1}{i}\right] - R$$

$$= 1000\left[\frac{(1 + 0.04075)^{19} - 1}{0.04075}\right] - 1000$$

$$\approx 26,874.97$$

The future value is $26,874.97.

$1000 is contributed in each of 18 periods. The total contribution is

$$\$1000(18) = \$18,000.$$

The amount from interest is

$$\$26,874.97 - 18,000 = \$8874.97.$$

45. $R = 250; \ i = \frac{0.042}{2} = 0.0105; \ n = 12(4) = 48$

$$S = R\left[\frac{(1 + i)^{n+1} - 1}{i}\right] - R$$

$$= 250\left[\frac{(1 + 0.0105)^{49} - 1}{0.0105}\right] - 250$$

$$\approx 15,662.40$$

The future value is $15,662.40.

$250 is contributed in each of 48 periods. The total contribution is

$$\$250(48) = \$12,000.$$

The amount from interest is

$$15,662.40 - 12,000 = \$3662.40.$$

47. (a) $R = 12,000; \ i = 0.08; \ n = 9$

$$S = R\left[\frac{(1+i)^n - 1}{i}\right]$$

$$= 12,000\left[\frac{(1 + 0.08)^9 - 1}{0.08}\right]$$

$$\approx 149,850.69$$

The final amount is $149,850.69.

(b) $R = 12,000; \ i = 0.06; \ n = 9$

$$S = R\left[\frac{(1+i)^n - 1}{i}\right]$$

$$= 12,000\left[\frac{(1 + 0.06)^9 - 1}{0.06}\right]$$

$$\approx 137,895.79$$

She will have $137,895.79.

(c) The amount that would be lost is the difference between the two amounts in parts (a) and (b), which is

$$\$149,850.69 - 137,895.79 = \$11,954.90.$$

49. This is a future value problem with $R = 136.50$, $i = 0.048/12 = 0.004$, and $n = (12)(40) = 480$.

$$S = R\left[\frac{(1+i)^n - 1}{i}\right],$$

$$S = 136.50\left[\frac{(1.004)^{480} - 1}{0.004}\right]$$

$$S = 197,750.47$$

The account would be worth $197,750.47.

51. From ages 50 to 60, we have an ordinary annuity with $R = 3000$, $i = \frac{0.05}{4} = 0.0125$, and $n = 10(4) = 40$. Use the formula for the future value of an ordinary annuity.

$$S = R\left[\frac{(1+i)^n - 1}{i}\right]$$

$$= 3000\left[\frac{(1.0125)^{40} - 1}{0.0125}\right]$$

$$\approx 154,468.67$$

At age 60, the value of the retirement account is $154,468.67. This amount now earns 6.9% interest compounded monthly for 5 yr. Use the formula for compound amount with $P = 154,468.67$, $i = \frac{0.069}{12} = 0.00575$, and

$n = 5(12) = 60$ to find the value of this amount after 5 yr.

$$A = P(1 + i)^n$$

$$= 154,468.67(1.00575)^{60}$$

$$\approx 217,892.80$$

The value of the amount she withdraws from the retirement account will be $217,892.80 when she reaches 65.

The deposits of $300 at the end of each month into the mutual fund form another ordinary annuity. Use the formula for the future value of an ordinary annuity with $R = 300$, $i = \frac{0.069}{12} = 0.00575$, and $n = 12(5) = 60$.

$$S = R\left[\frac{(1+i)^n - 1}{i}\right]$$

$$= 300\left[\frac{(1.00575)^{60} - 1}{0.00575}\right]$$

$$\approx 21,422.37$$

The value of this annuity after 5 yr is $21,422.37.

The total amount in the mutual fund account when the woman reaches age 65 will be

$$\$217,892.80 + 21,422.37 = \$239,315.17.$$

53. $R = 1000$, $i = \frac{0.08}{4} = 0.02$, and $n = 25(4) = 100$.

$$S = R\left[\frac{(1+i)^n - 1}{i}\right]$$

$$= 1000\left[\frac{(1 + 0.02)^{100} - 1}{0.02}\right]$$

$$\approx 312,232.31$$

There will be about $312,232.31 in the IRA.

The total amount deposited was $1000(100) = \$100,000$. Thus, the amount of interest earned was

$$\$312,232.31 - 100,000 = \$212,232.31.$$

55. $R = 1000$, $i = \frac{0.10}{4} = 0.025$, and $n = 100$.

$$S = R\left[\frac{(1+i)^n - 1}{i}\right]$$

$$= 1000\left[\frac{(1 + 0.025)^{100} - 1}{0.025}\right]$$

$$\approx 432,548.65$$

There will be about $432,548.65 in the IRA. The total amount deposited was $100,000. Thus, the amount of interest earned was

$$432,548.65 - 100,000 = \$332,548.65.$$

57. This is a sinking fund with $S = 12,000$, $i = \frac{0.06}{2} = 0.03$, and $n = 4(2) = 8$.

$$R = \frac{S}{s_{\overline{n}|i}}$$

$$= \frac{12,000}{s_{\overline{8}|0.03}}$$

$$= \frac{12,000(0.03)}{(1 + 0.03)^8 - 1}$$

$$\approx 1349.48$$

Each payment should be $1349.48.

59. $R = 80$; $i = \frac{0.025}{12}$; $n = 3(12) + 9 = 45$

Because the deposits are made at the beginning of each month, this is an annuity due.

$$S = R\left[\frac{(1 + i)^{n+1} - 1}{i}\right] - R$$

$$= 80\left[\frac{\left(1 + \frac{0.025}{12}\right)^{46} - 1}{\frac{0.025}{12}}\right] - 80$$

$$\approx 3777.89$$

The account will have $3777.89 in it.

61. For the first 8 yr, we have an annuity due with $R = 2435$, $i = \frac{0.06}{2} = 0.03$, and $n = 8(2) = 16$.

The amount on deposit after 8 yr is

$$S = R\left[\frac{(1 + i)^{n+1} - 1}{i}\right] - R$$

$$= 2435\left[\frac{(1 + 0.03)^{17} - 1}{0.03}\right] - 2435$$

$$\approx 50,554.47.$$

For the remaining 5 yr, this amount, $50,554.47, earns compound interest at 6% compounded semiannually. To find the final amount on deposit, use the formula for the compound amount with $P = 50,554.47$, $i = \frac{0.06}{2} = 0.03$, and $n = 5(2) = 10$.

$$A = P(1 + i)^n$$

$$= 50,554.47(1.03)^{10}$$

$$\approx 67,940.98$$

The final amount on deposit will be about $67,940.98.

63. Let $x =$ the annual interest rate.

$$n = 20(12) = 240$$

Graph $y_1 = 147,126$ and

$$y_2 = 300\left[\frac{\left(1 + \frac{x}{12}\right)^{240} - 1}{\frac{x}{12}}\right].$$

The x-coordinate of the point of intersection is 0.06499984. Thus, the annual interest rate was about 6.5%.

65. (a) Compare the future amounts for an ordinary annuity with $R = 1,350,000$ and $i = 0.08$ to compound amounts with $P = 7,000,000$ and $i = .08$ for different values of n, starting with $n = 1$.

n	$S = R\left[\dfrac{(1 + i)^n - 1}{i}\right]$	$A = P(1 + i)^n$
1	$1,350,000\left[\dfrac{1.08 - 1}{0.08}\right]$	
	$= \$1,350,000.00$	$\$7,560,000.00$
2	$1,350,000\left[\dfrac{(1.08)^2 - 1}{0.08}\right]$	
	$= \$2,808,000.00$	$\$8,164,800.00$
3	$1,350,000\left[\dfrac{(1.08)^3 - 1}{0.08}\right]$	
	$= \$4,382,640.00$	$\$8,817,984.00$
4	$1,350,000\left[\dfrac{(1.08)^4 - 1}{0.08}\right]$	
	$= \$6,083,251.20$	$\$9,523,422.72$
5	$1,350,000\left[\dfrac{(1.08)^5 - 1}{0.08}\right]$	
	$= \$7,919,911.30$	$\$10,285,296.54$
6	$1,350,000\left[\dfrac{(1.08)^6 - 1}{.08}\right]$	
	$= \$9,903,504.20$	$\$11,108,120.26$
7	$1,350,000\left[\dfrac{(1.08)^7 - 1}{0.08}\right]$	
	$= \$12,045,784.54$	$\$11,996,769.88$

After 7 yr, the investors would do better by winning the lottery.

(b) Repeat the calculations from part (a), but change the interest rate to $i = 0.12$.

n	$S = R\left[\dfrac{(1+i)^n - 1}{i}\right]$	$A = P(1+i)^n$
1	$\$ 1,350,000.00$	$\$ 7,840,000.00$
2	$\$ 2,862,000.00$	$\$ 8,780,800.00$
3	$\$ 4,555,440.00$	$\$ 9,834,496.00$
4	$\$ 6,452,092.80$	$\$11,014,635.52$
5	$\$ 8,576,343.94$	$\$12,336,391.78$
6	$\$10,955,505.21$	$\$13,816,758.80$
7	$\$13,620,165.83$	$\$15,474,769.85$
8	$\$16,604,585.73$	$\$17,331,742.23$
9	$\$19,947,136.02$	$\$19,411,551.30$

After 9 yr, the investors would do better by winning the lottery.

67. This exercise should be solved by graphing calculator or computer methods. The answers, which may vary slightly, are as follows.

(a) The amount of each interest payment is $120.

(b) The amount of each payment is $681.83, except the last payment, which is $681.80. A table showing the amount in the sinking fund after each deposit is as follows.

Payment Number	Amount of Deposit	Interest Earned	Total
1	$681.83	$ 0	$ 681.83
2	$681.83	$ 54.55	$1418.21
3	$681.83	$113.46	$2213.49
4	$681.83	$177.08	$3072.40
5	$681.81	$245.79	$4000.00

5.3 Present Value of an Annuity; Amortization

Your Turn 1

Use the formula for the present value of an annuity,

$$P = R\left[\frac{1 - (1+i)^{-n}}{i}\right],$$

with $R = 500$, $i = 0.048/12 = 0.004$, and $n = (12)(5) = 60$.

$$P = 120\left[\frac{1 - (1 + 0.004)^{-60}}{0.004}\right]$$

$$P = 6389.86$$

The present value is $6389.86.

Your Turn 2

Compute the monthly payment using the formula

$$R = \frac{Pi}{1 - (1+i)^{-n}}$$

with $P = 17,000$, $i = 0.054/12 = 0.0045$, and $n = 48$.

$$R = \frac{(17,000)(0.0045)}{1 - (1 + 0.0045)^{-48}}$$

$$R = 394.59$$

The monthly car payment will be $394.59.

Your Turn 3

The monthly payment will be given by

$$R = \frac{Pi}{1 - (1+i)^{-n}}$$

with $R = 220,000$, $i = 0.07/12$, and $n = (12)(15) = 180$.

$$R = \frac{(220,000)\left(\dfrac{0.07}{12}\right)}{1 - \left(1 + \dfrac{0.07}{12}\right)^{-180}} = 1977.42$$

The monthly payment will be $1977.42. The total of the 180 payments is

$$(1977.42)(180) = 355,935.60$$

To find the interest paid, subtract the principal from this payment total:

$$355,935.60 - 220,000 = 135,935.60$$

The total interest paid is $135,935.60.

Your Turn 4

The only change in the calculation from Example 4 is that number of months remaining is 8 instead of 9, which gives a present value for the remaining balance of

$$88.8488\left[\frac{1 - (1.01)^{-8}}{0.01}\right] = 679.84,$$

or $679.84.

5.3   Exercises

3.   Payments of $890 each year for 16 years at 6% compounded annually

Use the formula for present value of an annuity with $R = 890$, $i = 0.06$, and $n = 16$.

$$P = R\left[\frac{1 - (1 + i)^{-n}}{i}\right]$$

$$= 890\left[\frac{1 - (1 + 0.06)^{-16}}{0.06}\right]$$

$$\approx 8994.25$$

The present value is $8994.25

5.   Payments of $10,000 semiannually for 15 years at 5% compounded semiannually

Use the formula for present value of an annuity with $R = 10,000$, $i = \frac{0.05}{2} = 0.025$, and $n = 15(2) = 30$.

$$P = R\left[\frac{1 - (1 + i)^{-n}}{i}\right]$$

$$= 10,000\left[\frac{1 - (1 + 0.025)^{-30}}{0.025}\right]$$

$$\approx 209,302.93$$

The present value is $209,302.93.

7.   Payments of $15,806 quarterly for 3 years at 6.8% compounded quarterly

Use the formula for present value of an annuity with $R = 15,806$, $i = \frac{0.068}{4} = 0.017$, and $n = 3(4) = 12$.

$$P = R\left[\frac{1 - (1 + i)^{-n}}{i}\right]$$

$$= 15,806\left[\frac{1 - (1 + 0.017)^{-12}}{0.017}\right]$$

$$\approx 170,275.47$$

The present value is $170,275.47.

9.   4% compounded annually

We want the present value, P, of an annuity with $R = 10,000$, $i = 0.04$, and $n = 15$.

$$P = R\left[\frac{1 - (1 + i)^{-n}}{i}\right]$$

$$= 10,000\left[\frac{1 - (1.04)^{-15}}{0.04}\right]$$

$$\approx 111,183.87$$

The required lump sum is $111,183.87.

11. $P = 2500$, $i = \frac{0.06}{4} = 0.015$; $n = 6$

(a)   To find the payment amount, use the formula for amortization payments.

$$R = \frac{Pi}{1 - (1 + i)^{-n}}$$

$$R = \frac{2500(0.015)}{1 - (1 + 0.015)^{-6}}$$

$$\approx 438.81$$

Each payment is $438.81.

(b)   To find the total payments, multiply the amount of one payment by $n = 6$.

$$438.81(6) = 2632.86$$

The total payments come out to $2632.86.

To find the total amount of interest paid, subtract the original loan amount from the total payments.

$$2632.86 - 2500 = 132.86$$

The total amount of interest paid is $132.86.

(c)   Set $P = 2500$, $i = 0.015$, $n = 6$ and $R = 438.81$ and generate the following amortization table using software.

Payment Number	Amount of Payment	Interest for Period	Portion to Principal	Principal at End of Period
0	0.00	0.00	0.00	2500.00
1	438.81	37.50	401.31	2098.69
2	438.81	31.48	407.33	1691.36
3	438.81	25.37	413.44	1277.92
4	438.81	19.17	419.64	858.28
5	438.81	12.87	425.94	432.34
6	438.83	6.49	432.34	0.00

The sum of the Amount of Payment column gives the total payments, $2632.88.

The sum of the Interest column gives the total interest paid, $132.88.

13. $P = 90,000; i = 0.06; n = 12$

    (a) To find the payment amount, use the formula for amortization payments.

$$R = \frac{Pi}{1 - (1 + i)^{-n}}$$

$$R = \frac{90,000(0.06)}{1 - (1 + 0.06)^{-12}}$$

$$\approx 10,734.93$$

Each payment is $10,734.93.

    (b) To find the total payments, multiply the amount of one payment by $n = 12$.

$$10734.93(12) = 128,819.16$$

The total payments come out to $128,819.16.

To find the total amount of interest paid, subtract the original loan amount from the total payments.

$$128,819.16 - 90,000 = 38,819.16$$

The total amount of interest paid is $38,819.16.

    (c) Set $P = 90,000$, $i = 0.06$, $n = 12$ and $R = 10,734.93$ and generate the following amortization table using

Payment Number	Amount of Payment	Interest for Period	Portion to Principal	Principal at End of Period
0	0.00	0.00	0.00	90000.00
1	10734.93	5400.00	5334.93	84665.07
2	10734.93	5079.90	5655.03	79010.04
3	10734.93	4740.60	5994.33	73015.72
4	10734.93	4380.94	6353.99	66661.73
5	10734.93	3999.70	6735.23	59926.50
6	10734.93	3595.59	7139.34	52787.16
7	10734.93	3167.23	7567.70	45219.46
8	10734.93	2713.17	8021.76	37197.70
9	10734.93	2231.86	8503.07	28694.63
10	10734.93	1721.68	9013.25	19681.38
11	10734.93	1180.88	9554.05	10127.33
12	10734.97	607.64	10127.33	0.00

The sum of the Amount of Payment column gives the total payments, $128,819.20.

The sum of the Interest column gives the total interest paid, $38,819.20.

15. $P = 7400; i = \frac{0.062}{2} = 0.031; n = 18$

    (a) To find the payment amount, use the formula for amortization payments.

$$R = \frac{Pi}{1 - (1 + i)^{-n}}$$

$$R = \frac{7400(0.031)}{1 - (1 + 0.031)^{-18}}$$

$$\approx 542.60$$

Each payment is $542.60.

    (b) To find the total payments, multiply the amount of one payment by $n = 18$.

$$542.60(18) = 9766.80$$

The total payments come out to $9766.80.

To find the total amount of interest paid, subtract the original loan amount from the total payments

$$9766.80 - 7400 = 2366.80$$

The total amount of interest paid is $2366.80.

    (c) Set $P = 7400, i = 0.031, n = 18$ and $R = 542.60$ and generate the following amortization table using software.

Payment Number	Amount of Payment	Interest for Period	Portion to Principal	Principal at End of Period
0	0.00	0.00	0.00	7400.00
1	542.60	229.40	313.20	7086.80
2	542.60	219.69	322.91	6763.89
3	542.60	209.68	332.92	6430.97
4	542.60	199.36	343.24	6087.73
5	542.60	188.72	353.88	5733.85
6	542.60	177.75	364.85	5369.00
7	542.60	166.44	376.16	4992.84
8	542.60	154.78	387.82	4605.02
9	542.60	142.76	399.84	4205.17
10	542.60	130.36	412.24	3792.93
11	542.60	117.58	425.02	3367.91
12	542.60	104.41	438.19	2929.72
13	542.60	90.82	451.78	2477.94
14	542.60	76.82	465.78	2012.16
15	542.60	62.38	480.22	1531.93
16	542.60	47.49	495.11	1036.82
17	542.60	32.14	510.46	526.37
18	542.68	16.32	526.36	0.00

The sum of the Amount of Payment column gives the total payments, $9766.88.

The sum of the Interest column gives the total interest paid, $2366.88.

17. Using the first method in Example 4 and carrying more places in the payment amount we have

$$R = \frac{(90,000)(0.06)}{1 - (1 + 0.06)^{-12}}$$

$$R = 10,734.9326$$

There are $12 - 3 = 9$ payments left, so the amount to pay off the loan is

$$10,734.9326 \left[\frac{1 - (1.06)^{-9}}{0.06} \right] = 73,015.71$$

or \$73,015.71.

19. Using the first method in Example 4 and carrying more places in the payment amount we have

$$R = \frac{(7400)(0.031)}{1 - (1 + 0.031)^{-18}}$$

$$R = 542.6035$$

There are $18 - 6 = 12$ payments left, so the amount to pay off the loan is

$$542.6035 \left[\frac{1 - (1.031)^{-12}}{0.031} \right] = 5368.98$$

or \$5368.98.

21. Look at the entry for payment number 4 under the heading "Interest for Period." The amount of interest included in the fourth payment is \$7.61.

23. To find the amount of interest paid in the first 4 mo of the loan, add the entries for payment 1, 2, 3, and 4 under the heading "Interest for Period."

$$\$10.00 + 9.21 + 8.42 + 7.61 = \$35.24$$

In the first 4 mo of the loan, \$35.24 of interest is paid.

25. First, find the value of the annuity at the end of 8 yr. Use the formula for future value of an ordinary annuity.

$$S = R \left[\frac{(1 + i)^n - 1}{i} \right]$$

$$= 1000 \left[\frac{(1 + 0.06)^8 - 1}{0.06} \right]$$

$$\approx 9897.47$$

The future value of the annuity is \$9897.47.

Now find the present value of \$9897.47 at 5% compounded annually for 8 yr. Use the formula for present value for compound interest.

$$P = \frac{A}{(1 + i)^n} = \frac{9897.47}{(1.05)^8} \approx 6699.00$$

The required amount is \$6699.

27. $P = 199,000; \; i = \frac{0.0701}{12}; \; n = 25(12) = 300$

To find the payment amount, use the formula for amortization payments.

$$R = \frac{Pi}{1 - (1 + i)^{-n}}$$

$$R = \frac{199,000\left(\frac{0.0701}{12} \right)}{1 - \left(1 + \frac{0.0701}{12} \right)^{-300}}$$

$$\approx 1407.76$$

Each payment is \$1407.76.

To find the total payment, multiply the amount of one payment by $n = 300$.

$$1407.76(300) = 422,328$$

The total payments come out to \$422,328.

To find the total amount of interest paid, subtract the original loan amount from the total payments.

$$422,328 - 199,000 = 223,328$$

The total amount of interest paid is \$223,328.

29. $P = 253,000, \; i = \frac{0.0645}{12}, \; n = 30(12) = 360$

To find the payment amount, use the formula for amortization payments.

$$R = \frac{Pi}{1 - (1 + i)^{-n}}$$

$$R = \frac{253,000\left(\frac{0.0645}{12} \right)}{1 - \left(1 + \frac{0.0645}{12} \right)^{-360}}$$

$$\approx 1590.82$$

Each payment is \$1590.82.

To find the total payments, multiply the amount of one payment by $n = 360$.

$$1590.82(360) = 572,695.20$$

The total payments come out to \$572,695.20.

To find the total amount of interest paid, subtract the original loan amount from the total payments.

$$572,695.20 - 253,000 = 319,695.20$$

The total amount of interest paid is $319,695.20.

31. (a) Solve as in Example 6:

$$90,000 = 16,000\left(\frac{1 - 1.06^{-n}}{0.06}\right)$$

$$\left(\frac{90,000}{16,000}\right)(0.06) = 0.3375$$

$$0.3375 = 1 - 1.06^{-n}$$

$$1.06^{-n} = 1 - 0.3375 = 0.6625$$

$$n = -\frac{\log(0.6625)}{\log(1.06)} = 7.066$$

Rounding to the next whole year, the loan will take 8 years to pay off.

(b) Use software to build an amortization table with $P = 90,000$, $i = 0.06$, $n = 8$, and $R = 16,000$.

Payment Number	Amount of Payment	Interest for Period	Portion to Principal	Principal at End of Period
0	0.00	0.00	0.00	90000.00
1	16000.00	5400.00	10600.00	79400.00
2	16000.00	4764.00	11236.00	68164.00
3	16000.00	4089.84	11910.16	56253.84
4	16000.00	3375.23	12624.77	43629.07
5	16000.00	2617.74	13382.26	30246.81
6	16000.00	1814.81	14185.19	16061.62
7	16000.00	963.70	15036.30	1025.32
8	1086.84	61.52	1025.32	0.00

The amortization table shows that the total of payments is $113,068.84.

(c) Subtracting this value from the answer to Exercise 13(c), we find that the savings in interest is $128,819.20 - 113,086.84 = 15,732.36$ or $15,732.36.

33. (a) Solve as in Example 6:

$$7400 = 850\left(\frac{1 - 1.031^{-n}}{0.031}\right)$$

$$\left(\frac{7400}{850}\right)(0.031) = 0.270$$

$$0.270 = 1 - 1.031^{-n}$$

$$1.031^{-n} = 1 - 0.270 = 0.730$$

$$n = -\frac{\log(0.73)}{\log(1.031)} = 10.309$$

Rounding to the next half year, the loan will take 11 semiannual periods to pay off.

(b) Use software to build an amortization table with $P = 7400$, $i = 0.031$, $n = 11$, and $R = 850$.

Payment Number	Amount of Payment	Interest for Period	Portion to Principal	Principal at End of Period
0	0.00	0.00	0.00	7400.00
1	850.00	229.40	620.60	6779.40
2	850.00	210.16	639.84	6139.56
3	850.00	190.33	659.67	5479.89
4	850.00	169.88	680.12	4799.76
5	850.00	148.79	701.21	4098.56
6	850.00	127.06	722.94	3375.61
7	850.00	104.64	745.36	2630.26
8	850.00	81.54	768.46	1861.79
9	850.00	57.72	792.28	1069.51
10	850.00	33.15	816.85	252.66
11	260.50	7.83	252.67	0.00

The amortization table shows that the total of payments is $8760.50.

(c) Subtracting this value from the answer to Exercise 15(c), we find that the savings in interest is $9766.88 - 8760.50 = 1006.38$ or $1006.38.

35. From Example 3, $P = 220,000$ and $i = \frac{0.06}{12} = 0.005$. For a 15-year loan, use $n = 15(12) = 180$.

$$R = \frac{Pi}{1 - (1 + i)^{-n}}$$

$$= \frac{220,000(0.005)}{1 - (1 + 0.005)^{-180}}$$

$$\approx 1856.49$$

The monthly payments would be $1856.49. The family makes 180 payments of $1856.49 each, for a total of $334,168.20. Since the amount of the loan was $220,000, the total interest paid is

$$334,168.20 - 220,000 = 114,168.20.$$

The total amount of interest paid is $114,168.20.

The payments for the 15-year loan are

$$1856.49 - \$1319.01 = \$537.48$$

more than those for the 30-year loan in Example 3. However, the total interest paid is

$$254,843.60 - \$114,168.20 = \$140,675.40$$

less than for the 30-year loan in Example 3.

37. (a) $P = 14{,}000,\ i = \frac{0.07}{12},\ n = 4(12) = 48$

$$R = \frac{Pi}{1 - (1 + i)^{-n}}$$

$$= \frac{14{,}000\left(\frac{0.07}{12}\right)}{1 - \left(1 + \frac{0.07}{12}\right)^{-48}}$$

$$\approx 335.25$$

The amount of each payment is $335.25.

(b) 48 payments of $335.25 are made, and 48($335.25) = $16,092. The total amount of interest Le will pay is $16,092 − $14,000 = $2092.

39. (a) Compute the monthly payment using the formula

$$R = \frac{Pi}{1 - (1 + i)^{-n}}$$

with $P = 30{,}000,\ i = 0.009/12,$ and $n = 36.$

$$R = \frac{(30{,}000)\left(\frac{0.009}{12}\right)}{1 - \left(1 + \frac{0.009}{12}\right)^{-36}}$$

$$R = 844.95$$

The monthly payment is $844.95 and the total paid will be $(844.95)(36) = 30{,}418.20$ or $30,418.20.

(b) Compute the monthly payment with $P = 27{,}750,\ i = 0.0633/12,$ and $n = 48.$

$$R = \frac{(27{,}750)\left(\frac{0.0633}{12}\right)}{1 - \left(1 + \frac{0.0633}{12}\right)^{-48}}$$

$$R = 655.92$$

The monthly payment is $655.92 and the total paid will be $(655.92)(48) = 31{,}484.16$ or $31,484.16.

41. For parts (a) and (b), if $1 million is divided into 20 equal payments, each payment is $50,000.

(a) $i = 0.05, n = 20$

$$P = R\left[\frac{1 - (1 + i)^{-n}}{i}\right]$$

$$= 50{,}000\left[\frac{1 - (1 + 0.05)^{-20}}{0.05}\right]$$

$$\approx 623{,}110.52$$

The present value is $623,110.52.

(b) $i = 0.09, n = 20$

$$P = R\left[\frac{1 - (1 + i)^{-n}}{i}\right]$$

$$= 50{,}000\left[\frac{1 - (1 + 0.09)^{-20}}{0.09}\right]$$

$$\approx 456{,}427.28$$

The present value is $456,427.28.

For parts (c) and (d), if $1 million is divided into 25 equal payments, each payment is $40,000.

(c) $i = 0.05, n = 25$

$$P = R\left[\frac{1 - (1 + i)^{-n}}{i}\right]$$

$$= 40{,}000\left[\frac{1 - (1 + 0.05)^{-25}}{0.05}\right]$$

$$\approx 563{,}757.78$$

The present value is $563,757.78.

(d) $i = 0.09, n = 25$

$$P = R\left[\frac{1 - (1 + i)^{-n}}{i}\right]$$

$$= 40{,}000\left[\frac{1 - (1 + 0.09)^{-25}}{0.09}\right]$$

$$\approx 392{,}903.18$$

The present value is $392,903.18.

43. Compute the monthly payment using the formula

$$R = \frac{Pi}{1 - (1 + i)^{-n}}$$

with $P = 55{,}000,\ i = 0.068/12,$ and $n = 300.$

$$R = \frac{(55{,}000)\left(\frac{0.068}{12}\right)}{1 - \left(1 + \frac{0.068}{12}\right)^{-300}}$$

$$R = 381.74$$

The monthly payment will be $381.74 and the total paid will be $(381.74)(300) = 114{,}522.00$ or $114,522. The interest paid will be $114,522 − $55,000 = $59,522.

45. $P = 110{,}000$, $i = \frac{0.08}{2} = 0.04$, $n = 9$

$$R = \frac{110{,}000}{a_{\overline{9}|0.04}} \approx \$14{,}794.23$$

is the amount of each payment.

Of the first payment, the company owes interest of

$$I = Prt = 110{,}000(0.08)\left(\tfrac{1}{2}\right) = \$4400.$$

Therefore, from the first payment, \$4400 goes to interest, and the balance.

$$\$14{,}794.23 - 4400 = \$10{,}394.23,$$

goes to principal. The principal at the end of this period is

$$\$110{,}000 - 10{,}394.23 = \$99{,}605.77.$$

The interest for the second payment is

$$I = Prt = 99{,}605.77(0.08)\left(\tfrac{1}{2}\right) \approx \$3984.23$$

Of the second payment, \$3984.23 goes to interest and
$$\$14{,}794.23 - 3984.23 = \$10{,}810.00$$

goes to principal. Continue in this fashion to complete the amortization schedule for the first four payments.

Payment Number	Amount of Payment	Interest for Period	Portion to Principal	Principal at End of Period
0	—	—	—	\$110,000.00
1	\$14,794.23	\$4400.00	\$10,394.23	\$ 99,605.77
2	\$14,794.23	\$3984.23	\$10,810.00	\$ 88,795.77
3	\$14,794.23	\$3551.83	\$11,242.40	\$ 77,553.37
4	\$14,794.23	\$3102.13	\$11,692.10	\$ 65,861.27

47. \$150,000 is the future value of an annuity over 79 yr compounded quarterly. So, there are $79(4) = 316$ payment periods.

(a) The interest per quarter is $\frac{5.25\%}{4} = 1.3125\%$. Thus, $S = 150{,}000$, $n = 316$, $i = 0.013125$, and we must find the quarterly payment R in the formula

$$S = R\left[\frac{(1+i)^n - 1}{i}\right]$$

$$150{,}000 = R\left[\frac{(1.013125)^{316} - 1}{0.013125}\right]$$

$$R \approx 32.4923796$$

She would have to put \$32.49 into her savings at the end of every three months.

(b) For a 2% interest rate, the interest per quarter is $\frac{2\%}{4} = 0.5\%$. Thus, $S = 150{,}000$, $n = 316$, $i = 0.005$, and we must find the quarterly payment R in the formula

$$S = R\left[\frac{(1+i)^n - 1}{i}\right]$$

$$150{,}000 = R\left[\frac{(1.005)^{316} - 1}{0.005}\right]$$

$$R \approx 195.5222794$$

She would have to put \$195.52 into her savings at the end of every three months.

For a 7% interest rate, the interest per quarter is $\frac{7\%}{4} = 1.75\%$. Thus, $S = 150{,}000$, $n = 316$, $i = 0.0175$, and we must find the quarterly payment R in the formula

$$S = R\left[\frac{(1+i)^n - 1}{i}\right]$$

$$150{,}000 = R\left[\frac{(1.0175)^{316} - 1}{0.0175}\right]$$

$$R \approx 10.9663932$$

She would have to put \$10.97 into her savings at the end of every three months.

49. Throughout this exercise, $i = \frac{0.065}{12}$ and $P =$ the total amount financed, which is

$$\$285{,}000 - 60{,}000 = \$225{,}000.$$

(a) $n = 15(12) = 180$

$$R = \frac{Pi}{1 - (1+i)^{-n}}$$

$$= \frac{225{,}000\left(\frac{0.065}{12}\right)}{1 - \left(1 + \frac{0.065}{12}\right)^{-180}}$$

$$\approx 1959.99$$

The monthly payment is \$1959.99.

Total payments $= 180(\$1959.99) = \$352{,}798.20$

Total interest $= \$352{,}798.20 - 225{,}000$

$$= \$127{,}798.20$$

(b) $n = 20(12) = 240$

$$R = \frac{Pi}{1 - (1+i)^{-n}}$$

$$= \frac{225{,}000\left(\frac{0.065}{12}\right)}{1 - \left(1 + \frac{0.065}{12}\right)^{-240}}$$

$$\approx 1677.54$$

The monthly payment is $1677.54.

Total payments $= 240(\$1677.54) = \$402{,}609.60$

Total interest $= \$402{,}609.60 - 225{,}000$

$$= \$177{,}609.60$$

(c) $n = 25(12) = 300$

$$R = \frac{Pi}{1 - (1+i)^{-n}}$$

$$= \frac{225{,}000\left(\frac{0.065}{12}\right)}{1 - \left(1 + \frac{0.065}{12}\right)^{-300}}$$

$$\approx 1519.22$$

The monthly payment is $1519.22.

Total payments $= 300(\$1519.22) = \$455{,}766$

Total interest $= \$455{,}766 - 225{,}000$

$$= \$230{,}766$$

(d) Graph

$$y_1 = 1677.54\left[\frac{1 - \left(1 + \frac{0.065}{12}\right)^{-(240-x)}}{\frac{0.065}{12}}\right] \text{ and}$$

$$y_2 = \frac{285{,}000 - 60{,}000}{2}.$$

The x-coordinate of the point of intersection is 156.44167, which rounds up to 157. Half the loan will be paid after 157 payments.

51. $P = 150{,}000$, $i = \frac{0.082}{12}$, and $n = 30(12) = 360$.

$$R = \frac{Pi}{1 - (1+i)^{-n}}$$

$$= \frac{150{,}000\left(\frac{0.082}{12}\right)}{1 - \left(1 + \frac{0.082}{12}\right)^{-360}}$$

$$\approx 1121.63$$

The monthly payment is $1121.63.

Total payments $= 360(\$1121.63) = \$403{,}786.80$

Total interest $= \$403{,}786.80 - 150{,}000$

$$= \$253{,}786.80$$

(b) 15 years of payments means $15(12) = 180$ payments.

$$y_{15} = 1121.63\left[\frac{1 - \left(1 + \frac{0.082}{12}\right)^{-(360-180)}}{\frac{0.082}{12}}\right]$$

$$\approx 115{,}962.66$$

The unpaid balance after 15 years is approximately $115,962.66.

The total of the remaining 180 payments is

$$180(\$1121.63) = \$201{,}893.40.$$

(c) The unpaid balance from part (b) is the new loan amount. Now $P = 115{,}962.66$, $i = \frac{0.065}{12}$, and again $n = 30(12) = 360$.

$$R = \frac{Pi}{1 - (1+i)^{-n}}$$

$$= \frac{115{,}962.66\left(\frac{0.065}{12}\right)}{1 - \left(1 + \frac{0.065}{12}\right)^{-360}}$$

$$\approx 732.96$$

The new monthly payment would be $732.96.

Total payments $= 360(\$732.96) + \$3400$

$$= \$267{,}265.60$$

(d) Again the unpaid balance from part (b) is the new loan amount. Again $P = 115{,}962.66$ and $i = \frac{0.065}{12}$, and this time $n = 15(12) = 180$.

$$R = \frac{Pi}{1 - (1+i)^{-n}}$$

$$= \frac{115{,}962.66\left(\frac{0.065}{12}\right)}{1 - \left(1 + \frac{0.065}{12}\right)^{-180}}$$

$$\approx 1010.16$$

The new monthly payment would be $1010.16.

Total payments $= 180(\$1010.16) + \$4500$

$$= \$186{,}328.80$$

53. This is just like a sinking fund in reverse.

(a) $P = 150{,}000$, $i = \frac{0.06}{2} = 0.03$, $n = 2(5) = 10$

$$R = \frac{Pi}{1 - (1+i)^{-n}}$$

$$= \frac{150{,}000(.03)}{1 - (1 + 0.03)^{-10}}$$

$$\approx 17{,}584.58$$

The amount of each withdrawal is $17,584.58.

(b) $P = 150,000$, $i = \frac{0.06}{2} = 0.03$, $n = 2(6) = 12$

$$R = \frac{Pi}{1 - (1 + i)^{-n}}$$

$$= \frac{150,000(0.03)}{1 - (1 + 0.03)^{-12}}$$

$$\approx 15,069.31$$

If the money must last 6 yr, the amount of each withdrawal is $15,069.31.

55. This exercise should be solved by graphing calculator or computer methods. The amortization schedule, which may vary slightly, is as follows.

Payment Number	Amount of Payment	Interest for Period	Portion to Principal	Principal at End of Period
0	——	——	——	$4836.00
1	$585.16	$175.31	$409.85	$4426.15
2	$585.16	$160.45	$424.71	$4001.43
3	$585.16	$145.05	$440.11	$3561.32
4	$585.16	$129.10	$456.06	$3105.26
5	$585.16	$112.57	$472.59	$2632.67
6	$585.16	$ 95.43	$489.73	$2142.94
7	$585.16	$ 77.68	$507.48	$1635.46
8	$585.16	$ 59.29	$525.87	$1109.59
9	$585.16	$ 40.22	$544.94	$ 564.65
10	$585.12	$ 20.47	$564.65	$ 0.00

57. (a) Here $R = 1000$ and $i = 0.04$ and we have

$$P = \frac{R}{i} = \frac{100}{0.04} = 25,000$$

Therefore, the present value of the perpetuity is $25,000.

(b) Here $R = 600$ and $i = \frac{0.06}{4} = 0.015$ and we have

$$P = \frac{R}{i} = \frac{600}{0.015} = 40,000$$

Therefore, the present value of the perpetuity is $40,000.

Chapter 5 Review Exercises

1. True

2. False: The ratios of successive pairs of terms are not constant: For example, $\frac{4}{2} = 2$ but $\frac{6}{4} = 1.5$.

3. True

4. False: Both payments and interest on the accumulated value are added to a sinking fund at the end of each time period, so the value increases over time.

5. True

6. True

7. True

8. False: The effective rate formula gives an interest rate, not a present value.

9. False: The correct expression is

$$25,000\left[\frac{0.05/12}{1 - (1 + 0.05/12)^{-72}}\right].$$

10. True

11. $I = Prt$

$$= 15,903(0.06)\left(\frac{8}{12}\right)$$

$$= 636.12$$

The simple interest is $636.12.

13. $I = Prt$

$$= 42,368(0.0522)\left(\frac{7}{12}\right)$$

$$\approx 1290.11$$

The simple interest is $1290.11.

15. For a given amount of money at a given interest rate for a given time period greater than 1, compound interest produces more interest than simple interest.

17. $19,456.11 at 8% compounded semiannually for 7 yr

Use the formula for compound amount with $P = 19,456.11$, $i = \frac{0.08}{2} = 0.04$, and $n = 7(2) = 14$.

$$A = P(1 + i)^n$$

$$= 19,456.11(1.04)^{14}$$

$$\approx 33,691.69$$

The compound amount is $33,691.69.

19. $57,809.34 at 6% compounded quarterly for 5 yr

Use the formula for compound amount with $P = 57,809.34$, $i = \frac{0.06}{4} = 0.015$, and $n = 5(4) = 20$.

$$A = P(1 + i)^n$$
$$= 57,809.34(1.015)^{20}$$
$$\approx 77,860.80$$

The compound amount is $77,860.80.

21. $12,699.36 at 5% compounded semiannually for 7 yr

Here $P = 12,699.36$, $i = \frac{0.05}{2} = 0.025$, and $n = 7(2) = 14$. First find the compound amount.

$$A = P(1 + i)^n$$
$$= 12,699.36(1.025)^{14}$$
$$\approx 17,943.86$$

The compound amount is $17,943.86.

To find the amount of interest earned, subtract the initial deposit from the compound amount. The interest earned is

$$\$17,943.86 - 12,699.36 = \$5244.50.$$

23. $34,677.23 at 4.8% compounded monthly for 32 mo

Here $P = 34,677.23$, $i = \frac{0.048}{12} = 0.004$, and $n = 32$.

$$A = P(1 + i)^n$$
$$= 34,677.23(1.004)^{32}$$
$$\approx 39,402.45$$

The compound amount is $39,402.45

The interest earned is

$$\$39,402.45 - 34,677.23 = \$4725.22.$$

25. $42,000 in 7 yr, 6% compounded monthly

Use the formula for present value for compound interest with $A = 42,000$, $i = \frac{0.06}{12} = 0.005$, and $n = 7(12) = 84$.

$$P = \frac{A}{(1 + i)^n} = \frac{42,000}{(1.005)^{84}} \approx 27,624.86$$

The present value is $27,624.86.

27. $1347.89 in 3.5 yr, 6.77% compounded semiannually

Use the formula for present value for compound interest with $A = 1347.89$, $i = \frac{0.0677}{2} = 0.03385$, and $n = 3.5(2) = 7$.

$$P = \frac{A}{(1 + i)^n} = \frac{1347.89}{(1.03385)^7} \approx 1067.71$$

The present value is $1067.71.

29. $a = 2$; $r = 3$

The first five terms are

$$2, 2(3), 2(3)^2, 2(3)^3, \text{ and } 2(3)^4,$$

or

$$2, 6, 18, 54, \text{ and } 162.$$

31. $a = -3$; $r = 2$

To find the sixth term, use the formula $a_n = ar^{n-1}$ with $a = -3$, $r = 2$, and $n = 6$.

$$a_6 = ar^{6-1} = -3(2)^5 = -3(32) = -96$$

33. $a = -3$; $r = 3$

To find the sum of the first 4 terms of this geometric sequence, use the formula $S_n = \frac{a(r^n - 1)}{r - 1}$ with $n = 4$.

$$S_4 = \frac{-3(3^4 - 1)}{3 - 1} = \frac{-3(80)}{2} = \frac{-240}{2} = -120$$

35. $s_{\overline{n}|i} = \frac{(1+i)^n - 1}{i}$

$$s_{\overline{30}|0.02} = \frac{(1.02)^{30} - 1}{0.02} \approx 40.56808$$

39. $R = 1288$, $i = 0.04$, $n = 14$

This is an ordinary annuity.

$$S = Rs_{\overline{n}|i}$$
$$S = 1288s_{\overline{14}|0.04}$$
$$= 1288\left[\frac{(1 + 0.04)^{14} - 1}{0.04}\right]$$
$$\approx 23,559.98$$

The future value is $23,559.98.

The total amount deposited is $1288(14) = \$18,032$.

Thus, the amount of interest is

$$\$23,559.98 - 18,032 = \$5527.98.$$

41. $R = 233$, $i = \frac{0.048}{12} = 0.004$, $n = 4(12) = 48$

This is an ordinary annuity.

$$S = R\left[\frac{(1+i)^n - 1}{i}\right]$$

$$S = 233\left[\frac{(1.004)^{48} - 1}{0.004}\right]$$

$$\approx 12{,}302.78$$

The future value is $12,302.78.

The total amount deposited is $233(48) = \$11{,}184$.

Thus, the amount of interest is

$$\$12{,}302.78 - 11{,}184 = \$1118.78.$$

43. $R = 11{,}900$, $i = \frac{0.06}{12} = 0.005$, $n = 13$

This is an annuity due, so we use the formula for future value of an ordinary annuity, but include one additional time period and subtract the amount of one payment.

$$S = R\left[\frac{(1+i)^{n+1} - 1}{i}\right] - R$$

$$= 11{,}900\left[\frac{(1.005)^{14} - 1}{0.005}\right] - 11{,}900$$

$$\approx 160{,}224.29$$

The future value is $160,224.29.

The total amount deposited is $11,900(13) = $154,700.

Thus, the amount of interest is

$$\$160{,}224.29 - 154{,}700 = \$5524.29.$$

45. $6500; money earns 5% compounded annually; 6 annual payments

$$S = 6500, i = 0.05, n = 6$$

Let R be the amount of each payment.

$$S = Rs_{\overline{n}|i}$$

$$R = \frac{6500}{s_{\overline{6}|0.05}}$$

$$= \frac{6500(0.05)}{(1.05)^6 - 1}$$

$$\approx 955.61$$

The amount of each payment is $955.61.

47. $233,188; money earns 5.2% compounded quarterly for $7\frac{3}{4}$ years.

$$S = 233{,}188, i = \frac{0.052}{4} = 0.013, n = \left(7\frac{3}{4}\right)(4) = 31$$

Let R be the amount of each payment.

$$S = Rs_{\overline{n}|i}$$

$$R = \frac{233{,}188}{s_{\overline{31}|0.013}}$$

$$= \frac{233{,}188(0.013)}{(1.013)^{31} - 1}$$

$$\approx 6156.14$$

The amount of each payment is $6156.14.

49. Deposits of $850 annually for 4 years at 6% compounded annually

Use the formula for the present value of an annuity with $R = 850$, $i = 0.06$, and $n = 4$.

$$P = R\left[\frac{1 - (1+i)^{-n}}{i}\right]$$

$$= 850\left[\frac{1 - (1+0.06)^{-4}}{0.06}\right]$$

$$\approx 2945.34$$

The present value is $2945.34.

51. Deposits of $4210 semiannually for 8 years at 4.2% compounded annually

Use the formula for the present value of an annuity with $R = 4210$, $i = \frac{0.042}{2} = 0.021$, $n = 8(2) = 16$.

$$P = R\left[\frac{1 - (1+i)^{-n}}{i}\right]$$

$$= 4210\left[\frac{1 - (1.021)^{-16}}{0.021}\right]$$

$$\approx 56{,}711.93$$

The present value is $56,711.93.

53. Two types of loans that are commonly amortized are home loans and auto loans.

55. $P = 3200$, $i = \frac{0.08}{4} = 0.02$, $n = 12$

$$R = \frac{Pi}{1 - (1 + i)^{-n}}$$

$$= \frac{3200(0.02)}{1 - (1.02)^{-12}}$$

$$\approx 302.59$$

The amount of each payment is $302.59.

The total amount paid is $302.59(12) = \$3631.08$. Thus, the total interest paid is

$$\$3631.08 - 3200 = \$431.08.$$

57. $P = 51,607$, $i = \frac{0.08}{12} = 0.00\overline{6}$, $n = 32$

$$R = \frac{Pi}{1 - (1 + i)^{-n}}$$

$$= \frac{51,607(0.00\overline{6})}{1 - (1.00\overline{6})^{-32}}$$

$$\approx 1796.20$$

The amount of each payment is $1796.20.

The total amount paid is $1796.20(32) = \$57,478.40$. Thus, the total interest paid is

$$\$57,478.40 - 51,607 = \$5871.40.$$

59. $P = 177,110$, $i = \frac{0.0668}{12} = 0.00556$,

$n = 30(12) = 360$

$$R = \frac{Pi}{1 - (1 + i)^{-n}}$$

$$= \frac{177,110(0.00556)}{1 - (1.00556)^{-360}}$$

$$\approx 1140.50$$

The amount of each payment is $1140.50.

The total amount paid is $1140.50(360) = \$410,580$. Thus, the total interest paid is

$$\$410,580 - 177,110 = \$233,470.$$

61. The answer can be found in the table under payment number 12 in the column labeled "Portion to Principal." The amount of principal repayment included in the fifth payment is $132.99.

63. The last entry in the column "Principal at End of Period," $125,464.39, shows the debt remaining at the end of the first year (after 12 payments). Since the original debt (loan principal) was $127,000, the amount by which the debt has been reduced at the end of the first year is

$$\$127,000 - 125,464.39 = \$1535.61.$$

65. Here $P = 9820$, $r = 6.7\% = 0.067$, and $t = \frac{7}{12}$.

$$I = Prt$$

$$= 9820(0.067)\left(\frac{7}{12}\right)$$

$$\approx 383.80$$

The interest he will pay is $383.80. The total amount he will owe in 7 mo is

$$\$9820 + 383.80 = \$10,203.80.$$

67. $P = 84,720$, $t = \frac{7}{12}$, $I = 4055.46$

Substitute these values into the formula for simple interest to find the value of r.

$$I = Prt$$

$$4055.46 = 84,720r\left(\frac{7}{12}\right)$$

$$4055.46 = 49,420r$$

$$0.0821 \approx r$$

The interest rate is 8.21%.

69. In both cases use the formula for compound amount with $P = 500$ and $i = \frac{0.05}{4} = 0.0125$. For the investment at age 23 use $n = 42(4) = 168$.

$$A = P(1 + i)^n$$

$$= 500(1 + 0.0125)^{168}$$

$$\approx 4030.28$$

For the investment at age 40 use $n = 25(4) = 100$.

$$A = P(1 + i)^n$$

$$= 500(1 + 0.0125)^{100}$$

$$\approx 1731.70$$

The increased amount of money Tom will have if he invests now is

$$\$4030.28 - 1731.70 = \$2298.58.$$

71. $R = 5000$, $i = \frac{0.10}{2} = 0.05$, $n = 7\frac{1}{2}(2) = 15$

This is an ordinary annuity.

$$S = R\left[\frac{(1 + i)^n - 1}{i}\right]$$

$$S = 5000\left[\frac{(1 + 0.05)^{15} - 1}{0.05}\right]$$

$$\approx 107,892.82$$

The future value is \$107,892.82. The amount of interest earned is

$$\$107,892.82 - 15(5000) = \$32,892.82.$$

73. Use the formula for amortization payments with $P = 48,000$, $i = 0.065$, and $n = 7$.

$$R = \frac{Pi}{1 - (1+i)^{-n}}$$

$$= \frac{48,000(0.065)}{1 - (1.065)^{-7}}$$

$$\approx 8751.91$$

The owner should deposit \$8751.91 at the end of each year.

The total amount deposited is $\$8751.91(7) = \$61,263.37$. Thus, the total interest paid is

$$\$61,263.37 - 48,000 = \$13,263.37.$$

75. The effective rate paid by Ascencia would be

$$\left(1 + \frac{0.0149}{12}\right)^{12} - 1 = 0.015$$

or 1.50%.

The effective rate paid by giantbank.com would be

$$\left(1 + \frac{0.0145}{360}\right)^{360} - 1 = 0.0146$$

or 1.46%. Ascencia has the higher effective rate.

77. (a) For the 0% financing, the payments are simply 1/60 of the financed amount.

$$\frac{16,000}{60} = 266.67$$

Thus the rounded payments are \$266.67, and the total payments are equal to the financed amount of \$16,000. (In fact because of rounding the total payments are 20 cents more than this amount, but the final payment would be reduced by 20 cents to compensate.)

(b) For the 3.9% financing the monthly payment will be

$$\frac{(16,000)\left(\dfrac{0.039}{12}\right)}{1 - \left(1 + \dfrac{0.039}{12}\right)^{-72}} = 249.59$$

or \$249.59. The total payments will be 72 times this amount, or \$17,970.48.

(c) For the cash back option with a 6.33% interest rate for 489 months, the monthly payment will be

$$\frac{(12,000)\left(\dfrac{0.0633}{12}\right)}{1 - \left(1 + \dfrac{0.0633}{12}\right)^{-48}} = 283.64$$

or \$283.64. The total payments will be 48 times this amount, or \$13,614.72.

79. Amount of loan $= \$191,000 - 40,000$

$$= \$151,000$$

(a) Use the formula for amortization payments with $P = 151,000$, $i = \frac{0.065}{12} = 0.00541\overline{6}$, and $n = 30(12) = 360$.

$$R = \frac{Pi}{1 - (1+i)^{-n}}$$

$$= \frac{151,000(0.00541\overline{6})}{1 - (1.00541\overline{6})^{-360}}$$

$$\approx 954.42$$

The monthly payment for this mortgage is \$954.42.

(b) To find the amount of the first payment that goes to interest, use $I = Prt$ with $P = 151,000$, $i = 0.00541\overline{6}$, and $t = 1$.

$$I = 151,000(0.065)\left(\tfrac{1}{12}\right) = 817.92$$

Of the first payment, \$817.92 is interest.

(c) Using method 1, since 180 of 360 payments were made, there are 180 remaining payments. The present value is

$$954.42\left[\frac{1 - (1.00541\overline{6})^{-180}}{0.00541\overline{6}}\right] \approx 109,563.99,$$

so the remaining balance is \$109,563.99.

Using method 2, since 180 payments were already made, we have

$$954.42\left[\frac{1 - (1.00541\overline{6})^{-180}}{0.00541\overline{6}}\right] \approx 109,563.99.$$

She still owes

$$\$151,000 - 109,563.99 = \$41,436.01.$$

Furthermore, she owes the interest on this amount for 180 mo, for a total remaining balance of

$$41,436.01(1.00541\overline{6})^{180} = 109,565.13.$$

(d) Closing costs $= 3700 + 0.025(238,000)$

$$= 3700 + 5950$$

$$= 9650$$

Closing costs are $9650.

(e) Amount of money received

$=$ Selling price $-$ Closing costs $-$ Current mortgage balance

Using method 1, the amount received is

$238,000 - 9650 - 109,563.99 = \$118,786.01$.

Using method 2, the amount received is

$238,000 - 9650 - 109,565.13 = \$118,784.87$.

81. (a) Use the formula for effective rate with $r_E = 0.10$ and $m = 12$.

$$r_E = \left(1 + \frac{r}{m}\right)^m - 1$$

$$0.10 = \left(1 + \frac{r}{12}\right)^{12} - 1$$

$$1.10 = \left(1 + \frac{r}{12}\right)^{12}$$

$$(1.10)^{1/12} = 1 + \frac{r}{12}$$

$$1.007974 \approx 1 + \frac{r}{12}$$

$$0.007974 \approx \frac{r}{12}$$

$$0.095688 \approx r$$

The annual interest rate is 9.569%.

(b) Use the formula for amortization payments with $P = 140,000$, $i = \frac{0.06625}{12}$, and $n = 30(12) = 360$.

$$R = \frac{Pi}{1 - (1 + i)^{-n}}$$

$$= \frac{140,000\left(\frac{0.06625}{12}\right)}{1 - \left(1 + \frac{0.06625}{12}\right)^{-360}}$$

$$\approx 896.44$$

Her monthly payment is $896.44.

(c) This investment is an annuity with $R = 1200 - 896.44 = 303.56$, $i = \frac{0.09569}{12}$, and $n = 30(12) = 360$. The future value is

$$S = R\left[\frac{(1 + i)^n - 1}{i}\right]$$

$$= 303.56\left[\frac{\left(1 + \frac{0.09569}{12}\right)^{360} - 1}{\frac{0.09569}{12}}\right]$$

$$\approx 626,200.88$$

In 30 yr she will have $626,200.88 in the fund.

(d) Use the formula for amortization payments with $P = 14,000$, $i = \frac{0.0625}{12}$, and $n = 15(12) = 180$.

$$R = \frac{Pi}{1 - (1 + i)^{-n}}$$

$$= \frac{140,000\left(\frac{0.0625}{12}\right)}{1 - \left(1 + \frac{0.0625}{12}\right)^{-180}}$$

$$\approx 1200.39$$

His monthly payment is $1200.39.

(e) This investment is an annuity with $R = 1200$, $i = \frac{0.09569}{12}$, and $n = 15(12) = 180$. The future value is

$$S = R\left[\frac{(1 + i)^n - 1}{i}\right]$$

$$= 1200\left[\frac{\left(1 + \frac{0.09569}{12}\right)^{180} - 1}{\frac{0.09569}{12}}\right]$$

$$\approx 478,134.14$$

In 30 yr he will have $478,134.14.

(f) Sue is ahead by

$626,200.88 - 478,134.14 = \$148,066.74$.

LOGIC

6.1 Statements and Quantifiers

Your Turn 1
No; the statement "I bought Ben and Jerry's ice cream." is not compound because the "and" is not used as a logical connective that connects two simple statements.

Your Turn 2
The negative of the given statement is "Wal-Mart is the largest corporation in the USA."

Your Turn 3
The negation of $4x + 2y < 5$ is $4x + 2y \geq 5$.

Your Turn 4
$h \wedge \sim r$ represents "My backpack is heavy, and it is not going to rain."

Your Turn 5
p represents " $7 < 2$," which is false.

q represents " $4 > 3$," which is true.

$$\sim p \wedge q$$
$$\text{T} \wedge \text{T}$$
$$\text{T}$$

The statement $\sim p \wedge q$ is true.

Your Turn 6
p is false, q is true, and r is false.

$$(\sim p \wedge q) \vee r$$
$$(\sim \text{F} \wedge \text{T}) \vee \text{F}$$
$$(\text{T} \wedge \text{T}) \vee \text{F}$$
$$\text{T} \vee \text{F}$$
$$\text{T}$$

The statement $(\sim p \wedge q) \vee r$ is true.

Your Turn 7
p represents " $7 < 2$," q represents " $4 > 3$," and r represents " $2 > 8$."

$$p \vee (\sim q \wedge r)$$
$$\text{F} \vee (\sim \text{T} \wedge \text{F})$$
$$\text{F} \vee (\text{F} \wedge \text{F})$$
$$\text{F} \vee \text{F}$$
$$\text{F}$$

The statement $p \vee (\sim q \wedge r)$ is false.

6.1   Exercises

1. Because the declarative sentence "Montevideo is the capital of Uruguay" has the property of being true or false, it is considered a statement. It is not compound.

3. "Don't feed the animals" is not a declarative sentence and does not have the property of being true or false. Hence, it is not considered a statement.

5. " $2 + 2 = 5$ and $3 + 3 = 7$ " is a declarative sentence that is true or false and, therefore, is considered a statement. It is compound.

7. "Got milk?" is a question, not a declarative sentence, and, therefore, is not considered a statement.

9. "I am not a crook" is a compound statement because it contains the logical connective "not."

11. "She enjoyed the comedy team of Penn and Teller" is not compound because only one assertion is being made.

13. "If I get an A, I will celebrate" is a compound statement because it consists of two simple statements combined by the connective "if . . . then."

15. The negation of "My favorite flavor is chocolate" is "My favorite flavor is not chocolate."

17. A negation for "$y > 12$" (without using a slash sign) would be "$y \leq 12$."

19. A negation for "$q \geq 5$" would be "$q < 5$."

23. A translation of "$\sim b$" is "I'm not getting better."

25. A translation of "$\sim b \vee d$" is "I'm not getting better or my parrot is dead."

27. A translation of "$\sim(b \wedge \sim d)$" is "It is not the case that both I'm getting better and my parrot is not dead."

29. If q is false, then $(p \wedge \sim q) \wedge q$ is false, since both parts of the conjunction must be true for the compound statement to be true.

31. If the conjunction $p \wedge q$ is true, then both p and q must be true. Thus, q must be true.

33. If $\sim(p \vee q)$ is true, then $p \vee q$ must be false, since a statement and its negation have opposite truth values. In order for the disjunction $p \vee q$ to be false, both component statements must be false. Thus, p and q are both false.

35. Since p is false, $\sim p$ is true, since a statement and its negation have opposite truth values.

37. Since p is false and q is true, we may consider the statement $p \vee q$ as

$$F \vee T,$$

which is true by the *or* truth table. That is, $p \vee q$ is true.

39. Since p is false and q is true, we may consider $p \vee \sim q$ as

$$F \vee \sim T$$
$$F \vee F$$
$$F.$$

That is, $p \vee \sim q$ is false.

41. With the given truth values for p and q, we may consider $\sim p \vee \sim q$ as

$$\sim F \vee \sim T$$
$$T \vee F$$
$$T.$$

Thus, $\sim p \vee \sim q$ is true.

43. Replacing p and q with the given truth values, we have

$$\sim(F \wedge \sim T)$$
$$\sim(F \wedge F)$$
$$\sim F$$
$$T.$$

Thus, the compound statement $\sim(p \wedge \sim q)$ is true.

45. Replacing p and q with the given truth values, we have

$$\sim[\sim F \wedge (\sim T \vee F)]$$
$$\sim[T \wedge (F \vee F)]$$
$$\sim[T \wedge F]$$
$$\sim F$$
$$T.$$

Thus, the compound statement $\sim[\sim p \wedge (\sim q \vee p)]$ is true.

47. The statement $3 \geq 1$ is a disjunction since it means "$3 > 1$" or "$3 = 1$."

49. Replacing p, q, and r with the given truth values, we have

$$(T \wedge F) \vee \sim F$$
$$F \vee T$$
$$T.$$

Thus, the compound statement $(p \wedge r) \vee \sim q$ is true.

51. Replacing p, q, and r with the given truth values, we have

$$T \wedge (F \vee F)$$
$$T \wedge F$$
$$F.$$

Thus, the compound statement $p \wedge (q \vee r)$ is false.

53. Replacing p, q, and r with the given truth values, we have

$$\sim(T \wedge F) \wedge (F \vee \sim F)$$
$$\sim F \wedge (F \vee T)$$
$$T \wedge T$$
$$T.$$

Thus, the compound statement $\sim(p \wedge q) \wedge (r \vee \sim q)$ is true.

55. Replacing p, q, and r with the given truth values, we have

$$\sim[(\sim T \wedge F) \vee F]$$
$$\sim[(F \wedge F) \vee F]$$
$$\sim[F \vee F]$$
$$\sim F$$
$$T.$$

Thus, the compound statement $\sim[(p \wedge q) \vee r]$ is true.

57. Since p is false and r is true, we have

$$F \wedge T$$
$$F.$$

The compound statement $p \wedge r$ is false.

59. Since q is false and r is true, we have

$$\sim F \vee \sim T$$
$$T \vee F$$
$$T.$$

The compound statement $\sim q \vee \sim r$ is true.

61. Since p and q are false and r is true, we have

$$(F \wedge F) \vee T$$
$$F \vee T$$
$$T.$$

The compound statement $(p \wedge q) \vee r$ is true.

63. Since p and q are false and r is true, we have

$$(\sim T \wedge F) \vee \sim F$$
$$(F \wedge F) \vee T$$
$$F \vee T$$
$$T.$$

The compound statement $(\sim r \wedge q) \vee \sim p$ is true.

65. **(b)**, **(c)**, and **(d)** are declarative sentences that are true or false and are therefore statements.

(a) is a command, not a declarative sentence, and is therefore not a statement.

67. An individual has to be your biological child to be a "qualifying" child.

69. $a \wedge j$

71. $\sim a \vee j$

73. Statements a and j are both true.

$a \wedge j$	$\sim a \wedge \sim j$	$\sim a \vee j$	$a \vee j$
$T \wedge T$	$F \wedge F$	$F \vee T$	$T \vee T$
T	F	T	T

Exercises 69, 71, and 72 are true.

75. **(a)** This is a question, not a declarative sentence, and therefore not a statement.

(b), **(c)**, **(d)**, **(e)** There are declarative sentences that are true or false and therefore statements.

77. The negation of "You may find that exercise helps you cope with stress" is "You may not find that exercise helps you cope with stress."

79. **(c)**, and **(d)** are compound statements that are formed by the disjunction *or*.

83. "New Orleans won the Super Bowl but Peyton Manning is not the best quarterback" may be symbolized as $n \wedge \sim m$.

85. "New Orleans did not win the Super Bowl or Peyton Manning is the best quarterback" may be symbolized as $\sim n \vee m$.

87. "Neither did New Orleans win the Super Bowl nor is Peyton Manning the best quarterback" may be symbolized as $\sim n \wedge \sim m$ or $\sim(n \vee m)$.

89. Assume that n is true and m is true. Under these conditions, the statements in Exercises 83–88 have the following truth values:

83. $n \wedge \sim m$: False, because n is true and $\sim m$ is false.

84. $\sim n \vee \sim m$: False, because $\sim n$ is false and $\sim m$ is false.

85. ~n ∨ m: True, because ~n is false and m is true.

86. ~n ∧ m: False, because ~n is false and m is true.

87. ~n ∧ ~m: False, because ~n is false and ~m is false.

88. n ∨ m ∧ [~(n ∧ m)]: False, since (n ∨ m) is true but ~(n ∧ m) is false.

Therefore, only Exercise 85 is a true statement.

6.2   Truth Tables and Equivalent Statements

Your Turn 1

p	q	~p	~p ∨ q	p ∧ (~p ∨ q)
T	T	F	T	T
T	F	F	F	F
F	T	T	T	F
F	F	T	T	F

Your Turn 2

Let p represent "I order pizza" and d represent "You make dinner." Then the statement "I do not order pizza, or you do not make dinner and I order pizza" can be represented by ~p ∨ (~d ∧ p).

p	d	~p	~d	~d ∧ p	~p ∨ (~d ∧ p)
T	T	F	F	F	F
T	F	F	T	T	T
F	T	T	F	F	T
F	F	T	T	F	T

Your Turn 3

Let d represent "You do make dinner" and p represent "I order pizza." "You do not make dinner or I order pizza" is symbolically ~d ∨ p. Applying DeMorgan's first law, the negation is ~(~d ∨ p) = d ∧ ~p. In words this reads "You make dinner and I do not order pizza."

6.2   Exercises

1. Since there are two simple statements (p and r), we have $2^2 = 4$ rows in the truth table.

3. Since there are four simple statements (p, q, r, and s), we have $2^4 = 16$ rows in the truth table.

5. Since there are seven simple statements (p, q, r, s, t, u, and v), we have $2^7 = 128$ rows in the truth table.

7. If the truth table for a certain compound statement has 64 rows, then there must be six distinct component statements since $2^6 = 64$.

9. ~p ∧ q

p	q	~p	~p ∧ q
T	T	F	F
T	F	F	F
F	T	T	T
F	F	T	F

11. ~(p ∧ q)

p	q	p ∧ q	~(p ∧ q)
T	T	T	F
T	F	F	T
F	T	F	T
F	F	F	T

13. (q ∨ ~p) ∨ ~q

p	q	~p	~q	q ∨ ~p	(q ∨ ~p) ∨ ~q
T	T	F	F	T	T
T	F	F	T	F	T
F	T	T	F	T	T
F	F	T	T	T	T

In Exercises 15–23 to save space we are using the alternative method, filling in columns in the order indicated by the numbers. Observe that columns with the same number are combined (by the logical definition of the connective) to get the next numbered column. Note that <u>this is different</u> from the way the numbered columns are used in the textbook. Remember that the last column (highest numbered column) completed yields the truth values for the complete compound statement. Be sure to align truth values under the appropriate logical connective or simple statement.

15. $\sim q \land (\sim p \lor q)$

p	q	$\sim q$	$\land$	$(\sim p$	$\lor$	$q)$
T	T	F	F	F	T	T
T	F	T	F	F	F	F
F	T	F	F	T	T	T
F	F	T	T	T	T	F
		1	4	2	3	2

17. $(p \lor \sim q) \land (p \land q)$

p	q	$(p$	$\lor$	$\sim q)$	$\land$	$(p$	$\land$	$q)$
T	T	T	T	F	T	T	T	T
T	F	T	T	T	F	T	F	F
F	T	F	F	F	F	F	F	T
F	F	F	T	T	F	F	F	F
		1	2	1	5	3	4	3

19. $(\sim p \land q) \land r$

p	q	r	$(\sim p$	$\land$	$q)$	$\land$	r
T	T	T	F	F	T	F	T
T	T	F	F	F	T	F	F
T	F	T	F	F	F	F	T
T	F	F	F	F	F	F	F
F	T	T	T	T	T	T	T
F	T	F	T	T	T	F	F
F	F	T	T	F	F	F	T
F	F	F	T	F	F	F	F
			1	2	1	4	3

21. $(\sim p \land \sim q) \lor (\sim r \lor \sim p)$

p	q	r	$(\sim p$	$\land$	$\sim q)$	$\lor$	$(\sim r$	$\lor$	$\sim p)$
T	T	T	F	F	F	F	F	F	F
T	T	F	F	F	F	T	T	T	F
T	F	T	F	F	T	F	F	F	F
T	F	F	F	F	T	T	T	T	F
F	T	T	T	F	F	T	F	T	T
F	T	F	T	F	F	T	T	T	T
F	F	T	T	T	T	T	F	T	T
F	F	F	T	T	T	T	T	T	T
			1	2	1	5	3	4	3

23. $\sim(\sim p \land \sim q) \lor (\sim r \lor \sim s)$

p	q	r	s	$\sim$	$(\sim p$	$\land$	$\sim q)$	$\lor$	$(\sim r$	$\lor$	$\sim s)$
T	T	T	T	T	F	F	F	T	F	F	F
T	T	T	F	T	F	F	F	T	F	T	T
T	T	F	T	T	F	F	F	T	T	T	F
T	T	F	F	T	F	F	F	T	T	T	T
T	F	T	T	T	F	F	T	T	F	F	F
T	F	T	F	T	F	F	T	T	F	T	T
T	F	F	T	T	F	F	T	T	T	T	F
T	F	F	F	T	F	F	T	T	T	T	T
F	T	T	T	T	T	F	F	T	F	F	F
F	T	T	F	T	T	F	F	T	F	T	T
F	T	F	T	T	T	F	F	T	T	T	F
F	T	F	F	T	T	F	F	T	T	T	T
F	F	T	T	F	T	T	T	F	F	F	F
F	F	T	F	F	T	T	T	T	F	T	T
F	F	F	T	F	T	T	T	T	T	T	F
F	F	F	F	F	T	T	T	T	T	T	T
				3	1	2	1	6	4	5	4

25. "It's vacation and I am having fun" has the symbolic form $p \wedge q$. The negation, $\sim(p \wedge q)$, is equivalent, by one of DeMorgan's laws, to $\sim p \vee \sim q$. The corresponding word statement is "It's not vacation or I am not having fun."

27. "Either the door was unlocked or the thief broke a window" has the symbolic form $p \vee q$. The negation, $\sim(p \vee q)$, is equivalent, by one of DeMorgan's laws, to $\sim p \wedge \sim q$. The corresponding word statement is "The door was locked and the thief didn't break a window."

29. "I'm ready to go, but Naomi Bahary isn't" has the symbolic form $p \wedge \sim q$. (The connective "but" is logically equivalent to "and.") The negation, $\sim(p \wedge \sim q)$, is equivalent, by one of DeMorgan's laws, to $\sim p \vee q$. The corresponding word statement is "I'm not ready to go, or Naomi Bahary is."

31. "$12 > 4$ or $8 = 9$" has the symbolic form $p \vee q$. The negation, $\sim(p \vee q)$, is equivalent, by one of DeMorgan's laws, to $\sim p \wedge \sim q$. The corresponding statement is "$12 \leq 4$ and $8 \neq 9$." (Note that the inequality "$\leq$" is logically equivalent to "$\not>$.")

33. "Larry or Moe is out sick today" has the symbolic form $p \vee q$. The negation, $\sim(p \vee q)$, is equivalent, by one of DeMorgan's laws, to $\sim p \wedge \sim q$. The corresponding word statement is "Neither Larry nor Moe is out sick today."

35. $p \veebar q$

p	q	$p \veebar q$
T	T	F
T	F	T
F	T	T
F	F	F

Observe that it is only the first line in the truth table that changes for "exclusive disjunction" since the component statements can not both be true at the same time.

37. "$(3 + 1 = 4) \veebar (2 + 5 = 9)$" is <u>true</u> since the first component statement is true and the second is false.

39. Store the truth values of the statements p, q, and s as P, Q, and S, respectively.

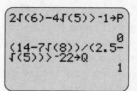
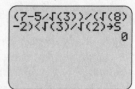

Use the stored values of P, Q, and S to find the truth values of each of the compound statements.

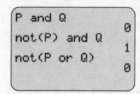
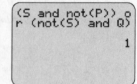

(a) P and Q returns 0, meaning $p \wedge q$ is false.

(b) not (P) and Q returns 1, meaning $\sim p \wedge q$ is true.

(c) not (P or Q) returns 0, meaning $\sim(p \vee q)$ is false.

(d) (S and not (P)) or (not (S) and Q) returns 1, meaning $(s \wedge \sim p) \vee (\sim s \wedge q)$ is true.

41. Service will not be performed at the location, and the store may not send the Covered Equipment to an Apple repair service location to be repaired.

43. Letting s represent "You will be completely satisfied," r represent "We will refund your money," and q represent "We will ask you questions," the guarantee translates into symbols as $s \vee (r \wedge \sim q)$. The truth table follows.

s	r	q	s	$\vee$	$(r$	$\wedge$	$\sim q)$
T	T	T	T	T	T	F	F
T	T	F	T	T	T	T	T
T	F	T	T	T	F	F	F
T	F	F	T	T	F	F	T
F	T	T	F	F	T	F	F
F	T	F	F	T	T	T	T
F	F	T	F	F	F	F	F
F	F	F	F	F	F	F	T
			1	1	4	2	3	2

The guarantee would be false in the three cases indicated as "F" in the column labeled "4." These would be if you are not completely satisfied, and they either don't refund your money or they ask you questions.

45. Inclusive, since the "and" case is allowed.

47. Letting p represent "Liberty without learning is always in peril" and q represent "Learning without liberty is always in vain," the quote translates symbolically as $p \wedge q$. Its negation is equivalent by DeMorgan's laws to $\sim p \vee \sim q$. A translation for the negation is "Liberty without learning is not always in peril, or learning without liberty is not always in vain."

49. p: You could reroll the die again for your Large Straight.

q: You could set aside the 2 Twos and roll for your Twos or for 3 of a Kind.
The statement is $p \vee q$.

Its negation is $\sim p \wedge \sim q$.

Negation: You cannot reroll the die again for your Large Straight and you cannot set aside the 2 Twos and roll for your Twos or for 3 of a Kind.

6.3   The Conditional and Circuits

Your Turn 1

If Little Rock is the capital of Arkansas, then New York City is the capital of New York.

The first component of this conditional is true, while the second is false. The given statement is false.

Your Turn 2

$$(4 > 5) \to F$$

The antecedent $4 > 5$ is false, while the consequent is false. The given statement is true.

Your Turn 3

p, q, and r are all false.

$$\sim q \to (p \to r)$$
$$T \to (F \to F)$$
$$T \to T$$
$$T$$

Your Turn 4

p: You do the homework.

q: You will pass the quiz.

$$p \to q \equiv \sim p \vee q$$

"If you do the homework, then you will pass the quiz" is equivalent to "You do not do the homework, or you will pass the quiz."

Your Turn 5

Since the negation of $p \to q$ is $p \wedge \sim q$, the negation of "If you are on time, then we will be on time" is "You are on time, and we will not be on time."

6.3   Exercises

1. The statement "If the antecedent of a conditional statement is false, the conditional statement is true" is <u>true</u>, since a false antecedent will always yield a true conditional statement.

3. The statement "If q is true, then $(p \wedge q) \to q$ is true" is <u>true</u>, since with a true consequent the conditional statement is always true (even though the antecedent may be false).

5. "Given that $\sim p$ is true and q is false, the conditional $p \to q$ is true" is a <u>true</u> statement since the antecedent, p, must be false.

9. "$F \to (4 \neq 7)$" is a <u>true</u>, statement, since a false antecedent always yields a conditional statement which is true.

11. "$(4 = 11 - 7) \to (8 > 0)$" is <u>true</u> since the antecedent and the consequent are both true.

13. $d \to (e \wedge s)$ expressed in words becomes "If she dances tonight, then I'm leaving early and he sings loudly."

15. $\sim s \to (d \vee \sim e)$ expressed in words becomes "If he doesn't sing loudly, then she dances tonight or I'm not leaving early."

17. The statement "My dog ate my homework, or if I receive a failing grade then I'll run for governor" can be symbolized as $d \vee (f \to g)$.

19. The statement "I'll run for governor if I don't receive a failing grade" can be symbolized as $\sim f \to g$.

21. Replacing r and p with the given truth values, we have

$$\sim F \to F$$
$$T \to F$$
$$F.$$

Thus, the statement $\sim r \to p$ is false.

23. Replacing p, q, and r with the given truth values, we have

$$\sim F \to (T \wedge F)$$
$$T \to F$$
$$F.$$

Thus, the statement $\sim p \to (q \wedge r)$ is false.

25. Replacing p, q, and r with the given truth values, we have

$$\sim T \to (F \wedge F)$$
$$F \to F$$
$$T.$$

Thus, the statement $\sim q \to (p \wedge r)$ is true.

27. Replacing p, q, and r with the given truth values, we have

$$(\sim F \to \sim T) \to (\sim F \wedge \sim F)$$
$$(T \to F) \to (T \wedge T)$$
$$F \to T$$
$$T.$$

Thus, the statement $(p \to \sim q) \to (\sim p \wedge \sim r)$ is true.

31. $\sim p \to \sim q \equiv \sim(\sim p) \vee \sim q$ By Equivalent Statement 8

$\sim(\sim p) \vee \sim q \equiv p \vee \sim q$ By Equivalent Statement 7

So, $\sim p \to \sim q \equiv p \vee \sim q$ and $\sim p \to \sim q$ and $p \vee \sim q$ have the same truth values.

The negation of $p \vee \sim q$ is $\sim(p \vee \sim q)$. Since $p \vee \sim q$ and $\sim(p \vee \sim q)$ cannot have the same truth value, $\sim p \to \sim q$ and $\sim(p \vee \sim q)$ cannot both be true. Thus, $(\sim p \to \sim q) \wedge \sim(p \vee \sim q)$ is a contradiction.

33. $\sim q \to p$

p	q	$\sim q$	$\to$	p
T	T	F	T	T
T	F	T	T	T
F	T	F	T	F
F	F	T	F	F

35. $(p \vee \sim p) \to (p \wedge \sim p)$

p	$(p$	$\vee$	$\sim p)$	$\to$	$(p$	$\wedge$	$\sim p)$
T	T	T	F	F	T	F	F
F	F	T	T	T	F	F	T
	1	2	1	5	3	4	3

Since the statement is always false (the truth values in column 5 are all false), it is a contradiction.

37. $(p \vee q) \to (q \vee p)$

p	q	$(p$	$\vee$	$q)$	$\to$	$(q$	$\vee$	$p)$
T	T	T	T	T	T	T	T	T
T	F	T	T	F	T	F	T	T
F	T	F	T	T	T	T	T	F
F	F	F	F	F	T	F	F	F
		1	2	1	5	3	4	3

Since this statement is always true (column 5), it is a tautology.

39. $r \to (p \wedge \sim q)$

p	q	r	r	$\to$	$(p$	$\wedge$	$\sim q)$
T	T	T	T	F	T	F	F
T	T	F	F	T	T	F	F
T	F	T	T	T	T	T	T
T	F	F	F	T	T	T	T
F	T	T	T	F	F	F	F
F	T	F	F	T	F	F	F
F	F	T	T	F	F	F	T
F	F	F	F	T	F	F	T
			1	4	2	3	2

41. $(\sim r \to s) \vee (p \to \sim q)$

p	q	r	s	$(\sim r$	$\to$	$s)$	$\vee$	$(p$	$\to$	$\sim q)$
T	T	T	T	F	T	T	T	T	F	F
T	T	T	F	F	T	F	T	T	F	F
T	T	F	T	T	T	T	T	T	F	F
T	T	F	F	T	F	F	F	T	F	F
T	F	T	T	F	T	T	T	T	T	T
T	F	T	F	F	T	F	T	T	T	T
T	F	F	T	T	T	T	T	T	T	T
T	F	F	F	T	F	F	T	T	T	T
F	T	T	T	F	T	T	T	F	T	F
F	T	T	F	F	T	F	T	F	T	F
F	T	F	T	T	T	T	T	F	T	F
F	T	F	F	T	F	F	T	F	T	F
F	F	T	T	F	T	T	T	F	T	T
F	F	T	F	F	T	F	T	F	T	T
F	F	F	T	T	T	T	T	F	T	T
F	F	F	F	T	F	F	T	F	T	T
				1	2	1	5	3	4	3

43. Let *p* represent "your eyes are bad" and *q* represent "your whole body will be full of darkness." The statement has the form $p \rightarrow q$. In words, the equivalent form $\sim p \lor q$ becomes "Your eyes are not bad or your whole body will be full of darkness."

45. Let *p* represent "I have the money" and *q* represent "I'd buy that car." The statement has the form $p \rightarrow q$. In words, the equivalent form $\sim p \lor q$ becomes "I don't have the money or I'd buy that car."

47. "If you ask me, I will do it" has the form $p \rightarrow q$. The negation has the form $p \land \sim q$, which translates as "You ask me, and I will not do it."

49. "If you don't love me, I won't be happy" has the form $\sim p \rightarrow \sim q$. The negation has the form $\sim p \land q$, which translates as "You don't love me and I will be happy."

51. The statements $p \rightarrow q$ and $\sim p \lor q$ are equivalent if they have the same truth tables.

p	*q*	*p*	$\rightarrow$	*q*	$\sim p$	$\lor$	*q*
T	T	T	T	T	F	T	T
T	F	T	F	F	F	F	F
F	T	F	T	T	T	T	T
F	F	F	T	F	T	T	F
		1	2	1	1	2	1

Since the truth values in the final columns for each statement are the same, the statements are equivalent.

53.

p	*q*	$p \rightarrow q$	$q \rightarrow p$
T	T	T T T	T T T
T	F	T F F	F T T
F	T	F T T	T F F
F	F	F T F	F T F
		1 2 1	1 2 1

Since the truth values in the final columns for each statement are not the same, the statements are not equivalent.

55.

p	*q*	*p*	$\rightarrow$	$\sim q$	$\sim p$	$\lor$	$\sim q$
T	T	T	F	F	F	F	F
T	F	T	T	T	F	T	T
F	T	F	T	F	T	T	F
F	F	F	T	T	T	T	T
		1	2	1	1	2	1

Since the truth values in the final columns for each statement are the same, the statements are equivalent.

57.

p	*q*	*p*	$\land$	$\sim q$	$\sim q$	$\rightarrow$	$\sim p$
T	T	T	F	F	F	T	F
T	F	T	T	T	T	F	F
F	T	F	F	F	F	T	T
F	F	F	F	T	T	T	T
		1	2	1	1	2	1

Since the truth values in the final columns for each statement are not the same, the statements are not equivalent. Observe that since they have opposite truth values, each statement is the negation of the other.

59.

p	*q*	*p*	$\land$	*q*	$\sim$	(*p*	$\rightarrow$	$\sim q$)
T	T	T	T	T	T	T	F	F
T	F	T	F	F	F	T	T	T
F	T	F	F	T	F	F	T	F
F	F	F	F	F	F	F	T	T
		1	2	1	5	3	4	3

The columns labeled 2 and 5 are identical.

61.

p	*q*	*p*	$\lor$	*q*	*q*	$\lor$	*p*
T	T	T	T	T	T	T	T
T	F	T	T	F	F	T	T
F	T	F	T	T	T	T	F
F	F	F	F	F	F	F	F
		1	2	1	3	4	3

The columns labeled 2 and 4 are identical.

63.

p	q	r	(p	∨	q	) ∨	r	p	∨	(q	∨	r)
T	T	T	T	T	T	T	T	T	T	T	T	T
T	T	F	T	T	T	T	F	T	T	T	T	F
T	F	T	T	T	F	T	T	T	T	F	T	T
T	F	F	T	T	F	T	F	T	T	F	F	F
F	T	T	F	T	T	T	T	F	T	T	T	T
F	T	F	F	T	T	T	F	F	T	T	T	F
F	F	T	F	F	F	T	T	F	T	F	T	T
F	F	F	F	F	F	F	F	F	F	F	F	F
			1	2	1	4	3	5	8	6	7	6

The columns labeled 4 and 8 are identical.

65.

p	q	r	p	∨	(q	∧	r)	(p	∨	q)	∧	(p	∨	r)
T	T	T	T	T	T	T	T	T	T	T	T	T	T	T
T	T	F	T	T	T	F	F	T	T	T	T	T	T	F
T	F	T	T	T	F	F	T	T	T	F	T	T	T	T
T	F	F	T	T	F	F	F	T	T	F	T	T	T	F
F	T	T	F	T	T	T	T	F	T	T	T	F	T	T
F	T	F	F	F	T	F	F	F	T	T	F	F	F	F
F	F	T	F	F	F	F	T	F	F	F	F	F	T	T
F	F	F	F	F	F	F	F	F	F	F	F	F	F	F
			1	4	2	3	2	5	6	5	9	7	8	7

The columns labeled 4 and 9 are identical.

67.

p	q	(p	∧	q)	∨	p
T	T	T	T	T	T	T
T	F	T	F	F	T	T
F	T	F	F	T	F	F
F	F	F	F	F	F	F
		1	2	1	4	3

The p column and the column labeled 4 are identical.

69. In the diagram, two series circuits are shown, which correspond to $p \wedge q$ and $p \wedge {\sim}q$. These circuits, in turn, form a parallel circuit. Thus, the logical statement is

$$(p \wedge q) \vee (p \wedge {\sim}q).$$

One pair of equivalent statements listed in the text includes

$$(p \wedge q) \vee (p \wedge {\sim}q) \equiv p \wedge (q \vee {\sim}q)$$

Since $(q \vee {\sim}q)$ is always true, $p \wedge (q \vee {\sim}q)$ simplifies to

$$p \wedge \mathrm{T} \equiv p.$$

71. In the diagram, a series circuit is shown, which corresponds to ${\sim}q \wedge r$. This circuit, in turn, forms a parallel circuit with p. Thus, the logical statement is

$$p \vee ({\sim}q \wedge r).$$

73. In the diagram, a parallel circuit corresponds to $p \vee q$. This circuit is parallel to ${\sim}p$. Thus, the total circuit corresponds to the logical statement

$$(p \vee q) \vee {\sim}p.$$

This statement, in turn, is equivalent to

$$({\sim}p \vee p) \vee q.$$

Since ${\sim}p \vee p$ is always true, we have

$$\mathrm{T} \vee q \equiv \mathrm{T}.$$

75. In the diagram, series circuits corresponding to $p \wedge q$ and $p \wedge p$ form a parallel circuit. This parallel circuit is parallel to the series circuit corresponding to $r \wedge {\sim}r$. Thus, the logical statement is

$$[(p \wedge q) \vee (p \wedge p)] \vee (r \wedge {\sim}r).$$

This statement simplifies to p as follows:

$$[(p \wedge q) \vee (p \wedge p)] \vee (r \wedge {\sim}r)$$
$$\equiv [(p \wedge q) \vee p] \vee (r \wedge {\sim}r)$$
$$\equiv p \vee (r \wedge {\sim}r)$$
$$\equiv p \vee \mathrm{F}$$
$$\equiv p.$$

77. The logical statement $p \wedge (q \vee \sim p)$ can be represented by the following circuit.

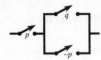

The statement $p \wedge (q \vee \sim p)$ simplifies to $p \wedge q$ as follows:

$$p \wedge (q \vee \sim p) \equiv (p \wedge q) \vee (p \wedge \sim p)$$
$$\equiv (p \wedge q) \vee F$$
$$\equiv p \wedge q.$$

79. The logical statement $(p \vee q) \wedge (\sim p \wedge \sim q)$ can be represented by the following circuit.

The statement $(p \vee q) \wedge (\sim p \wedge \sim q)$ simplifies to F as follows.

$$(p \vee q) \wedge (\sim p \wedge \sim q)$$
$$\equiv [p \wedge (\sim p \wedge \sim q)] \vee [q \wedge (\sim p \wedge \sim q)]$$
$$\equiv [(p \wedge \sim p) \wedge \sim q)] \vee [q \wedge (\sim q \wedge \sim p)]$$
$$\equiv [F \wedge \sim q)] \vee [(q \wedge \sim q) \wedge \sim p]$$
$$\equiv F \vee (F \wedge \sim p)$$
$$\equiv F \vee F$$
$$\equiv F$$

81. The logical statement $[(p \vee q) \wedge r] \wedge \sim p$ can be represented by the following circuit.

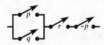

The statement $[(p \vee q) \wedge r] \wedge \sim p$ simplifies to $(r \wedge \sim p) \wedge q$ as follows:

$$[(p \vee q) \wedge r] \wedge \sim p$$
$$\equiv [(p \wedge r) \vee (q \wedge r)] \wedge \sim p$$
$$\equiv [(p \wedge r) \wedge \sim p] \vee [(q \wedge r) \wedge \sim p]$$
$$\equiv [p \wedge r \wedge \sim p] \vee [q \wedge r \wedge \sim p]$$
$$\equiv [(p \wedge \sim p) \wedge r] \vee [(r \wedge \sim p) \wedge q]$$
$$\equiv (F \wedge r) \vee [(r \wedge \sim p) \wedge q]$$
$$\equiv F \vee [(r \wedge \sim p) \wedge q]$$
$$\equiv (r \wedge \sim p) \wedge q$$
$$\equiv r \wedge (\sim p \wedge q).$$

83. The logical statement $\sim q \rightarrow (\sim p \rightarrow q)$ can be represented by the following circuit.

The statement $\sim q \rightarrow (\sim p \rightarrow q)$ simplifies to $p \vee q$ as follows:

$$\sim q \rightarrow (\sim p \rightarrow q) \equiv \sim q \rightarrow (p \vee q)$$
$$\equiv q \vee (p \vee q)$$
$$\equiv q \vee p \vee q$$
$$\equiv p \vee q \vee q$$
$$\equiv p \vee (q \vee q)$$
$$\equiv p \vee q.$$

85. The logical statement $[(p \wedge q) \vee p] \wedge [(p \vee q) \wedge q]$ can be represented by the following circuit.

The statement simplifies to $p \wedge q$ as follows:

$$[(p \wedge q) \vee p] \wedge [(p \vee q) \wedge q]$$
$$\equiv p \wedge [(p \vee q) \wedge q]$$
$$\equiv p \wedge q.$$

89. Each statement has the form $p \rightarrow q$. The equivalent form using *or* is $\sim p \vee q$.

(a) You are not married at the end of the year, or you may file a joint return with your spouse.

(b) A bequest received by an executor from an estate is compensation for services, or it is tax free.

(c) A course does not improve your current job skills or does not lead to qualification for a new profession, or the course is not deductable.

91. (a) $(v \vee p) \rightarrow (s \wedge g)$

(b) The portfolio being worth \$80,0000 means v is F, selling the Ford Motor stock at \$56 per share means s is T and p is F, and keeping the proceeds means g is F.

$$(v \vee p) \rightarrow (s \wedge g)$$
$$(F \vee F) \rightarrow (T \wedge F)$$
$$F \rightarrow F$$
$$T$$

The statement is true.

(d) $\sim[(v \lor p) \to (s \land g)]$

$\equiv (v \lor p) \land \sim(s \land g)$

$\equiv (v \lor p) \land (\sim s \lor \sim g)$

In words, the negation of the given statement is:

The value of my portfolio exceeds \$100,000 or the price of my stock in Ford Motor Company falls below \$50 per share, and I will not sell all my shares of Ford stock or I will not give the proceeds to United Way.

93. $p \lor q \equiv \sim p \to q$

(a) If you cannot file a civil lawsuit yourself, then your attorney can do it for you.

(b) If your driver's license does not come with restrictions, then restrictions may sometimes be added on later.

(c) If you can marry when you're not at least 18 years old, then you have the permission of your parents or guardian.

6.4 More on the Conditional

Your Turn 1

(a) Being happy is sufficient for you to clap your hands is equivalent to If you are happy, then you clap your hands.

(b) All who seek shall find is equivalent to If you seek, then you shall find.

Your Turn 2

Given Statement: If I get another ticket, then I lose my license.

Converse: If I lose my license, then I get another ticket.

Inverse: If I didn't get another ticket, then I didn't lose my license.

Contrapositive: If I didn't lose my license, then I didn't get another ticket. The contrapositive is equivalent.

Your Turn 3

New Your City is the capital of the United States if and only if Paris is the capital of France. The statement is false because the two simple statements have different truth values.

6.4 Exercises

Wording may vary in the answers to Exercises 1–27.

1. *The direct statement:* If the exit is ahead, then I don't see it.

(a) *Converse:* If I don't see it, then the exit is ahead.

(b) *Inverse:* If the exit is not ahead, then I see it.

(c) *Contrapositive:* If I see it, then the exit is not ahead.

3. *The direct statement:* If I knew you were coming, I'd have cleaned the house.

(a) *Converse:* If I cleaned the house, then I knew you were coming.

(b) *Inverse:* If I didn't know you were coming, I wouldn't have cleaned the house.

(c) *Contrapositive:* If I didn't clean the house, then I didn't know you were coming.

5. *It is helpful to reword the given statement.*

The direct statement: If you are a mathematician, then you wear a pocket protector.

(a) *Converse:* If you wear a pocket protector, then you are a mathematician.

(b) *Inverse:* If you are not a mathematician, then you do not wear a pocket protector.

(c) *Contrapositive:* If you don't wear a pocket protector, then you are not a mathematician.

7. The direct statement: $p \to \sim q$.

(a) *Converse:* $\sim q \to p$.

(b) *Inverse:* $\sim p \to q$.

(c) *Contrapositive:* $q \to \sim p$.

9. The direct statement: $p \to (q \lor r)$.

(a) *Converse:* $(q \lor r) \to p$.

(b) *Inverse:* $\sim p \to \sim(q \lor r)$
or $\sim p \to (\sim q \land \sim r)$.

(c) *Contrapositive:* $(\sim q \land \sim r) \to \sim p$.

13. The statement "Your signature implies that you accept the conditions" becomes "If you sign, then you accept the conditions."

15. The statement "You can take this course pass/fail only if you have prior permission" becomes "If you can take this course pass/fail, then you have prior permission."

17. The statement "You can skate on the pond when the temperature is below 10°" becomes "If the temperature is below 10°, then you can skate on the pond."

19. The statement "Eating ten hot dogs is sufficient to make someone sick" becomes "If someone eats ten hot dogs, then he or she will get sick."

21. The statement "A valid passport is necessary for travel to France" becomes "If you travel to France, then you have a valid passport."

23. The statement "For a number to have a real square root, it is necessary that it be nonnegative" becomes "If a number has a real square root, then it is nonnegative."

25. The statement "All brides are beautiful" becomes "If someone is a bride, then she is beautiful."

27. The statement "A number is divisible by 3 if the sum of its digits is divisible by 3" becomes "If the sum of a number's digits is divisible by 3, then it is divisible by 3."

29. Option d is the answer since "r is necessary for s" represents the converse, $s \rightarrow r$, of all of the other statements.

33. The statement "$5 = 9 - 4$ if and only if $8 + 2 = 10$" is <u>true</u>, since this is a biconditional composed of two true statements.

35. The statement "$8 + 7 \neq 15$ if and only if $3 \times 5 \neq 9$" is <u>false</u>, since this is a biconditional consisting of one false statement and one true statement.

37. The statement "China is in Asia if and only if Mexico is in Europe" is <u>false</u>, since it is a biconditional consisting of a true statement and a false statement.

39.

p	q	$(\sim p$	$\wedge$	$q)$	$\leftrightarrow$	$(p$	$\rightarrow$	$q)$
T	T	F	F	T	F	T	T	T
T	F	F	F	F	T	T	F	F
F	T	T	T	T	T	F	T	T
F	F	T	F	F	F	F	T	F
		1	2	1	5	3	4	3

41. **(a)** If it is an employee contribution, then it must be reported on Form 8889.

  (b) If certain tax benefits may be claimed by married persons, then they file jointly.

(c) If a child provides over half of his or her own support, then the child is not a qualifying child.

43. *Given statement:*

If your account is in default, then we may close your account without notice.

Converse:

If we close your account without notice, then your account is in default.

Inverse:

If your account is not in default, then we may not close your account without notice.

Contrapositive:

If we do not close your account without notice, then your account is not in default.

  The original statement and the contrapositive are equivalent, and the converse and inverse are equivalent.

45. **(a)** Let p represent "there are triplets," let q represent "the most persistent stands to gain an extra meal," and let r represent "it may eat at the expense of another." Then the statement can be written as $p \rightarrow (q \wedge r)$.

  (b) The contrapositive is $\sim(q \wedge r) \rightarrow p$, which is equivalent to $(\sim q \vee \sim r) \rightarrow p$: If the most persistent does not stand to gain an extra meal or it does not eat at the expense of another, then there are not triplets.

47. **(a)** *Converse:* If you can't get married again, then you are married.

    Inverse: If you aren't married, then you can get married again.

    Contrapositive: If you can get married again, then you are not married.

  (b) *Converse:* If you are protected by the Fair Credit Billing Act, then you pay for your purchase with a credit card.

    Inverse: If you do not pay for your purchase with a credit card, then you are not protected by the Fair Credit Billing Act.

    Contrapositive: If you are not protected by the Fair Credit Billing Act, then you do not pay for your purchase with a credit card.

  (c) *Converse:* If you're expected to make a reasonable effort to locate the owner, then you hit a parked car.

    Inverse: If you did not hit a parked car, then you are not expected to make a reasonable effort to locate the owner.

Contrapositive: If you are not expected to make a reasonable effort to locate the owner, then you did not hit a parked car.

The original statement and the contrapositive are equivalent, and the converse and inverse are equivalent.

49. (a) Let *d* represent "political development in Western Europe will increase" and let *a* represent "social assimilation is increasing." Then the statement can be written as $d \leftrightarrow a$. The truth table for the statement is as follows.

d	a	$d \leftrightarrow a$
T	T	T
T	F	F
F	T	F
F	F	T

(b) If *a* is true and *d* is false then $d \leftrightarrow a$ is false.

51. If a country has democracy, then it has a high level of education.

Converse: If a country has a high level of education, then it has democracy.

Inverse: If a country does not have democracy, then it does not have a high level of education.
Contrapositive: If a country does not have a high level of education, then it does not have democracy.

The contrapositive is equivalent to the original.

53. The rule "If a card has a D on one side, then it must have a 3 on the other side" is violated when a card has a D on one side and the number on the other side is not 3. Thus, we only need to turn over cards that have a D or a number other than 3.

D card: This card must be turned over to see whether the rule has been violated. If the number on the other side is 3, the rule has not been violated; however, if the number is not 3, the rule has been violated.

F card: Since the premise of the rule is false for this card, the rule automatically holds. Thus, this card does not need to be turned over.

3 card: Since the conclusion of the rule is true, the rule automatically holds. Thus, this card does not need to be turned over.

7 card: This card must be turned over to see whether the rule has been violated. If the letter is

D, then the rule has been violated; however, if the letter on the other side is not D, the rule has not been violated.

55. (a) If … then *form:* If nothing is ventured, then nothing is gained.

Contrapositive: If something is gained, then something is ventured.

Statement using or*:* Something is ventured or nothing is gained.

(b) If … then *form:* If something is one of the best things in life, then it is free.

Contrapositive: If something is not free, then it is not one of the best things in life.

Statement using or*:* Something is not one of the best things in life or it is free.

(c) If … then *form:* If something is a cloud, then it has a silver lining.

Contrapositive: If something doesn't have a silver lining, then it isn't a cloud.

Statement using or*:* Something is not a cloud or it doesn't have a silver lining.

57. (a) If you can score in this box, then the dice show any sequence of four numbers. You cannot score in this box, or the dice show any sequence of four numbers.

(b) If two or more words are formed in the same play, then each is scored. Two or more words are not formed in the same play, or each is scored.

(c) If words are labeled as a part of speech, then they are permitted. Words are not labeled as parts of speech, or they are permitted.

6.5 Analyzing Arguments and Proofs

Your Turn 1

Let t represent "You watch television tonight," p represent "You write your paper tonight," and g represent "You get a good grade."

The given argument written symbolically is

1. $t \lor p$ T
2. $p \rightarrow g$ T
3. $\dfrac{g}{\sim t}$ $\begin{array}{c}\text{T}\\\text{F}\end{array}$

Invalid argument; If t is true, p is true, and g is true, the premises are true, but the conclusion is false.

Your Turn 2

Let m represent "You put money in the parking meter", c represent "You buy a cup of coffee", and t represent "You get a ticket".

The given argument written symbolically as

1. $\sim m \lor \sim c$ Premise
2. $\sim m \rightarrow t$ Premise
3. $\sim t$ Premise
4. m Statements 2, 3, Modus Tollens
5. $\sim c$ Statements 1, 4, Disjunctive Syllogism

Conclusion: You did not buy a cup of coffee. Valid argument.

6.5 Exercises

1. Let p represent "she weighs the same as a duck," q represent "she's made of wood," and r represent "she's a witch." The argument is then represented symbolically by:

$$\begin{array}{c} p \rightarrow q \\ q \rightarrow r \\ \hline p \rightarrow r. \end{array}$$

This is the valid argument form Reasoning by Transitivity.

3. Let p represent "I had the money" and q represent "I'd go on vacation." The argument is then represented symbolically by:

$$\begin{array}{c} p \rightarrow q \\ p \\ \hline q. \end{array}$$

This is the valid argument form Modus Ponens.

5. Let p represent "you want to make trouble" and q represent "the door is that way." The argument is then represented symbolically by:

$$\begin{array}{c} p \rightarrow q \\ q \\ \hline p. \end{array}$$

Since this is the form Fallacy of the Converse, it is invalid and considered a fallacy.

7. Let p represent "Andrew Crowley plays" and q represent "the opponent gets shut out." The argument is then represented symbolically by:

$$\begin{array}{c} p \rightarrow q \\ \sim q \\ \hline \sim p. \end{array}$$

This is the valid argument form Modus Tollens.

9. Let p represent "we evolved a race of Isaac Newtons" and q represent "that would not be progress." The argument is then represented symbolically by:

$$\begin{array}{c} p \rightarrow q \\ \sim p \\ \hline \sim q. \end{array}$$

Note that since we let q represent "that would not be progress," $\sim q$ represents "that is progress."

Since this is the form Fallacy of the Inverse, it is invalid and considered a fallacy.

11. Let p represent "Something is rotten in the state of Denmark" and q represent "my name isn't Hamlet." The argument is then represented symbolically by:

$$\begin{array}{c} p \lor q \text{ (or } q \lor p) \\ \sim q \\ \hline \sim p. \end{array}$$

Since this is the form Disjunctive Syllogism, it is a valid argument.

To show validity for the arguments in the following exercises we must show that the conjunction of the premises implies the conclusion. That is, the conditional statement $[P_1 \wedge P_2 \wedge ... \wedge P_n] \rightarrow C$ must be a tautology.

13. 1. $p \vee q$ T

2. $\underline{p \qquad}$ T

$\qquad \sim q$ F

The argument is <u>invalid</u>. When $p = $ T and $q = $ T, the premises are true but the conclusion is false.

15. 1. $p \rightarrow q$ T

2. $\underline{q \rightarrow p}$ T

$\qquad p \wedge q$ F

The argument is <u>invalid</u>. When $p = $ F and $q = $ F , the premises are true but the conclusion is false.

17. 1. $\sim p \rightarrow \sim q$ Premise
   2. q Premise
   3. p 1, 2, Modus Tollens

The argument is <u>valid</u>.

19. 1. $p \rightarrow q$ Premise
   2. $\sim q$ Premise
   3. $\sim p \rightarrow r$ Premise
   4. $\sim p$ 1, 2, Modus Tollens
   5. r 3, 4, Modus Ponens

The argument is <u>valid</u>.

21. 1. $p \rightarrow q$ Premise
   2. $q \rightarrow r$ Premise
   3. $\sim r$ Premise
   4. $p \rightarrow r$ 1, 2, Transitivity
   5. $\sim p$ 3, 4, Modus Tollens

The argument is <u>valid</u>.

23. 1. $p \rightarrow q$ Premise
   2. $q \rightarrow \sim r$ Premise
   3. p Premise
   4. $r \vee s$ Premise
   5. q 1, 3, Modus Ponens
   6. $\sim r$ 2, 5, Modus Ponens
   7. s 4, 6, Disjunctive Syllogism

The argument is <u>valid</u>.

25. Make a truth table for the statement $(p \wedge q) \rightarrow p$.

p	q	$(p$	$\wedge$	$q)$	$\rightarrow$	p
T	T	T	T	T	T	T
T	F	T	F	F	T	T
F	T	F	F	T	T	F
F	F	F	F	F	T	F
		1	2	1	3	2

Since the final column, 3, indicates that the conditional statement that represents the argument is true for all possible truth values of p and q, the statement is a tautology.

27. Make a truth table for the statement $(p \wedge q) \rightarrow (p \wedge q)$.

p	q	$(p$	$\wedge$	$q)$	$\rightarrow$	$(p$	$\wedge$	$q)$
T	T	T	T	T	T	T	T	T
T	F	T	F	F	T	T	F	F
F	T	F	F	T	T	F	F	T
F	F	F	F	F	T	F	F	F
		1	2	1	5	3	4	3

Since the final column, 3, indicates that the conditional statement that represents the argument is true for all possible truth values of p and q, the statement is a tautology.

29. Let a represent "Alex invests in AT&T," s represent "Sophia invests in Sprint Nextel," and v represent "Victor invests in Verizon".

The given argument written symbolically is

$$a \rightarrow s$$
$$v \vee a$$
$$\underline{\sim v}$$
$$s$$

1. $a \rightarrow s$ Premise
2. $v \vee a$ Premise
3. $\sim v$ Premise
4. a 2, 3, Disjunctive Syllogism
5. s 1, 4, Modus Ponens

The argument is <u>valid</u>.

31. Let b represent "it is a bearish market," p represent "prices are rising," and i represent "the investor sells stocks."

The given argument written symbolically is

$$\begin{array}{ll} b & \text{T} \\ p \to {\sim}b & \text{T} \\ \underline{{\sim}p \to i} & \text{T} \\ {\sim}i & \text{F} \end{array}$$

The argument is <u>invalid</u>. When $b = \text{T}$, $p = \text{F}$, and $i = \text{T}$, the premises are true but the conclusion is false.

33. Let s represent "animal is a spider," i represent "animal is an insect," l represent "animal has eight legs," and b represent "animal has two main body parts."

The given argument written symbolically is

$$\begin{array}{l} s \lor i \\ s \to (l \land b) \\ \underline{{\sim}l \lor {\sim}b} \\ i \end{array}$$

1. $s \lor i$ Premise
2. $s \to (l \land b)$ Premise
3. ${\sim}l \lor {\sim}b$ Premise
4. ${\sim}(l \land b)$ 3, DeMorgan's Law
5. ${\sim}s$ 2, 4, Modus Tollens
6. i 1, 5, Disjunctive Syllogism

The argument is <u>valid</u>.

35. Let m represent "I am married to you," o represent "we are one," and r represent "you are really a part of me."

The given argument written symbolically is

$$\begin{array}{ll} m \to o & \text{T} \\ o \lor {\sim}r & \text{T} \\ \underline{m \lor r} & \text{T} \\ m \to r & \text{F} \end{array}$$

The argument is <u>invalid</u>. When $m = \text{T}$, $o = \text{T}$, and $r = \text{F}$, the premises are true but the conclusion is false.

37. Let y represent "the Yankees will be in the World Series," p represent "the Phillies will be in the World Series," and n represent "the National League wins." The argument is then represented symbolically by:

$$\begin{array}{l} y \lor {\sim}p \\ {\sim}p \to {\sim}n \\ n \\ \hline y. \end{array}$$

The argument is <u>valid</u>.

1. $y \lor {\sim}p$ Premise
2. ${\sim}p \to {\sim}n$ Premise
3. n Premise
4. p 2, 3, Modus Tollens
5. y 1, 4, Disjunctive Syllogism

39. (a) $d \to {\sim}w$ **(b)** $o \to w$ or $w \to {\sim}o$

(c) $p \to d$ **(d)** $p \to {\sim}o$,

Conclusion: If it is my poultry, then it is not an officer. In Lewis Carroll's words, "My poultry are not officers."

41. (a) $b \to {\sim}t$ or $t \to {\sim}b$

(b) $w \to c$

(c) ${\sim}b \to h$

(d) ${\sim}w \to {\sim}p$ or $p \to w$

(e) $c \to t$

(f) $p \to h$,

Conclusion: If one is a pawnbroker, then one is honest. In Lewis Carroll's words, "No pawnbroker is dishonest."

43. (a) $d \to p$

(b) ${\sim}t \to {\sim}i$

(c) $r \to {\sim}f$ or $f \to {\sim}r$

(d) $o \to d$ or ${\sim}d \to {\sim}o$

(e) ${\sim}c \to i$

(f) $b \to s$

(g) $p \to f$

(h) ${\sim}o \to {\sim}c$ or $c \to o$

(i) $s \to {\sim}t$ or $t \to {\sim}s$

(j) $b \to {\sim}r$,

Conclusion: If it is written by Brown, then I can't read it. In Lewis Carroll's words, "I cannot read any of Brown's letters."

6.6 Analyzing Arguments with Quantifiers

Your Turn 1

(a) All college students study.

Let $c(x)$ represent "x is a college student" and $s(x)$ represent "x studies."

The statement can be written as

$$\forall x\,[c(x) \rightarrow s(x)]$$

Its negation is

$$\exists x\{\sim[c(x) \rightarrow s(x)]\},$$

which is equivalent to $\exists x\,[c(x) \wedge \sim s(x)]$.

In words, "Some college students do not study."

(b) Some professors are not organized.

Let $p(x)$ represent "x is a professor" and $o(x)$ represent "x is organized."

The statement can be written as

$$\exists x\,[p(x) \wedge \sim o(x)]$$

Its negation is

$$\forall x\,\{\sim[p(x) \wedge \sim o(x)]\},$$

which is equivalent to $\forall x\,[p(x) \rightarrow o(x)]$.

In words, "All professors are organized."

Your Turn 2

All insects are arthropods.

<u>A bee is an insect.</u>

A bee is an arthropod.

Let $i(x)$ represent "x is an insect," $a(x)$ represent "x is an arthropod," and b represent a "bee."

The argument becomes

$$\forall x[i(x) \rightarrow a(x)]$$
$$\frac{i(b)}{a(b)}$$

The argument resembles Modus Ponens, so the argument is <u>valid</u>.

Your Turn 3

All birds have wings.

<u>Rover does not have wings.</u>

Rover is not a bird.

Let $b(x)$ represent "x is a bird," $w(x)$ represent "x has wings," and r represent "Rover."

The argument becomes

$$\forall x\,[b(x) \rightarrow w(x)]$$
$$\frac{\sim w(r)}{\sim b(r)}$$

The argument resembles Modus Tollens, so the argument is <u>valid</u>.

Your Turn 4

Every man has his price.

<u>Sam has a price.</u>

Sam is a man.

Let $m(x)$ represent "x is a man," $p(x)$ represent "x has a price," and s represent "Sam."

The argument becomes

$$\forall x[m(x) \rightarrow p(x)]$$
$$\frac{p(s)}{m(s)}$$

Draw an Euler diagram.

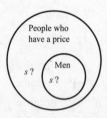

It is possible that Sam is not a man.

The argument resembles the Fallacy of the Converse, so the argument is <u>invalid</u>.

Your Turn 5

Some vegetarians eat eggs.

<u>Sarah is a vegetarian.</u>

Sarah eats eggs.

Let $v(x)$ represent "x is a vegetarian," $e(x)$ represent "x eats eggs," and s represent "Sarah."

The argument becomes

$$\exists x[v(x) \wedge e(x)]$$
$$\frac{v(s)}{e(s)}$$

Draw an Euler diagram.

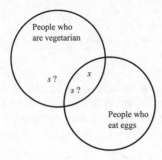

It's possible that Sarah is a vegetarian but does not eat eggs.

The argument is <u>invalid</u>.

6.6   Exercises

1. Let $b(x)$ represent "x is a book" and $s(x)$ represent" "x is a bestseller."

   (a) $\exists x[b(x) \wedge s(x)]$

   (b) $\forall x[b(x) \rightarrow \sim s(x)]$

   (c)   No books are bestsellers.

3. Let $c(x)$ represent "x is a CEO" and $s(x)$ represent "x sleeps well at night."

   (a) $\forall x[c(x) \rightarrow \sim s(x)]$

   (b) $\exists x[c(x) \wedge s(x)]$

   (c)   There is a CEO who sleeps well at night.

5. Let $l(x)$ represent "x is a leaf" and $b(x)$ represent "x is brown."

   (a) $\forall x[l(x) \rightarrow b(x)]$

   (b) $\exists x[l(x) \wedge \sim b(x)]$

   (c)   There is a leaf that's not brown.

7. **(a)** Let $g(x)$ represent "x is a graduate" and $f(x)$ represent "x wants to find a good job." Let t represent Theresa Cortesini. We can represent the argument symbolically as follows.

   $$\forall x[g(x) \rightarrow f(x)]$$
   $$\dfrac{g(t)}{f(t)}$$

   (b) Draw an Euler diagram where the region representing "graduates" must be inside the region representing "people who want to find good jobs" so that the first premise is true.

By the second premise, t must lie in the "graduates" region. Since this forces the conclusion to be true, the argument is <u>valid</u>.

9. **(a)** Let $p(x)$ represent "x is a professor" and $c(x)$ represent "x is covered with chalk dust." Let o represent Otis Taylor. We can represent the argument symbolically as follows.

   $$\forall x[p(x) \rightarrow c(x)]$$
   $$\dfrac{c(o)}{p(o)}$$

   (b) Draw an Euler diagram where the region representing "professors" must be inside the region representing "those who are covered with chalk dust" so that the first premise is true.

By the second premise, o must lie in the "those who are covered with chalk dust" region. Thus, o could be inside or outside the inner region "professors." Since this allows for a false conclusion (Otis doesn't have to be a professor to be covered with chalk dust), the argument is <u>invalid</u>.

11. **(a)** Let $c(x)$ represent "x is an accountant" and $p(x)$ represent "x uses a spreadsheet." Let n represent Nancy Hart. We can represent the argument symbolically as follows.

   $$\forall x[c(x) \rightarrow p(x)]$$
   $$\dfrac{\sim p(n)}{\sim c(n)}$$

   (b) Draw an Euler diagram where the region representing "accountants" must be inside the region representing "those who use spreadsheets" so that the first premise is true.

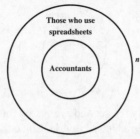

By the second premise, n must lie outside the region representing "those who use spreadsheets." Since this forces the conclusion to be true, the argument is <u>valid</u>.

13. **(a)** Let $t(x)$ represent "x is turned down for a mortgage," $s(x)$ represent "x has a second income," and $b(x)$ represent "x needs a mortgage broker." We can represent the argument symbolically as follows.

$$\exists x[t(x) \wedge s(x)]$$
$$\underline{\forall x[t(x) \rightarrow b(x)]}$$
$$\exists x[s(x) \wedge b(x)]$$

(b) Draw an Euler diagram where the region representing "Those who are turned down for a mortgage" intersects the region representing "Those with a 2nd income." This keeps the first premise true.

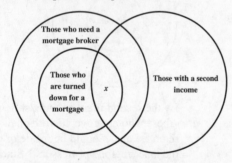

By the second premise the region representing "Those who are turned down for a mortgage" must be inside the region "Those who need a mortgage broker." Since x lies in both regions "These with a 2nd income" and "Those who need a mortgage broker," the conclusion is true and so the argument is <u>valid</u>.

15. **(a)** Let $w(x)$ represent "x wanders" and $l(x)$ represent "x is lost." Let m represent Marty McDonald. We can represent the argument symbolically as follows.

$$\exists x[w(x) \wedge l(x)]$$
$$\underline{w(m)}$$
$$l(m)$$

(b) Draw an Euler diagram where the region representing "those who wander" intersects the region representing "those who are lost." This keeps the first premise true.

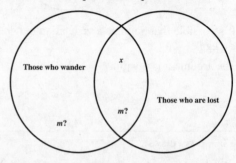

By the second premise, m must lie in the region representing "those who wander." But, m could be inside or outside the region representing "those who are lost." Since this allows for a false conclusion, the argument is <u>invalid</u>.

17. **(a)** Let $p(x)$ represent "x is a psychologist," $u(x)$ represent "x is a university professor," and $r(x)$ represent "x has a private practice." We can represent the argument symbolically as follows.

$$\exists x[p(x) \wedge u(x)]$$
$$\underline{\exists x[p(x) \wedge r(x)]}$$
$$\exists x[u(x) \wedge r(x)]$$

(b) Draw an Euler diagram where the region representing "psychologists" and "university professors" intersect each other to keep the first premise true. Then add a region representing "those with a private practice" intersecting the region "university professors" to keep the second premise true. In the most general case, this region should also intersect the region representing "psychologists."

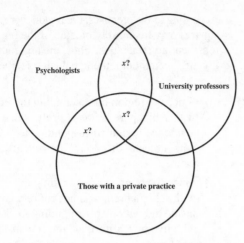

By the first premise, x must lie in the region shared by "psychologists" and "university professors"; by the second premise, x must lie in the region shared by "psychologists" and "those with a private practice." But x may not lie in the region shared by all three regions. Since this diagram shows true premises but a false conclusion, the argument is <u>invalid</u>.

19. **(a)** Let $a(x)$ represent "x is a saint" and $i(x)$ represent "x is a sinner." We can represent the argument symbolically as follows.

$$\forall x[a(x) \vee i(x)]$$
$$\underline{\exists x[\sim a(x)]}$$
$$\exists x[i(x)]$$

(b) Draw an Euler diagram where the region representing "saints" intersects the region representing "sinners" to keep the first premise true.

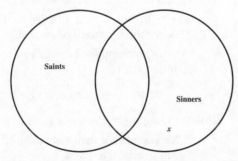

By the second premise, x must lie outside the region representing "saints." But that means x must lie in the part of the "sinners" region not shared by the "saints" region. Hence, the conclusion is true and so the argument is <u>valid</u>.

21. Interchanging the second premise and the conclusion of Example 4 yields the following argument.

All well-run businesses generate profits.

<u>Monsters, Inc. is a well-run business.</u>

Monsters, Inc. generates profits.

Draw an Euler diagram where the region representing "well-run businesses" must be inside the region representing "things that generate profits" so that the first premise is true.

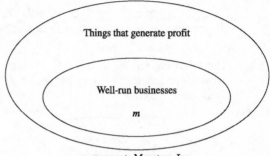

m represents Monsters, Inc.

Let m represent Monsters, Inc. By the second premise, m must lie inside the region representing "well-run businesses." Since this forces the conclusion to be true, the argument is <u>valid</u>, which makes the answer to the question "yes."

23. Since the region representing "major league baseball players" lies entirely inside the region representing "people who earn at least $300,000 a year," a possible first premise is

All major league baseball players earn at least $300,000 a year.

And if r represents Ryan Howard and r is inside the region representing "major league baseball players," then a possible second premise is

Ryan Howard is a major league baseball player.

A valid conclusion drawn from these two premises is that

Ryan Howard earns at least $300,000 a year.

Therefore, a valid argument based on the Euler diagram is as follows.

All major league baseball players earn at least $300,000 a year.

<u>Ryan Howard is a major league baseball player.</u>

Ryan Howard earns at least $300,000 a year.

25. The following diagram yields true premises. It also forces the conclusion to be true.

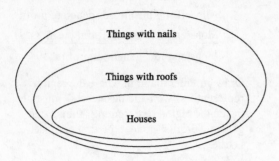

Thus, the argument is <u>valid</u>. Observe that the diagram is the only way to show true premises.

27. The following represents one way to diagram the premises so that they are true but does not lead to a true conclusion.

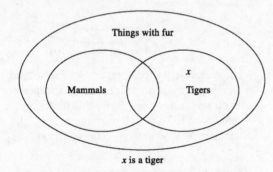

x is a tiger

If we let *x* be a tiger, according to the premises, *x* could also be a tiger but not a mammal. Thus, the argument is <u>invalid</u>.

29. The following Euler diagram illustrates that the conclusion is not forced to be true.

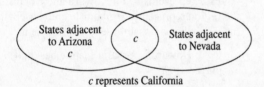

c represents California

The argument is <u>invalid</u> even though the conclusion is true.

31. The following Euler diagram represents true premises.

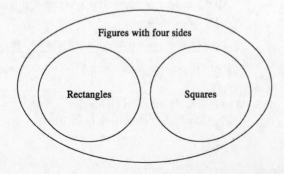

According to the premises, the region representing squares may lie entirely outside of the region representing rectangles. Thus, the argument is <u>invalid</u>, even though the conclusion is true.

37. (a) Since the region corresponding to "People with schizophrenia" is entirely contained inside the region corresponding to "People with a mental disorder," the conclusion is valid.

(b) Since the region corresponding to "People with schizophrenia" is not entirely contained inside the region corresponding to "People who live in California," the conclusion is invalid.

(c) Since some of the region corresponding to "People with schizophrenia" also lies in the region corresponding to "People who live in California," the conclusion is valid.

(d) Since some of the region corresponding to "People who live in California" lies outside the region corresponding to "People with schizophrenia" the conclusion is valid.

(e) Since some of the region corresponding to "People with a mental disorder" lies outside the region corresponding to "People with schizophrenia," the conclusion is invalid.

Thus, the answer is a, c, and d.

39. (a) Let $r(x)$ represent "*x* is a Representative,"

$a(x)$ represent "*x* has attained to the age of twenty-five years,"

$c(x)$ represent "*x* has been seven years a citizen of the United States,"

and $i(x)$ represent "*x* is an inhabitant of that State in which he shall be chosen."

We can represent the passage symbolically as follows.

$$\forall x \, \{r(x) \rightarrow [a(x) \wedge c(x) \wedge i(x)]\}$$

(b) We make the following argument.

A Representative shall have attained to the age of twenty-five years.

A Representative shall have been seven years a citizen of the United states.

A Representative shall, when elected, be an inhabitant of that State in which he shall be chosen.

John Boehner is a Representative.

John Boehner has attained to the age of twenty-five years, and been seven years a citizen of the United States, and was, when elected, an inhabitant of that State in which he was chosen.

The conclusion is true. Thus, the argument is <u>valid</u>.

(c) Draw an Euler diagram where the region representing "representative" must be inside the three other regions and the three other regions intersect each other. Let *b* represent John Boehner. By the fourth premise, *b* must lie in the region representing "Representative" and, thus, be in all of the regions.

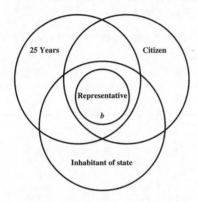

41. (a) Let $s(x)$ represent "x is a State,"

$t(x)$ represent "x enters into a treaty,"

$a(x)$ represent "x enters into an alliance,"

and $c(x)$ represent "x enters into a confederation."

We can represent the passage symbolically as follows.

$$\forall x\ \{s(x) \rightarrow \sim[t(x) \lor a(x) \lor c(x)]\}$$

(b) We make the following argument.

No State shall enter into any treaty, alliance, or confederatrion.

Texas is a state.

Texas shall not enter into any treaty, alliance, or confederation.

The conclusion is true. Thus, the argument is <u>valid</u>.

(c) Draw an Euler diagram where the region representing "states" must be outside the other three regions and the three other regions intersect. Let *t* represent Texas. By the second premise, *t* must lie in the region representing "states" and, thus, must be outside the other three regions.

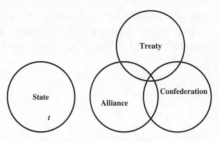

In Exercises 43–47, the premises marked A, B, and C are followed by several possible conclusions. Take each conclusion in turn, and check the resulting argument as valid *or* invalid.

    A. All kittens are cute animals.

    B. All cute animals are admired by animal lovers.

    C. Some dangerous animals are admired by animal lovers.

       Diagram the first two premises to be true. Then, notice that premise C is correctly represented by Case I, Case II, or Case III in the diagram.

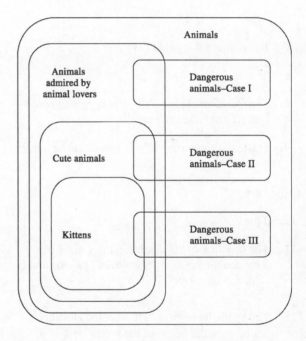

43. We are not forced into the conclusion "Some kittens are dangerous animals" since Case I and Case II represent true premises where this conclusion is false. Thus, the argument is <u>invalid</u>.

45. We are not forced into the conclusion "Some dangerous animals are cute" since Case I represents true premises where this conclusion is false. Thus, the argument is <u>invalid</u>.

47. The conclusion "All kittens are admired by animal lovers" yields a <u>valid</u> argument since premises A and B force the conclusion to be true.

Chapter 6 Review Exercises

1. True

2. False; a truth table with 5 variables has $2^5 = 32$ rows.

3. False; the number of rows any truth table has is always a power of 2, which is an even number.

4. False; the negation of a disjunction results in a conjunction.

5. False; the negation of a conditional statement is a conjunction.

6. True

7. False; a tautology is always true.

8. True

9. False; the conclusion of a valid argument could be a true or false statement.

10. False; the conclusion of a fallacy could be a true statement.

11. True

12. True

13. The negation of "If she doesn't pay me, I won't have enough cash" is "She doesn't pay me and I have enough cash."

15. The symbolic form of "He loses the election, but he wins the hearts of the voters" is $l \wedge w$.

17. The symbolic form of "He loses the election only if he doesn't win the hearts of the voters" is $l \rightarrow {\sim}w$.

19. Writing the symbolic form ${\sim}l \wedge w$ in words, we get "He doesn't lose the election and he wins the hearts of the voters."

21. Replacing q and r with the given truth values, we have

$${\sim}F \wedge {\sim}F$$
$$T \wedge T$$
$$T$$

The compound statement ${\sim}q \wedge {\sim}r$ is true.

23. Replacing r with the given truth value (s not known), we have

$$F \rightarrow (s \wedge F)$$
$$T$$

since a conditional statement with a false antecedent is true.

The compound statement $r \rightarrow (s \vee r)$ is true.

27.

p	q	p	$\wedge$	$({\sim}p$	$\vee$	$q)$
T	T	T	T	F	T	T
T	F	T	F	F	F	F
F	T	F	F	T	T	T
F	F	F	F	T	T	F
		1	4	2	3	2

The statement is not a tautology.

29. "All mathematicians are loveable" can be restated as "If someone is a mathematician, then that person is loveable."

31. "Having at least as many equations as unknowns is necessary for a system to have a unique solution" can be restated as "If a system has a unique solution, then it has at least as many equations as unknowns."

33. *The direct statement*: If the proposed regulations have been approved, then we need to change the way we do business.

(a) *Converse:* If we need to change the way we do business, then the proposed regulations have been approved.

(b) *Inverse:* If the proposed regulations have not been approved, then we do not need to change the way we do business.

(c) *Contrapositive:* If we do not need to change the way we do business, then the proposed regulations have not been approved.

35. In the diagram, a series circuit corresponding to $p \wedge p$ is followed in series by a parallel circuit represented by $\sim p \vee q$. The logical statement is $(p \wedge p) \wedge (\sim p \vee q)$. This statement is equivalent to $p \wedge q$.

p	q	$(p$	$\wedge$	$p)$	$\wedge$	$(\sim p$	$\vee$	$q)$	p	$\wedge$	q
T	T	T	T	T	T	F	T	T	T	T	T
T	F	T	T	T	F	F	F	F	T	F	F
F	T	F	F	F	F	T	T	T	F	F	T
F	F	F	F	F	F	T	T	F	F	F	F
		1	2	1	6	3	5	4	7	9	8

Columns 6 and 9 are identical.

37. The logical statement $(p \wedge q) \vee (p \wedge p)$ can be represented by the following circuit.

The statement simplifies to p as follows:

$$(p \wedge q) \vee (p \wedge p) = p \wedge (q \wedge p)$$
$$= p$$

39.

p	q	$(p$	$\underline{\vee}$	$q)$	$(p$	$\vee$	$q)$	$\wedge$	$\sim$	$(p$	$\wedge$	$q)$
T	T	T	F	T	T	T	T	F	F	T	T	T
T	F	T	T	F	T	T	F	T	T	T	F	F
F	T	F	T	T	F	T	T	T	T	F	F	T
F	F	F	F	F	F	F	F	T	F	F	F	F
		1	2	1	3	4	3	8	7	5	6	5

The columns labeled 2 and 8 are identical.

41. (a) Yes, the statement is true because "this year is 2010" is false and "$1 + 1 = 3$" is false and $F \rightarrow F$ is true.

(b) No, the statement was not true in 2010 because then the statement had the value $T \rightarrow F$, which is false.

43. Let l represent "you're late one more time" and d represent "you'll be docked." The argument is then presented symbolically as follows.

$$l \rightarrow d$$
$$\underline{l}$$
$$d$$

The argument is valid by Modus Ponens.

45. Let l represent "the instructor is late" and w represent "my watch is wrong." The argument is then presented symbolically as follows.

$$l \vee w$$
$$\underline{\sim w}$$
$$l$$

The argument is valid by Disjunctive Syllogism.

47. Let p represent "you play that song one more time" and n represent "I'm going nuts." The argument is then presented symbolically as follows.

$$p \rightarrow n$$
$$\underline{n}$$
$$p$$

The argument is invalid by Fallacy of the Converse.

49. Let h represent "we hire a new person," t represent "we'll spend more on training," and r represent "we rewrite the manual." The argument is then presented symbolically as follows.

$$h \rightarrow t$$
$$\underline{r \rightarrow \sim t}$$
$$\sim h$$

1. $h \rightarrow t$ Premise
2. $r \rightarrow \sim t$ Premise
3. r Premise
4. $\sim t$ 2, 3, Modus Ponens
5. $\sim h$ 1, 4, Modus Tollens

The argument is valid.

51. 1. $\sim p \rightarrow \sim q$ T
2. $\underline{q \rightarrow p}$ T
      $p \vee q$ F.

The argument is underline{invalid}. When $p = $ F and $q = $ F, the premises are true but the conclusion is false.

53. Let $d(x)$ represent "x is a dog" and $\ell(x)$ represent "x has a license".

(a) $\forall x \, [d(x) \rightarrow \ell(x)]$

(b) $\exists x \, [d(x) \wedge \sim\ell(x)]$

(c) There is a dog that doesn't have a license.

55. (a) Let $f(x)$ represent "x is a member of that fraternity," $w(x)$ represent "x does well academically," and j represent Jordan Enzor We can represent the argument symbolically as follows.

$$\forall x\,[f(x) \rightarrow w(x)]$$
$$\underline{f(j)}$$
$$w(j)$$

(b) Because of the first premise, the region representing "members of that fraternity" must be inside the region representing "those who do well academically." And j must be within the region representing "members of that fraternity" because of the second premise. Complete the Euler diagram as follows.

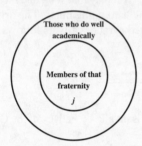

Since, when the premises are diagrammed as being true, we are forced into a true conclusion, the argument is <u>valid</u>.

57.

p	q	r	p	$\rightarrow$	$(q$	$\rightarrow$	$r)$	$(p$	$\rightarrow$	$q)$	$\rightarrow$	r
T	T	T	T	T	T	T	T	T	T	T	T	T
T	T	F	T	F	T	F	F	T	T	T	F	F
T	F	T	T	T	F	T	T	T	F	F	T	T
T	F	F	T	T	F	T	F	T	F	F	T	F
F	T	T	F	T	T	T	T	F	T	T	T	T
F	T	F	F	T	T	F	F	F	T	T	F	F
F	F	T	F	T	F	T	T	F	T	F	T	T
F	F	F	F	T	F	T	F	F	T	F	F	F
			1	4	2	3	2	5	6	5	8	7

To determine if the statements are equivalent, compare columns 4 and 8. Since they are not identical, the statements are not equivalent.

59. (a)

p	q	$(p$	$\wedge$	$\sim p)$	$\rightarrow$	q
T	T	T	F	F	T	T
T	F	T	F	F	T	F
F	T	F	F	T	T	T
F	F	F	F	T	T	F
		1	2	1	4	3

61. **(b)** and **(c)** are compound statements.

63. "If you use the Tax Table, then you do not have to compute your tax mathematically" is equivalent to "You do not use the Tax Table or you do not have to compute your tax mathematically."

65. (a) If you exercise regularly, then your heart becomes stronger and more efficient.

(b) If you are a teenager, then you need to be aware of the risks of drinking and driving.

(c) If you are visiting a country that has a high incidence of infectious diseases, then you may need extra immunizations.

(d) If you have good health, then you have food.

67. $(w \rightarrow d) \rightarrow v$

69. **(b), (c), (d)** These are declarative sentences and therefore are statements.

71. (a) $\sim s \rightarrow g$

(b) $l \rightarrow \sim g$

(c) $w \rightarrow l \equiv \sim l \rightarrow \sim w$

(d) $\sim s \rightarrow \sim w$

If the puppy does not lie still, it does not care to do worsted work. In Lewis Carroll's words, "Puppies that will not lie still never care to do worsted work."

73. (a) $f \rightarrow t \equiv \sim t \rightarrow \sim f$

(b) $\sim a \rightarrow \sim g \equiv g \rightarrow a$

(c) $w \rightarrow f$

(d) $t \rightarrow \sim g \equiv g \rightarrow \sim t$

(e) $a \rightarrow w \equiv \sim w \rightarrow \sim a$

(f) $g \rightarrow \sim e$

If the kitten will play with a gorilla, it does not have green eyes. In Lewis Carroll's words, "No kitten with green eyes will play with a gorilla."

Chapter 7

SETS AND PROBABILITY

7.1 Sets

Your Turn 1

There are three states whose names begin with O, Ohio Oklahoma, and Oregon Thus
$\{x \mid x$ is a state that begins with the letter O$\}$
$= \{$Ohio, Oklahoma, Oregon$\}$.

Your Turn 2

$\{2,4,6\} \subseteq \{6,2,4\}$ is true because each element of the first set is an element of the second set. (In this example the sets are in fact equal.)

Your Turn 3

A set of k distinct elements has 2^k subsets, so since there are four seasons, $\{x \mid x$ is a season of the year$\}$ has $2^4 = 16$ subsets.

Your Turn 4

$$U = \{0,1,2,...,10\}$$
$$A = \{3,6,9\}$$
$$B = \{2,4,6,8\}$$

Then $A' = \{0,1,2,4,5,7,8,10\}$. The elements common to A' and B are 2,4, and 8. Thus $A' \cap B = \{2,4,8\}$.

Your Turn 5

$$U = \{0,1,2,...,12\}$$
$$A = \{1,3,5,7,9,11\}$$
$$B = \{3,6,9,12\}$$
$$C = \{1,2,3,4,5\}$$

To find $A \cup (B \cap C')$ begin with the expression in parentheses and find C', which includes all the elements of the universal set that are not in C.

$$C' = \{6,7,8,9,10,11,12\}$$

The elements in C' that are also in B are 6, 9, and 12, so $B \cap C' = \{6,9,12\}$.

Now we list the elements of A and include any elements of $B \cap C'$ that are not already listed:

$$A \cup (B \cap C') = \{1,3,5,6,7,9,11,12\}$$

7.1 Exercises

1. $3 \in \{2,5,7,9,10\}$

   The number 3 is not an element of the set, so the statement is false.

3. $9 \notin \{2,1,5,8\}$

   Since 9 is not an element of the set, the statement is true.

5. $\{2,5,8,9\} = \{2,5,9,8\}$

   The sets contain exactly the same elements, so they are equal. The statement is true.

7. $\{$All whole numbers greater than 7 and less than 10$\} = \{8,9\}$

   Since 8 and 9 are the only such numbers, the statement is true.

9. $0 \in \emptyset$

   The empty set has no elements. The statement is false.

In Exercises 11–22,

$$A = \{2,4,6,8,10,12\},$$
$$B = \{2,4,8,10\},$$
$$C = \{4,8,12\},$$
$$D = \{2,10\},$$
$$E = \{6\},$$
and $$U = \{2,4,6,8,10,12,14\}.$$

11. Since every element of A is also an element of U, A is a subset of U, written $A \subseteq U$.

13. A contains elements that do not belong to E, namely 2, 4, 8, 10, and 12, so A is not a subset of E, written $A \nsubseteq E$.

15. The empty set is a subset of every set, so $\emptyset \subseteq A$.

17. Every element of D is also an element of B, so D is a subset of B, $D \subseteq B$.

19. Since every element of A is also an element of U, and $A \neq U$, $A \subset U$.

    Since every element of E is also an element of A, and $E \neq A$, $E \subset A$.

Since every element of A is not also an element of E, $A \not\subset E$.

Since every element of B is not also an element of C, $B \not\subset C$.

Since $\emptyset$ is a subset of every set, and $\emptyset \neq A$, $\emptyset \subset A$.

Since every element of $\{0,2\}$ is not also an element of D, $\{0,2\} \not\subset D$.

Since every element of D is not also an element of B, and $D \neq B$, $D \subset B$.

Since every element of A is not also an element of C, $A \not\subset C$.

21. A set with n distinct elements has 2^n subsets. A has $n = 6$ elements, so there are exactly $2^6 = 64$ subsets of A.

23. A set with n distinct elements has 2^n subsets, and C has $n = 3$ elements. Therefore, there are exactly $2^3 = 8$ subsets of C.

25. Since $\{7,9\}$ is the set of elements belonging to both sets, which is the intersection of the two sets, we write
$$\{5,7,9,19\} \cap \{7,9,11,15\} = \{7,9\}.$$

27. Since $\{1,2,5,7,9\}$ is the set of elements belonging to one or the other (or both) of the listed sets, it is their union.
$$\{2,1,7\} \cup \{1,5,9\} = \{1,2,5,7,9\}$$

29. Since $\emptyset$ contains no elements, there are no elements belonging to both sets. Thus, the intersection is the empty set, and we write
$$\{3,5,9,10\} \cap \emptyset = \emptyset.$$

31. $\{1,2,4\}$ is the set of elements belonging to both sets, and $\{1,2,4\}$ is also the set of elements in the first set or in the second set or possibly both. Thus,
$$\{1,2,4\} \cap \{1,2,4\} = \{1,2,4\}$$
and
$$\{1,2,4\} \cup \{1,2,4\} = \{1,2,4\}$$
are both true statements.

In Exercises 35–44,
$$U = \{1,2,3,4,5,6,7,8,9\},$$
$$X = \{2,4,6,8\},$$
$$Y = \{2,3,4,5,6\},$$
and $\quad Z = \{1,2,3,8,9\}.$

35. $X \cap Y$, the intersection of X and Y, is the set of elements belonging to both X and Y. Thus,
$$X \cap Y = \{2,4,6,8\} \cap \{2,3,4,5,6\}$$
$$= \{2,4,6\}.$$

37. X', the complement of X, consists of those elements of U that are not in X. Thus,
$$X' = \{1,3,5,7,9\}.$$

39. From Exercise 37, $X' = \{1,3,5,7,9\}$; from Exercise 38, $Y' = \{1,7,8,9\}$. There are no elements common to both X' and Y' so
$$X' \cap Y' = \{1,7,9\}.$$

41. First find $X \cup Z$.
$$X \cup Z = \{2,4,6,8\} \cup \{1,2,3,8,9\}$$
$$= \{1,2,3,4,6,8,9\}$$

Now find $Y \cap (X \cup Z)$.
$$Y \cap (X \cup Z) = \{2,3,4,5,6\} \cap \{1,2,3,4,6,8,9\}$$
$$= \{2,3,4,6\}$$

43. $U = \{1,2,3,4,5,6,7,8,9\}$ and $Z = \{1,2,3,8,9\}$, so $Z' = \{4,5,6,7\}$.

From Exercise 38, $Y' = \{1,7,8,9\}$.
$$(X \cap Y') \cup (Z' \cap Y') = (\{2,4,6,8\} \cap \{1,7,8,9\})$$
$$\cup (\{4,5,6,7\} \cap \{1,7,8,9\})$$
$$= \{8\} \cup \{7\}$$
$$= \{7,8\}$$

45. $(A \cap B) \cup (A \cap B') = (\{3,6,9\} \cap \{2,4,6,8\})$
$$\cup (\{3,6,9\} \cap \{0,1,3,5,7,9,10\})$$
$$= \{6\} \cup \{3,9\}$$
$$= \{3,6,9\}$$
$$= A$$

47. M' consists of all students in U who are not in M, so M' consists of all students in this school not taking this course.

49. $N \cap P$ is the set of all students in this school taking both accounting and zoology.

51. $A = \{2,4,6,8,10,12\}$,

$B = \{2,4,8,10\}$,

$C = \{4,8,12\}$,

$D = \{2,10\}$,

$E = \{6\}$,

$U = \{2,4,6,8,10,12,14\}$

A pair of sets is disjoint if the two sets have no elements in common. The pairs of these sets that are disjoint are B and E, C and E, D and E, and C and D.

53. B' is the set of all stocks on the list with a closing price below \$60 or above \$70.

$B' = \{$AT&T, Coca-Cola, FedEx, Disney$\}$

55. $(A \cap B)'$ is the set of all stocks on the list that do not have both a high price greater than \$50 and a closing price between \$60 and \$70.

$(A \cap B)' = \{$AT&T, Coca-Cola, FedEx, Disney$\}$

57. $A = \{1,2,3,\{3\},\{1,4,7\}\}$

(a) $1 \in A$ is true.

(b) $\{3\} \in A$ is true.

(c) $\{2\} \in A$ is false. $(\{2\} \subseteq A)$

(d) $4 \in A$ is false. $(4 \in \{1,4,7\})$

(e) $\{\{3\}\} \subset A$ is true.

(f) $\{1,4,7\} \in A$ is true.

(g) $\{1,4,7\} \subseteq A$ is false. $(\{1,4,7\} \in A)$

For Exercises 59 through 61 refer to this abbreviated version of the table:

Vanguard 500 (*V*)	Fidelity New Millennium Fund (*F*)	Janus Perkins Large Cap Value (*J*)	Templeton Large Cap Value Fund (T)
Exxon	Pfizer	Exxon	IBM
Apple	Cisco	GE	GE
GE	Wal-Mart	Wal-Mart	HP
IBM	Apple	AT&T	Home Depot
JPMorgan	JPMorgan	JPMorgan	Aflac

59. $V \cap J$

$= \{$Exxon, Apple, GE, IBM, JPMorgan$\}$

$\cap \{$Exxon, GE, Wal-Mart, AT&T, JPMorgan$\}$

$= \{$Exxon, GE, JPMorgan$\}$

61.

$J \cup F$

$= \{$Exxon, GE, Wal-Mart, AT&T, JPMorgan$\}$

$\cup \{$Pfizer, Cisco,Wal-Mart, Apple, JPMorgan$\}$

$= \{$Exxon, GE, Wal-Mart, AT&T, JPMorgan, Pfizer, Cisco, Apple$\}$

$(J \cup F)' = \{$IBM, HP, Home Depot, Aflac$\}$

63. The number of subsets of a set with k elements is 2^k, so the number of possible sets of customers (including the empty set) is $2^9 = 512$.

65. $U = \{s,d,c,g,i,m,h\}$ and $N = \{s,d,c,g\}$, so

$$N' = \{i,m,h\}.$$

67. $N \cup O = \{s,d,c,g\} \cup \{i,m,h,g\}$

$= \{s,d,c,g,i,m,h\} = U$

69. The number of subsets of a set with 51 elements (50 states plus the District of Columbia) is

$$2^{51} \approx 2.252 \times 10^{15}.$$

For Exercises 71 through 75 refer to this table:

Network	Subscribers (millions)	Launch	Content
The Discovery Channel	98.0	1985	Nonfiction, nature, science
TNT	98.0	1988	Movies, sports, original programming
USA Network	97.5	1980	Sports, family entertainment
TLC	97.3	1980	Original programming, family entertainment
TBS	97.3	1976	Movies, sports, original programming

71. $F = \{$USA, TLC, TBS$\}$

73. $H = \{$Discovery, TNT$\}$

75. $G \cup H = \{$TNT, USA, TBS$\} \cup \{$Discovery, TNT$\}$

$= \{$TNT, USA, TBS, Discovery$\}$;

the set of networks that feature sports or that have more than 97.6 million viewers.

77. Joe should always first choose the complement of what Dorothy chose. This will leave only two sets to choose from, and Joe will get the last choice.

79. (a) $(A \cup B') \cap C$

$A \cup B$ is the set of states whose name contains the letter e or which have a population over 4,000,000. Therefore, $(A \cup B)'$ is the set of states which are not among those whose name contains the letter e or which have a population over 4,000,000. As a result, $(A \cup B)' \cap C$ is the set of states which are not among those whose name contains the letter e or which have a population over 4,000,000 and which also have an area over 40,000 square miles.

(b)

$(A \cup B)' = \{($Kentucky, Maine, Nebraska, New Jersey$\}$
$\cup \{$Alabama, Colorado, Florida, Indiana,
Kentucky, New Jersey$\})'$
$= \{$Alabama, Colorado, Florida, Indiana,
Kentucky, Maine, Nebraska, New Jersey$\}'$
$= \{$Alaska, Hawaii$\}$

$(A \cup B)' \cap C = \{$Alaska, Hawaii$\} \cap \{$Alabama, Alaska,
Colorado, Florida, Kentucky, Nebraska$\}$
$= \{$Alaska$\}$

7.2 Applications of Venn Diagrams

Your Turn 1

$A \cup B'$ is the set of elements in A or not in B or both in A and not in B.

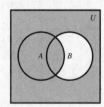

Your Turn 2

$A' \cap (B \cup C)$

First find A'.

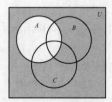

Then find $B \cup C$.

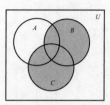

Then intersect these regions.

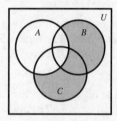

Your Turn 3

Start with $n(M \cap T \cap W) = 13$ and label the corresponding region with 13.

Since $n(M \cap T) = 17$, there are an additional 4 elements in $M \cap T$ but not in $M \cap T \cap W$. Label the corresponding region with 4.

Since $n(T \cap W) = 19$, there are an additional 6 elements in $T \cap W$ but not in $M \cap T \cap W$. Label the corresponding region with 6. The Venn diagram now looks like this:

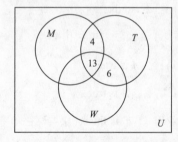

Since $n(T) = 46$, the number who ate only at Taco Bell is $46 - 4 - 13 - 6 = 23$.

Your Turn 4

Let T represent the set of those texting and M represent the set of those listening to music. The number in the lounge is $n(T \cup M) = n(T) + n(M) - n(T \cap M)$. We know that $n(T) = 15$, $n(M) = 11$, and $n(T \cap M) = 8$. Then $n(T \cup M) = 15 + 11 - 8 = 18$.

7.2 Exercises

1. $B \cap A'$ is the set of all elements in B *and* not in A.

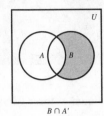

$B \cap A'$

3. $A' \cup B$ is the set of all elements that do not belong to A *or* that do belong to B, or both.

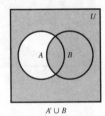

$A' \cup B$

5. $B' \cup (A' \cap B')$

First find $A' \cap B'$, the set of elements not in A *and* not in B.

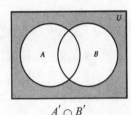

$A' \cap B'$

For the union, we want those elements in B' *or* $(A' \cap B')$, or both.

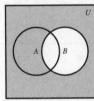

$B' \cup (A' \cap B')$

7. U' is the empty set $\emptyset$.

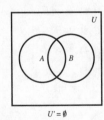

$U' = \emptyset$

9. Three sets divide the universal set into at most 8 regions. (Examples of this situation will be seen in Exercises 11–17.)

11. $(A \cap B) \cap C$

First form the intersection of A with B.

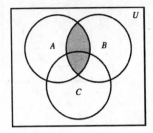

$A \cap B$

Now form the intersection of $A \cap B$ with C. The result will be the set of all elements that belong to all three sets.

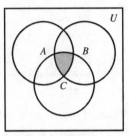

$(A \cap B) \cap C$

13. $A \cap (B \cup C')$

C' is the set of all elements in U that are not elements of C.

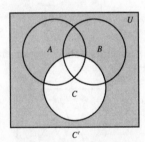

C'

Now form the union of C' with B.

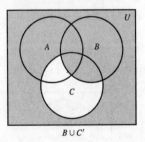

$B \cup C'$

Finally, find the intersection of this region with A.

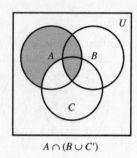

$$A \cap (B \cup C')$$

15. $(A' \cap B') \cap C'$

$A' \cap B'$ is the part of the universal set not in A *and* not in B:

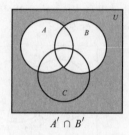

$$A' \cap B'$$

C' is the part of the universal set not in C:

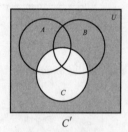

$$C'$$

Now intersect the shaded regions in these two diagrams:

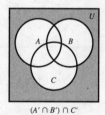

$$(A' \cap B') \cap C'$$

17. $(A \cap B') \cup C'$

First find $A \cap B'$, the region in A and not in B:

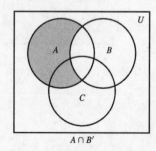

$$A \cap B'$$

C' is the region of the universal set not in C:

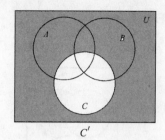

$$C'$$

Now form the union of these two regions.

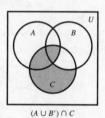

$$(A \cap B') \cup C'$$

19. $(A \cup B') \cap C$

First find $A \cup B'$, the region in A or B' or both.

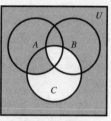

$$A \cup B'$$

Intersect this with C.

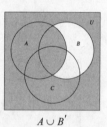

$$(A \cup B') \cap C$$

21. $n(A \cup B) = n(A) + n(B) - n(A \cap B)$
$$= 5 + 12 - 4$$
$$= 13$$

23. $n(A \cup B) = n(A) + n(B) - n(A \cap B)$
$$22 = n(A) + 9 - 5$$
$$22 = n(A) + 4$$
$$18 = n(A)$$

25. $n(U) = 41$

$n(A) = 16$

$n(A \cap B) = 12$

$n(B') = 20$

First put 12 in $A \cap B$. Since $n(A) = 16$, and 12 are in $A \cap B$, there must be 4 elements in A that are not in $A \cap B$. $n(B') = 20$, so there are 20 not in B. We already have 4 not in B (but in A), so there must be another 16 outside B *and* outside A. So far we have accounted for 32, and $n(U) = 41$, so 9 must be in B but not in any region yet identified. Thus $n(A' \cap B) = 9$.

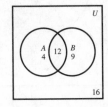

27. $n(A \cup B) = 24$

$n(A \cap B) = 6$

$n(A) = 11$

$n(A' \cup B') = 25$

Start with $n(A \cap B) = 6$. Since $n(A) = 11$, there must be 5 more in A not in B. $n(A \cup B) = 24$; we already have 11, so 13 more must be in B not yet counted. $A' \cup B'$ consists of all the region not in $A \cap B$, where we have 6. So far $5 + 13 = 18$ are in this region, so another $25 - 18 = 7$ must be outside both A and B.

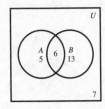

29. $n(A) = 28$ $n(B) = 34$ $n(C) = 25$

$n(A \cap B) = 14$ $n(B \cap C) = 15$ $n(A \cap C) = 11$

$n(A \cap B \cap C) = 9$

$n(U) = 59$

We start with $n(A \cap B \cap C) = 9$. If $n(A \cap B) = 14$, an additional 5 are in $A \cap B$ but not in $A \cap B \cap C$. Similarly, $n(B \cap C) = 15$, so $15 - 9 = 6$ are in $B \cap C$ but not in $A \cap B \cap C$. Also, $n(A \cap C) = 11$, so $11 - 9 = 2$ are in $A \cap C$ but not in $A \cap B \cap C$.

Now we turn our attention to $n(A) = 28$. So far we have $2 + 9 + 5 = 16$ in A; there must be another $28 - 16 = 12$ in A not yet counted. Similarly, $n(B) = 34$; we have $5 + 9 + 6 = 20$ so far, and $34 - 20 = 14$ more must be put in B.

For C, $n(C) = 25$; we have $2 + 9 + 6 = 17$ counted so far. Then there must be 8 more in C not yet counted. The count now stands at 56, and $n(U) = 59$, so 3 must be outside the three sets.

31. $n(A \cap B) = 6$ $n(A \cap B \cap C) = 4$

$n(A \cap C) = 7$ $n(B \cap C) = 4$

$n(A \cap C') = 11$ $n(B \cap C') = 8$

$n(C) = 15$ $n(A' \cap B' \cap C') = 5$

Start with $n(A \cap B) = 6$ and $n(A \cap B \cap C) = 4$ to get $6 - 4 = 2$ in that portion of $A \cap B$ outside of C. From $n(B \cap C) = 4$, there are $4 - 4 = 0$ elements in that portion of $B \cap C$ outside of A. Use $n(A \cap C) = 7$ to get $7 - 4 = 3$ elements in that portion of $A \cap C$ outside of B.

Since $n(A \cap C') = 11$, there are $11 - 2 = 9$ elements in that part of A outside of B and C. Use $n(B \cap C') = 8$ to get $8 - 2 = 6$ elements in that part of B outside of A and C. Since $n(C) = 15$, there are $15 - 3 - 4 - 0 = 8$ elements in C outside of A and B. Finally, 5 must be outside all three sets, since $n(A' \cap B' \cap C') = 5$.

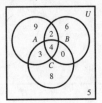

33. $(A \cup B)' = A' \cap B'$

For $(A \cup B)'$, first find $A \cup B$.

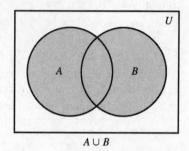

$A \cup B$

Now find $(A \cup B)'$, the region outside $A \cup B$.

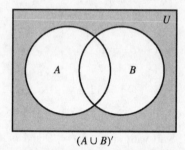

$(A \cup B)'$

For $A' \cap B'$, first find A' and B' individually.

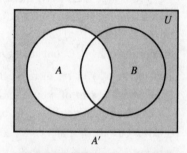

A'

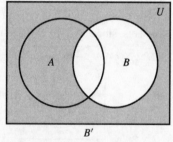

B'

Then $A' \cap B'$ is the region where A' and B' overlap, which is the entire region outside $A \cup B$ (the same result as in the second diagram). Therefore,

$$(A \cup B)' = A' \cap B'.$$

35. $A \cap (B \cup C) = (A \cap B) \cup (A \cap C)$

First find A and $B \cup C$ individually.

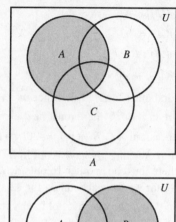

A

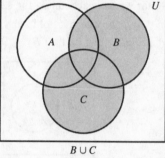

$B \cup C$

Then $A \cap (B \cup C)$ is the region where the above two diagram overlap.

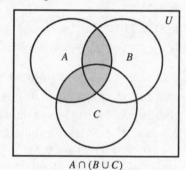

$A \cap (B \cup C)$

Next find $A \cap B$ and $A \cap C$ individually.

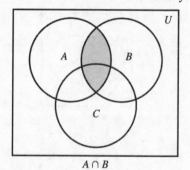

$A \cap B$

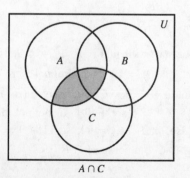

$A \cap C$

Then $(A \cap B) \cup (A \cap C)$ is the union of the above two diagrams.

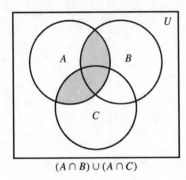

$$(A \cap B) \cup (A \cap C)$$

The Venn diagram for $A \cap (B \cup C)$ is identical to the Venn diagram for $(A \cap B) \cup (A \cap C)$, so conclude that

$$A \cap (B \cup C) = (A \cap B) \cup (A \cap C).$$

37. Prove

$n(A \cup B \cup C)$
$\quad = n(A) + n(B) + n(C) - n(A \cap B) - n(A \cap C)$
$\qquad\qquad - n(B \cap C) + n(A \cap B \cap C)$

$n(A \cup B \cup C)$
$\quad = n[A \cup (B \cup C)]$
$\quad = n(A) + n(B \cup C) - n[A \cap (B \cup C)]$
$\quad = n(A) + n(B) + n(C) - n(B \cap C)$
$\qquad\qquad - n[(A \cap B) \cup (A \cap C)]$
$\quad = n(A) + n(B) + n(C) - n(B \cap C)$
$\qquad\quad - \{n(A \cap B) + n(A \cap C)$
$\qquad\qquad - n[(A \cap B) \cap (A \cap C)]\}$
$\quad = n(A) + n(B) + n(C) - n(B \cap C) - n(A \cap B)$
$\qquad\qquad - n(A \cap C) + n(A \cap B \cap C)$

39. Let A be the set of trucks that carried early peaches, B be the set of trucks that carried late peaches, and C be the set of trucks that carried extra late peaches. We are given the following information.

$n(A) = 34 \qquad n(B) = 61 \qquad n(C) = 50$
$n(A \cap B) = 25$
$n(B \cap C) = 30$
$n(A \cap C) = 8$
$n(A \cap B \cap C) = 6$
$n(A' \cap B' \cap C') = 9$

Start with $A \cap B \cap C$.

We know that $n(A \cap B \cap C) = 6$.

Since $n(A \cap B) = 25$, the number in $A \cap B$ but not in C is $25 - 6 = 19$.

Since $n(B \cap C) = 30$, the number in $B \cap C$ but not in A is $30 - 6 = 24$.

Since $n(A \cap C) = 8$, the number in $A \cap C$ but not in B is $8 - 6 = 2$.

Since $n(A) = 34$, the number in A but not in B or C is $34 - (19 + 6 + 2) = 7$.

Since $n(B) = 61$, the number in B but not in A or C is $61 - (19 + 6 + 24) = 12$.

Since $n(C) = 50$, the number in C but not in A or B is $50 - (24 + 6 + 2) = 18$.

Since $n(A' \cap B' \cap C') = 9$, the number outside $A \cup B \cup C$ is 9.

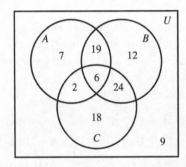

(a) From the Venn diagram, 12 trucks carried only late peaches.

(b) From the Venn diagram, 18 trucks carried only extra late peaches.

(c) From the Venn diagram, $7 + 12 + 18 = 37$ trucks carried only one type of peach.

(d) From the Venn diagram, $6 + 2 + 19 + 24 + 7 + 12 + 18 + 9 = 97$ trucks went out during the week.

41. (a) $n(Y \cap B) = 2$ since 2 is the number in the table where the Y row and the B column meet.

(b) $n(M \cup A) = n(M) + n(A) - n(M \cap A)$
$\qquad\qquad\quad = 33 + 41 - 14 = 60$

(c) $n[Y \cap (S \cup B)] = 6 + 2 = 8$

(d)

$n[O' \cup (S \cup A)]$
$\quad = n(O') + n(S \cup A) - n[O' \cap (S \cup A)]$
$\quad = (23 + 33) + (52 + 41) - (6 + 14 + 15 + 14)$
$\quad = 100$

(e) Since is $M' \cup O'$ is the entire set,
$(M' \cup O') \cap B = B$. Therefore,

$$n[(M' \cup O') \cap B] = n(B) = 27.$$

(f) $Y \cap (S \cup B)$ is the set of all bank customers who are of age 18–29 and who invest in stocks or bonds.

43. Let T be the set of all tall pea plants, G be the set of plants with green peas, and S be the set of plants with smooth peas. We are given the following information.

$n(U) = 50 \quad n(T) = 22 \quad n(G) = 25 \quad n(S) = 39$

$n(T \cap G) = 9$

$n(G \cap S) = 20$

$n(T \cap G \cap S) = 6$

$n(T' \cap G' \cap S') = 4$

Start by filling in the Venn Diagram with the numbers for the last two regions, $T \cap G \cap S$ and $T' \cap G' \cap S'$, as shown below. With $n(T \cap G) = 9$, this leaves

$n(T \cap G \cap S') = 9 - 6 = 3$.

With $n(G \cap S) = 20$, this leaves

$n(T' \cap G \cap S) = 20 - 6 = 14$.

Since $n(G) = 25$, $n(T' \cap G \cap S')$

$= 25 - 3 - 6 - 14 = 2$.

With no other regions that we can calculate, denote by x the number in $T \cap G' \cap S'$. Then $n(T \cap G' \cap S') = 22 - 3 - 6 - x = 13 - x$, and $n(T' \cap G' \cap S) = 39 - 6 - 14 - x = 19 - x$, as shown. Summing the values for all eight regions,

$(13 - x) + 3 + 2 + x + 6 + 14 + (19 - x) + 4 = 50$

$61 - x = 50$

$x = 11$

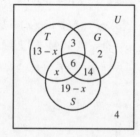

(a) $n(T \cap S) = 11 + 6 = 17$

(b) $n(T \cap G' \cap S') = 13 - x = 13 - 11 = 2$

(c) $n(T' \cap G \cap S) = 14$

45. First fill in the Venn diagram, starting with the region common to all three sets.

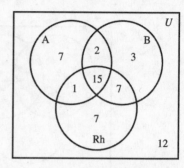

(a) The total of these numbers in the diagram is 54.

(b) $7 + 3 + 7 = 17$ had only one antigen.

(c) $1 + 2 + 7 = 10$ had exactly two antigens.

(d) A person with O-positive blood has only the Rh antigen, so this number is 7.

(e) A person with AB-positive blood has all three antigens, so this number is 15.

(f) A person with B-negative blood has only the B antigen, so this number is 3.

(g) A person with O-negative blood has none of the antigens. There are 12 such people.

(h) A person with A-positive blood has the A and Rh antigens, but not the B-antigen. The number is 1.

47. Extend the table to include totals for each row and each column.

	H	F	Total
A	95	34	129
B	41	38	79
C	9	7	16
D	202	150	352
Total	347	229	576

(a) $n(A \cap F)$ is the entry in the table that is in both row A and column F. Thus, there are 34 players in the set $A \cap F$.

(b) Since all players in the set C are either in set H or set F, $C \cap (H \cup F) = C$. Thus, $n(C \cap (H \cup F)) = n(C) = 16$, the total for row C. There are 16 players in the set $C \cap (H \cup F)$.

(c) $n(D \cup F) = n(D) + n(F) - n(D \cap F)$

$= 352 + 229 - 150$

$= 431$

(d) $B' \cap C'$ is the set of players who are both *not* in B and *not* in C. Thus, $B' \cap C' = A \cup D$, and since A and D are disjoint, $n(A \cup D) = n(A) + n(D) = 129 + 352 = 481$.

There are 481 players in the set $B' \cap C'$.

49. Reading directly from the table, $n(A \cap B) = 110.6$. Thus, there are 110.6 million people in the set $A \cap B$.

51. $n(G \cup (C \cap H)) = n(G) + n(C \cap H)$
$$= 80.4 + 5.0$$
$$= 85.4$$

There are 85.4 million people in the set $G \cup (C \cap H)$.

53. $n(H \cup D) = n(H) + n(D) - n(H \cap D)$
$$= 53.6 + 19.6 - 2.2$$
$$= 71.0$$

There are 71.0 million people in the set $H \cup D$.

For Exercises 55 through 58, use the following table, where the numbers are in thousands and the table has been extended to include row and column totals.

	W	B	H	A	Totals
N	54,205	13,547	12,021	3518	83,291
M	106,517	9577	16,111	6741	138,946
I	11,968	1740	1068	507	15,283
D	23,046	4590	3477	665	31,778
Totals	195,736	29,454	32,677	11,431	269,298

55. $N \cap (B \cup H)$ is the set of Blacks or Hispanics who never married. These people are located in the first row of the table, in the B and H columns, so $n(N \cap (B \cup H)) = 13,547 + 12,021 = 25,568$; since the table values are in thousands, this set contains 25,568,000 people.

57. $(D \cup W) \cap A'$ is the set of Whites or Divorced/ separated people who are not Asian/Pacific Islanders. The number of these people is found by adding the total of the W column to the total of the D row and then subtracting the number of Divorced/separated Whites (so we don't count them twice) and subtracting the number of Divorced/separated Asians/Pacific Islanders (whom we don't want to count).

$n((D \cup W) \cap A') = 195,736 + 31,778 - 23,046 - 665$
$$= 203,803;$$

since the table values are in thousands, this set contains 203,803,000 people.

59. Let W be the set of women, C be the set of those who speak Cantonese, and F be the set of those who set off firecrackers. We are given the following information.

$n(W) = 120 \qquad n(C) = 150 \qquad n(F) = 170$

$n(W' \cap C) = 108 \qquad n(W' \cap F') = 100$

$n(W \cap C' \cap F) = 18$

$n(W' \cap C' \cap F') = 78$

$n(W \cap C \cap F) = 30$

Note that

$n(W' \cap C \cap F') = n(W' \cap F') - n(W' \cap C' \cap F')$
$$= 100 - 78$$
$$= 22.$$

Furthermore,

$n(W' \cap C \cap F) = n(W' \cap C) - n(W' \cap C \cap F')$
$$= 108 - 22$$
$$= 86.$$

We now have

$n(W \cap C \cap F')$
$$= n(C) - n(W' \cap C \cap F) - n(W \cap C \cap F)$$
$$\quad - n(W' \cap C \cap F')$$
$$= 150 - 86 - 30 - 22 = 12.$$

With all of the overlaps of W, C, and F determined, we can now compute $n(W \cap C' \cap F') = 60$ and $n(W' \cap C' \cap F) = 36$.

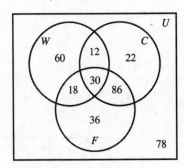

(a) Adding up the disjoint components, we find the total attendance to be

$60 + 12 + 18 + 30 + 22 + 86 + 36 + 78 = 342$.

(b) $n(C') = 342 - n(C) = 342 - 150 = 192$

(c) $n(W \cap F') = 60 + 12 = 72$

(d) $n(W' \cap C \cap F) = 86$

61. Let F be the set of fat chickens (so F' is the set of thin chickens), R be the set of red chickens (so R' is the set of brown chickens), and M be the set of male chickens, or roosters (so M' is the set of female chickens, or hens). We are given the following information.

$$n(F \cap R \cap M) = 9$$
$$n(F' \cap R' \cap M') = 13$$
$$n(R \cap M) = 15$$
$$n(F' \cap R) = 11$$
$$n(R \cap M') = 17$$
$$n(F) = 56$$
$$n(M) = 41$$
$$n(M') = 48$$

First, note that $n(M) + n(M') = n(U) = 89$, the total number of chickens.

Since $n(R \cap M) = 15, n(F' \cap R \cap M)$
$= 15 - 9 = 6$.

Since $n(F' \cap R) = 11, n(F' \cap R \cap M')$
$= 11 - 6 = 5$.

Since $n(R \cap M') = 17, n(F \cap R \cap M')$
$= 17 - 5 = 12$.

Since $n(M') = 48, n(F \cap R' \cap M')$
$= 48 - (12 + 5 + 13) = 18$.

Since $n(F) = 56, n(F \cap R' \cap M)$
$= 56 - (18 + 12 + 9) = 17$.

And, finally, since $n(M) = 41$,

$$n(F' \cap R' \cap M) = 41 - (17 + 9 + 6)$$
$$= 9.$$

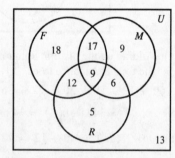

(a) $n(U) = 18 + 9 + 5 + 17 + 6$
$+ 12 + 9 + 13$
$= 89$

(b) $n(R) = n(R \cap M) + n(R \cap M')$
$= 15 + 17$
$= 32$

(c) $n(F \cap M) = n(F \cap R \cap M)$
$+ n(F \cap R' \cap M)$
$= 17 + 9$
$= 26$

(d) $n(F \cap M') = n(F) - n(F \cap M)$
$= 56 - 26$
$= 30$

(e) $n(F' \cap R') = n(F' \cap R' \cap M)$
$+ n(F' \cap R' \cap M')$
$= 9 + 13$
$= 22$

(f) $n(F \cap R)$
$= n(F \cap R \cap M) + n(F \cap R \cap M')$
$= 9 + 12 = 21$

7.3 Introduction to Probability

Your Turn 1

The sample space of equally likely outcomes is $\{hh, ht, th, tt\}$.

Your Turn 2

The sample space for tossing two coins is $\{hh, ht, th, tt\}$. The event E: "the coins show exactly one head" is $E = \{ht, th\}$.

Your Turn 3

E: worker is under 20, so E': worker is 20 or over.

F: worker is white, so F': worker is not white.

Thus, $E' \cap F'$ is the event that the worker is 20 or over and is not white.

Your Turn 4

The sample space S is $S = \{1, 2, 3, 4, 5, 6\}$.

The event H that the die shows a number less than 5 is $H = \{1, 2, 3, 4\}$. The probability of H is

$$P(H) = \frac{n(H)}{n(S)} = \frac{4}{6} = \frac{2}{3}.$$

Your Turn 5

In a standard 52-card deck there are 4 jacks and 4 kings, so $P(\text{jack or king}) = \frac{8}{52} = \frac{2}{13}$.

7.3 Exercises

3. The sample space is the set of the twelve months, {January, February, March, . . ., December}.

5. The possible number of points earned could be any whole number from 0 to 80. The sample space is the set

$$\{0,1,2,3,\ldots, 79,80\}.$$

7. The possible decisions are to go ahead with a new oil shale plant or to cancel it. The sample space is the set {go ahead, cancel}.

9. Let h = heads and t = tails for the coin; the die can display 6 different numbers. There are 12 possible outcomes in the sample space, which is the set

$$\{(h,1), (h,2), (h,3), (h,4), (h,5), (h,6),$$
$$(t,1), (t,2), (t,3), (t,4), (t,5), (t,6)\}.$$

13. Use the first letter of each name. The sample space is the set

$$S = \{AB, AC, AD, AE, BC, BD, BE, CD, CE, DE\}.$$

$n(S) = 10.$ Assuming the committee is selected at random, the outcomes are equally likely.

(a) One of the committee members must be Chinn. This event is {AC, BC, CD, CE}.

(b) Alam, Bartolini, and Chinn may be on any committee; Dickson and Ellsberg may not be on the same committee. This event is

{AB, AC, AD, AE, BC, BD, BE, CD, CE}.

(c) Both Alam and Chinn are on the committee. This event is {AC}.

15. Each outcome consists of two of the numbers 1, 2, 3, 4, and 5, without regard for order. For example, let (2, 5) represent the outcome that the slips of paper marked with 2 and 5 are drawn. There are ten equally likely outcomes in this sample space, which is

$$S = \{(1,2), (1,3), (1,4), (1,5), (2,3),$$
$$(2,4), (2,5), (3,4), (3,5), (4,5)\}.$$

(a) Both numbers in the outcome pair are even. This event is {(2, 4)}, which is called a simple event since it consists of only one outcome.

(b) One number in the pair is even and the other number is odd. This event is

$$\{(1,2), (1,4), (2,3), (2,5), (3,4), (4,5)\}.$$

(c) Each slip of paper has a different number written on it, so it is not possible to draw two slips marked with the same number. This event is ∅, which is called an impossible event since it contains no outcomes.

17. $S = \{HH, THH, HTH, TTHH, THTH, HTTH,$
$$TTTH, TTHT, THTT, HTTT, TTTT\}$$

$n(S) = 11.$ The outcomes are not equally likely.

(a) The coin is tossed four times. This event is written {TTHH, THTH, HTTH, TTTH, TTHT, THTT, HTTT, TTTT}.

(b) Exactly two heads are tossed. This event is written {HH, THH, HTH, TTHH, THTH, HTTH}.

(c) No heads are tossed. This event is written {TTTT}.

For Exercises 19–24, use the sample space

$$S = \{1,2,3,4,5,6\}.$$

19. "Getting a 2" is the event $E = \{2\}$, so $n(E) = 1$ and $n(S) = 6.$

If all the outcomes in a sample space S are equally likely, then the probability of an event E is

$$P(E) = \frac{n(E)}{n(S)}.$$

In this problem,

$$P(E) = \frac{n(E)}{n(S)} = \frac{1}{6}.$$

21. "Getting a number less than 5" is the event $E = \{1,2,3,4\}$, so $n(E) = 4.$

$$P(E) = \frac{4}{6} = \frac{2}{3}.$$

23. "Getting a 3 or a 4" is the event $E = \{3,4\}$, so $n(E) = 2.$

$$P(E) = \frac{2}{6} = \frac{1}{3}.$$

For Exercises 25–34, the sample space contains all 52 cards in the deck, so $n(S) = 52.$

25. Let E be the event "a 9 is drawn." There are four 9's in the deck, so $n(E) = 4.$

$$P(9) = P(E) = \frac{n(E)}{n(S)} = \frac{4}{52} = \frac{1}{13}$$

27. Let F be the event "a black 9 is drawn." There are two black 9's in the deck, so $n(F) = 2.$

$$P(\text{black } 9) = P(F) = \frac{n(F)}{n(S)} = \frac{2}{52} = \frac{1}{26}$$

29. Let G be the event "a 9 of hearts is drawn." There is only one 9 of hearts in a deck of 52 cards, so $n(G) = 1$.

$$P(9 \text{ of hearts}) = P(G) = \frac{n(G)}{n(S)} = \frac{1}{52}$$

31. Let H be the event "a 2 or a queen is drawn." There are four 2's and four queens in the deck, so $n(H) = 8$.

$$P(2 \text{ or queen}) = P(H) = \frac{n(H)}{n(S)} = \frac{8}{52} = \frac{2}{13}$$

33. Let E be the event "a red card or a ten is drawn." There are 26 red cards and 4 tens in the deck. But 2 tens are red cards and are counted twice. Use the result from the previous section.

$$n(E) = n(\text{red cards}) + n(\text{tens}) - n(\text{red tens})$$
$$= 26 + 4 - 2$$
$$= 28$$

Now calculate the probability of E.

$$P(\text{red cards or ten}) = \frac{n(E)}{n(S)}$$
$$= \frac{28}{52}$$
$$= \frac{7}{13}$$

For Exercises 35–40, the sample space consists of all the marbles in the jar. There are $3 + 4 + 5 + 8 = 20$ marbles, so $n(S) = 20$.

35. 3 of the marbles are white, so

$$P(\text{white}) = \frac{3}{20}.$$

37. 5 of the marbles are yellow, so

$$P(\text{yellow}) = \frac{5}{20} = \frac{1}{4}.$$

39. $3 + 4 + 5 = 12$ of the marbles are not black, so

$$P(\text{not black}) = \frac{12}{20} = \frac{3}{5}.$$

41. It is possible to establish an exact probability for this event, so it is not an empirical probability.

43. It is not possible to establish an exact probability for this event, so this an empirical probability.

45. It is not possible to establish an exact probability for this event, so this is an empirical probability.

47. The gambler's claim is a mathematical fact, so this is not an empirical probability.

49. The outcomes are not equally likely.

51. E: worker is female

F: worker has worked less than 5 yr

G: worker contributes to a voluntary retirement plan

(a) E' occurs when E does not, so E' is the event "worker is male."

(b) $E \cap F$ occurs when both E and F occur, so $E \cap F$ is the event "worker is female and has worked less than 5 yr."

(c) $E \cup G'$ is the event "worker is female or does not contribute to a voluntary retirement plan."

(d) F' occurs when F does not, so F' is the event "worker has worked 5 yr or more."

(e) $F \cup G$ occurs when F or G occurs or both, so $F \cup G$ is the event "worker has worked less than 5 yr or contributes to a voluntary retirement plan."

(f) $F' \cap G'$ occurs when F does not and G does not, so $F' \cap G'$ is the event "worker has worked 5 yr or more and does not contribute to a voluntary retirement plan."

53. (a) From the solution to Exercise 42 in Section 7.2 we know that 80 investors made all three types of investments, so the probability that a randomly chosen professor invested in stocks and bonds and certificates of deposit is $\frac{80}{150} = \frac{8}{15}$.

(b) From the solution to Exercise 42 in Section 7.2 we find that $80 - 65$ or 15 professors invested only in bonds, so the probability that a randomly chosen professor invested only in bonds is $\frac{15}{150} = \frac{1}{10}$.

55. E: person smokes

F: person has a family history of heart disease

G: person is overweight

(a) G': "person is not overweight."

(b) $F \cap G$: "person has a family history of heart disease and is overweight."

(c) $E \cup G'$: "person smokes or is not overweight."

57. (a) $P(\text{heart disease}) = \dfrac{615{,}651}{2{,}424{,}059} = 0.2540$

(b) $P(\text{cancer or heart disease}) = \dfrac{1{,}175{,}838}{2{,}424{,}059}$
$$= 0.4851$$

(c) P(not accident and not diabetes mellitus)

$$= \frac{2,424,059 - 117,075 - 70,905}{2,424,059}$$

$$= \frac{2,236,079}{2,424,059}$$

$$= 0.9225$$

59. P(served $20-29$ years) $= \dfrac{17}{100} = 0.17$

61. (a) P(III Corps) $= \dfrac{22,083}{70,076} \approx 0.3151$

(b) P(lost in battle) $= \dfrac{22,557}{70,076} \approx 0.3219$

(c) P(I Corps lost in battle) $= \dfrac{7661}{20,706} \approx 0.3700$

(d) P(I Corps not lost in battle)

$$= \frac{20,706 - 7661}{20,706} \approx 0.6300$$

P(II Corps not lost in battle)

$$= \frac{20,666 - 6603}{20,666} \approx 0.6805$$

P(III Corps not lost in battle)

$$= \frac{22,083 - 8007}{22,083} \approx 0.6374$$

P(Cavalry not lost in battle)

$$= \frac{6621 - 286}{6621} \approx 0.9568$$

The Cavalry had the highest probability of not being lost in battle.

(e) P(I Corps loss) $= \dfrac{7661}{20,706} \approx 0.3700$

P(II Corps loss) $= \dfrac{6603}{20,666} \approx 0.3195$

P(III Corps loss) $= \dfrac{8007}{22,083} \approx 0.3626$

P(Cavalry loss) $= \dfrac{286}{6621} \approx 0.0432$

I Corps had the highest probability of loss.

63. There were 342 in attendance.

(a) P(speaks Cantonese) $= \dfrac{150}{342} = \dfrac{25}{57}$

(b) P(does not speaks Cantonese) $= \dfrac{192}{342} = \dfrac{32}{57}$

(c) P(woman who did not light firecracker).

$$= \frac{72}{342} = \frac{4}{19}$$

7.4   Basic Concepts of Probability

Your Turn 1

Let A stand for the event "ace" and C stand for the event "club."

$$P(A \cup C) = P(A) + P(C) - P(A \cap C)$$

$$= \frac{4}{52} + \frac{13}{52} - \frac{1}{52}$$

$$= \frac{16}{52} = \frac{4}{13}$$

Your Turn 2

Let E stand for the event "eight" and B stand for the event "both dice show the same number." From Figure 18 we see that E contains the 5 events 6-2, 5-3, 4-4, 3-5, and 2-6. B contains the 6 events 1-1, 2-2, 3-3, 4-4, 5-5, and 6-6. Only the even 4-4 belongs to both E and B. The sample space contains 36 equally likely events.

$$P(E \cup B) = P(E) + P(B) - P(E \cap B)$$

$$= \frac{5}{36} + \frac{6}{36} - \frac{1}{36}$$

$$= \frac{10}{36} = \frac{5}{18}$$

Your Turn 3

The complement of the event "sum < 11" is "sum $= 11$ or sum $= 12$." Figure 18 shows that "sum $= 11$ or sum $= 12$" contains the 3 events 5-6, 6-5, and 6-6.

$$P(\text{sum} < 11) = 1 - P(\text{sum} = 11 \text{ or sum} = 12)$$

$$= 1 - \frac{3}{36}$$

$$= \frac{33}{36} = \frac{11}{12}$$

Your Turn 4

Let E be the event "snow tomorrow." Since $P(E) = \frac{3}{10}$,

$P(E') = 1 - P(E) = \frac{7}{10}$. The odds in favor of snow

tomorrow are

$$\frac{P(E)}{P(E')} = \frac{3/10}{7/10} = \frac{3}{7}.$$

We can write these odds as 3 to 7 or 3:7.

Your Turn 5

Let E be the event "package delivered on time." The

odds in favor of E are 17 to 3, so $P(E') = \frac{3}{17+3} =$

$\frac{3}{20}$. The probability that the package will not be

delivered on time is $\frac{3}{20}$.

Your Turn 6

If the odds against the horse winning are 7 to 3,

$$P(\text{loses}) = \frac{7}{7+3} = \frac{7}{10}, \text{ so } P(\text{wins}) = 1 - \frac{7}{10} = \frac{3}{10}.$$

7.4 Exercises

3. A person can own a dog and own an MP3 player at the same time. No, these events are not mutually exclusive.

5. A person can be retired and be over 70 years old at the same time. No, these events are not mutually exclusive.

7. A person cannot be one of the ten tallest people in the United States and be under 4 feet tall at the same time. Yes, these events are mutually exclusive.

9. When two dice are rolled, there are 36 equally likely outcomes.

   (a) Of the 36 ordered pairs, there is only one for which the sum is 2, namely $\{(1,1)\}$. Thus,

$$P(\text{sum is 2}) = \frac{1}{36}.$$

   (b) $\{(1,3), (2,2), (3,1)\}$ comprise the ways of getting a sum of 4. Thus,

$$P(\text{sum is 4}) = \frac{3}{36} = \frac{1}{12}.$$

   (c) $\{(1,4), (2,3), (3,2), (4,1)\}$ comprise the ways of getting a sum of 5. Thus,

$$P(\text{sum is 5}) = \frac{4}{36} = \frac{1}{9}.$$

   (d) $\{(1,5), (2,4), (3,3), (4,2), (5,1)\}$ comprise the ways of getting a sum of 6. Thus,

$$P(\text{sum is 6}) = \frac{5}{36}.$$

11. Again, when two dice are rolled there are 36 equally likely outcomes.

   (a) Here, the event is the union of four mutually exclusive events, namely, the sum is 9, the sum is 10, the sum is 11, and the sum is 12. Hence,

$$\begin{aligned}
P(\text{sum is 9 or more}) &= P(\text{sum is 9}) + P(\text{sum is 10}) \\
&\quad + P(\text{sum is 11}) + (\text{sum is 12}) \\
&= \frac{4}{36} + \frac{3}{36} + \frac{2}{36} + \frac{1}{36} \\
&= \frac{10}{36} = \frac{5}{18}.
\end{aligned}$$

   (b) $P(\text{sum is less than 7})$

$$\begin{aligned}
&= P(2) + P(3) + P(4) + P(5) + P(6) \\
&= \frac{1}{36} + \frac{2}{36} + \frac{3}{36} + \frac{4}{36} + \frac{5}{36} \\
&= \frac{15}{36} \\
&= \frac{5}{12}
\end{aligned}$$

   (c) $P(\text{sum is between 5 and 8})$

$$\begin{aligned}
&= P(\text{sum is 6}) + P(\text{sum is 7}) \\
&= \frac{5}{36} + \frac{6}{36} \\
&= \frac{11}{36}
\end{aligned}$$

13. $P(\text{first die is 3 or sum is 8})$

$$\begin{aligned}
&= P(\text{first die is 3}) + P(\text{sum is 8}) \\
&\quad - P(\text{first die is 3 and sum is 8}) \\
&= \frac{6}{36} + \frac{5}{36} - \frac{1}{36} \\
&= \frac{10}{36} \\
&= \frac{5}{18}
\end{aligned}$$

15. **(a)** The events E, "9 is drawn," and F, "10 is drawn," are mutually exclusive, so $P(E \cap F) = 0$. Using the union rule,

$$P(9 \text{ or } 10) = P(9) + P(10)$$
$$= \frac{4}{52} + \frac{4}{52}$$
$$= \frac{8}{52} = \frac{2}{13}.$$

(b)

$$P(\text{red or } 3) = P(\text{red}) + P(3) - P(\text{red and } 3)$$
$$= \frac{26}{52} + \frac{4}{52} - \frac{2}{52}$$
$$= \frac{28}{52} = \frac{7}{13}$$

(c) Since these events are mutually exclusive,
$$P(9 \text{ or black } 10) = P(9) + P(\text{black } 10)$$
$$= \frac{4}{52} + \frac{2}{52}$$
$$= \frac{6}{52}$$
$$= \frac{3}{26}.$$

(d) $P(\text{heart or black }) = \dfrac{13}{52} + \dfrac{26}{52} = \dfrac{39}{52} = \dfrac{3}{4}.$

(e) $P(\text{face card or diamond})$
$$= P(\text{face card}) + P(\text{diamond})$$
$$- P(\text{face card and diamond})$$
$$= \frac{12}{52} + \frac{13}{52} - \frac{3}{52}$$
$$= \frac{22}{52} = \frac{11}{26}$$

17. (a) Since these events are mutually exclusive,
$$P(\text{brother or uncle}) = P(\text{brother}) + P(\text{uncle})$$
$$= \frac{2}{13} + \frac{3}{13} = \frac{5}{13}.$$

(b) Since these events are mutually exclusive,
$$P(\text{brother or cousin}) = P(\text{brother}) + P(\text{cousin})$$
$$= \frac{2}{13} + \frac{5}{13}$$
$$= \frac{7}{13}.$$

(c) Since these events are mutually exclusive,
$$P(\text{brother or mother}) = P(\text{brother}) + P(\text{mother})$$
$$= \frac{2}{13} + \frac{1}{13}$$
$$= \frac{3}{13}.$$

19. (a) There are 5 possible numbers on the first slip drawn, and for each of these, 4 possible numbers on the second, so the sample space contains $5 \cdot 4 = 20$ ordered pairs. Two of these ordered pairs have a sum of 9: $(4, 5)$ and $(5, 4)$. Thus,

$$P(\text{sum is } 9) = \frac{2}{20} = \frac{1}{10}.$$

(b) The outcomes for which the sum is 5 or less are $(1, 2), (1, 3), (1, 4), (2, 1), (2, 3), (3, 1),$ $(3, 2),$ and $(4,1)$. Thus,

$$P(\text{sum is 5 or less}) = \frac{8}{20} = \frac{2}{5}.$$

(c) Let A be the event "the first number is 2" and B the event "the sum is 6." Use the union rule.
$$P(A \cup B) = P(A) + P(B) - P(A \cap B)$$
$$= \frac{4}{20} + \frac{4}{20} - \frac{1}{20}$$
$$= \frac{7}{20}$$

21. Since $P(E \cap F) = 0.16$, the overlapping region $E \cap F$ is assigned the probability 0.16 in the diagram. Since $P(E) = 0.26$ and $P(E \cap F) = 0.16$, the region in E but not F is given the label 0.10. Similarly, the remaining regions are labeled.

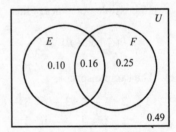

(a) $P(E \cup F) = 0.10 + 0.16 + 0.25$
$$= 0.51$$

Consequently, the part of U outside $E \cup F$ receives the label
$$1 - 0.51 = 0.49.$$

(b) $P(E' \cap F) = P(\text{in } F \text{ but not in } E)$
$$= 0.25$$

(c) The region $E \cap F'$ is that part of E which is not in F. Thus,
$$P(E \cap F') = 0.10.$$

(d) $P(E' \cup F') = P(E') + P(F') - P(E' \cap F')$
$$= 0.74 + 0.59 - 0.49$$
$$= 0.84$$

23. **(a)** The sample space is

$$3 - 1 \quad 3 - 1 \quad 3 - 5 \quad 3 - 5 \quad 3 - 9 \quad 3 - 9$$
$$3 - 1 \quad 3 - 1 \quad 3 - 5 \quad 3 - 5 \quad 3 - 9 \quad 3 - 9$$
$$4 - 1 \quad 4 - 1 \quad 4 - 5 \quad 4 - 5 \quad 4 - 9 \quad 4 - 9$$
$$4 - 1 \quad 4 - 1 \quad 4 - 5 \quad 4 - 5 \quad 4 - 9 \quad 4 - 9$$
$$8 - 1 \quad 8 - 1 \quad 8 - 5 \quad 8 - 5 \quad 8 - 9 \quad 8 - 9$$
$$8 - 1 \quad 8 - 1 \quad 8 - 5 \quad 8 - 5 \quad 8 - 9 \quad 8 - 9$$

where the first number in each pair is the number that appears on A and the second the number that appears on B. B beats A in 20 of 36 possible outcomes. Thus,

$$P(B \text{ beats } A) = \frac{20}{36} = \frac{5}{9}.$$

(b) The sample space is

$$1 - 2 \quad 1 - 2 \quad 1 - 6 \quad 1 - 6 \quad 1 - 7 \quad 1 - 7$$
$$1 - 2 \quad 1 - 2 \quad 1 - 6 \quad 1 - 6 \quad 1 - 7 \quad 1 - 7$$
$$5 - 2 \quad 5 - 2 \quad 5 - 6 \quad 5 - 6 \quad 5 - 7 \quad 5 - 7$$
$$5 - 2 \quad 5 - 2 \quad 5 - 6 \quad 5 - 6 \quad 5 - 7 \quad 5 - 7$$
$$9 - 2 \quad 9 - 2 \quad 9 - 6 \quad 9 - 6 \quad 9 - 7 \quad 9 - 7$$
$$9 - 2 \quad 9 - 2 \quad 9 - 6 \quad 9 - 6 \quad 9 - 7 \quad 9 - 7$$

where the first number in each pair is the number that appears on B and the second the number that appears on C. C beats B in 20 of 36 possible outcomes. Thus,

$$P(C \text{ beats } B) = \frac{20}{36} = \frac{5}{9}.$$

(c) The sample space is

$$3 - 2 \quad 3 - 2 \quad 3 - 6 \quad 3 - 6 \quad 3 - 7 \quad 3 - 7$$
$$3 - 2 \quad 3 - 2 \quad 3 - 6 \quad 3 - 6 \quad 3 - 7 \quad 3 - 7$$
$$4 - 2 \quad 4 - 2 \quad 4 - 6 \quad 4 - 6 \quad 4 - 7 \quad 4 - 7$$
$$4 - 2 \quad 4 - 2 \quad 4 - 6 \quad 4 - 6 \quad 4 - 7 \quad 4 - 7$$
$$8 - 2 \quad 8 - 2 \quad 8 - 6 \quad 8 - 6 \quad 8 - 7 \quad 8 - 7$$
$$8 - 2 \quad 8 - 2 \quad 8 - 6 \quad 8 - 6 \quad 8 - 7 \quad 8 - 7$$

where the first number in each pair is the number that appears on A and the second the number that appears on C. A beats C in 20 of 36 possible outcomes. Thus,

$$P(A \text{ beats } C) = \frac{20}{36} = \frac{5}{9}.$$

27. Let E be the event "a 3 is rolled."

$$P(E) = \frac{1}{6} \text{ and } P(E') = \frac{5}{6}.$$

The odds in favor of rolling a 3 are

$$\frac{P(E)}{P(E')} = \frac{\frac{1}{6}}{\frac{5}{6}} = \frac{1}{5},$$

which is written "1 to 5."

29. Let E be the event "a 2, 3, 4, or 5 is rolled." Here $P(E) = \frac{4}{6} = \frac{2}{3}$ and $P(E') = \frac{1}{3}$. The odds in favor of E are

$$\frac{P(E)}{P(E')} = \frac{\frac{2}{3}}{\frac{1}{3}} = \frac{2}{1},$$

which is written "2 to 1."

31. **(a)** Yellow: There are 3 ways to win and 15 ways to lose. The odds in favor of drawing yellow are 3 to 15, or 1 to 5.

(b) Blue: There are 11 ways to win and 7 ways to lose; the odds in favor of drawing blue are 11 to 7.

(c) White: There are 4 ways to win and 14 ways to lose; the odds in favor of drawing white are 4 to 14, or 2 to 7.

(d) Not white: Since the odds in favor of white are 2 to 7, the odds in favor of not white are 7 to 2.

35. Each of the probabilities is between 0 and 1 and the sum of all the probabilities is

$$0.09 + 0.32 + 0.21 + 0.25 + 0.13 = 1,$$

so this assignment is possible.

$$0.92 + 0.03 + 0 + 0.02 + 0.03 = 1.$$

37. The sum of the probabilities

$$\frac{1}{3} + \frac{1}{4} + \frac{1}{6} + \frac{1}{8} + \frac{1}{10} = \frac{117}{120} < 1,$$

so this assignment is not possible.

39. This assignment is not possible because one of the probabilities is -0.08, which is not between 0 and 1. A probability cannot be negative.

41. The answers that are given are theoretical. Using the Monte Carlo method with at least 50 repetitions on a graphing calculator should give values close to these.

(a) 0.2778

(b) 0.4167

43. The answers that are given are theoretical. Using the Monte Carlo method with at least 100 repetitions should give values close to these.

(a) 0.0463

(b) 0.2963

47. Let C be the event "the calculator has a good case," and let B be the event "the calculator has good batteries."

$$P(C \cap B)$$
$$= 1 - P[(C \cap B)']$$
$$= 1 - P(C' \cup B')$$
$$= 1 - [P(C') + P(B') - P(C' \cap B')]$$
$$= 1 - (0.08 + 0.11 - 0.03)$$
$$= 0.84$$

Thus, the probability that the calculator has a good case and good batteries is 0.84.

49. (a) $P(\$500 \text{ or more}) = 1 - P(\text{less than } \$500)$
$$= 1 - (0.21 + 0.17)$$
$$= 1 - 0.38$$
$$= 0.62$$

(b) $P(\text{less than } \$1000) = 0.21 + 0.17 + 0.16$
$$= 0.54$$

(c) $P(\$500 \text{ to } \$2999) = 0.16 + 0.15 + 0.12$
$$= 0.43$$

(d) $P(\$3000 \text{ or more}) = 0.08 + 0.07 + 0.04$
$$= 0.19$$

51. (a) The probability of Female and 16 to 24 years old is the entry in the first row in the Female column, which is 0.061.

(b) 16 to 54 years old includes the first two rows of the table, so adding the totals for these two rows we find that the probability is $0.127 + 0.634 = 0.761$.

(c) For Male or 25 to 54, we add the totals of the second row and the Male column and then subtract the value for Male and 25 to 54 in order not to count it twice. Thus the probability is $0.634 + 0.531 - 0.343 = 0.822$.

(d) For Female or 16 to 24 we add the totals of the first row and the Female column and then subtract the value for Female and 16 to 24 in order not to count it twice. Thus the probability is $0.127 + 0.469 - 0.061 = 0.535$.

53. $P(C) = 0.039, P(M \cap C) = 0.035,$
$P(M \cup C) = 0.491$

Place the given information in a Venn diagram by starting with 0.035 in the intersection of the regions for M and C.

Since $P(C) = 0.039, 0.039 - 0.035 = 0.004$

goes inside region C, but outside the intersection of C and M. Thus,

$$P(C \cap M') = 0.004.$$

Since

$P(M \cup C) = 0.491, 0.491 - 0.035 - 0.004 = 0.452$

goes inside region M, but outside the intersection of C and M. Thus, $P(M \cap C') = 0.452$. The labeled regions have probability

$$0.452 + 0.035 + 0.004 = 0.491.$$

Since the entire region of the Venn diagram must have probability 1, the region outside M and C, or $M' \cap C'$, has probability

$$1 - 0.491 = 0.509.$$

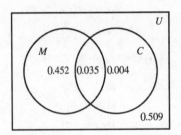

(a) $P(C') = 1 - P(C)$
$$= 1 - 0.039$$
$$= 0.961$$

(b) $P(M) = 0.452 + 0.035$
$$= 0.487$$

(c) $P(M') = 1 - P(M)$
$$= 1 - 0.487$$
$$= 0.513$$

(d) $P(M' \cap C') = 0.509$

(e) $P(C \cap M') = 0.004$

(f) $P(C \cup M')$
$$= P(C) + P(M') - P(C \cap M')$$
$$= 0.039 + 0.513 - 0.004$$
$$= 0.548$$

55. (a) Now red is no longer dominant, and RW or WR results in pink, so

$$P(\text{red}) = P(RR) = \frac{1}{4}.$$

(b) $P(\text{pink}) = P(RW) + P(WR)$

$$= \frac{1}{4} + \frac{1}{4} = \frac{1}{2}$$

(c) $P(\text{white}) = P(WW) = \frac{1}{4}$

57. Let L be the event "visit results in lab work" and R be the event "visit results in referral to specialist." We are given the probability a visit results in neither is 35%, so $P((L \cup R)') = 0.35$. Since $P(L \cup R) = 1 - P((L \cup R)')$, we have

$$P(L \cup R) = 1 - 0.35 = 0.65.$$

We are also given $P(L) = 0.40$ and $P(R) = 0.30$. Using the union rule for probability,

$$P(L \cup R) = P(L) + P(R) - P(L \cap R)$$
$$0.65 = 0.40 + 0.30 - P(L \cap R)$$
$$0.65 = 0.50 - P(L \cap R)$$
$$P(L \cap R) = 0.05$$

The correct answer choice is **a**.

59. Let $x = P(A \cap B)$,

$y = P(B \cap C)$,

$z = P(A \cap C)$,

and $w = P((A \cup B \cup C)')$.

If an employee must choose exactly two or none of the supplementary coverages A, B, and C, then $P(A \cap B \cap C) = 0$ and the probabilities of the region representing a single choice of coverages A, B, or C are also 0. We can represent the choices and probabilities with the following Venn diagram.

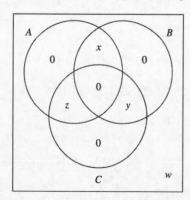

The information given leads to the following system of equations.

$$x + y + z + w = 1$$
$$x \quad + z \quad = \frac{1}{4}$$
$$x + y \quad = \frac{1}{3}$$
$$y + z \quad = \frac{5}{12}$$

Using a graphing calculator or computer program, the solution to the system is $x = 1/2$, $y = 1/4$, $z = 1/6$, $w = 1/2$. Since the probability that a randomly chosen employee will choose no supplementary coverage is w, the correct answer choice is **c**.

61. Since 55 of the workers were women, $130 - 55 = 75$ were men. Since 3 of the women earned more than \$40,000, $55 - 3 = 52$ of them earned \$40,000 or less. Since 62 of the men earned \$40,000 or less, $75 - 62 = 13$ earned more than \$40,000. These data for the 130 workers can be summarized in the following table.

	Men	Women
\$40,000 or less	62	52
Over \$40,000	13	3

(a) $P(\text{a woman earning \$40,000 or less})$

$$= \frac{52}{130} = 0.4$$

(b) $P(\text{a man earning more than \$40,000})$

$$= \frac{13}{130} = 0.1$$

(c) $P(\text{a man or is earning more than \$40,000})$

$$= \frac{62 + 13 + 3}{130}$$

$$= \frac{78}{130} = 0.6$$

(d) $P(\text{a woman or is earning \$40,000 or less})$

$$= \frac{52 + 3 + 62}{130}$$

$$= \frac{117}{130} = 0.9$$

63. Let A be the set of refugees who came to escape abject poverty and B be the set of refugees who came to escape political oppression. Then $P(A) = 0.80$, $P(B) = 0.90$, and $P(A \cap B) = 0.70$.

$$P(A \cup B) = P(A) + P(B) - P(A \cap B)$$
$$= 0.80 + 0.90 - 0.70 = 1$$
$$P(A' \cap B') = 1 - P(A \cap B)$$
$$= 1 - 1 = 0$$

The probability that a refugee in the camp was neither poor nor seeking political asylum is 0.

65. The odds of winning are 3 to 2; this means there are 3 ways to win and 2 ways to lose, out of a total of $2 + 3 = 5$ ways altogether. Hence, the probability of losing is $\frac{2}{5}$.

67. (a) P(somewhat or extremely intolerant of Facists)

= P(somewhat intolerant of Facists)

+ P(extremely intolerant of Facists)

$= \frac{27.1}{100} + \frac{59.5}{100} = \frac{86.6}{100} = 0.866$

(b) P(completely tolerant of Communists)

= P(no intolerance at all of Communists)

$= \frac{47.8}{100} = 0.478$

69. (a) There are $67 + 25 = 92$ possible judging combinations with Sasha Cohen finishing in first place. The probability of this outcome is, therefore, $92/220 = 23/55$.

(b) There are 67 possible judging combinations with Irina Slutskaya finishing in second place. The probability of this outcome is, therefore, $67/220$.

(c) There are $92 + 67 = 159$ possible judging combinations with Shizuka Arahawa finishing in third place. The probability of this outcome is, therefore, $159/220$.

71. The probabilities are as follows:

$$\frac{1}{1 + 195,199,999} = \frac{1}{195,200,000}$$
$$= 0.0000000051$$

$$\frac{1}{1 + 835,499} = \frac{1}{835,500} = 0.0000012$$

$$\frac{1}{1 + 157.6} = \frac{1}{158.6} = 0.0063$$

$$\frac{1}{1 + 59.32} = \frac{1}{60.32} = 0.0166$$

7.5 Conditional Probability; Independent Events

Your Turn 1

Reduce the sample space to B' and then find $n(A \cap B')$ and $n(B')$.

$$P(A|B') = \frac{P(A \cap B')}{P(B')} = \frac{n(A \cap B')}{n(B')} = \frac{30}{55} = \frac{6}{11}$$

Your Turn 2

$$P(E \cup F) = P(E) + P(F) - P(E \cap F)$$
$$0.80 = 0.56 + 0.64 - P(E \cap F)$$
$$P(E \cap F) = 0.56 + 0.64 - 0.80$$
$$P(E \cap F) = 0.40$$

$$P(E \mid F) = \frac{P(E \cap F)}{P(F)}$$
$$= \frac{0.40}{0.64} = 0.625$$

Your Turn 3

Since at least one coin is a tail, the sample space is reduced to $\{ht, th, tt\}$. Two of these equally likely outcomes have exactly one head, so P(one head|at least one tail) $= \frac{2}{3}$.

Your Turn 4

Let C represent "lives on campus" and A represent "has a car on campus." Using the given information, $P(A|C) = \frac{1}{4}$ and $P(C) = \frac{4}{5}$. By the product rule,

$$P(A \cap C) = P(A|C) \cdot P(C)$$
$$= \frac{1}{4} \cdot \frac{4}{5} = \frac{1}{5}.$$

Your Turn 5

Using Figure 23, we follow the A branch and then the U branch. Multiplying along the tree we find the probability of the composite branch, which represents $A \cap U$ or the event that A is in charge of the campaign and it produces unsatisfactory results.

$$P(A \cap U) = \frac{2}{3} \cdot \frac{1}{4} = \frac{1}{6}$$

Your Turn 6

There are two paths that result in one NY plant and one Chicago plant: C first and NY second, and NY first and C second. From the tree, $P(C, NY) = \frac{1}{2} \cdot \frac{1}{5} = \frac{1}{10}$, and $P(NY, C) = \frac{1}{6} \cdot \frac{3}{5} = \frac{1}{10}$. So

P(one NY plant and one Chicago plant)

$$= P(C, NY) + P(NY, C) = \frac{1}{10} + \frac{1}{10} = \frac{1}{5}.$$

Your Turn 7

Successive rolls of a die are independent events, so

$$P(\text{two fives in a row}) = P(\text{five}) \cdot P(\text{five})$$

$$= \frac{1}{6} \cdot \frac{1}{6}$$

$$= \frac{1}{36}.$$

Your Turn 8

Let M represent the event "you do your math homework" and H represent the event "you do your history assignment."

$$P(M) = 0.8$$

$$P(H) = 0.7$$

$$P(M \cup H) = 0.9$$

Now solve for $P(M \cap H)$.

$$P(M \cup H) = P(M) + P(H) - P(M \cap H)$$

$$0.9 = 0.8 + 0.7 - P(M \cap H)$$

$$P(M \cap H) = 0.8 + 0.7 - 0.9$$

$$P(M \cap H) = 0.6$$

However, $P(M)P(H) = (0.8)(0.7) = 0.56$, so $P(M \cap H) \neq P(M)P(H)$. Since M and H do not satisfy the product rule for independent events, they are not independent.

7.5 Exercises

1. Let A be the event "the number is 2" and B be the event "the number is odd."

   The problem seeks the conditional probability $P(A|B)$. Use the definition

   $$P(A|B) = \frac{P(A \cap B)}{P(B)}.$$

   Here, $P(A \cap B) = 0$ and $P(B) = \frac{1}{2}$. Thus,

   $$P(A|B) = \frac{0}{\frac{1}{2}} = 0.$$

3. Let A be the event "the number is even" and B be the event "the number is 6." Then

   $$P(A|B) = \frac{P(A \cap B)}{P(B)} = \frac{\frac{1}{6}}{\frac{1}{6}} = 1.$$

5. $P(\text{sum of } 8 \,|\, \text{greater than } 7)$

   $$= \frac{P(8 \cap \text{greater than } 7)}{P(\text{greater than } 7)}$$

   $$= \frac{n(8 \cap \text{greater than } 7)}{n(\text{greater than } 7)}$$

   $$= \frac{5}{15} = \frac{1}{3}$$

7. The event of getting a double given that 9 was rolled is impossible; hence,

   $$P(\text{double} \,|\, \text{sum of } 9) = 0.$$

9. Use a reduced sample space. After the first card drawn is a heart, there remain 51 cards, of which 12 are hearts. Thus,

   $$P(\text{heart on 2nd} \,|\, \text{heart on 1st}) = \frac{12}{51} = \frac{4}{17}.$$

11. Use a reduced sample space. After the first card drawn is a jack, there remain 51 cards, of which 11 are face cards. Thus,

    $$P(\text{face card on 2nd} \,|\, \text{jack on 1st}) = \frac{11}{51}.$$

13. $P(\text{a jack and a 10})$

    $$= P(\text{jack followed by } 10)$$
    $$+ P(10 \text{ followed by jack})$$
    $$= \frac{4}{52} \cdot \frac{4}{51} + \frac{4}{52} \cdot \frac{4}{51}$$
    $$= \frac{16}{2652} + \frac{16}{2652}$$
    $$= \frac{32}{2652} = \frac{8}{663}$$

15. $P(\text{two black cards})$

    $$= P(\text{black on 1st})$$
    $$\cdot P(\text{black on 2nd} \,|\, \text{black on 1st})$$
    $$= \frac{26}{52} \cdot \frac{25}{51}$$
    $$= \frac{650}{2652} = \frac{25}{102}$$

19. Examine a table of all possible outcomes of rolling a red die and rolling a green die (such as Figure 18 in Section 7.4). There are 9 outcomes of the 36 total outcomes that correspond to rolling "red die comes up even and green die comes up even"—in other words, corresponding to $A \cap B$. Therefore,

$$P(A \cap B) = \frac{9}{36} = \frac{1}{4}.$$

We also know that $P(A) = 1/2$ and $P(B) = 1/2$. Since

$$P(A \cap B) = \frac{1}{4} = \frac{1}{2} \cdot \frac{1}{2} = P(A) \cdot P(B),$$

the events A and B are independent.

21. Notice that $P(F|E) \neq P(F)$: the knowledge that a person lives in Dallas affects the probability that the person lives in Dallas or Houston. Therefore, the events are dependent.

23. **(a)** The events that correspond to "sum is 7" are (2, 5), (3, 4), (4, 3), and (5, 2), where the first number is the number on the first slip of paper and the second number is the number on the second. Of these, only (3, 4) corresponds to "first is 3," so

$$P(\text{first is 3}|\text{sum is 7}) = \frac{1}{4}.$$

(b) The events that correspond to "sum is 8" are (3, 5) and (5, 3). Of these, only (3, 5) corresponds to "first is 3," so

$$P(\text{first is 3}|\text{sum is 8}) = \frac{1}{2}.$$

25. **(a)** Many answers are possible; for example, let B be the event that the first die is a 5. Then

$$P(A \cap B) = P(\text{sum is 7 and first is 5}) = \frac{1}{36}$$

$$P(A) \cdot P(B) = P(\text{sum is 7}) \cdot P(\text{first is 5})$$

$$= \frac{6}{36} \cdot \frac{1}{6} = \frac{1}{36}$$

so, $P(A \cap B) = P(A) \cdot P(B)$.

(b) Many answers are possible; for example, let B be the event that at least one die is a 5.

$$P(A \cap B) = P(\text{sum is 7 and at least one is a 5})$$

$$= \frac{2}{36}$$

$$P(A) \cdot P(B) = P(\text{sum is 7}) \cdot P(\text{at least one is a 5})$$

$$= \frac{6}{36} \cdot \frac{11}{6}$$

so, $P(A \cap B) \neq P(A) \cdot P(B)$.

29. Since A and B are independent events,

$$P(A \cap B) = P(A) \cdot P(B) = \frac{1}{4} \cdot \frac{1}{5} = \frac{1}{20}.$$

Thus,

$$P(A \cup B) = P(A) + P(B) - P(A \cap B)$$

$$= \frac{1}{4} + \frac{1}{5} - \frac{1}{20}$$

$$= \frac{2}{5}.$$

31. At the first booth, there are three possibilities: shaker 1 has heads and shaker 2 has heads; shaker 1 has tails and shaker 2 has heads; shaker 1 has heads and shaker 2 has tails. We restrict ourselves to the condition that at least one head has appeared. These three possibilities are equally likely so the probability of two heads is $\frac{1}{3}$.

At the second booth we are given the condition of one head in one shaker. The probability that the second shaker has one head is $\frac{1}{2}$. Therefore, you stand the best chance at the second booth.

33. No, these events are not independent.

35. Assume that each box is equally likely to be drawn from and that within each box each marble is equally likely to be drawn. If Laura does not redistribute the marbles, then the probability of winning the Porsche is $\frac{1}{2}$, since the event of a pink marble being drawn is equivalent to the event of choosing the first of the two boxes.

If however, Laura puts 49 of the pink marbles into the second box with the 50 blue marbles, the probability of a pink marble being drawn increases to $\frac{74}{99}$. The probability of the first box being chosen is $\frac{1}{2}$, and the probability of drawing a pink marble from this box is 1. The probability of the second box being chosen is $\frac{1}{2}$, and the probability of drawing a pink marble from this box is $\frac{49}{99}$. Thus, the probability of drawing a pink marble is $\frac{1}{2} \cdot 1 + \frac{1}{2} \cdot \frac{49}{99} = \frac{74}{99}$. Therefore Laura increases her chances of winning by redistributing some marbles.

37. The probability that a customer cashing a check will fail to make a deposit is

$$P(D'|C) = \frac{n(D' \cap C)}{n(C)} = \frac{30}{90} = \frac{1}{3}.$$

39. The probability that a customer making a deposit will not cash a check is

$$P(C'|D) = \frac{n(C' \cap D)}{n(D)} = \frac{20}{80} = \frac{1}{4}.$$

41. **(a)** Since the separate flights are independent, the probability of 4 flights in a row is

$$(0.773)(0.773)(0.773)(0.773) \approx 0.3570$$

43. Let W be the event "withdraw cash from ATM" and C be the event "check account balance at ATM."

$$P(C \cup W) = P(C) + P(W) - P(C \cap W)$$
$$0.96 = 0.32 + 0.92 - P(C \cap W)$$
$$-0.28 = -P(C \cap W)$$
$$P(C \cap W) = 0.28$$

$$P(W|C) = \frac{P(C \cap W)}{P(C)}$$
$$= \frac{0.28}{0.32}$$
$$\approx 0.875$$

The probability that she uses an ATM to get cash given that she checked her account balance is 0.875.

Use the following tree diagram for Exercise 45.

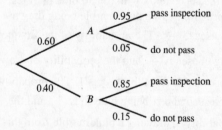

45. Since 40% of the production comes off B, $P(B) = 0.40$. Also, $P(\text{pass}|B) = 0.85$, so $P(\text{not pass}|B) = 0.15$. Therefore,

$$P(\text{not pass} \cap B) = P(B) \cdot P(\text{not pass}|B)$$
$$= 0.40(0.15)$$
$$= 0.06$$

47. The sample space is

$$\{RW, WR, RR, WW\}.$$

The event "red" is $\{RW, WR, RR\}$, and the event "mixed" is $\{RW, WR\}$.

$$P(\text{mixed}|\text{red}) = \frac{n(\text{mixed and red})}{n(\text{red})}$$
$$= \frac{2}{3}.$$

Use the following tree diagram for Exercises 49 through 53.

1st child	2nd child	3rd child	Branch	Probability
1/2	1/2 B	1/2 B	1	1/8
		1/2 G	2	1/8
B	1/2 G	1/2 B	3	1/8
		1/2 G	4	1/8
1/2	1/2 B	1/2 B	5	1/8
G		1/2 G	6	1/8
	1/2 G	1/2 B	7	1/8
		1/2 G	8	1/8

49. $P(\text{all girls}|\text{first is a girl})$
$$= \frac{P(\text{all girls and first is a girl})}{P(\text{first is a girl})}$$
$$= \frac{n(\text{all girls and first is a girl})}{n(\text{first is a girl})}$$
$$= \frac{1}{4}$$

51. $P(\text{all girls}|\text{second is a girl})$
$$= \frac{P(\text{all girls and second is a girl})}{P(\text{second is a girl})}$$
$$= \frac{n(\text{all girls and second is a girl})}{n(\text{second is a girl})}$$
$$= \frac{1}{4}$$

53. $P(\text{all girls}|\text{at least 1 girl})$
$$= \frac{P(\text{all girls and at least 1 girl})}{P(\text{at least 1 girl})}$$
$$= \frac{n(\text{all girls and at least 1 girl})}{n(\text{at least 1 girl})}$$
$$= \frac{1}{7}$$

55. $P(C) = 0.039$, the total of the C row.

57. $P(M \cup C) = P(M) + P(C) - P(M \cap C)$
$= 0.487 + 0.039 - 0.035$
$= 0.491$

59. $P(C|M) = \dfrac{P(C \cap M)}{P(M)}$
$= \dfrac{0.035}{0.487}$
≈ 0.072

61. By the definition of independent events, C and M are independent if

$$P(C|M) = P(C).$$

From Exercises 55 and 59,

$$P(C) = 0.039$$

and $\qquad P(C|M) = 0.072.$

Since $P(C|M) \neq P(C)$, events C and M are not independent, so we say that they are dependent. This means that red-green color blindness does not occur equally among men and women.

63. First draw a tree diagram.

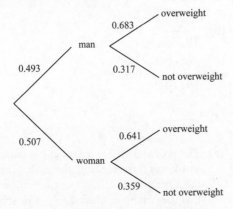

(a) $P(\text{overweight man}) = (0.493)(0.683)$
$= 0.3367$

(b) $P(\text{overweight}) = 0.3367 + (0.507)(0.641)$
$= 0.3367 + 0.3250$
$= 0.6617$

(c) The two events "man" and "overweight" are independent if

$P(\text{overweight man}) = P(\text{overweight})P(\text{man}).$
$P(\text{overweight})P(\text{man}) = (0.6617)(0.493) = 0.3262.$

Since $0.3367 \neq 0.3262$, the events are not independent

65. $P(C|F) = \dfrac{n(C \cap F)}{n(F)} = \dfrac{7}{229}$

67. $P(B'|H') = P((A \cup C \cup D) \mid F)$
$= \dfrac{n((A \cup C \cup D) \cap F)}{n(F)}$
$= \dfrac{34 + 7 + 150}{229}$
$= \dfrac{191}{229}$

69. Let H be the event "patient has high blood pressure,"

N be the event "patient has normal blood pressure,"

L be the event "patient has low blood pressure,"

R be the event "patient has a regular heartbeat,"

and I be the event "patient has an irregular heartbeat.

We wish to determine $P(R \cap L)$.

Statement (i) tells us $P(H) = 0.14$ and statement (ii) tells us $P(L) = 0.22$. Therefore,

$$P(H) + P(N) + P(L) = 1$$
$$0.14 + P(N) + 0.22 = 1$$
$$P(N) = 0.64.$$

Statement (iii) tells us $P(I) = 0.15$. This and statement (iv) lead to

$$P(I \cap H) = \frac{1}{3}P(I) = \frac{1}{3}(0.15) = 0.05.$$

Statement (v) tells us

$$P(N \cap I) = \frac{1}{8}P(N) = \frac{1}{3}(0.64) = 0.08.$$

Make a table and fill in the data just found.

	H	N	L	Totals
R	–	–	–	–
I	0.05	0.08	–	0.15
Totals	0.14	0.64	0.22	1.00

To determine $P(R \cap L)$, find $P(I \cap L)$.

$$P(I) = P(I \cap H) + P(I \cap N) + P(I \cap L)$$
$$0.15 = 0.05 + 0.08 + P(I \cap L)$$
$$0.15 = 0.13 + P(I \cap L)$$
$$P(I \cap L) = 0.02$$

Now calculate $P(R \cap L)$.

$$P(L) = P(R \cap L) + P(I \cap L)$$
$$0.22 = P(R \cap L) + 0.02$$
$$P(I \cap L) = 0.20$$

The correct answer choice is **e**.

71. (a) The total number of males is 2(5844)
$+ 6342 = 18{,}030$. The total number of
infants is $2(17{,}798) = 35{,}596$. So among
infants who are part of a twin pair, the
proportion of males is $\frac{18{,}030}{35{,}596} = 0.5065$.

To answer parts (b) through (g) we make the
assumption that the event "twin comes from
an identical pair" is independent of the event
"twin is male." We can then construct the
following tree diagram.

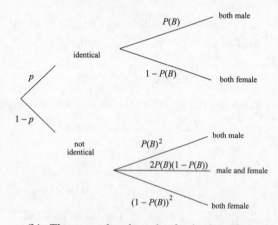

(b) The event that the pair of twins is male happens
along two compound branches. Multiplying
the probabilities along these branches and
adding the products we get

$$pP(B) + (1 - p)(P(B))^2.$$

(c) Using the values from (a) and (b) together
with the fraction of mixed twins, we solve
this equation:

$$\frac{5844}{17{,}798} = p(0.506518) + (1 - p)(0.506518)^2$$
$$0.328352 = 0.506518p + 0.256560 - 0.256560p$$
$$0.071792 = 0.249958p$$
$$p = \frac{0.071792}{0.249958}$$
$$p = 0.2872$$

So our estimate for p is 0.2872. Note that if
you use fewer places in the value for $P(B)$
you may get a slightly different answer.

(d) Multiplying along the two branches that result
in two female twins and adding the products
gives

$$p(1 - P(B)) + (1 - p)(1 - P(B))^2.$$

(e) The equation to solve will now be the following:

$$\frac{5612}{17{,}798}$$
$$= p(1 - 0.506518) + (1 - p)(1 - 0.506518)^2$$

The answer will be the same as in part (c)

(f) Now only the "not identical" branch is
involved, and the expression is

$$2(1 - p)P(B)(1 - P(B)).$$

(g) The equation to solve will now be the
following:

$$\frac{6342}{17{,}798} = 2(1 - p)(0.506518)(1 - 0.506518)$$

Again the answer will be the same as in part
(c). Note that because of our independence
assumption these three estimates for p
must agree.

73. (a) $\frac{39.66}{192.64} = 0.2059$

(b) $\frac{28.68}{192.64} = 0.1489$

(c) $\frac{7.90}{192.64} = 0.0410$

(d) $\frac{7.90}{28.68} = 0.2755$

(e) The probability that a person is a current
smoker is different from the probability that
the a person is a current smoker *given* that
that the person has less than a high school
diploma. Thus knowing that a person has less
than a high school diploma changes our
estimate of the probability that the person is
a smoker, so these events are not
independent. Alternatively, we could note
that the product of the answers to (a) and (b),
which is
$P(\text{smoker}) \cdot P(\text{less than HS diploma})$
is equal to $(0.2059)(0.1489) = 0.0307$,
while according to (c), $P(\text{smoker } and \text{ less than}$
$\text{HS diploma}) = 0.0410$. Since these values
are different, the two events "current smoker"
and "less than a high school diploma" are not
independent.

75. (a) In this exercise, it is easier to work with complementary events. Let E be the event "at least one of the faults erupts." Then the complementary event E' is "none of the faults erupts," and we can use $P(E) = 1 - P(E')$.

Consider the event E': "none of the faults erupts." This means "the first fault does not erupt <u>and</u> the second fault does not erupt <u>and</u> . . . <u>and</u> the seventh fault does not erupt." Letting F_i denote the event "the ith fault erupts," we wish to find

$$P(E') = P(F_1' \cap F_2' \cap F_3' \cap F_4' \cap F_5' \cap F_6' \cap F_7').$$

Since we are assuming the events are independent, we have

$$P(E') = P(F_1' \cap F_2' \cap F_3' \cap F_4' \cap F_5' \cap F_6' \cap F_7')$$
$$= P(F_1') \cdot P(F_2') \cdot P(F_3') \cdot P(F_4') \cdot P(F_5') \cdot P(F_6') \cdot P(F_7')$$

Now use $P(F_i') = 1 - P(F_i)$ and perform the calculations.

$$P(E') = P(F_1') \cdot P(F_2') \cdot P(F_3') \cdot P(F_4') \cdot P(F_5') \cdot P(F_6') \cdot P(F_7')$$
$$= (1 - 0.27) \cdot (1 - 0.21) \cdot \ldots \cdot (1 - 0.03)$$
$$= (0.73)(0.79)(0.89)(0.90)(0.96)(0.97)(0.97)$$
$$\approx 0.42$$

Therefore,

$$P(E) = 1 - P(E')$$
$$\approx 1 - 0.42 \approx 0.58.$$

77. (a) $P(\text{second class}) = \dfrac{357}{1316} \approx 0.2713$

(b) $P(\text{surviving}) = \dfrac{499}{1316} \approx 0.3792$

(c) $P(\text{surviving} | \text{first class}) = \dfrac{203}{325} \approx 0.6246$

(d) $P(\text{surviving} | \text{child and third class}) = \dfrac{27}{79}$
≈ 0.3418

(e) $P(\text{woman} | \text{first class and survived}) = \dfrac{140}{203}$
≈ 0.6897

(f) $P(\text{third class} | \text{man and survived}) = \dfrac{75}{146}$
≈ 0.5137

(g) $P(\text{survived} | \text{man}) = \dfrac{146}{805} \approx 0.1814$

$P(\text{survived} | \text{man and third class}) = \dfrac{75}{462}$
≈ 0.1623

No, the events are not independent.

79. First draw the tree diagram.

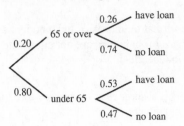

(a) $P(\text{person is 65 or over and has a loan})$
$= P(65 \text{ or over}) \cdot P(\text{has loan} | 65 \text{ or over})$
$= 0.20(0.26) = 0.052$

(b)

$P(\text{person has a loan}) = P(65 \text{ or over and has loan})$
$\qquad\qquad + P(\text{under 65 and has loan})$
$= 0.20(0.26) + 0.80(0.53)$
$= 0.052 + 0.424$
$= 0.476$

81.

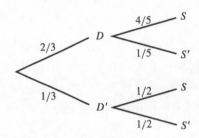

From the tree diagram, we see that the probability that a person

(a) drinks diet soft drinks is

$$\frac{2}{3}\left(\frac{4}{5}\right) + \frac{1}{3}\left(\frac{1}{2}\right) = \frac{8}{15} + \frac{1}{6}$$
$$= \frac{21}{30} = \frac{7}{10};$$

(b) diets, but does not drink diet soft drinks is

$$\frac{2}{3}\left(\frac{1}{5}\right) = \frac{2}{15}.$$

83. Let F_i be the event "the ith burner fails." The event "all four burners fail" is equivalent to the event "the first burner fails <u>and</u> the second burner fails <u>and</u> the third burner fails <u>and</u> the fourth burner fails"—that is, the event $F_1 \cap F_2 \cap F_3 \cap F_4$. We are told that the burners are independent. Therefore

$$P(F_1 \cap F_2 \cap F_3 \cap F_4) = P(F_1) \cdot P(F_2) \cdot P(F_3) \cdot P(F_4)$$
$$= (0.001)(0.001)(0.001)(0.001)$$
$$= 0.000000000001 = 10^{-12}.$$

85. $P(\text{luxury car}) = 0.04$ and $P(\text{luxury car} \mid \text{CPA})$
$= 0.17$

Use the formal definition of independent events. Since these probabilities are not equal, the events are not independent.

87. (a) We will assume that successive free throws are independent.

$$P(0) = 0.4$$
$$P(1) = (0.6)(0.4) = 0.24$$
$$P(2) = (0.6)(0.6) = 0.36$$

(b) Let p be her season free throw percentage.

$$P(0) = 1 - p$$
$$P(2) = p^2$$

Solve:

$$p^2 = 1 - p$$
$$p^2 + p - 1 = 0$$
$$p = \frac{-1 + \sqrt{5}}{2} \approx 0.618$$

89. (a)

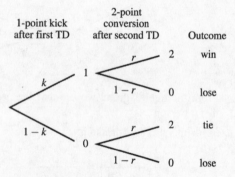

From the tree diagram,

$$P(\text{win}) = kr$$
$$P(\text{tie}) = (1 - k)r$$
$$P(\text{lose}) = k(1 - r) + (1 - k)(1 - r)$$
$$= k - kr + 1 - r - k + kr$$
$$= 1 - r$$

(b)

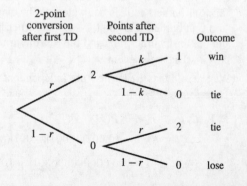

From the tree diagram,

$$P(\text{win}) = rk$$
$$P(\text{tie}) = r(1 - k) + (1 - r)r$$
$$= r - rk + r - r^2$$
$$= 2r - rk - r^2$$
$$= r(2 - k - r)$$
$$P(\text{lose}) = (1 - r)(1 - r)$$
$$= (1 - r)^2$$

(c) $P(\text{win})$ is the same under both strategies.

(d) If $r < 1, (1 - r) > (1 - r)^2$. The probability of losing is smaller for the 2-point first strategy.

7.6   Bayes' Theorem

Your Turn 1

Let E be the event "passes math exam" and F be the event "attended review session."

$$P(E|F) = 0.8$$
$$P(E|F') = 0.65$$
$$P(F) = 0.6 \text{ so } P(F') = 0.4$$

Now apply Bayes' Theorem.

$$P(F|E) = \frac{P(F)P(E|F)}{P(F)P(E|F) + P(F')P(E|F')}$$

$$= \frac{(0.6)(0.8)}{(0.6)(0.8) + (0.4)(0.65)}$$

$$= 0.6486$$

So the probability that the student attended the review session given that the student passed the exam is 0.6486.

Your Turn 2

$$\text{Let} \quad F_1 = \text{in English I}$$
$$F_2 = \text{in English II}$$
$$F_3 = \text{in English III}$$
$$E = \text{received help from writing center}$$

Express the given information in terms of these variables.

$$P(F_1) = 0.12 \quad P(E|F_1) = 0.80$$
$$P(F_2) = 0.68 \quad P(E|F_2) = 0.40$$
$$P(F_3) = 0.20 \quad P(E|F_3) = 0.11$$

$$P(F_1|E) = \frac{P(F_1)P(E|F_1)}{P(F_1)P(E|F_1) + P(F_2)P(E|F_2) + P(F_3)P(E|F_3)}$$

$$= \frac{(0.12)(0.80)}{(0.12)(0.80) + (0.68)(0.40) + (0.20)(0.11)}$$

$$= 0.2462$$

Given that a student received help from the writing center, the probability that the student is in English I is 0.2462.

7.6   Exercises

1. Use Bayes' theorem with two possibilities M and M'.

$$P(M|N) = \frac{P(M) \cdot P(N|M)}{P(M) \cdot P(N|M) + P(M') \cdot P(N|M')}$$

$$= \frac{0.4(0.3)}{0.4(0.3) + 0.6(0.4)}$$

$$= \frac{0.12}{0.12 + 0.24}$$

$$= \frac{0.12}{0.36} = \frac{12}{36} = \frac{1}{3}$$

3. Using Bayes' theorem,

$$P(R_1|Q) = \frac{P(R_1) \cdot P(Q|R_1)}{P(R_1) \cdot P(Q|R_1) + P(R_2) \cdot P(Q|R_2) + P(R_3) \cdot P(Q|R_3)}$$

$$= \frac{0.15(0.40)}{(0.15)(0.40) + 0.55(0.20) + 0.30(0.70)}$$

$$= \frac{0.06}{0.38} = \frac{6}{38} = \frac{3}{19}.$$

5. Using Bayes' theorem,

$$P(R_3|Q) = \frac{P(R_3) \cdot P(Q|R_3)}{P(R_1) \cdot P(Q|R_1) + P(R_2) \cdot P(Q|R_2) + P(R_3) \cdot P(Q|R_3)}$$

$$= \frac{0.30(0.70)}{(0.15)(0.40) + 0.55(0.20) + 0.30(0.70)}$$

$$= \frac{0.21}{0.38} = \frac{21}{38}.$$

7. We first draw the tree diagram and determine the probabilities as indicated below.

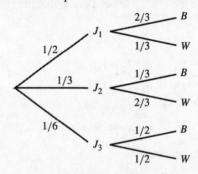

We want to determine the probability that if a white ball is drawn, it came from the second jar. This is $P(J_2 | W)$. Use Bayes' theorem.

$$P(J_2|W) = \frac{P(J_2) \cdot P(W|J_2)}{P(J_2) \cdot P(W|J_2) + P(J_1) \cdot P(W|J_1) + P(J_3) \cdot P(W|J_3)} = \frac{\frac{1}{3} \cdot \frac{2}{3}}{\frac{1}{3} \cdot \frac{2}{3} + \frac{1}{2} \cdot \frac{1}{3} + \frac{1}{6} \cdot \frac{1}{2}}$$

$$= \frac{\frac{2}{9}}{\frac{2}{9} + \frac{1}{6} + \frac{1}{12}} = \frac{\frac{2}{9}}{\frac{17}{36}} = \frac{8}{17}$$

9. Let G represent "good worker," B represent "bad worker," S represent "pass the test," and F represent "fail the test." The given information if $P(G) = 0.70, P(B) = P(G') = 0.30, P(S|G) = 0.85$ (and therefore $P(F|G) = 0.15$), and $P(S|B) = 0.35$ (and therefore $P(F|B) = 0.65$). If passing the test is made a requirement for employment, then the percent of the new hires that will turn out to be good workers is

$$P(G|S) = \frac{P(G) \cdot P(S|G)}{P(G) \cdot P(S|G) + P(B) \cdot P(S|B)}$$

$$= \frac{0.70(0.85)}{0.70(0.85) + 0.30(0.35)}$$

$$= \frac{0.595}{0.700}$$

$$= 0.85.$$

85% of new hires become good workers.

11. Let Q represent "qualified" and A represent "approved by the manager." Set up the tree diagram.

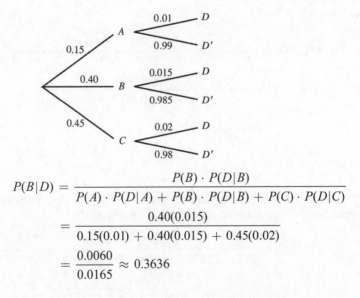

$$P(Q'|A) = \frac{P(Q') \cdot P(A|Q')}{P(Q) \cdot P(A|Q) + P(Q') \cdot P(A|Q')}$$

$$= \frac{0.25(0.20)}{0.75(0.85) + 0.25(0.20)}$$

$$= \frac{0.05}{0.6875} = \frac{4}{55} \approx 0.0727$$

13. Let D represent "damaged," A represent "from supplier A," and B represent "from supplier B." Set up the tree diagram.

$$P(B|D) = \frac{P(B) \cdot P(D|B)}{P(B) \cdot P(D|B) + P(A) \cdot P(D|A)}$$

$$= \frac{0.30(0.05)}{0.30(0.05) + 0.70(0.10)}$$

$$= \frac{0.015}{0.015 + 0.07}$$

$$= \frac{0.015}{0.085} \approx 0.1765$$

15. Start with the tree diagram, where the first state refers to the companies and the second to a defective appliance.

$$P(B|D) = \frac{P(B) \cdot P(D|B)}{P(A) \cdot P(D|A) + P(B) \cdot P(D|B) + P(C) \cdot P(D|C)}$$

$$= \frac{0.40(0.015)}{0.15(0.01) + 0.40(0.015) + 0.45(0.02)}$$

$$= \frac{0.0060}{0.0165} \approx 0.3636$$

17. Let H represent "high rating," F_1 represent "sponsors college game," F_2 represent "sponsors baseball game," and F_3 represent "sponsors pro football game."

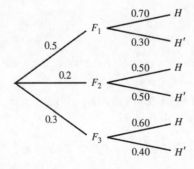

$$P(F_3|H) = \frac{P(F_3) \cdot P(H|F_3)}{P(F_1) \cdot P(H|F_1) + P(F_2) \cdot P(H|F_2) + P(F_3) \cdot P(H|F_3)}$$

$$= \frac{0.3(0.60)}{0.5(0.70) + 0.2(0.50) + 0.3(0.60)}$$

$$= \frac{0.18}{0.35 + 0.10 + 0.18}$$

$$= \frac{0.18}{0.63} = \frac{18}{63} = \frac{2}{7}$$

19. Using the given information, construct a table similar to the one in the previous exercise.

Category of Policyholder	Portion of Policyholders	Probability of Dying in the Next Year
Standard	0.50	0.010
Preferred	0.40	0.005
Ultra-preferred	0.10	0.001

Let S represent "standard policyholder,"

R represent "preferred policyholder,"

U represent "ultra-preferred policyholder,"

and D represent "policyholder dies in the next year."

We wish to find $P(U|D)$.

$$P(U|D) = \frac{P(U) \cdot P(D|U)}{P(S) \cdot P(D|S) + P(R) \cdot P(D|R) + P(U) \cdot P(D|U)}$$

$$= \frac{0.10(0.001)}{0.50(0.010) + 0.40(0.005) + 0.10(0.001)}$$

$$= \frac{0.0001}{0.0071} \approx 0.141$$

The correct answer choice is **d**.

21. Let L be the event "the object was shipped by land," A be the event "the object was shipped by air," S be the event "the object was shipped by sea," and E be the event "an error occurred."

$$P(L|E) = \frac{P(L) \cdot P(E|L)}{P(L) \cdot P(E|L) + P(A) \cdot P(E|A) + P(S) \cdot P(E|S)}$$

$$= \frac{0.50(0.02)}{0.50(0.02) + 0.40(0.04) + 0.10(0.14)}$$

$$= \frac{0.0100}{0.0400} = 0.25$$

The correct response is **c**.

23. Let E represent the event "hemoccult test is positive," and let F represent the event "has colorectal cancer." We are given

$$P(F) = 0.003, P(E|F) = 0.5,$$

$$\text{and } P(E|F') = 0.03$$

and we want to find $P(F|E)$. Since $P(F) = 0.003, P(F') = 0.997$. Therefore,

$$P(F|E) = \frac{P(F) \cdot P(E|F)}{P(F) \cdot P(E|F) + P(F') \cdot P(E|F')} = \frac{0.003 \cdot 0.5}{0.003 \cdot 0.5 + 0.997 \cdot 0.03} \approx 0.0478.$$

25. $P(T^+ \,|\, D^+) = 0.796$

$P(T^- \,|\, D^-) = 0.902$

$P(D^+) = 0.005$ so $P(D^-) = 0.995$

We can now fill in the complete tree.

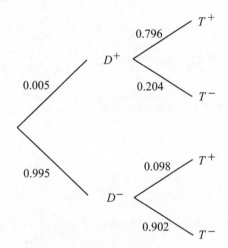

(a) $P(D^+ \,|\, T^+) = \dfrac{P(D^+)P(T^+ \,|\, D^+)}{P(D^+)P(T^+ \,|\, D^+) + P(D^-)P(T^+ \,|\, D^-)}$

$ = \dfrac{(0.005)(0.796)}{(0.005)(0.796) + (0.995)(0.098)}$

$ = 0.039$

(b) $P(D^- \,|\, T^-) = \dfrac{P(D^-)P(T^- \,|\, D^-)}{P(D^-)P(T^- \,|\, D^-) + P(D^+)P(T^- \,|\, D^+)}$

$ = \dfrac{(0.995)(0.902)}{(0.995)(0.902) + (0.005)(0.204)}$

$ = 0.999$

(c) $P(D^+ \,|\, T^-) = \dfrac{P(D^+)P(T^- \,|\, D^+)}{P(D^+)P(T^- \,|\, D^+) + P(D^-)P(T^- \,|\, D^-)}$

$ = \dfrac{(0.005)(0.204)}{(0.005)(0.204) + (0.995)(0.902)}$

$ = 0.001$

Alternatively, since $P(D^- \,|\, T^-) + P(D^+ \,|\, T^-) = 1$, we can subtract the answer to (b) from 1.

(d) The right half of the tree stays the same, but $P(D^+)$ is now 0.015 and $P(D^-)$ is 0.985.

$$P(D^+ \,|\, T^+) = \frac{P(D^+)P(T^+ \,|\, D^+)}{P(D^+)P(T^+ \,|\, D^+) + P(D^-)P(T^+ \,|\, D^-)}$$

$$= \frac{(0.015)(0.796)}{(0.015)(0.796) + (0.985)(0.098)}$$

$$= 0.110$$

27. (a) Let H represent "heavy smoker," L be "light smoker," N be "nonsmoker," and D be "person died."
Let $x = P(D|N)$, that is, let x be the probability that a nonsmoker died. Then $P(D|L) = 2x$
and $P(D|H) = 4x$. Create a table.

Level of Smoking	Probability of Level	Probability of Death for Level
H	0.2	$4x$
L	0.3	$2x$
N	0.5	x

We wish to find $P(H|D)$.

$$P(H|D) = \frac{P(H) \cdot P(D|H)}{P(H) \cdot P(D|H) + P(L) \cdot P(D|L) + P(N) \cdot P(D|N)}$$

$$= \frac{0.2(4x)}{0.2(4x) + 0.3(2x) + 0.5(x)}$$

$$= \frac{0.8x}{1.9x}$$

$$\approx 0.42$$

The correct answer choice is **d**.

29. Let H represent "person has the disease" and R be "test indicates presence of the disease." We wish to
determine $P(H|R)$.

Construct a table as before.

Category of Person	Probability of Population	Probability of Presence of Disease
H	0.01	0.950
H'	0.99	0.005

$$P(H|R) = \frac{P(H) \cdot P(R|H)}{P(H) \cdot P(R|H) + P(H') \cdot P(R|H')}$$

$$= \frac{0.01(0.950)}{0.01(0.950) + 0.99(0.005)}$$

$$= \frac{0.00950}{0.01445} \approx 0.657$$

The correct answer choice is **b**.

31. We start with a tree diagram based on the given information. Let W stand for "woman" and A stand for "abstains
from alcohol."

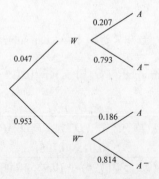

(a) $P(A) = P(A \mid W)P(W) + P(A \mid W^-)P(W^-)$

$\qquad\qquad = (0.047)(0.207) + (0.953)(0.186)$

$\qquad\qquad = 0.1870$

(b) $P(W^- \mid A) = \dfrac{P(W^-)P(A \mid W^-)}{P(W^-)P(A \mid W^-) + P(W)P(A \mid W)}$

$\qquad\qquad\quad = \dfrac{(0.953)(0.186)}{(0.953)(0.186) + (0.047)(0.207)}$

$\qquad\qquad\quad = 0.9480$

33. $P(\text{between 35 and 44} \mid \text{never married})$ (for a randomly selected man)

$\qquad = \dfrac{(0.186)(0.204)}{(0.186)(0.204) + (0.132)(0.901) + (0.186)(0.488) + (0.348)(0.118) + (0.148)(0.044)}$

$\qquad = 0.1285$

35. $P(\text{between 45 and 64} \mid \text{never married})$ (for a randomly selected woman)

$\qquad = \dfrac{(0.345)(0.092)}{(0.345)(0.092) + (0.121)(0.825) + (0.172)(0.366) + (0.178)(0.147) + (0.184)(0.040)}$

$\qquad = 0.1392$

In Exercise 37, let S stand for "smokes" and S^- for "does not smoke."

37. $P(18 - 44 \mid S)$

$\qquad = \dfrac{P(18 - 44)P(S \mid 18 - 44)}{P(18 - 44)P(S \mid 18 - 44) + P(45 - 64)P(S \mid 45 - 65) + P(65 - 74)P(S \mid 65 - 74) + P(>75)P(S \mid >75)}$

$\qquad = \dfrac{(0.49)(0.23)}{(0.49)(0.23) + (0.34)(0.22) + (0.09)(0.12) + (0.08)(0.06)}$

$\qquad = 0.5549$

39.

Category	Proportion of Population	Probability of Being Picked Up
Has terrorist ties	$\dfrac{1}{1,000,000}$	0.99
Does not have terrorists ties	$\dfrac{999,999}{1,000,000}$	0.01

$P(\text{Has terrorist ties} \mid \text{Picked up}) = \dfrac{\frac{1}{1,000,000}(0.99)}{\frac{1}{1,000,000}(0.99) + \frac{999,999}{1,000,000}(0.01)}$

$\qquad\qquad\qquad\qquad\qquad = \dfrac{\frac{1}{1,000,000}(0.99)}{\frac{1}{1,000,000}(0.99) + \frac{999,999}{1,000,000}(0.01)} \cdot \dfrac{1,000,000}{1,000,000}$

$\qquad\qquad\qquad\qquad\qquad = \dfrac{0.99}{10,000.98}$

$\qquad\qquad\qquad\qquad\qquad \approx 9.9 \times 10^{-5}$

Chapter 7   Review Exercises

1. True

2. True

3. False: The union of a set with itself has the same number of elements as the set.

4. False: The intersection of a set with itself has the same number of elements as the set.

5. False: If the sets share elements, this procedure gives the wrong answer.

6. True

7. False: This procedure is correct only if the two events are mutually exclusive.

8. False: We can calculate this probability by assuming a sample space in which each card in the 52-card deck is equally likely to be drawn.

9. False: If two events A and B are mutually exclusive, then $P(A \cap B) = 0$ and this will not be equal to $P(A)P(B)$ if $P(A)$ and $P(B)$ are greater than 0.

10. True

11. False: In general these two probabilities are different. For example, for a draw from a 52-card deck, $P(\text{heart}|\text{queen}) = 1/4$ and $P(\text{queen}|\text{heart}) = 1/13$.

12. True

13. $9 \in \{8, 4, -3, -9, 6\}$

Since 9 is not an element of the set, this statement is false.

15. $2 \notin \{0, 1, 2, 3, 4\}$

Since 2 is an element of the set, this statement is false.

17. $\{3, 4, 5\} \subseteq \{2, 3, 4, 5, 6\}$

Every element of $\{3, 4, 5\}$ is an element of $\{2, 3, 4, 5, 6\}$, so this statement is true.

19. $\{3, 6, 9, 10\} \subseteq \{3, 9, 11, 13\}$

10 is an element of $\{3, 6, 9, 10\}$, but 10 is not an element of $\{3, 9, 11, 13\}$. Therefore, $\{3, 6, 9, 10\}$ is not a subset of $\{3, 9, 11, 13\}$. The statement is false.

21. $\{2, 8\} \nsubseteq \{2, 4, 6, 8\}$

Since both 2 and 8 are elements of $\{2, 4, 6, 8\}$, $\{2, 8\}$ is a subset of $\{2, 4, 6, 8\}$. This statement is false.

In Exercises 23–32

$$U = \{a, b, c, d, e, f, g, h\},$$
$$K = \{c, d, e, f, h\},$$
$$\text{and } R = \{a, c, d, g\}.$$

23. K has 5 elements, so it has $2^5 = 32$ subsets.

25. K' (the complement of K) is the set of all elements of U that do *not* belong to K.

$$K' = \{a, b, g\}$$

27. $K \cap R$ (the intersection of K and R) is the set of all elements belonging to both set K and set R.

$$K \cap R = \{c, d\}$$

29. $(K \cap R)' = \{a, b, e, f, g, h\}$ since these elements are in U but not in $K \cap R$. (See Exercise 27.)

31. $\emptyset' = U$

33. $A \cap C$ is the set of all female employees in the K.O. Brown Company who are in the accounting department.

35. $A \cup D$ is the set of all employees in the K.O. Brown Company who are in the accounting department *or* have MBA degrees or both.

37. $B' \cap C'$ is the set of all male employees who are not in the sales department.

39. $A \cup B'$ is the set of all elements which belong to A or do not belong to B, or both.

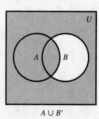

$A \cup B'$

41. $(A \cap B) \cup C$

First find $A \cap B$.

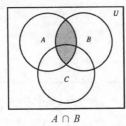

$A \cap B$

Now find the union of this region with C.

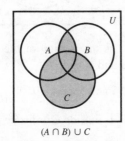

$(A \cap B) \cup C$

43. The sample space for rolling a die is

$$S = \{1, 2, 3, 4, 5, 6\}.$$

45. The sample space of the possible weights is

$$S = \{0, 0.5, 1, 1.5, 2, \ldots, 299.5, 300\}.$$

47. The sample space consists of all ordered pairs (a, b) where a can be 3, 5, 7, 9, or 11, and b is either R (red) or G (green). Thus,

$$S = \{(3, R), (3, G), (5, R), (5, G), (7, R),$$
$$(7, G), (9, R), (9, G), (11, R), (11, G)\}.$$

49. The event F that the second ball is green is

$$F = \{(3, G), (5, G), (7, G), (9, G), (11, G)\}.$$

51. There are 13 hearts out of 52 cards in a deck. Thus,

$$P(\text{heart}) = \frac{13}{52} = \frac{1}{4}.$$

53. There are 3 face cards in each of the four suits.

$$P(\text{face card}) = \frac{12}{52}$$

$$P(\text{heart}) = \frac{13}{52}$$

$$P(\text{face card and heart}) = \frac{3}{52}$$

$$P(\text{face card or heart}) = P(\text{face card}) + P(\text{heart})$$
$$- P(\text{face card and heart})$$

$$= \frac{12}{52} + \frac{13}{52} - \frac{3}{52}$$

$$= \frac{22}{52} = \frac{11}{26}$$

55. There are 4 queens of which 2 are red, so

$$P(\text{red} | \text{queen}) = \frac{n(\text{red and queen})}{n(\text{queen})}$$

$$= \frac{2}{4} = \frac{1}{2}.$$

57. There are 4 kings of which all 4 are face cards. Thus,

$$P(\text{face card} | \text{king}) = \frac{n(\text{face card and king})}{n(\text{king})}$$

$$= \frac{4}{4} = 1.$$

63. If A and B are nonempty and independent, then

$$P(A \cap B) = P(A) \cdot P(B).$$

For mutually exclusive events, $P(A \cap B) = 0$, which would mean $P(A) = 0$ or $P(B) = 0$. So independent events with nonzero probabilities are not mutually exclusive. But independent events one of which has zero probability are mutually exclusive.

65. Let C be the event "a club is drawn." There are 13 clubs in the deck, so $n(C) = 13$,
$P(C') = \frac{13}{52} = \frac{1}{4}$, and $P(C') = 1 - P(C) = \frac{3}{4}$.
The odds in favor of drawing a club are

$$\frac{P(C)}{P(C')} = \frac{\frac{1}{4}}{\frac{3}{4}} = \frac{1}{3},$$

which is written "1 to 3."

67. Let R be the event "a red face card is drawn" and Q be the event "a queen is drawn." Use the union rule for probability to find $P(R \cup Q)$.

$$P(R \cup Q) = P(R) + P(Q) - P(R \cap Q)$$

$$= \frac{6}{52} + \frac{4}{52} - \frac{2}{52}$$

$$= \frac{8}{52} = \frac{2}{13}$$

$$P(R \cup Q)' = 1 - P(R \cup Q)$$

$$= 1 - \frac{2}{13} = \frac{11}{13}$$

The odds in favor of drawing a red face card or a queen are

$$\frac{P(R \cup Q)}{P(R \cup Q)'} = \frac{\frac{2}{13}}{\frac{11}{13}} = \frac{2}{11},$$

which is written "2 to 11."

69. The sum is 8 for each of the 5 outcomes 2-6, 3-5, 4-4, 5-3, and 6-2. There are 36 outcomes in all in the sample space.

$$P(\text{sum is } 8) = \frac{5}{36}$$

71. $P(\text{sum is at least } 10)$

$$= P(\text{sum is } 10) + P(\text{sum is } 11)$$

$$+ P(\text{sum is } 12)$$

$$= \frac{3}{36} + \frac{2}{36} + \frac{1}{36}$$

$$= \frac{6}{36} = \frac{1}{6}$$

73. The sum can be 9 or 11. $P(\text{sum is } 9) = \frac{4}{36}$ and $P(\text{sum is } 11) = \frac{2}{36}$.

$P(\text{sum is odd number greater than } 8)$

$$= \frac{4}{36} + \frac{2}{36}$$

$$= \frac{6}{36} = \frac{1}{6}$$

75. Consider the reduced sample space of the 11 outcomes in which at least one die is a four. Of these, 2 have a sum of 7, 3-4 and 4-3. Therefore,

$P(\text{sum is } 7 \,|\, \text{at least one die is a } 4)$

$$= \frac{2}{11}.$$

77. $P(E) = 0.51, \quad P(F) = 0.37, \quad P(E \cap F) = 0.22$

(a) $P(E \cup F) = P(E) + P(F) - P(E \cap F)$

$$= 0.51 + 0.37 - 0.22$$

$$= 0.66$$

(b) Draw a Venn diagram.

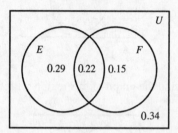

$E \cap F'$ is the portion of the diagram that is inside E and outside F.

$$P(E \cap F') = 0.29$$

(c) $E' \cup F$ is outside E or inside F, or both.

$$P(E' \cup F) = 0.22 + 0.15 + 0.34 = 0.71.$$

(d) $E' \cap F'$ is outside E and outside F.

$$P(E' \cap F') = 0.34$$

79. First make a tree diagram. Let A represent "box A" and K represent "black ball."

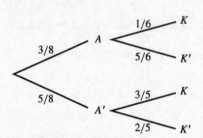

Use Bayes' theorem.

$$P(A|K) = \frac{P(A) \cdot P(K|A)}{P(A) \cdot P(K|A) + P(A') \cdot P(K|A')}$$

$$= \frac{\frac{3}{8} \cdot \frac{1}{6}}{\frac{3}{8} \cdot \frac{1}{6} + \frac{5}{8} \cdot \frac{3}{5}}$$

$$= \frac{\frac{1}{16}}{\frac{7}{16}} = \frac{1}{7}$$

81. First make a tree diagram letting C represent "a competent shop" and R represent "an appliance is repaired correctly."

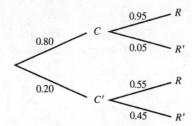

To obtain $P(C|R)$, use Bayes' theorem.

$$P(C|R) = \frac{P(C) \cdot P(R|C)}{P(C) \cdot P(R|C) + P(C') \cdot P(R|C')}$$

$$= \frac{0.80(0.95)}{0.80(0.95) + 0.20(0.55)}$$

$$= \frac{0.76}{0.87} \approx 0.8736$$

83. Refer to the tree diagram for Exercise 81. Use Bayes' theorem.

$$P(C|R') = \frac{P(C) \cdot P(R'|C)}{P(C) \cdot P(R'|C) + P(C') \cdot P(R'|C')}$$

$$= \frac{0.80(0.05)}{0.80(0.05) + 0.20(0.45)}$$

$$= \frac{0.04}{0.13} \approx 0.3077$$

85. To find $P(R)$, use

$$P(R) = P(C) \cdot P(R|C) + P(C') \cdot P(R|C')$$
$$= 0.80(0.95) + 0.20(0.55) = 0.87.$$

87. (a) "A customer buys neither machine" may be written $(E \cup F)'$ or $E' \cap F'$.

(b) "A customer buys at least one of the machines" is written $E \cup F$.

89. Use Bayes' theorem to find the required probabilities.

(a) Let D be the event "item is defective" and E_k be the event "item came from supplier k," $k = 1, 2, 3, 4$.

$$P(D) = P(E_1) \cdot P(D|E_1) + P(E_2) \cdot P(D|E_2)$$
$$+ P(E_3) \cdot P(D|E_3) + P(E_4) \cdot P(D|E_4)$$
$$= 0.17(0.01) + 0.39(0.02) + 0.35(0.05)$$
$$+ 0.09(0.03)$$
$$= 0.0297$$

(b) Find $P(E_4|D)$. Using Bayes' theorem, the numerator is

$$P(E_4) \cdot P(D|E_4) = 0.09(0.03) = 0.0027.$$

The denominator is $P(E_1) \cdot P(D|E_1) +$ $P(E_2) \cdot P(D|E_2) + P(E_3) \cdot P(D|E_3) +$ $P(E_4) \cdot P(D|E_4)$, which, from part (a), equals 0.0297.

Therefore,

$$P(E_4|D) = \frac{0.0027}{0.0297} \approx 0.0909.$$

(c) Find $P(E_2|D)$. Using Bayes' theorem with the same denominator as in part (a),

$$P(E_2|D) = \frac{P(E_2) \cdot P(D|E_2)}{0.0418}$$
$$= \frac{0.39(0.02)}{0.0297}$$
$$= \frac{0.0078}{0.0297}$$
$$\approx 0.2626.$$

(d) Since $P(D) = 0.0297$ and $P(D|E_4) = 0.03$,

$$P(D) \neq P(D|E_4)$$

Therefore, the events are not independent.

91. Let E represent "customer insures exactly one car" and S represent "customer insures a sports car." Let x be the probability that a customer who insures exactly one car insures a sports car, or $P(S|E)$. Make a tree diagram.

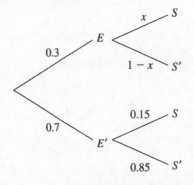

We are told that 20% of the customers insure a sports car, or $P(S) = 0.20$.

$$P(S) = P(E) \cdot P(S|E) + P(E') \cdot P(S|E')$$
$$0.20 = 0.30(x) + 0.70(0.15)$$
$$0.20 = 0.3x + 0.105$$
$$0.3x = 0.095$$
$$x \approx 0.316666667$$

Therefore, the probability that a customer insures a car other than a sports car is

$$P(S') = 1 - P(S)$$
$$\approx 1 - 0.316666667$$
$$= 0.683333333.$$

Finally, the probability that a randomly selected customer insures exactly one car and that car is not a sports car is

$$P(E \cap S') = P(E) \cdot P(S')$$
$$\approx 0.3 \cdot 0.683333333$$
$$\approx 0.21.$$

The correct answer choice is **b**.

93. Let C represent "the automobile owner purchases collision coverage" and D represent "the automobile owner purchases disability coverage." We want to find $P(C' \cap D') = P[(C \cup D)'] = 1 - P(C \cup D)$. We are given that $P(C) = 2 \cdot P(D)$ and that $P(C \cap D) = 0.15$. Let $x = P(D)$.

$$P(C \cap D) = P(C) \cdot P(D)$$
$$0.15 = 2x \cdot x$$
$$0.075 = x^2$$
$$x = \sqrt{0.075}$$
$$x \approx 0.2739$$

So $P(D) = x \approx 0.2739$ and $P(C) = 2x$.

$$P(C \cup D) = P(C) + P(D) - P(C \cap D)$$
$$= 2(0.2739) + 0.2739 - 0.15$$
$$= 0.6720$$

$$P(C' \cap D') = 1 - P(C \cup D)$$
$$= 1 - 0.6720$$
$$\approx 0.33$$

The correct choice is answer **b**.

95. (a)

	N_2	T_2
N_1	N_1N_2	N_1T_2
T_1	T_1N_2	T_1T_2

Since the four combinations are equally likely, each has probability $\frac{1}{2}$.

(b) $P(\text{two trait cells}) = P(T_1T_2) = \frac{1}{4}$

(c) $P(\text{one normal cell and one trait cell})$
$$= P(N_1T_2) + P(T_1N_2)$$
$$= \frac{1}{4} + \frac{1}{4} = \frac{1}{2}$$

(d) $P(\text{not a carrier and does not have disease})$
$$= P(N_1N_2) = \frac{1}{4}$$

97. We want to find $P(A' \cap B' \cap C' | A')$. Use a Venn diagram, fill in the information given, and use the diagram to help determine the missing values.

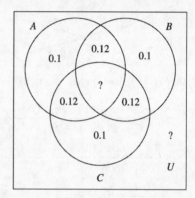

To determine $P(A \cap B \cap C)$, we are told that "The probability that a woman has all three risk factors, given that she has A and B, is 1/3." Therefore, $P(A \cap B \cap C | A \cap B) = 1/3$.

Let $x = P(A \cap B)$; then, using the diagram as a guide,

$$P(A \cap B \cap C) + P(A \cap B \cap C') = P(A \cap B)$$
$$\frac{1}{3}x + 0.12 = x$$
$$0.12 = \frac{2}{3}x$$
$$x = 0.18$$

So, $P(A \cap B \cap C) = (1/3)(0.16) = 0.06$. By DeMorgan's laws, we have

$$A' \cap B' \cap C' = (A \cup B \cup C)'$$

so that

$$P(A' \cap B' \cap C') = P[(A \cup B \cup C)']$$
$$= 1 - P(A \cup B \cup C)$$
$$= 1 - [3(0.10) + 3(0.12) + 0.06]$$
$$= 0.28.$$

Therefore,

$$P(A' \cap B' \cap C' | A') = \frac{P(A' \cap B' \cap C' \cap A')}{P(A')}$$
$$= \frac{P(A' \cap B' \cap C')}{P(A')}$$
$$= \frac{0.28}{0.6} \approx 0.467.$$

The correct answer choice is **c**.

99. Let C be the set of viewers who watch situation comedies,
G be the set of viewers who watch game shows,
and M be the set of viewers who watch movies.

We are given the following information.

$$n(C) = 20 \quad n(G) = 19 \quad n(M) = 27$$

$n(M \cap G') = 19$

$n(C \cap G') = 15$

$n(C \cap M) = 10$

$n(C \cap G \cap M) = 3$

$n(C' \cap G' \cap M') = 7$

Start with $C \cap G \cap M$: $n(C \cap G \cap M) = 3$.

Since $n(C \cap M) = 10$, the number of people who watched comedies and movies but not game shows, or $n(C \cap G' \cap M)$, is $10 - 3 = 7$.

Since $n(M \cap G') = 19, n(C' \cap G' \cap M)$
$$= 19 - 7 = 12.$$

Since $n(M) = 27, n(C' \cap G \cap M)$
$$= 27 - 3 - 7 - 12 = 5.$$

Since $n(C \cap G') = 15, n(C \cap G' \cap M')$
$$= 15 - 7 = 8.$$

Since $n(C) = 20$,
$$n(C \cap G \cap M') = 20 - 8 - 3 - 7 = 2.$$

Finally, since $n(G) = 19, n(C' \cap G \cap M')$
$$= 19 - 2 - 3 - 5 = 9.$$

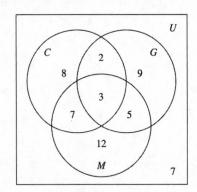

(a) $n(U) = 8 + 2 + 9 + 7$
$$+3 + 5 + 12 + 7 = 53$$

(b) $n(C \cap G' \cap M) = 7$

(c) $n(C' \cap G' \cap M) = 12$

(d) $n(M') = n(U) - n(M)$
$$= 53 - 27 = 26$$

101. Let C be the event "the culprit penny is chosen." Then

$$p(C|HHH) = \frac{P(C \cap HHH)}{P(HHH)}.$$

These heads will result two different ways. The culprit coin is chosen $\frac{1}{3}$ of the time and the probability of a head on any one flip is $\frac{3}{4}$: $P(C \cap HHH) = \frac{1}{3}\left(\frac{3}{4}\right)^3 \approx 0.1406$. If a fair (innocent) coin is chosen, the probability of a head on any one flip is $\frac{1}{2}$: $P(C'|HHH) = \frac{2}{3}\left(\frac{1}{2}\right)^3 \approx 0.0833$. Therefore,

$$
\begin{aligned}
P(C|HHH) &= \frac{P(C \cap HHH)}{P(HHH)} \\
&= \frac{P(C \cap HHH)}{P(C \cap HHH) + P(C' \cap HHH)} \\
&\approx \frac{0.1406}{0.1406 + 0.0833} \\
&\approx 0.6279
\end{aligned}
$$

103. $P(\text{earthquake}) = \dfrac{9}{9+1} = \dfrac{9}{10} = 0.90$

105. Let W be the set of western states,
S be the set of small states, and
E be the set of early states.

We are given the following information.

$$n(W) = 24 \qquad\qquad n(S) = 22 \qquad\quad n(E) = 26$$
$$n(W' \cap S' \cap E') = 9 \qquad\qquad n(W \cap S) = 14$$
$$n(S \cap E) = 11 \qquad\qquad\qquad n(W \cap S \cap E) = 7$$

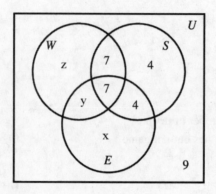

First, put 7 in $W \cap S \cap E$ and 9 in $W' \cap S' \cap E'$.

Complete $S \cap E$ with 4 for a total of 11.
Complete $W \cap S$ with 7 for a total of 14.
Complete S with 4 for a total of 22.

To complete the rest of the diagram requires solving some equations. Let the incomplete region of E be x, the incomplete region of $W \cap E$ be y, and the incomplete region of W be z. Then, using the given values and the fact that $n(U) = 50$,

$$x + y \qquad = 15$$
$$y + z = 10$$
$$x + y + z = 50 - 22 - 9 = 19.$$

The solution to the system is $x = 9$, $y = 6$, and $z = 4$.

Complete $W \cap E$ with.
Complete E with 9 for a total of 26.
Complete W with 4 for a total of 24.

The completed diagram is as follows.

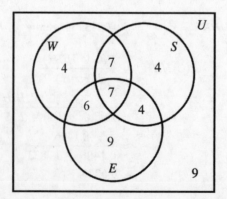

(a) $n(W \cap S' \cap E') = 4$

(b) $n(W' \cap S') = n((W \cup S)') = 18$

107. Let R be "a red side is facing up" and
RR be "the 2-sided red card is chosen."

If a red side is facing up, we want to find $P(RR|R)$ since the other possibility would be a green side is facing down.

$$P(RR|R) = \frac{P(RR)}{P(R)} = \frac{\frac{1}{3}}{\frac{1}{2}} = \frac{2}{3}$$

No, the bet is not a good bet.

109. Let G be the set of people who watched gymnastics,
B be the set of people who watched baseball,
and S be the set of people who watched soccer.

We want to find $P(G' \cap B' \cap S')$ or, by DeMorgan's laws, $P[(G \cup B \cup S)']$.

We are given the following information.

$$P(G) = 0.28 \qquad P(B) = 0.29 \qquad P(S) = 0.19 \qquad P(G \cap B) = 0.14$$
$$P(B \cap S) = 0.12 \qquad P(G \cap S) = 0.10 \qquad P(G \cap B \cap S) = 0.08$$

Start with $P(G \cap B \cap S) = 0.08$ and work from the inside out.

Since $P(G \cap S) = 0.10$, $P(G \cap B' \cap S) = 0.02$.

Since $P(B \cap S) = 0.12$, $P(G' \cap B \cap S) = 0.04$.

Since $P(G \cap B) = 0.14$, $P(G \cap B \cap S') = 0.06$.

Since $P(S) = 0.19, P(G' \cap B' \cap S) = 0.19 - 0.14 = 0.05$.

Since $P(B) = 0.29, P(G' \cap B \cap S') = 0.29 - 0.18 = 0.11$.

Since $P(G) = 0.28, P(G \cap B' \cap S') = 0.28 - 0.16 = 0.12$.

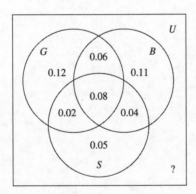

Therfore,

$$P(G' \cap B' \cap S) = P[(G \cup B \cup S)']$$
$$= 1 - P(G \cup B \cup S)$$
$$= 1 - (0.12 + 0.06 + 0.11 + 0.02 + 0.08 + 0.04 + 0.05) = 0.52.$$
$$P(G' \cap B' \cap S) = 1 - (0.12 + 0.06 + 0.11 + 0.02 + 0.08 + 0.04 + 0.05) = 0.52.$$

The correct answer choice is **d**.

Chapter 8

COUNTING PRINCIPLES; FURTHER PROBABILITY TOPICS

8.1 The Multiplication Principle; Permutations

Your Turn 1

Each of the four digits can be one of the ten digits $0, 1, 2, \ldots, 10,$ so there are $10 \cdot 10 \cdot 10 \cdot 10$ or $10,000$ possible sequences. If no digit is repeated, there are 10 choices for the first place, 9 for the second, 8 for the third, and 7 for the fourth, so there are $10 \cdot 9 \cdot 8 \cdot 7 = 5040$ possible sequences.

Your Turn 2

Any of the 8 students could be first in line. Then there are 7 choices for the second spot, 6 for the third, and so on, with only one student remaining for the last spot in line. There are $8 \cdot 7 \cdot 6 \cdot 5 \cdot 4 \cdot 3 \cdot 2 \cdot 1 = 40,320$ different possible lineups.

Your Turn 3

The teacher has 8 ways to fill the first space (say the one on the left), 7 choices for the next book, and so on, leaving 4 choices for the last book on the right. So the number of possible arrangements is $8 \cdot 7 \cdot 6 \cdot 5 \cdot 4 = 6720.$

Your Turn 4

There are 6 letters. If we use 3 of the 6, the number of permutations is

$$P(6, 3) = \frac{6!}{(6-3)!} = \frac{6 \cdot 5 \cdot 4 \cdot 3 \cdot 2 \cdot 1}{3 \cdot 2}$$
$$= 6 \cdot 5 \cdot 4 = 120.$$

Your Turn 5

If the panel sits in a row, there are $4!$ or 24 ways of arranging the four class groups. Within the groups, there are $2!$ ways of arranging the freshmen, $2!$ ways of arranging the sophomores, $2!$ ways of arranging the juniors, and $3!$ ways of arranging the seniors. Using the multiplication principle, the number of ways of seating the panel with the classes together is $24 \cdot 2! \cdot 2! \cdot 2! \cdot 3! = 24 \cdot 2 \cdot 2 \cdot 2 \cdot 6 = 1152.$

Your Turn 6

The word *Tennessee* contains 9 letters, consisting of 1 t, 4 e's, 2 n's and 2 s's. Thus the number of possible arrangements is

$$\frac{9!}{1! \, 4! \, 2! \, 2!} = 3780.$$

Your Turn 7

The student has $4 + 5 + 3 + 2 = 14$ pairs of socks, so if the pairs were distinguishable there would be $14!$ possible selections for the next two weeks. But since the pairs of each color are identical, the number of distinguishable selections is

$$\frac{14!}{4! \, 5! \, 3! \, 2!} = 2,522,520.$$

8.1 Exercises

1. $6! = 6 \cdot 5 \cdot 4 \cdot 3 \cdot 2 \cdot 1 = 720$

3. $15! = 15 \cdot 14 \cdot 13 \cdot 12 \cdot 11 \cdot 10 \cdot 9 \cdot 8 \cdot 7$
    $\cdot 6 \cdot 5 \cdot 4 \cdot 3 \cdot 2 \cdot 1$
    $\approx 1.308 \times 10^{12}$

5. $P(13, 2) = \dfrac{13!}{(13-2)!} = \dfrac{13!}{11!}$
    $= \dfrac{13 \cdot 12 \cdot 11!}{11!}$
    $= 156$

7. $P(38, 17) = \dfrac{38!}{(38-17)!} = \dfrac{38!}{21!}$
    $\approx 1.024 \times 10^{25}$

9. $P(n, 0) = \dfrac{n!}{(n-0)!} = \dfrac{n!}{n!} = 1$

11. $P(n, 1) = \dfrac{n!}{(n-1)!} = \dfrac{n(n-1)!}{(n-1)!} = n$

13. By the multiplication principle, there will be $6 \cdot 3 \cdot 2 = 36$ different home types available.

15. There are 4 choices for the first name and 5 choices for the middle name, so, by the multiplication principle, there are $4 \cdot 5 = 20$ possible arrangements.

19. There is exactly one 3-letter subset of the letters A, B, and C, namely A, B, and C.

21. **(a)** initial

This word contains 3 i's, 1 n, 1 t, 1 a, and $1\ \ell$. Use the formula for distinguishable permutations with $n = 7, n_1 = 3,\ n_2 = 1,$ $n_3 = 1, n_4 = 1,$ and $n_5 = 1$.

$$\frac{n!}{n_1!n_2!n_3!n_4!n_5!} = \frac{7!}{3!1!1!1!1!}$$
$$= \frac{7 \cdot 6 \cdot 5 \cdot 4 \cdot 3!}{3!}$$
$$= 840$$

There are 840 distinguishable permutations of the letters.

(b) little

Use the formula for distinguishable permutetions with $n = 6, n_1 = 2,\ n_2 = 1,$ $n_3 = 2,$ and $n_4 = 1$.

$$\frac{6!}{2!1!2!1!} = \frac{6!}{2!2!}$$
$$= \frac{6 \cdot 5 \cdot 4 \cdot 3 \cdot 2 \cdot 1}{2 \cdot 1 \cdot 2 \cdot 1}$$
$$= 180$$

There are 180 distinguishable permutations.

(c) decreed

Use the formula for distinguishable permutations with $n = 7,\ n_1 = 2,$ $n_2 = 3,\ n_3 = 1,$ and $n_4 = 1$.

$$\frac{7!}{2!3!1!1!} = \frac{7!}{2!3!}$$
$$= \frac{7 \cdot 6 \cdot 5 \cdot 4 \cdot 3!}{2 \cdot 1 \cdot 3!}$$
$$= 420$$

There are 420 distinguishable permutations.

23. **(a)** The 9 books can be arranged in
$P(9, 9) = 9! = 362,880$ ways.

(b) The blue books can be arranged in 4! ways, the green books can be arranged in 3! ways, and the red books can be arranged in 2! ways. There are 3! ways to choose the order of the 3 groups of books. Therefore, using

the multiplication principle, the number of possible arrangements is

$$4!3!2!3! = 24 \cdot 6 \cdot 2 \cdot 6 = 1728.$$

(c) Use the formula for distinguishable permutations with $n = 9, n_1 = 4, n_2 = 3,$ and $n_3 = 2$. The number of distinguishable arrangements is

$$\frac{9!}{4!3!2!} = \frac{9 \cdot 8 \cdot 7 \cdot 6 \cdot 5 \cdot 4!}{4! \cdot 6 \cdot 2}$$
$$= 1260.$$

(d) There are 4 choices for the blue book, 3 for the green book, and 2 for the red book. The total number of arrangements is

$$4 \cdot 3 \cdot 2 = 24.$$

(e) From part (d) there are 24 ways to select a blue, red, and green book if the order does not matter. There 3! ways to choose the order. Using the multiplication principle, the number of possible ways is

$$24 \cdot 3! = 24 \cdot 6 = 144.$$

25. $10! = 10 \cdot 9!$

To find the value of 10!, multiply the value of 9! by 10.

27. **(a)** The number 13! has 2 factors of five so there must be 2 ending zeros in the answer.

(b) The number 27! has 6 factors of five (one each in 5, 10, 15, and 20 and two factors in 25), so there must be 6 ending zeros in the answer.

(c) The number 75! has $15 + 3 = 18$ factors of five (one each in $5, 10, \ldots, 75$ and two factors each in 25, 50, and 75), so there must be 18 ending zeros in the answer.

29. $P(4, 4) = \dfrac{4!}{(4 - 4)!} = \dfrac{4!}{0!}$

If $0! = 0$, then $P(4, 4)$ would be undefined.

31. **(a)** By the multiplication principle, since there are 7 pastas and 6 sauces, the number of different bowls is $7 \cdot 6 = 42$.

(b) If we exclude the two meat sauces there are 4 sauces left and the number of bowls is now $7 \cdot 4 = 28$.

33. (a) Since there are 11 slots, the 11 commercials can be arranged in $11! = 39,916,800$ ways.

(b) Use the multiplication principle. We can put either stores or restaurants first (2 choices). Then there are 6! orders for the stores and 5! orders for the restaurants, so the number of groupings is $2 \cdot 6! \cdot 5! = 172,800$.

(c) Since the number of restaurants is one more than the number of stores, a restaurant must come first. This eliminates the first choice in part (b), but we still can order the restaurants and the stores freely within each category, so the answer is $6! \cdot 5! = 86,400$.

35. If each species were to be assigned 3 initials, since there are 26 different letters in the alphabet, there could be $26^3 = 17,576$ different 3-letter designations. This would not be enough. If 4 initials were used, the biologist could represent $26^4 = 456,976$ different species, which is more than enough. Therefore, the biologist should use at least 4 initials.

37. The number of ways to seat the people is
$$P(6,6) = \frac{6!}{0!} = \frac{6!}{1}$$
$$= 6 \cdot 5 \cdot 4 \cdot 3 \cdot 2 \cdot 1$$
$$= 720.$$

39. The number of possible batting orders is
$$P(19,9) = \frac{19!}{(19-9)!} = \frac{19!}{10!}$$
$$= 33,522,128,640$$
$$\approx 3.352 \times 10^{10}.$$

41. (a) The number of ways 5 works can be arranged is
$$P(5,5) = 5! = 120.$$

(b) If one of the 2 overtures must be chosen first, followed by arrangements of the 4 remaining pieces, then
$$P(2,1) \cdot P(4,4) = 2 \cdot 24 = 48$$
is the number of ways the program can be arranged.

43. (a) There are 4 tasks to be performed in selecting 4 letters for the call letters. The first task may be done in 2 ways, the second in 25, the third in 24, and the fourth in 23. By the multiplication principle, there will be

$$2 \cdot 25 \cdot 24 \cdot 23 = 27,600$$

different call letter names possible.

(b) With repeats possible, there will be

$$2 \cdot 26 \cdot 26 \cdot 26 = 2 \cdot 26^3 \quad \text{or} \quad 35,152$$

call letter names possible.

(c) To start with W or K, make no repeats, and end in R, there will be

$$2 \cdot 24 \cdot 23 \cdot 1 = 1104$$

possible call letter names.

45. (a) Our number system has ten digits, which are 1 through 9 and 0.

There are 3 tasks to be performed in selecting 3 digits for the area code. The first task may be done in 8 ways, the second in 2, and the third in 10. By the multiplication principle, there will be

$$8 \cdot 2 \cdot 10 = 160$$

different area codes possible.

There are 7 tasks to be performed in selecting 7 digits for the telephone number. The first task may be done in 8 ways, and the other 6 tasks may each be done in 10 ways. By the multiplication principle, there will be

$$8 \times 10^6 = 8,000,000$$

different telephone numbers possible within each area code.

(b) Some numbers, such as 911, 800, and 900, are reserved for special purposes and are therefore unavailable for use as area codes.

47. (a) There were

$$26^3 \cdot 10^3 = 17,576,000$$

license plates possible that had 3 letters followed by 3 digits.

(b) There were

$$10^3 \cdot 26^3 = 17,576,000$$

new license plates possible when plates were also issued having 3 digits followed by 3 letters.

(c) There were

$$26 \cdot 10^3 \cdot 26^3 = 456,976,000$$

new license plates possible when plates were also issued having 1 letter followed by 3 digits and then 3 letters.

49. If there are no restrictions on the digits used, there would be

$$10^5 = 100,000$$

different 5-digit zip codes possible.

If the first digit is not allowed to be 0, there would be

$$9 \cdot 10^4 = 90,000$$

zip codes possible.

51. There are 3 possible identical shapes on each card.

There are 3 possible shapes for the identical shapes.

There are 3 possible colors.

There are 3 possible styles.

Therefore, the total number of cards is
$3 \cdot 3 \cdot 3 \cdot 3 = 81$.

53. There are 3 possible answers for the first question and 2 possible answers for each of the 19 other questions. The number of possible objects is

$$3 \cdot 2^{19} = 1,572,864.$$

20 questions are not enough.

55. (a) Since the starting seat is not counted, the number of arrangements is

$$P(19, 19) = 19! \approx 1.216451 \times 10^{17}.$$

(b) Since the starting bead is not counted and the necklace can be flipped, the number of arrangements is

$$\frac{P(14, 14)}{2} = \frac{14!}{2} = 43,589,145,600.$$

8.2 Combinations

Your Turn 1

Use the combination formula.

$$C(10, 4) = \frac{10!}{6!4!} = 210$$

Your Turn 2

Since the group of students contains either 3 or 4 students out of 15, it can be selected in $C(15,3) + C(15,4)$ ways.

$$\begin{aligned} C(15, 3) + C(15, 4) &= \frac{15!}{12!3!} + \frac{15!}{11!4!} \\ &= 455 + 1365 \\ &= 1820 \end{aligned}$$

Your Turn 3

(a) Use permutations.

$$P(10, 4) = 10 \cdot 9 \cdot 8 \cdot 7 = 5040$$

(b) Use combinations.

$$C(15, 3) = \frac{15!}{12!3!} = 455$$

(c) Use combinations.

$$C(8, 2) = \frac{8!}{6!2!} = 28$$

(d) Use combinations and permutations. First pick 4 rooms; this is an unordered selection:

$$C(6, 4) = \frac{6!}{2!4!} = 15$$

Now assign the patients to the rooms; this is an ordered selection:

$$P(4, 4) = 4! = 24$$

The number of possible assignments is
$15 \cdot 24 = 360$.

Your Turn 4

The committee is an unordered selection.

$$C(20, 3) = \frac{20!}{17!3!} = 1140$$

If the selection includes assignment to one of the three offices we have an ordered selection.

$$P(20, 3) = 20 \cdot 19 \cdot 18 = 6840$$

Your Turn 5

There are $C(4,2)$ ways to select 2 aces from the 4 aces in the deck and $C(48,3)$ ways to select the 3 remaining cards from the 48 non-aces. Now use the multiplication principle.

$$C(4, 2) \cdot C(48, 3) = 6 \cdot 17,296 = 103,776$$

8.2 Exercises

3. To evaluate $C(8, 3)$, use the formula

$$C(n, r) = \frac{n!}{(n - r)!r!}$$

with $n = 8$ and $r = 3$.

$$\begin{aligned} C(8, 3) &= \frac{8!}{(8 - 3)!3!} \\ &= \frac{8!}{5!3!} \\ &= \frac{8 \cdot 7 \cdot 6 \cdot 5!}{5! \cdot 3 \cdot 2 \cdot 1} = 56 \end{aligned}$$

5. To evaluate $C(44, 20)$, use the formula

$$C(n, r) = \frac{n!}{(n - r)!r!}$$

with $n = 44$ and $r = 20$.

$$C(44, 20) = \frac{44!}{(44 - 20)!20!}$$

$$= \frac{44!}{24!20!}$$

$$= 1.761 \times 10^{12}$$

7. $C(n, 0) = \dfrac{n!}{(n - 0)!0!}$

$$= \frac{n!}{n! \cdot 1}$$

$$= 1$$

9. $C(n, 1) = \dfrac{n!}{(n - 1)!1!}$

$$= \frac{n(n - 1)!}{(n - 1)! \cdot 1}$$

$$= n$$

11. There are 13 clubs, from which 6 are to be chosen. The number of ways in which a hand of 6 clubs can be chosen is

$$C(13, 6) = \frac{13!}{7!6!} = 1716.$$

13. **(a)** There are

$$C(5, 2) = \frac{5!}{3!2!} = \frac{5 \cdot 4 \cdot 3!}{3! \cdot 2 \cdot 1} = 10$$

different 2-card combinations possible.

(b) The 10 possible hands are

$$\{1, 2\}, \{2, 3\}, \{3, 4\}, \{4, 5\}, \{1, 3\},$$
$$\{2, 4\}, \{3, 5\}, \{1, 4\}, \{2, 5\}, \{1, 5\}.$$

Of these, 7 contain a card numbered less than 3.

15. Choose 2 letters from {L, M, N}; order is important.

(a)

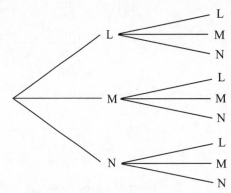

There are 9 ways to choose 2 letters if repetition is allowed.

(b)

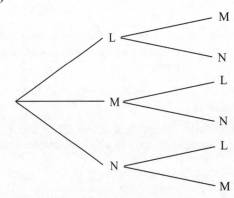

There are 6 ways to choose 2 letters if no repeats are allowed.

(c) The number of combinations of 3 elements taken 2 at a time is

$$C(3, 2) = \frac{3!}{1!2!} = 3.$$

This answer differs from both parts (a) and (b).

17. Order does not matter in choosing members of a committee, so use combinations rather than permutations.

(a) The number of committees whose members are all men is

$$C(9, 5) = \frac{9!}{4!5!}$$

$$= \frac{9 \cdot 8 \cdot 7 \cdot 6 \cdot 5!}{4 \cdot 3 \cdot 2 \cdot 1 \cdot 5!}$$

$$= 126.$$

(b) The number of committees whose members are all women is

$$C(11, 5) = \frac{11!}{6!5!}$$
$$= \frac{11 \cdot 10 \cdot 9 \cdot 8 \cdot 7 \cdot 6!}{6! \cdot 5 \cdot 4 \cdot 3 \cdot 2 \cdot 1}$$
$$= 462.$$

(c) The 3 men can be chosen in

$$C(9, 3) = \frac{9!}{6!3!} = \frac{9 \cdot 8 \cdot 7 \cdot 6!}{6! \cdot 3 \cdot 2 \cdot 1}$$
$$= 84 \text{ ways.}$$

The 2 women can be chosen in

$$C(11, 2) = \frac{11!}{9!2!} = \frac{11 \cdot 10 \cdot 9!}{9! \cdot 2 \cdot 1}$$
$$= 55 \text{ ways.}$$

Using the multiplication principle, a committee of 3 men and 2 women can be chosen in

$$84 \cdot 55 = 4620 \text{ ways.}$$

19. Order is important, so use permutations. The number of ways in which the children can find seats is

$$P(12, 11) = \frac{12!}{(12 - 11)!} = \frac{12!}{1!}$$
$$= 12!$$
$$= 479{,}001{,}600.$$

21. Since order does not matter, the answers are combinations.

(a) $C(16, 2) = \frac{16!}{14!2!} = \frac{16 \cdot 15 \cdot 14!}{14! \cdot 2 \cdot 1} = 120$

120 samples of 2 marbles can be drawn.

(b) $C(16, 4) = 1820$

1820 samples of 4 marbles can be drawn.

(c) Since there are 9 blue marbles in the bag, the number of samples containing 2 blue marbles is

$$C(9, 2) = 36.$$

23. Since order does not matter, use combinations.

(a) $C(5, 3) = \frac{5!}{2!3!} = \frac{5 \cdot 4 \cdot 3!}{2 \cdot 1 \cdot 3!} = 10$

There are 10 possible samples with all black jelly beans.

(b) There is only 1 red jelly bean, so there are no samples in which all 3 are red.

(c) $C(3, 3) = 1$

There is 1 sample with all yellow.

(d) $C(5, 2)C(1, 1) = 10 \cdot 1 = 10$

There are 10 samples with 2 black and 1 red.

(e) $C(5, 2)C(3, 1) = 10 \cdot 3 = 30$

There are 30 samples with 2 black and 1 yellow.

(f) $C(3, 2)C(5, 1) = 3 \cdot 5 = 15$

There are 15 samples with 2 yellow and 1 black.

(g) There is only 1 red jelly bean, so there are no samples containing 2 red jelly beans.

25. Show that $C(n, r) = C(n, n - r)$.

Work with each side of the equation separately.

$$C(n, r) = \frac{n!}{r!(n - r)!}$$

$$C(n, n - r) = \frac{n!}{(n - r)![n - (n - r)]!}$$
$$= \frac{n!}{(n - r)!r!}$$

Since both results are the same, we have shown that

$$C(n, r) = C(n, n - r).$$

27. Use combinations since order does not matter.

(a) First consider how many pairs of circles there are. This number is

$$C(6, 2) = \frac{6!}{2!4!} = 15.$$

Each pair intersects in two points. The total number of intersection points is $2 \cdot 15 = 30$.

(b) The number of pairs of circles is

$$C(n, 2) = \frac{n!}{(n - 2)!2!}$$
$$= \frac{n(n - 1)(n - 2)!}{(n - 2)! \cdot 2}$$
$$= \frac{1}{2}n(n - 1).$$

Each pair intersects in two points. The total number of points is

$$2 \cdot \frac{1}{2}n(n - 1) = n(n - 1).$$

29. Since the assistants are assigned to different managers, this amounts to an ordered selection of 3 from 8.

$$P(8, 3) = 8 \cdot 7 \cdot 6 = 336$$

31. Order is important in arranging a schedule, so use permutations.

    (a) $P(6,6) = \dfrac{6!}{0!} = 6! = 720$

    She can arrange her schedule in 720 ways if she calls on all 6 prospects.

    (b) $P(6,4) = \dfrac{6!}{2!} = 360$

    She can arrange her schedule in 360 ways if she calls on only 4 of the 6 prospects.

33. There are 2 types of meat and 6 types of extras. Order does not matter here, so use combinations.

    (a) There are $C(2,1)$ ways to choose one type of meat and $C(6,3)$ ways to choose exactly three extras. By the multiplication principle, there are

$$C(2,1)C(6,3) = 2 \cdot 20 = 40$$

    different ways to order a hamburger with exactly three extras.

    (b) There are

$$C(6,3) = 20$$

    different ways to choose exactly three extras.

    (c) "At least five extras" means "5 extras or 6 extras." There are $C(6,5)$ different ways to choose exactly 5 extras and $C(6,6)$ ways to choose exactly 6 extras, so there are

$$C(6,5) + C(6,6) = 6 + 1 = 7$$

    different ways to choose at least five extras.

35. Select 8 of the 16 smokers and 8 of the 22 non-smokers; order does not matter in the group, so use combinations. There are

$$C(16,8)C(22,8) = 4,115,439,900$$

different ways to select the study group.

37. Order does not matter in choosing a delegation, so use combinations. This committee has $5 + 4 = 9$ members.

    (a) There are

$$C(9,3) = \frac{9!}{6!\,3!}$$
$$= \frac{9 \cdot 8 \cdot 7 \cdot 6!}{6! \cdot 3 \cdot 2 \cdot 1}$$
$$= 84 \text{ possible delegations.}$$

    (b) To have all Democrats, the number of possible delegations is

$$C(5,3) = 10.$$

    (c) To have 2 Democrats and 1 Republican, the number of possible delegations is

$$C(5,2)C(4,1) = 10 \cdot 4 = 40.$$

    (d) We have previously calculated that there are 84 possible delegations, of which 10 consist of all Democrats. Those 10 delegations are the only ones with no Republicans, so the remaining $84 - 10 = 74$ delegations include at least one Republican.

39. Order does not matter in choosing the panel, so use combinations.

$$C(45,3) = \frac{45!}{42!\,3!}$$
$$= \frac{45 \cdot 44 \cdot 43 \cdot 42!}{3 \cdot 2 \cdot 1 \cdot 42!}$$
$$= 14,190$$

The publisher was wrong. There are 14,190 possible three judge panels.

41. Since the cards are chosen at random, that is, order does not matter, the answers are combinations.

    (a) There are 4 queens and 48 cards that are not queens. The total number of hands is

$$C(4,4)C(48,1) = 1 \cdot 48 = 48.$$

    (b) Since there are 12 face cards (3 in each suit), there are 40 nonface cards. The number of ways to choose no face cards (all 5 nonface cards) is

$$C(40,5) = \frac{40!}{35!\,5!} = 658,008.$$

    (c) If there are exactly 2 face cards, there will be 3 nonface cards. The number of ways in which the face cards can be chosen is $C(12,2)$, while the number of ways in which the nonface cards can be chosen is $C(40,3)$. Using the multiplication principle, the number of ways to get this result is

$$C(12,2)C(40,3) = 66 \cdot 9880$$
$$= 652,080.$$

    (d) If there are at least 2 face cards, there must be either 2 face cards and 3 nonface cards, 3 face cards and 2 nonface cards, 4 face cards and 1 nonface card, or 5 face cards. Use the multiplication principle as in part (c) to find the number of ways to obtain each of these possibilities. Then add these numbers. The total number of ways to get at least 2 face cards is

$$C(12,2)C(40,3) + C(12,3)C(40,2)$$
$$+ C(12,4)C(40,1) + C(12,5)$$
$$= 66 \cdot 9880 + 220 \cdot 780$$
$$+ 495 \cdot 40 + 792$$
$$= 652{,}080 + 171{,}600 + 19{,}800 + 792$$
$$= 844{,}272.$$

(e) The number of ways to choose 1 heart is $C(13, 1)$, the number of ways to choose 2 diamonds is $C(13, 2)$, and the number of ways to choose 2 clubs is $C(13, 2)$. Using the multiplication principle, the number of ways to get this result is

$$C(13,1)C(13,2)C(13,2) = 13 \cdot 78 \cdot 78$$
$$= 79{,}092.$$

43. Since order does not matter, use combinations.

2 good hitters: $C(5,2)C(4,1) = 10 \cdot 4 = 40$

2 good hitters: $C(5,3)C(4,0) = 10 \cdot 1 = 10$

The total number of ways is $40 + 10 = 50$.

45. Since order does not matter, use combinations.

(a) There are
$$C(20,5) = 15{,}504$$
different ways to select 5 of the orchids.

(b) If 2 special orchids must be included in the show, that leaves 18 orchids from which the other 3 orchids for the show must be chosen. This can be done in
$$C(18,3) = 816$$
different ways.

47. Since order is not important, use combinations. To pick 5 of the 6 winning numbers, we must also pick 1 of the 93 losing numbers. Therefore, the number of ways to pick 5 of the 6 winning numbers is

$$C(6,5)C(93,1) = 6 \cdot 93 = 558.$$

49. **(a)** The number of different committees possible is

$$C(5,2) + C(5,3) + C(5,4) + C(5,5)$$
$$= 10 + 10 + 5 + 1 = 26.$$

(b) The total number of subsets is

$$2^5 = 32.$$

The number of different committees possible is

$$2^5 - C(5,1) - C(5,0)$$
$$= 32 - 5 - 1$$
$$= 26.$$

51. **(a)** If the letters can be repeated, there are 26^6 choices of 6 letters, and there are 10 choices for the digit, giving
$$26^6 \cdot 10 = 3{,}089{,}157{,}760 \text{ passwords.}$$

(b) For nonrepeating letters we have
$P(26, 6) \cdot 10$ or $1{,}657{,}656{,}000$ passwords.

53. **(a)** A pizza can have 3, 2, 1, or no toppings. The number of possibilities is

$$C(17,3) + C(17,2) + C(17,1) + C(17,0)$$
$$= 680 + 136 + 17 + 1$$
$$= 834.$$

There are also four speciality pizzas, so the number of different pizzas is
$834 + 4 = 838$.

(b) The number of 4forAll Pizza possibilities if all four pizzas are different is

$$C(838,4) = 20{,}400{,}978{,}015.$$

The number of 4forAll Pizza possibilities if there are three different pizzas (2 pizzas are the same and the other 2 are different) is

$$838 \cdot C(837,2) = 838 \cdot 349{,}866$$
$$= 293{,}187{,}708.$$

The number of 4forAll Pizza possibilities if there are two different pizzas (3 pizzas are the same or 2 pizzas and 2 pizzas are the same) is

$$838 \cdot 837 + C(838,2)$$
$$= 701{,}406 + 350{,}703$$
$$= 1{,}052{,}109.$$

The number of 4forAll Pizza possibilities if all four are the same is 838. The total number of 4forAll Pizza possibilities is

$$20{,}400{,}978{,}015 + 293{,}187{,}708$$
$$+ 1{,}052{,}109 + 838$$
$$= 20{,}695{,}218{,}670$$

(c) Using the described method, there would be 837 vertical lines and 4 X's or 841 objects, so the total number is

$$C(841,4) = 20{,}695{,}218{,}670.$$

55. (a) $C(9,3) = \dfrac{9!}{6!\,3!} = 84$

   (b) $9 \cdot 9 \cdot 9 = 729$

   (d) $C(17,3) = \dfrac{17!}{14!\,3!} = 680$

   (e) First pick the two boneless buffalo wing flavors; there are $C(5,2) = 10$ ways of doing this. Then we still have 7 non-wing options, plus the 5 buffalo chicken wings available for the third item. So our total is $10 \cdot 12 = 120$.

57. (a) The number of ways the names can be arranged is
   $$18! \approx 6.402 \times 10^{15}.$$

   (b) 4 lines consist of a 3 syllable name repeated, followed by a 2 syllable name and then a 4 syllable name. Including order, the number of arrangements is
   $$10 \cdot 4 \cdot 4 \cdot 9 \cdot 3 \cdot 3 \cdot 8 \cdot 2 \cdot 2 \cdot 7 \cdot 1 \cdot 1$$
   $$= 2{,}903{,}040.$$

   2 lines consist of a 3 syllable name repeated, followed by two more 3 syllable names. Including order, the number of arrangements is
   $$6 \cdot 5 \cdot 4 \cdot 3 \cdot 2 \cdot 1 = 720.$$

   The number of ways the similar 4 lines can be arranged among the 6 total lines is
   $$C(6,4) = 15.$$

   The number of arrangements that fit the pattern is
   $$2{,}903{,}040 \cdot 720 \cdot 15 \approx 3.135 \times 10^{10}.$$

8.3 Probability Applications of Counting Principles

Your Turn 1

Using method 2,
$$P(1 \text{ NY}, 1 \text{ Chicago}) = \frac{C(3,1)\,C(1,1)}{C(6,2)} = \frac{1}{5}.$$

Your Turn 2

Since 8 of the nurses are men, $22 - 8 = 14$ of them are women. We choose 4 nurses, 2 men and 2 women.

$P(2 \text{ men among 4 selected})$
$$= \frac{C(8,2)\,C(14,2)}{C(22,4)} = \frac{2548}{7315} \approx 0.3483$$

Your Turn 3

The probability that the container will be shipped is the probability of selecting 3 working engines for testing when there are $12 - 4 = 8$ working engines in the container. This is
$$P(\text{all 3 work}) = \frac{C(8,3)}{C(12,3)} = \frac{56}{220} = \frac{14}{55}.$$

The probability that at least one defective engine is in the batch is
$$P(\text{at least one defective}) = 1 - \frac{14}{55} = \frac{41}{55} \approx 0.7455.$$

Your Turn 4

$$P(2 \text{ aces}, 2 \text{ kings}, 1 \text{ other}) = \frac{C(4,2)\,C(4,2)\,C(44,1)}{C(52,5)}$$
$$= \frac{6 \cdot 6 \cdot 44}{2{,}598{,}960}$$
$$\approx 0.0006095$$

Your Turn 5

If the slips are chosen without replacement, there are $P(7,3) = 7 \cdot 6 \cdot 5 = 210$ ordered selections. Only one of these spells "now" so $P(\text{now}) = \frac{1}{210}$. If the slips are chosen with replacement there are 7 choices for each and thus $7 \cdot 7 \cdot 7 = 343$ selections. Again, only one of these spells "now" so in this case $P(\text{now}) = \frac{1}{343}$.

Your Turn 6

Using method 1 we compute the number of ways to arrange 14 pieces of fruit which come in 4 kinds: 2 kiwis, 3 apricots, 4 pineapples and 5 coconuts. Assuming the pieces of each kind of fruit are indistinguishable (all kiwis look alike, and so on), the number of arrangements is $\frac{14!}{2!\,3!\,4!\,5!} = 2{,}522{,}520$. If we keep the four kinds together (all kiwis next to each other, and so on) there are $4! = 24$ arrangements of the kinds that keep them together. So
$$P(\text{all of same kind together}) = \frac{24}{2{,}522{,}520}$$
$$\approx 9.514 \times 10^{-6}.$$

8.3 Exercises

1. There are $C(11,3)$ samples of 3 apples.
$$C(11,3) = \frac{11 \cdot 10 \cdot 9}{3 \cdot 2 \cdot 1} = 165$$

There are $C(7,3)$ samples of 3 red apples.

$$C(7,3) = \frac{7 \cdot 6 \cdot 5}{3 \cdot 2 \cdot 1} = 35$$

Thus,

$$P(\text{all red apples}) = \frac{35}{165} = \frac{7}{33}.$$

3. There are $C(4,2)$ samples of 2 yellow apples.

$$C(4,2) = \frac{4 \cdot 3}{2 \cdot 1} = 6$$

There are $C(7,1) = 7$ samples of 1 red apple. Thus, there are $6 \cdot 7 = 42$ samples of 3 in which 2 are yellow and 1 red. Thus,

$$P(\text{2 yellow and 1 red apple}) = \frac{42}{165} = \frac{14}{55}.$$

For Exercises 5 through 9 the number of possible 5-member committees is $C(20,5) = 15{,}504.$

5. $P(\text{all men}) = \dfrac{C(9,5)}{C(20,5)}$

$$= \frac{126}{15{,}504} \approx 0.008127$$

7. $P(\text{3 men, 2 women}) = \dfrac{C(9,3)\,C(11,2)}{C(20,5)}$

$$= \frac{4620}{15{,}504} \approx 0.2980$$

9. $P(\text{at least 4 women})$

$$= P(\text{4 women}) + P(\text{5 women})$$

$$= \frac{C(11,4)\,C(9,1) + C(11,5)\,C(9,0)}{C(20,5)}$$

$$= \frac{3532}{15{,}504} \approx 0.2214$$

11. The number of 2-card hands is

$$C(52,2) = \frac{52 \cdot 51}{2 \cdot 1} = 1326.$$

13. There are $C(52,2) = 1326$ different 2-card hands. The number of 2-card hands with exactly one ace is

$$C(4,1)\,C(48,2) = 4 \cdot 48 = 192.$$

The number of 2-card hands with two aces is

$$C(4,2) = 6.$$

Thus there are 198 hands with at least one ace. Therefore,

$P(\text{the 2-card hand contains an ace})$

$$= \frac{198}{1326} = \frac{33}{221} \approx 0.149.$$

15. There are $C(52,2) = 1326$ different 2-card hands. There are $C(13,2) = 78$ ways to get a 2-card hand where both cards are of a single named suit, but there are 4 suits to choose from. Thus,

$P(\text{two cards of same suit})$

$$= \frac{4 \cdot C(13,2)}{C(52,2)} = \frac{312}{1326} = \frac{52}{221} \approx 0.235.$$

17. There are $C(52,2) = 1326$ different 2-card hands. There are 12 face cards in a deck, so there are 40 cards that are not face cards. Thus,

$P(\text{no face cards})$

$$= \frac{C(40,2)}{C(52,2)} = \frac{780}{1326} = \frac{130}{221} \approx 0.588.$$

19. There are 26 choices for each slip pulled out, and there are 5 slips pulled out, so there are

$$26^5 = 11{,}881{,}376$$

different "words" that can be formed from the letters. If the "word" must be "chuck," there is only one choice for each of the 5 letters (the first slip must contain a "c," the second an "h," and so on). Thus,

$P(\text{word is "chuck"})$

$$= \frac{1^5}{26^5} = \left(\frac{1}{26}\right)^5 \approx 8.417 \times 10^{-8}.$$

21. There are $26^5 = 11{,}881{,}376$ different "words" that can be formed. If the "word" is to have no repetition of letters, then there are 26 choices for the first letter, but only 25 choices for the second (since the letters must all be different), 24 choices for the third, and so on. Thus,

$P(\text{all different letters})$

$$= \frac{26 \cdot 25 \cdot 24 \cdot 23 \cdot 22}{26^5}$$

$$= \frac{1 \cdot 25 \cdot 24 \cdot 23 \cdot 22}{26^4}$$

$$= \frac{303{,}600}{456{,}976}$$

$$= \frac{18{,}975}{28{,}561} \approx 0.664.$$

25. P(at least 2 presidents have the same birthday)

$= 1 - P$(no 2 presidents have the

same birthday)

The number of ways that 43 people can have the same or different birthdays is $(365)^{43}$. The number of ways that 43 people can have all different birthdays is the number of permutations of 365 things taken 43 at a time or $P(365, 43)$. Thus,

P(at least 2 presidents have the same birthday)

$= 1 - \dfrac{P(365, 43)}{365^{43}}$.

(Be careful to realize that the symbol P is sometimes used to indicate permutations and sometimes used to indicate probability; in this solution, the symbol is used both ways.)

27. Since there are 435 members of the House of Representatives, and there are only 365 days in a year, it is a certain event that at least 2 people will have the same birthday. Thus,

P(at least 2 members have the same birthday) $= 1$.

29. P(matched pair)

$= P$(2 black or 2 brown or 2 blue)

$= P$(2 black) $+ P$(2 brown) $+ 2$(2 blue)

$= \dfrac{C(9, 2)}{C(17, 2)} + \dfrac{C(6, 2)}{C(17, 2)} + \dfrac{C(2, 2)}{C(17, 2)}$

$= \dfrac{36}{136} + \dfrac{15}{136} + \dfrac{1}{136}$

$= \dfrac{52}{136} = \dfrac{13}{34}$

31. There are 6 letters so the number of possible spellings (counting duplicates) is $6! = 720$. Since the letter l is repeated 2 times and the letter t is repeated 2 times, the spelling "little" will occur $2! 2! = 4$ times. The probability that "little" will be spelled is $\frac{4}{720} = \frac{1}{180}$.

33. Each of the 4 people can choose to get off at any one of the 7 floors, so there are 7^4 ways the four people can leave the elevator. The number of ways the people can leave at different floors is the number of permutations of 7 things (floors) taken 4 at a time or

$P(7, 4) = 7 \cdot 6 \cdot 5 \cdot 4 = 840.$

The probability that no 2 passengers leave at the same floor is

$\dfrac{P(7, 4)}{7^4} = \dfrac{840}{2401} \approx 0.3499.$

Thus, the probability that at least 2 passengers leave at the same floor is

$1 - 0.3499 = 0.6501.$

(Note the similarity of this problem and the "birthday problem.")

35. P(at least one \$100-bill)

$= P$(1 \$100-bill) $+ P$(2 \$100-bills)

$= \dfrac{C(2, 1) C(4, 1)}{C(6, 2)} + \dfrac{C(2, 2) C(4, 0)}{C(6, 2)}$

$= \dfrac{8}{15} + \dfrac{1}{15} = \dfrac{9}{15} = \dfrac{3}{5}$

P(no \$100-bill) $= \dfrac{C(2, 0) C(4, 2)}{C(6, 2)} = \dfrac{6}{15} = \dfrac{2}{5}$

It is more likely to get at least one \$100-bill.

37. There are $C(9, 2)$ possible ways to choose 2 nondefective typewriters out of the $C(11, 2)$ possible ways of choosing any 2. Thus,

P(no defective) $= \dfrac{C(9, 2)}{C(11, 2)} = \dfrac{36}{55}.$

39. There are $C(9, 4)$ possible ways to choose 4 nondefective typewriters out of the $C(11, 4)$ possible ways of choosing any 4. Thus,

P(no defective) $= \dfrac{C(9, 4)}{C(11, 4)} = \dfrac{126}{330} = \dfrac{21}{55}.$

41. There are $C(12, 5) = 792$ ways to pick a sample of 5. It will be shipped if all 5 are good. There are $C(10, 5) = 252$ ways to pick 5 good ones, so

P(all good) $= \dfrac{252}{792} = \dfrac{7}{22} \approx 0.318.$

43. P(not Scottsdale customer)

$= \dfrac{\left[\begin{array}{l}\text{number of choices of 4 out of the}\\ \text{5 non-Scottsdale customers}\end{array}\right]}{\text{number of choices of 4 out of the 6 customers}}$

$= \dfrac{C(5, 4)}{C(6, 4)} = \dfrac{5}{15} = \dfrac{1}{3}$

45. There are 20 people in all, so the number of possible 5-person committees is $C(20, 5) = 15,504$. Thus, in parts (a)-(g), $n(S) = 15,504$.

(a) There are $C(10, 3)$ ways to choose the 3 men and $C(10, 2)$ ways to choose the 2 women. Thus,

$$P(\text{3 men and 2 women})$$
$$= \frac{C(10, 3)C(10, 2)}{C(20, 5)} = \frac{120 \cdot 45}{15,504}$$
$$= \frac{225}{646} \approx 0.348.$$

(b) There are $C(6, 3)$ ways to chose the 3 Miwoks and $C(9, 2)$ ways to choose the 2 Pomos. Thus,

$$P(\text{exactly 3 Miwoks and 2 Pomos})$$
$$= \frac{C(6, 3)C(9, 2)}{C(20, 5)} = \frac{20 \cdot 36}{15,504}$$
$$= \frac{15}{323} \approx 0.046.$$

(c) Choose 2 of the 6 Miwoks, 2 of the 5 Hoopas, and 1 of the 9 Pomos. Thus,

$$P(\text{2 Miwoks, 2 Hoopas, and a Pomo})$$
$$= \frac{C(6, 2)C(5, 2)C(9, 1)}{C(20, 5)} = \frac{15 \cdot 10 \cdot 9}{15,504}$$
$$= \frac{225}{2584} \approx 0.087.$$

(d) There cannot be 2 Miwoks, 2 Hoopas, and 2 Pomos, since only 5 people are to be selected. Thus,

$$P(\text{2 Miwoks, 2 Hoopas, and 2 Pomos}) = 0.$$

(e) Since there are more women then men, there must be 3, 4, or 5 women.

$$P(\text{more women than men})$$
$$= \frac{C(10, 3)C(10, 2) + C(10, 4)C(10, 1) + C(10, 5)C(10, 0)}{C(20, 5)}$$
$$= \frac{7752}{15,504} = \frac{1}{2}$$

(f) Choose 3 of 5 Hoopas and any 2 of the 15 non-Hoopas.

$$P(\text{exactly 3 Hoopas})$$
$$= \frac{C(5, 3)C(15, 2)}{C(20, 5)}$$
$$= \frac{175}{2584} \approx 0.068$$

(g) There can be 2 to 5 Pomos, the rest chosen from the 11 non-Pomos.

$$P(\text{at least 2 Pomos})$$
$$= \frac{C(9, 2)C(11, 3) + C(9, 3)C(11, 2) + C(9, 4)C(11, 1) + C(9, 5)C(11, 0)}{C(20, 5)}$$
$$= \frac{503}{646} \approx 0.779$$

47. There are 21 books, so the number of selection of any 6 books is

$$C(21, 6) = 54,264.$$

(a) The probability that the selection consisted of 3 Hughes and 3 Morrison books is

$$\frac{C(9, 3)C(7, 3)}{C(21, 6)} = \frac{85 \cdot 35}{54,264}$$
$$= \frac{2940}{54,264} \approx 0.0542.$$

(b) A selection containing exactly 4 Baldwin books will contain 2 of the 16 books by the other authors, so the probability is

$$\frac{C(5, 4)C(16, 2)}{C(21, 6)} = \frac{5 \cdot 120}{54,264}$$
$$= \frac{600}{54,264} \approx 0.0111.$$

(c) The probability of a selection consisting of 2 Hughes, 3 Baldwin, and 1 Morrison book is

$$\frac{C(9, 2)C(5, 3)C(7, 1)}{C(21, 6)} = \frac{30 \cdot 10 \cdot 7}{54,264}$$
$$= \frac{2520}{54,264}$$
$$\approx 0.0464.$$

(d) A selection consisting of at least 4 Hughes books may contain 4, 5, or 6 Hughes books, with any remaining books by the other authors. Therefore, the probability is

$$\frac{\left(\begin{array}{c} C(9, 4)C(12, 2) + C(9, 5)C(12, 1) \\ + C(9, 6)C(12, 0) \end{array} \right)}{C(21, 6)}$$
$$= \frac{126 \cdot 66 + 126 \cdot 12 + 84}{54,264}$$
$$= \frac{8316 + 1512 + 84}{54,264}$$
$$= \frac{9912}{54,264} \approx 0.1827.$$

(e) Since there are 9 Hughes books and 5 Baldwin books, there are 14 books written by males. The probability of a selection with exactly 4 books written by males is

$$\frac{C(14,2)C(7,2)}{C(21,6)} = \frac{1001 \cdot 21}{54,264}$$

$$= \frac{21,021}{54,264} \approx 0.3874.$$

(f) A selection with no more than 2 books written by Baldwin may contain 0, 1, or 2 books by Baldwin, with the remaining books by the other authors. Therefore, the probability is

$$\frac{C(5,0)C(16,6) + C(5,1)C(16,5) + C(5,2)C(16,4)}{C(21,6)}$$

$$= \frac{8008 + 5 \cdot 4368 + 10 \cdot 1820}{54,264}$$

$$= \frac{8008 + 21,840 + 18,200}{54,264}$$

$$= \frac{48,048}{54,264} \approx 0.8854.$$

49. A straight flush could start with an ace, 2, 3, 4, …, 7, 8, or 9. This gives 9 choices in each of 4 suits, so there are 36 choices in all. Thus,

$$P(\text{straight flush}) = \frac{36}{C(52,2)} = \frac{36}{2,598,960}$$

$$\approx 0.00001385$$

$$= 1.385 \times 10^{-5}.$$

51. A straight could start with an ace, 2, 3, 4, 5, 6, 7, 8, 9, or 10 as the low card, giving 40 choices. For each succeeding card, only the suit may be chosen. Thus, the number of straights is

$$40 \cdot 4^4 = 10,240.$$

But this also counts the straight flushes, of which there are 36 (see Exercise 49), and the 4 royal flushes. There are thus 10,200 straights that are not also flushes, so

$$P(\text{straight}) = \frac{10,200}{2,598,960} \approx 0.0039.$$

53. There are 13 different values of cards and 4 cards of each value. Choose 2 values out of the 13 for the values of the pairs. The number of ways to select the 2 values is $C(13,2)$. The number of ways to select a pair for each value is $C(4,2)$. There are $52 - 8 = 44$ cards that are neither of these 2 values, so the number of ways to select the fifth card is $C(44,1)$. Thus,

$$P(\text{two pairs}) = \frac{C(13,2)\, C(4,2)\, C(4,2)\, C(44,1)}{C(52,5)}$$

$$= \frac{123,552}{2,598,960} \approx 0.0475.$$

55. There are $C(52,13)$ different 13-card bridge hands. Since there are only 13 hearts, there is exactly one way to get a bridge hand containing only hearts. Thus,

$$P(\text{only hearts}) = \frac{1}{C(52,13)} \approx 1.575 \cdot 10^{-12}.$$

57. There are $C(4,2)$ ways to obtain 2 aces, $C(4,2)$ ways to obtain 2 kings, and $C(44,9)$ ways to obtain the remaining 9 cards. Thus,

$$P(\text{exactly 2 aces and exactly 2 kings})$$
$$= \frac{C(4,2)\, C(4,2)\, C(44,9)}{C(52,13)} \approx 0.0402.$$

For Exercises 59 through 65, use the fact that the number of 7-card selections is $C(52,7)$.

59. Pick a kind for the pair: 13 choices
Choose 2 suits out of the 4 suits for this kind: $C(4,2)$
Then pick 5 kinds out of the 12 remaining: $C(12,5)$
For each of these 5 kinds, pick one of the 4 suits: 4^5
The product of these factors gives the numerator and the denominator is $C(52,7)$.

$$\frac{13 \cdot C(4,2) \cdot C(12,5) \cdot 4^5}{C(52,7)} \approx 0.4728$$

61. Pick a kind for the three-of-a-kind: 13 choices
Choose 3 suits out of the 4 suits for this kind: $C(4,3)$
There are now 4 cards remaining which must be 4 of the 12 kinds remaining: $C(12,4)$
For each of these 4 kinds we pick one of the 4 suits: 4^4
The product of these factors gives the numerator and the denominator is $C(52,7)$.

$$\frac{13 \cdot C(4,3) \cdot C(12,4) \cdot 4^4}{C(52,7)} \approx 0.0493$$

63. Pick a suit: 4

Now we either get exactly 5 of this suit and 2 of other suits: $C(13,5) \cdot C(39,2)$

…or exactly 6 of this suit and 1 of another suit: $C(13,6) \cdot C(39,1)$

…or all 7 of our chosen suit: $C(13,7)$. We now add these options over our usual denominator.

$$\frac{4 \cdot [C(13,5) \cdot C(39,2) + C(13,6) \cdot C(39,1) + C(13,7)]}{C(52,7)}$$

$$\approx 0.0306$$

65. We need at least 3 hearts out of 5 cards. This can happen in three ways:

3 hearts, 2 non-hearts: $C(11,3) \cdot C(39,2)$

4 hearts, 1 non-heart: $C(11,4) \cdot C(39,1)$

5 hearts: $C(11,5)$

Add these options over the usual denominator.

$$\frac{C(11,3) \cdot C(39,2) + C(11,4) \cdot C(39,1) + C(11,5)}{C(52,7)}$$

$$\approx 0.0640$$

67. To find the probability of picking 5 of the 6 lottery numbers correctly, we must recall that the total number of ways to pick the 6 lottery numbers is $C(99, 6) = 1,120,529,256$. To pick 5 of the 6 winning numbers, we must also pick 1 of the 93 losing numbers. Therefore, the number of ways of picking 5 of the 6 winning numbers is

$$C(6,5)\,C(93,1) = 558.$$

Thus, the probability of picking 5 of the 6 numbers correctly is

$$\frac{C(6,5)\,C(93,1)}{C(99,6)} \approx 4.980 \times 10^{-7}.$$

69. The probability of picking six numbers out of 49 is

$$P(6 \text{ out of } 49) = \frac{1}{C(49,6)} = \frac{1}{13,983,816}$$

The probability of picking of picking five numbers out of 52 is

$$P(5 \text{ out of } 52) = \frac{1}{C(52,5)} = \frac{1}{2,598,960}$$

The probability of winning the lottery when picking five out of 52 is higher.

71. (a) The number of ways to select 6 numbers between 1 and 49 is $C(49,6) = 13,983,816$. The number of ways to select 3 of the 6 numbers, while not selecting the bonus number is

$$C(6,3)\,C(42,3) = 20 \cdot 11,480$$
$$= 229,600.$$

The probability of winning fifth prize is

$$\frac{229,600}{13,983,816} \approx 0.01642.$$

(b) The number of ways to select 2 of the 6 numbers plus the bonus number is

$$C(6,2)\,C(1,1)\,C(42,3) = 15 \cdot 1 \cdot 11,480$$
$$= 172,200.$$

The probability of winning sixth prize is

$$\frac{172,200}{13,983,816} \approx 0.01231.$$

73. (a) There were 28 games played in the season, since the numbers in the "Won" column have a sum of 28 (and the numbers in the "Lost" column have a sum of 28).

(b) Assuming no ties, each of the 28 games had 2 possible outcomes; either Team A won and Team B lost, or else Team A lost and Team B won. By the multiplication principle, this means that there were

$$2^{28} = 268,435,456$$

different outcomes possible.

(c) Any one of the 8 teams could have been the one that won all of its games, any one of the remaining 7 teams could have been the one that won all but one of its games, and so on, until there is only one team left, and it is the one that lost all of its games. By the multiplication principle, this means that there were

$$8! = 8 \cdot 7 \cdot 6 \cdot 5 \cdot 4 \cdot 3 \cdot 2 \cdot 1 = 40,320$$

different "perfect progressions" possible.

(d) Thus,

$P(\text{"perfect progression" in an 8-team league})$

$$= \frac{8!}{2^{28}} \approx 0.0001502 = 1.502 \times 10^{-4}.$$

(e) If there are n teams in the league, then the "Won" column will begin with $n-1$, followed by $n-2$, then $n-3$, and so on down to 0. It can be shown that the sum of these n numbers is $\frac{n(n-1)}{2}$, so there are $2^{n(n-1)/2}$ different win/lose progressions possible. The n teams can be ordered in $n!$

different ways, so there are $n!$ different "perfect progressions" possible. Thus,

P("perfect progression" in an n-team league)

$$= \frac{n!}{2^{n(n-1)/2}}.$$

75. (a) There are only 4 ways to win in just 4 calls: the 2 diagonals, the center column, and the center row. There are $C(75,4)$ combinations of 4 numbers that can occur. The probability that a person will win bingo after just 4 numbers are called is

$$\frac{4}{C(75,4)} \approx 3.291 \times 10^{-6}.$$

(b) There is only 1 way to get an L. It can occur in as few as 9 calls. There are $C(75,9)$ combinations of 9 numbers that can occur in 9 calls, so the probability of an L in 9 calls is $\frac{1}{C(75,9)} \approx 7.962 \times 10^{-12}$.

(c) There is only 1 way to get an X-out. It can ocur in as few as 8 calls. There are $C(75,8)$ combinations of 8 numbers that can occur. The probability that an X-out occurs in 8 calls is $\frac{1}{C(75,8)} \approx 5.927 \times 10^{-11}$.

(d) Four columns contain a permutation of 15 numbers taken 5 at a time. One column contains a permutation of 15 numbers taken 4 at a time. The number of distinct cards is

$$P(15,5)^4 \cdot P(15,4) \approx 5.524 \times 10^{26}.$$

8.4 Binomial Probability

Your Turn 1

$$P(\text{exactly 2 of 6}) = C(6,2)(0.59)^2(0.41)^4$$
$$= 15(0.3481)(0.0283)$$
$$\approx 0.1475$$

Your Turn 2

$$P(\text{4 heads in 8 tosses}) = C(8,4)\left(\frac{1}{2}\right)^4\left(\frac{1}{2}\right)^4$$

$$= 70\left(\frac{1}{2}\right)^8$$

$$\approx 0.2734$$

Your Turn 3

$$P(\text{2 or 3 defective of 15})$$
$$= C(15,2)(0.01)^2(0.99)^{13} + C(15,3)(0.01)^3(0.99)^{12}$$
$$\approx 0.009214 + 0.000403$$
$$= 0.009617$$

Your Turn 4

$$P(\text{at least one incorrect charge})$$
$$= 1 - P(\text{no incorrect charges in 4})$$
$$= 1 - C(4,0)(0.29)^0(0.71)^4$$
$$\approx 1 - 0.2541$$
$$= 0.7459$$

Your Turn 5

$$P(\text{at most 3 incorrect charges in 6})$$
$$= P(0) + P(1) + P(2) + P(3)$$
$$= C(6,0)(0.29)^0(0.71)^6 + C(6,1)(0.29)^1(0.71)^5$$
$$\quad + C(6,2)(0.29)^2(0.71)^4 + C(6,3)(0.29)^3(0.71)^3$$
$$\approx 0.9372$$

8.4 Exercises

1. This is a Bernoulli trial problem with

$P(\text{success}) = P(\text{girl}) = \frac{1}{2}$. The probability of exactly x successes in n trials is

$$C(n,x)p^x(1-p)^{n-x},$$

where p is the probability of success in a single trial. We have $n = 5, x = 2$, and $p = \frac{1}{2}$

Note that

$$1 - p = 1 - \frac{1}{2} = \frac{1}{2}.$$

$P(\text{exactly 2 girls and 3 boys})$

$$= C(5,2)\left(\frac{1}{2}\right)^2\left(\frac{1}{2}\right)^3$$

$$= \frac{10}{32} = \frac{5}{16} \approx 0.313$$

3. We have
$$n = 5, x = 0, p = \tfrac{1}{2}, \text{ and } 1 - p = \tfrac{1}{2}.$$

$$P(\text{no girls}) = C(5,0)\left(\frac{1}{2}\right)^0\left(\frac{1}{2}\right)^5$$

$$= \frac{1}{32} \approx 0.031$$

5. "At least 4 girls" means either 4 or 5 girls.

$$P(\text{at least 4 girls})$$

$$= C(5,4)\left(\frac{1}{2}\right)^4\left(\frac{1}{2}\right)^1 + C(5,5)\left(\frac{1}{2}\right)^5\left(\frac{1}{2}\right)^0$$

$$= \frac{5}{32} + \frac{1}{32} = \frac{6}{32} = \frac{3}{16} \approx 0.188$$

7. $P(\text{no more than 3 boys})$
$$= 1 - P(\text{at least 4 boys})$$
$$= 1 - P(\text{4 boys or 5 boys})$$
$$= 1 - [P(\text{4 boys}) + P(\text{5 boys})]$$
$$= 1 - \left(\frac{5}{32} + \frac{1}{32}\right)$$
$$= 1 - \frac{6}{32}$$
$$= 1 - \frac{3}{16} = \frac{13}{16} \approx 0.813$$

9. On one roll, $P(1) = \tfrac{1}{6}$. We have $n = 12$,
$x = 12$, and $p = \tfrac{1}{6}$. Note that $1 - p = \tfrac{5}{6}$.
Thus,

$$P(\text{exactly 12 ones}) = C(12,12)\left(\frac{1}{6}\right)^{12}\left(\frac{5}{6}\right)^0$$

$$\approx 4.594 \times 10^{-10}.$$

11. $$P(\text{exactly 1 one }) = C(12,1)\left(\frac{1}{6}\right)^1\left(\frac{5}{6}\right)^{11}$$

$$\approx 0.2692$$

13. "No more than 3 ones" means 0, 1, 2, or 3 ones.
Thus,

$P(\text{no more than 3 ones})$
$$= P(\text{0 ones}) + P(\text{1 one}) + P(\text{2 ones}) + P(\text{3 ones})$$

$$= C(12,0)\left(\frac{1}{6}\right)^0\left(\frac{5}{6}\right)^{12} + C(12,1)\left(\frac{1}{6}\right)^1\left(\frac{5}{6}\right)^{11}$$

$$+ C(12,2)\left(\frac{1}{6}\right)^2\left(\frac{5}{6}\right)^{10} + C(12,3)\left(\frac{1}{6}\right)^3\left(\frac{5}{6}\right)^9$$

$$\approx 0.8748.$$

For Exercises 15 and 17 we have
$$n = 6, p = \tfrac{1}{4}, \text{ and } 1 - p = \tfrac{3}{4}.$$

15. $$P(\text{all heads}) = C(6,6)\left(\frac{1}{4}\right)^6\left(\frac{3}{4}\right)^0$$

$$= \frac{1}{4096} \approx 0.0002441$$

17. $P(\text{no more than 3 heads})$
$$= P(\text{0 heads}) + P(\text{1 head}) + P(\text{2 heads}) + P(\text{3 heads})$$

$$= C(6,0)\left(\frac{1}{4}\right)^0\left(\frac{3}{4}\right)^6 + C(6,1)\left(\frac{1}{4}\right)^1\left(\frac{3}{4}\right)^5$$

$$+ C(6,2)\left(\frac{1}{4}\right)^2\left(\frac{3}{4}\right)^4 + C(6,3)\left(\frac{1}{4}\right)^3\left(\frac{3}{4}\right)^3$$

$$= \frac{3942}{4096} \approx 0.9624$$

21. $C(n,r) + C(n,r+1)$
$$= \frac{n!}{r!(n-r)!} + \frac{n!}{(r+1)![n-(r+1)]!}$$
$$= \frac{n!(r+1)}{r!(r+1)(n-r)!}$$
$$+ \frac{n!(n-r)}{(r+1)![n-(r+1)]!(n-r)}$$
$$= \frac{rn! + n!}{(r+1)!(n-r)!} + \frac{n(n!) - rn!}{(r+1)!(n-r)!}$$
$$= \frac{rn! + n! + n(n!) - rn!}{(r+1)!(n-r)!}$$
$$= \frac{n!(n+1)}{(r+1)!(n-r)!}$$
$$= \frac{(n+1)!}{(r+1)![(n+1)-(r+1)]!}$$
$$= C(n+1, r+1)$$

23. Since the potential callers are not likely to have birthdates that are distributed evenly throughout the twentieth century, the use of binomial probabilities is not applicable and thus, the probabilities that are computed are not correct.

For Exercises 25 and 27 we define a success to be the event that a customer is charged incorrectly. In this situation, $n = 15, p = \tfrac{1}{30}$ and $1 - p = \tfrac{29}{30}$.

25. $P(\text{0 incorrect charges})$

$$= C(15,0)\left(\frac{1}{30}\right)^0\left(\frac{29}{30}\right)^{15}$$

$$\approx 0.6014$$

27. P(at least 2 incorrect charges)

$= 1 - P(0 \text{ or } 1 \text{ incorrect charges})$

$= 1 - C(15,0)\left(\dfrac{1}{30}\right)^0\left(\dfrac{29}{30}\right)^{15}$

$\qquad - C(15,1)\left(\dfrac{1}{30}\right)^1\left(\dfrac{29}{30}\right)^{14}$

≈ 0.0876

For Exercises 29 and 31, we define a success to be the event that the family hardly ever pays off the balance. In this situation, $n = 20, p = 0.254$ and $1 - p = 0.746$.

29. $P(6) = C(20,6)(0.254)^6(0.746)^{14}$

$\qquad \approx 0.1721$

31.

P(at least 4)

$= 1 - C(20,0)(0.254)^0(0.746)^{20} - C(20,1)(0.254)^1(0.746)^{19}$

$\quad - C(20,2)(0.254)^2(0.746)^{18} - C(20,3)(0.254)^3(0.746)^{17}$

≈ 0.7868

33. We have $n = 6, x = 2, p = \frac{1}{5}$, and $1 - p = \frac{4}{5}$. Thus,

$$P(\text{exactly 2 correct}) = C(6,2)\left(\frac{1}{5}\right)^2\left(\frac{4}{5}\right)^4$$

$$\approx 0.2458.$$

35. We have

P(at least 4 correct)

$= P(4 \text{ correct}) + P(5 \text{ correct}) + P(6 \text{ correct})$

$= C(6,4)\left(\dfrac{1}{5}\right)^4\left(\dfrac{4}{5}\right)^2 + C(6,5)\left(\dfrac{1}{5}\right)^5\left(\dfrac{4}{5}\right)^1$

$\qquad\qquad + C(6,6)\left(\dfrac{1}{5}\right)^6\left(\dfrac{4}{5}\right)^0$

$\approx 0.0170.$

37. $n = 20, p = 0.05, x = 0$

$P(0 \text{ defective transistors}) = C(20,0)(0.05)^0(0.95)^{20}$

$\qquad \approx 0.3585$

39. Let success mean producing a defective item. Then we have $n = 75, p = 0.05$, and $1 - p = 0.95$.

(a) If there are exactly 5 defective items, then $x = 5$. Thus,

P(exactly 5 defective)

$= C(75,5)(0.05)^5(0.95)^{70}$

$\approx 0.1488.$

(b) If there are no defective items, then $x = 0$. Thus,

P(none defective)

$= C(75,0)(0.05)^0(0.95)^{75}$

$\approx 0.0213.$

(c) If there is at least 1 defective item, then we are interested in $x \geq 1$. We have

P(at least one defective)

$= 1 - P(x = 0)$

$\approx 1 - 0.021$

$= 0.9787.$

41. (a) Since 80% of the "good nuts" are good, 20% of the "good nuts" are bad. Let's let success represent "getting a bad nut." Then 0.2 is the probability of success in a single trial. The probability of 8 successes in 20 trials is

$C(20,8)(0.2)^8(1 - 0.2)^{20-8}$

$= C(20,8)(0.2)^8(0.8)^{12}$

≈ 0.0222

(b) Since 60% of the "blowouts" are good, 40% of the "blowouts" are bad. Let's let success represent "getting a bad nut." Then 0.4 is the probability of success in a single trial. The probability of 8 successes in 20 trials is

$C(20,8)(0.4)^8(1 - 0.4)^{20-8}$

$= C(20,8)(0.4)^8(0.6)^{12}$

≈ 0.1797

(c) The probability that the nuts are "blowouts" is

$$\dfrac{\left(\begin{array}{l}\text{Probability of "Blowouts"}\\ \text{having 8 bad nuts out of 20}\end{array}\right)}{\left(\begin{array}{l}\text{Probability of "Good Nuts" or "Blowouts"}\\ \text{having 8 bad nuts of 20}\end{array}\right)}$$

$$= \dfrac{0.3\left[C(20,8)(0.4)^8(0.6)^{12}\right]}{0.7\left[C(20,8)(0.2)^8(0.8)^{12}\right] + 0.3\left[C(20,8)(0.4)^8(0.6)^{12}\right]}$$

$\approx 0.7766.$

43. $n = 15, p = 0.85$

$$P(\text{all } 15) = C(15,15)(0.85)^{15}(0.15)^0$$
$$\approx 0.0874$$

45. $n = 15, p = 0.85$

$$P(\text{not all}) = 1 - P(\text{all } 15)$$
$$= 1 - C(15,15)(0.85)^{15}(0.15)^0$$
$$\approx 0.9126$$

47. $n = 100, p = 0.012, x = 2$

$$P(\text{exactly 2 sets of twins})$$
$$= C(100,2)(0.012)^2(0.988)^{98}$$
$$\approx 0.2183$$

49.   We have $n = 10{,}000$,

$p = 2.5 \cdot 10^{-7} = 0.00000025$, and
$1 - p = 0.99999975$. Thus,

$$P(\text{at least 1 mutation occurs})$$
$$= 1 - P(\text{none occurs})$$
$$= 1 - C(10{,}000, 0)p^0(1 - p)^{10{,}000}$$
$$= 1 - (0.99999975)^{10{,}000}$$
$$\approx 0.0025.$$

51. $n = 53, p = 0.042$

(a) The probability that exactly 5 men are color-blind is

$$P(5) = C(53,5)(0.042)^5(0.958)^{48}$$
$$\approx 0.0478.$$

(b) The probability that no more than 5 men are color-blind is

$P(\text{no more than 5 men are color-blind})$
$$= C(53,0)(0.042)^0(0.958)^{53}$$
$$+ C(53,1)(0.042)^1(0.958)^{52}$$
$$+ C(53,2)(0.042)^2(0.958)^{51}$$
$$+ C(53,3)(0.042)^3(0.958)^{50}$$
$$+ C(53,4)(0.042)^4(0.958)^{49}$$
$$+ C(53,5)(0.042)^5(0.958)^{48}$$
$$\approx 0.9767.$$

(c) The probability that at least 1 man is color-blind is

$1 - P(0 \text{ men are color-blind})$
$$= 1 - C(53,0)(0.042)^0(0.958)^{53}$$
$$\approx 0.8971.$$

53.   **(a)** Since the probability of a particular band matching is 1 in 4 or $\frac{1}{4}$, the probability that 5 bands match is $\left(\frac{1}{4}\right)^5 = \frac{1}{1024}$ or 1 chance in 1024.

(b) The probability that 20 bands match is $\left(\frac{1}{4}\right)^{20} \approx \frac{1}{1.1 \times 10^{12}}$ or about 1 chance in 1.1×10^{12}.

(c) If 20 bands are compared, the probability that 16 or more bands match is

$P(\text{at least 16})$
$$= P(16) + P(17) + P(18) + P(19) + P(20)$$
$$= C(20,16)\left(\frac{1}{4}\right)^{16}\left(\frac{3}{4}\right)^4 + C(20,17)\left(\frac{1}{4}\right)^{17}\left(\frac{3}{4}\right)^3$$
$$+ C(20,18)\left(\frac{1}{4}\right)^{18}\left(\frac{3}{4}\right)^2 + C(20,19)\left(\frac{1}{4}\right)^{19}\left(\frac{3}{4}\right)^1$$
$$+ C(20,20)\left(\frac{1}{4}\right)^{20}\left(\frac{3}{4}\right)^0$$
$$= 4845\left(\frac{1}{4}\right)^{16}\left(\frac{3}{4}\right)^4 + 1140\left(\frac{1}{4}\right)^{17}\left(\frac{3}{4}\right)^3$$
$$+ 190\left(\frac{1}{4}\right)^{18}\left(\frac{3}{4}\right)^2 + 20\left(\frac{1}{4}\right)^{19}\left(\frac{3}{4}\right)^1 + \left(\frac{1}{4}\right)^{20}\cdot 1$$
$$= \left(\frac{1}{4}\right)^{16}\left[4845\left(\frac{81}{256}\right) + 1140\left(\frac{1}{4}\right)\left(\frac{27}{64}\right) + 190\left(\frac{1}{16}\right)\left(\frac{9}{16}\right)\right.$$
$$\left. + 20\left(\frac{1}{64}\right)\left(\frac{3}{4}\right) + \frac{1}{256}\right]$$
$$= \left(\frac{1}{4}\right)^{16}\left(\frac{392{,}445 + 30{,}780 + 1710 + 60 + 1}{256}\right)$$
$$= \frac{424{,}996}{4^{20}} = \frac{1}{\frac{4^{20}}{424{,}996}}$$
$$\approx \frac{1}{2{,}587{,}110}$$

or about 1 chance in 2.587×10^6.

55. $n = 4800, p = 0.001$

$P(\text{more than } 1)$

$= 1 - P(1) - P(0)$

$= 1 - C(4800,1)(0.001)^1(0.999)^{4799}$

$\qquad - C(4800,0)(0.001)^0(0.999)^{4800}$

≈ 0.9523

57.   First, find the probability that one group of ten has at least 9 participants complete the study.
$n = 10, P = 0.8,$

$P(\text{at least 9 complete}) = P(9) + P(10)$

$\qquad = C(10,9)(0.8)^9(0.2)^1$

$\qquad\qquad + C(10,10)(0.8)^{10}(0.2)^0$

$\qquad \approx 0.3758$

The probability that 2 or more drop out in one group is $1 - 0.3758 = 0.6242$. Thus, the probability that at least 9 participants complete the study in one of the two groups, but not in both groups, is

$(0.3758)(0.6242) + (0.6242)(0.3758) \approx 0.469.$

The answer is e.

59. $n = 12, x = 7, p = 0.83$

$P(7) = C(12,7)(0.83)^7(0.17)^5 \approx 0.0305$

61. $n = 12, p = 0.83$

$P(\text{at least 9}) = P(9) + P(10) + P(11) + P(12)$

$= C(12,9)(0.83)^9(0.17)^3 + C(12,10)(0.83)^{10}(0.17)^2$

$\qquad + C(12,11)(0.83)^{11}(0.17)^1 + C(12,12)(0.83)^{12}(0.17)^0$

≈ 0.8676

63. $n = 10, p = 0.322, 1 - p = 0.678$

(a) $P(2) = C(10,2)(0.322)^2(0.678)^8 \approx 0.2083$

(b) $P(3 \text{ or fewer}) = C(10,0)(0.322)^0(0.678)^{10}$

$\qquad\qquad + C(10,1)(0.322)^1(0.678)^9$

$\qquad\qquad + C(10,2)(0.322)^2(0.678)^8$

$\qquad\qquad + C(10,3)(0.322)^3(0.678)^7$

$\qquad \approx 0.5902$

(c) If exactly 5 *do not* belong to a minority, then exactly $10 - 5 = 5$ *do* belong to a minority, and this probability is

$P(5) = C(10,5)(0.322)^5(0.678)^5 \approx 0.1250.$

(d) If 6 or more *do not* belong to a minority, then at most 4 *do* belong to a minority, and this probability is $P(\text{at most } 4)$

$P(\text{at most } 4) = C(10,0)(0.322)^0(0.678)^{10}$

$\qquad\qquad + C(10,1)(0.322)^1(0.678)^9$

$\qquad\qquad + C(10,2)(0.322)^2(0.678)^8$

$\qquad\qquad + C(10,3)(0.322)^3(0.678)^7$

$\qquad\qquad + C(10,4)(0.322)^4(0.678)^6$

≈ 0.8095

65. (a) Using the binomcdf function on a graphing calculator, we find

$P(\text{at least } 30) = 1 - P(29 \text{ or fewer})$

$\qquad = 1 - \text{binomcdf}(40, 0.74, 29)$

$\qquad \approx 1 - 0.4740$

$\qquad = 0.5260$

(b) Using the binomcdf function on a graphing calculator, we find

$P(\text{at least } 30) = 1 - P(29 \text{ or fewer})$

$\qquad = 1 - \text{binomcdf}(40, 0.83, 29)$

$\qquad \approx 1 - 0.0657$

$\qquad = 0.9343$

67. (a) Suppose the National League wins the series in four games. Then they must win all four games and $P = C(4,4)(0.5)^4(0.5)^0 = 0.0625$. Since the probability that the American League wins the series in four games is equally likely, the probability the series lasts four games is $2(0.0625) = 0.125$.

Suppose the National League wins the series in five games. Then they must win exactly three of the previous four games and $P = C(4,3)(0.5)^3(0.5)^1 \cdot (0.5) = 0.125$. Since the probability that the American League wins the series in five games is equally likely, the probability the series lasts five games is $2(0.125) = 0.25$. Suppose the National League wins the series in six games. Then they must win exactly three of the previous five games and $P = C(5,3)(0.5)^3(0.5)^2 \cdot (0.5) = 0.15625$. Since the probability that the American League wins the series in six games is equally likely, the probability the series lasts six games is $2(0.15625) = 0.3125$. Suppose the National League wins the series in seven games. Then they must win exactly three of the previous six games and

$$P = C(6,3)(0.5)^3(0.5)^3 \cdot (0.5) = 0.15625.$$

Since the Probability that the American League wins the series in seven games is equally likely, the probability the series last seven games is $2(0.15625) = 0.3125.$

(b) Suppose the better team wins the series in four games. Then they must win all four games and $P = C(4,4)(0.73)^4(0.27)^0 \approx$ 0.2840. Suppose the other team wins the series in four games. Then they must win all four games and

$$P = C(4,4)(0.27)^4(0.73)^0 \approx 0.0053.$$ The probability the series lasts four games is the sum of two probabilities, 0.2893.

Suppose the better team wins the series in five games. Then they must win exactly three of the previous four games and

$$P = C(4,3)(0.73)^3(0.27)^1 \cdot (0.73) \approx 0.3067.$$
Suppose the other team wins the series in five games. Then they must win exactly three of the previous four games and

$$C(4,3)(0.27)^3(0.73)^1 \cdot (0.27) \approx 0.0155.$$ The probability the series lasts five games is the sum of the two probabilities, 0.3222.

Suppose the better team wins the series in six games. Then they must win exactly three of the previous five games and

$$P = C(5,3)(0.73)^3(0.27)^2 \cdot (0.73) \approx 0.2070.$$
Suppose the other team wins the series in six games. Then they must win exactly three of the previous five games and

$$P = C(5,3)(0.27)^3(0.73)^2 \cdot (0.27) \approx 0.0283.$$
The probability the series lasts six games is the sum of the two probabilities, 0.2353.

Suppose the better team wins the series in seven games. Then they must win exactly three of the previous six games and

$$P = C(6,3)(0.73)^3(0.27)^3 \cdot (0.73) \approx 0.1118.$$
Suppose the other team wins the series in seven games. Then they must win exactly three of the previous six games and

$$P = C(6,3)(0.27)^3(0.73)^3 \cdot (0.27) \approx 0.0413.$$
The probability the series lasts seven games is the sum of the two probabilities, 0.1531.

8.5 Probability Distributions; Expected Value

Your Turn 1

$$P(x = 0) = \frac{C(3,0)C(9,2)}{C(12,2)} = \frac{6}{11}$$

$$P(x = 1) = \frac{C(3,1)C(9,1)}{C(12,2)} = \frac{9}{22}$$

$$P(x = 2) = \frac{C(3,2)C(9,0)}{C(12,2)} = \frac{1}{22}$$

The distribution is shown in the following table:

x	0	1	2
$P(x)$	6/11	9/22	1/22

Your Turn 2

Let the random variable x represent the number of tails.

$$P(x = 0) = C(3,0)\left(\frac{1}{2}\right)^0\left(\frac{1}{2}\right)^3 = \frac{1}{8}$$

$$P(x = 1) = C(3,1)\left(\frac{1}{2}\right)^1\left(\frac{1}{2}\right)^2 = \frac{3}{8}$$

$$P(x = 2) = C(3,2)\left(\frac{1}{2}\right)^2\left(\frac{1}{2}\right)^1 = \frac{3}{8}$$

$$P(x = 3) = C(3,3)\left(\frac{1}{2}\right)^3\left(\frac{1}{2}\right)^0 = \frac{1}{8}$$

The distribution is shown n the following table:

x	0	1	2	3
$P(x)$	1/8	3/8	3/8	1/8

Here is the histogram.

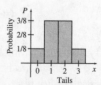

Your Turn 3

The expected payback is

$$955\left(\frac{1}{1000}\right) + 495\left(\frac{1}{1000}\right) + 245\left(\frac{1}{1000}\right) + (-5)\left(\frac{997}{1000}\right)$$

$$= \frac{-3250}{1000}$$

$$= -3.25 \quad \text{or} \quad -\$3.25.$$

Your Turn 4

Let the random variable x represents the number of male engineers in the sample of 5.

$$P(x = 0) = C(5,0)(0.816)^0(0.184)^5 \approx 0.00021$$

$$P(x = 1) = C(5,1)(0.816)^1(0.184)^4 \approx 0.00468$$

$$P(x = 2) = C(5,2)(0.816)^2(0.184)^3 \approx 0.04148$$

$$P(x = 3) = C(5,3)(0.816)^3(0.184)^2 \approx 0.18395$$

$$P(x = 4) = C(5,4)(0.816)^4(0.184)^1 \approx 0.40790$$

$$P(x = 5) = C(5,5)(0.816)^5(0.184)^0 \approx 0.36179$$

$$E(x) \approx (0)(0.0021) + (1)(0.00468) + (2)(0.04148)$$
$$+ (3)(0.18395) + (4)(0.40790) + (5)(0.36179)$$
$$= 4.0803$$

In fact the exact value of the expectation can be computed quickly using the formula $E(x) = np$. For this example, $n = 5$ and $p = 0.816$ so $np = (5)(0.816) = 4.08$.

Your Turn 5

The expected number of girls in a family of a dozen children is $12\left(\frac{1}{2}\right) = 6$.

8.5 Exercises

1. Let x denote the number of heads observed. Then x can take on 0, 1, 2, 3, or 4 as values. The probabilities are as follows.

$$P(x = 0) = C(4,0)\left(\frac{1}{2}\right)^0\left(\frac{1}{2}\right)^4 = \frac{1}{16}$$

$$P(x = 1) = C(4,1)\left(\frac{1}{2}\right)^1\left(\frac{1}{2}\right)^3 = \frac{4}{16} = \frac{1}{4}$$

$$P(x = 2) = C(4,2)\left(\frac{1}{2}\right)^2\left(\frac{1}{2}\right)^2 = \frac{6}{16} = \frac{3}{8}$$

$$P(x = 3) = C(4,3)\left(\frac{1}{2}\right)^3\left(\frac{1}{2}\right)^1 = \frac{4}{16} = \frac{1}{4}$$

$$P(x = 4) = C(4,4)\left(\frac{1}{2}\right)^4\left(\frac{1}{2}\right)^0 = \frac{1}{16}$$

Therefore, the probability distribution is as follows.

Number of Heads	0	1	2	3	4
Probability	$\frac{1}{16}$	$\frac{1}{4}$	$\frac{3}{8}$	$\frac{1}{4}$	$\frac{1}{16}$

3. Let x denote the number of aces drawn. Then x can take on values 0, 1, 2, or 3. The probabilities are as follows.

$$P(x = 0) = C(3,0)\left(\frac{48}{52}\right)\left(\frac{47}{51}\right)\left(\frac{46}{50}\right) \approx 0.7826$$

$$P(x = 1) = C(3,1)\left(\frac{4}{52}\right)\left(\frac{48}{51}\right)\left(\frac{47}{50}\right) \approx 0.2042$$

$$P(x = 2) = C(3,2)\left(\frac{4}{52}\right)\left(\frac{3}{51}\right)\left(\frac{48}{50}\right) \approx 0.0130$$

$$P(x = 3) = C(3,3)\left(\frac{4}{52}\right)\left(\frac{3}{51}\right)\left(\frac{2}{50}\right) \approx 0.0002$$

Therefore, the probability distribution is as follows.

Number of Aces	0	1	2	3
Probability	0.7826	0.2042	0.0130	0.0002

5. Use the probabilities that were calculated in Exercise 1. Draw a histogram with 5 rectangles, corresponding to $x = 0, x = 1, x = 2, x = 3$, and $x = 4$. $P(x \le 2)$ corresponds to

$$P(x = 0) + P(x = 1) + P(x = 2),$$

so shade the first 3 rectangles in the histogram.

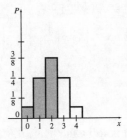

7. Use the probabilities that were calculated in Exercise 3. Draw a histogram with 4 rectangles, corresponding to $x = 0, x = 1, x = 2$, and

$x = 3$. $P(\text{at least one ace}) = P(x \geq 1)$ corresponds to

$$P(x = 1) + P(x = 2) + P(x = 3),$$

so shade the last 3 rectangles.

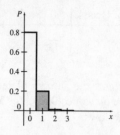

9. $E(x) = 2(0.1) + 3(0.4) + 4(0.3) + 5(0.2)$

             $= 3.6$

11. $E(z) = 9(0.14) + 12(0.22) + 15(0.38)$

                $+ 18(0.19) + 21(0.07)$

             $= 14.49$

13.   It is possible (but not necessary) to begin by writing the histogram's data as a probability distribution, which would look as follows.

x	1	2	3	4
$P(x)$	0.2	0.3	0.1	0.4

The expected value of x is

$$E(x) = 1(0.2) + 2(0.3) + 3(0.1) + 4(0.4)$$

             $= 2.7.$

15.   The expected value of x is

$$E(x) = 6(0.1) + 12(0.2) + 18(0.4)$$

                $+ 24(0.2) + 30(0.1)$

             $= 18.$

17.   Using the data from Example 5, the expected winnings for Mary are

$$E(x) = -1.2\left(\frac{1}{4}\right) + 1.2\left(\frac{1}{4}\right)$$

$$+ 1.2\left(\frac{1}{4}\right) + (-1.2)\left(\frac{1}{4}\right)$$

             $= 0.$

Yes, it is still a fair game if Mary tosses and Donna calls.

19.   **(a)**

Number of Yellow Marbles	Probability
0	$\dfrac{C(3,0)\,C(4,3)}{C(7,3)} = \dfrac{4}{35}$
1	$\dfrac{C(3,1)\,C(4,2)}{C(7,3)} = \dfrac{18}{35}$
2	$\dfrac{C(3,2)\,C(4,1)}{C(7,3)} = \dfrac{12}{35}$
3	$\dfrac{C(3,3)\,C(4,0)}{C(7,3)} = \dfrac{1}{35}$

Draw a histogram with four rectangles corresponding to $x = 0, 1, 2,$ and 3.

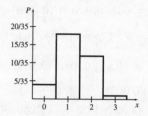

   (b)   Expected number of yellow marbles

$$= 0\left(\frac{4}{35}\right) + 1\left(\frac{18}{35}\right) + 2\left(\frac{12}{35}\right) + 3\left(\frac{1}{35}\right)$$

$$= \frac{45}{35} = \frac{9}{7} \approx 1.286$$

21.   **(a)**   Let x be the number of times 1 is rolled. Since the probability of getting a 1 on any single roll is $\frac{1}{6}$, the probability of any other outcome is $\frac{5}{6}$. Use combinations since the order of outcomes is not important.

$$P(x = 0) = C(4,0)\left(\frac{1}{6}\right)^0\left(\frac{5}{6}\right)^4 = \frac{625}{1296}$$

$$P(x = 1) = C(4,1)\left(\frac{1}{6}\right)^1\left(\frac{5}{6}\right)^3 = \frac{125}{324}$$

$$P(x = 2) = C(4,2)\left(\frac{1}{6}\right)^2\left(\frac{5}{6}\right)^2 = \frac{25}{216}$$

$$P(x = 3) = C(4,3)\left(\frac{1}{6}\right)^3\left(\frac{5}{6}\right)^1 = \frac{5}{324}$$

$$P(x = 4) = C(4,4)\left(\frac{1}{6}\right)^4\left(\frac{5}{6}\right)^0 = \frac{1}{1296}$$

x	0	1	2	3	4
$P(x)$	$\frac{625}{1296}$	$\frac{125}{324}$	$\frac{25}{216}$	$\frac{5}{324}$	$\frac{1}{1296}$

(b)
$$E(x) = 0\left(\frac{625}{1296}\right) + 1\left(\frac{125}{324}\right) + 2\left(\frac{25}{216}\right)$$
$$+ 3\left(\frac{5}{324}\right) + 4\left(\frac{1}{1296}\right)$$
$$= \frac{2}{3}$$

23. Set up the probability distribution.

Number of Women	0	1	2
Probability	$\frac{C(3,0)\,C(5,2)}{C(8,2)}$	$\frac{C(3,1)\,C(5,1)}{C(8,2)}$	$\frac{C(3,2)\,C(5,0)}{C(8,2)}$
Simplified	$\frac{5}{14}$	$\frac{15}{28}$	$\frac{3}{28}$

$$E(x) = 0\left(\frac{5}{14}\right) + 1\left(\frac{15}{28}\right) + 2\left(\frac{3}{28}\right)$$
$$= \frac{21}{28} = \frac{3}{4} = 0.75$$

25. Set up the probability distribution as in Exercise 20.

Number of Women	Probability	Simplified
0	$\frac{C(13,0)\,C(39,2)}{C(52,2)}$	$\frac{741}{1326}$
1	$\frac{C(13,1)\,C(39,1)}{C(52,2)}$	$\frac{507}{1326}$
2	$\frac{C(13,2)\,C(39,0)}{C(52,2)}$	$\frac{78}{1326}$

$$E(x) = 0\left(\frac{741}{1326}\right) + 1\left(\frac{507}{1326}\right) + 2\left(\frac{78}{1326}\right)$$
$$= \frac{663}{1326} = \frac{1}{2}$$

29. (a) First list the possible sums, 5, 6, 7, 8, and 9, and find the probabilities for each. The total possible number of results are $4 \cdot 3 = 12$. There are two ways to draw a sum of 5 (2 then 3, and 3 then 2). The probability of 5 is $\frac{2}{12} = \frac{1}{6}$. There are two ways to draw a sum of 6 (2 then 4, and 4 then 2). The

probability of 6 is $\frac{2}{12} = \frac{1}{6}$. There are four ways to draw a sum of 7 (2 then 5, 3 then 4, 4 then 3, and 5 then 2). The probability of 7 is $\frac{4}{12} = \frac{1}{3}$. There are two ways to draw a sum of 8 (3 then 5, and 5 then 3). The probability of 8 is $\frac{2}{12} = \frac{1}{6}$. There are two ways to draw a sum of 9 (4 then 5, and 5 then 4). The probability of 9 is $\frac{2}{12} = \frac{1}{6}$. The distribution is as follows.

Sum	5	6	7	8	9
Probability	$\frac{1}{6}$	$\frac{1}{6}$	$\frac{1}{3}$	$\frac{1}{6}$	$\frac{1}{6}$

(b)

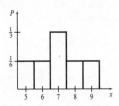

(c) The probability that the sum is even is $\frac{1}{6} + \frac{1}{6} = \frac{1}{3}$. Thus the odds are 1 to 2.

(d)
$$E(x) = \frac{1}{6}(5) + \frac{1}{6}(6) + \frac{1}{3}(7)$$
$$+ \frac{1}{6}(8) + \frac{1}{6}(9) = 7$$

31. We first compute the amount of money the company can expect to pay out for each kind of policy. The sum of these amounts will be the total amount the company can expect to pay out. For a single $100,000 policy, we have the following probability distribution.

	Pay	Don't Pay
Outcome	$100,000	$100,000
Probability	0.0012	0.9998

$$E(\text{payoff}) = 100,000(0.0012) + 0(0.9998)$$
$$= \$120$$

For all 100 such policies, the company can expect to pay out

$$100(120) = \$12,000.$$

For a single $50,000 policy,

$$E(\text{payoff}) = 50,000(0.0012) + 0(0.9998)$$
$$= \$60.$$

For all 500 such policies, the company can expect to pay out

$500(60) = \$30,000.$

Similarly, for all 1000 policies of \$10,000, the company can expect to pay out

$1000(12) = \$12,000.$

Thus, the total amount the company can expect to pay out is

$\$12,000 + \$30,000 + \$12,000 = \$54,000.$

33. (a) Expected number of good nuts in 50 "blow outs" is

$$E(x) = 50(0.60) = 30.$$

(b) Since 80% of the "good nuts" are good, 20% are bad. Expected number of bad nuts in 50 "good nuts" is

$$E(x) = 50(0.20) = 10.$$

35. The tour operator earns \$1050 if 1 or more tourists do not show up. The tour operator earns \$950 if all tourists show up. The probability that all tourists show up is $(0.98)^{21} \approx 0.6543.$ The expected revenue is
$1050(0.3457) + 950(0.6543) = 984.57$

The answer is e.

37. (a) Expected cost of Amoxicillin:
$E(x) = 0.75(\$59.30) + 0.25(\$96.15) = \$68.51$

Expected cost of Cefaclor:
$E(x) = 0.90(\$69.15) + 0.10(\$106.00) = \$72.84$

(b) Amoxicillin should be used to minimize total expected cost.

39. $E(x) = 250(0.74) = 185$

We would expect 38 low-birth-weight babies to graduate from high school.

41. (a) Using binomial probability, $n = 48, x = 0,$ $p = 0.0976.$

$$P(0) = C(48,0)(0.0976)^0(0.9024)^{48} \approx 0.007230$$

(b) Using combinations, the probability is
$$\frac{C(74,48)}{C(82,48)} \approx 5.094 \times 10^{-4}.$$

(c) Using binomial probability, $n = 6, x = 5,$ $p = 0.1.$

$$P(0) = C(6,5)(0.1)^5(0.9)^1 + (0.1)^6$$
$$= 5.5 \times 10^{-5}$$

(d) Using binomial probability,
$n = 6, p = 0.1.$

$P(\text{at least } 2)$
$$= 1 - C(6,0)(0.1)^0(0.9)^6 - C(6,1)(0.1)^1(0.9)^5$$
$$\approx 0.1143$$

43. (a) We define a success to be a cat sitting in the chair with Kimberly. For this situation, $n = 4; x = 0, 1, 2, 3,$ or $4; p = 0.3;$ and $1 - p = 0.7.$

Number of Cats	Probability
0	$C(4,0)(0.3)^0(0.7)^4 = 0.2401$
1	$C(4,1)(0.3)^1(0.7)^3 = 0.4116$
2	$C(4,2)(0.3)^2(0.7)^2 = 0.2646$
3	$C(4,3)(0.3)^3(0.7)^1 = 0.0756$
4	$C(4,4)(0.3)^4(0.7)^0 = 0.0081$

(b) Expected number of cats
$$= 0(0.2401) + 1(0.4116) + 2(0.2646)$$
$$+ 3(0.0756) + 4(0.0081)$$
$$= 1.2$$

(c) Expected number of cats
$$= np = 4(0.3) = 1.2$$

45. Below is the probability distribution of x, which stands for the person's payback.

x	\$398	\$78	$-\$2$
$P(x)$	$\frac{1}{500} = 0.002$	$\frac{3}{500} = 0.006$	$\frac{497}{500} = 0.994$

The expected value of the person's winnings is

$$E(x) = 398(0.002) + 78(0.006) + (-2)(0.994)$$
$$\approx -\$0.72 \text{ or } -72\cent.$$

Since the expected value of the payback is not 0, this is not a fair game.

47. There are 13 possible outcomes for each suit. That would make $13^4 = 28,561$ total possible outcomes. In one case, you win $5000 (minus the $1 cost to play the game). In the other 28,560, cases, you lose your dollar.

$$E(x) = 4999\left(\frac{1}{28,561}\right) + (-1)\left(\frac{28,560}{28,561}\right)$$

$$= -82\cancel{c}$$

49. There are $18 + 20 = 38$ possible outcomes. In 18 cases you win a dollar and in 20 you lose a dollar; hence,

$$E(x) = 1\left(\frac{18}{38}\right) + (-1)\left(\frac{20}{38}\right)$$

$$= -\frac{1}{19}, \text{ or about } -5.3\cancel{c}.$$

51. You have one chance in a thousand of winning $500 on a $1 bet for a net return of $499. In the 999 other outcomes, you lose your dollar.

$$E(x) = 499\left(\frac{1}{1000}\right) + (-1)\left(\frac{999}{1000}\right)$$

$$= -\frac{500}{1000} = -50\cancel{c}$$

53. Let x represent the payback. The probability distribution is as follows.

x	$P(x)$
100,000	$\frac{1}{2,000,000}$
40,000	$\frac{2}{2,000,000}$
10,000	$\frac{2}{2,000,000}$
0	$\frac{1,999,995}{2,000,000}$

The expected value is

$$E(x) = 100,000\left(\frac{1}{2,000,000}\right) + 40,000\left(\frac{2}{2,000,000}\right)$$

$$+ 10,000\left(\frac{2}{2,000,000}\right) + 0\left(\frac{1,999,995}{2,000,000}\right)$$

$$= 0.05 + 0.04 + 0.01 + 0$$

$$= \$0.10 = 10\cancel{c}.$$

Since the expected payback is $10\cancel{c}$, if entering the context costs $100\cancel{c}$, then it would be worth it to enter. The expected net return is $-\$0.90$.

55. (a) The possible scores are 0, 2, 3, 4, 5, 6. Each score has a probability of $\frac{1}{6}$.

$$E(x) = 0\left(\frac{1}{6}\right) + 2\left(\frac{1}{6}\right) + 3\left(\frac{1}{6}\right)$$

$$+ 4\left(\frac{1}{6}\right) + 5\left(\frac{1}{6}\right) + 6\left(\frac{1}{6}\right)$$

$$= \frac{1}{6}(20) = \frac{10}{3}$$

(b) The possible scores are

0 which has a probability of $\frac{11}{36}$,

4 which has a probability of $\frac{1}{36}$,

5 which has a probability of $\frac{2}{36}$,

6 which has a probability of $\frac{3}{36}$,

7 which has a probability of $\frac{4}{36}$,

8 which has a probability of $\frac{5}{36}$,

9 which has a probability of $\frac{4}{36}$,

10 which has a probability of $\frac{3}{36}$,

11 which has a probability of $\frac{2}{36}$,

12 which has a probability of $\frac{1}{36}$.

$$E(x) = 0\left(\frac{11}{36}\right) + 4\left(\frac{1}{36}\right) + 5\left(\frac{2}{36}\right) + 6\left(\frac{3}{36}\right) + 7\left(\frac{4}{36}\right)$$

$$+ 8\left(\frac{5}{36}\right) + 9\left(\frac{4}{36}\right) + 10\left(\frac{3}{36}\right) + 11\left(\frac{2}{36}\right) + 12\left(\frac{1}{36}\right)$$

$$= \frac{4}{36} + \frac{10}{36} + \frac{18}{36} + \frac{28}{36} + \frac{40}{36} + \frac{36}{36} + \frac{30}{36} + \frac{22}{36} + \frac{12}{36}$$

$$= \frac{200}{36} = \frac{50}{9}$$

(c) If a single die does not result in a score of zero, the possible scores are 2, 3, 4, 5, 6 with each of these having a probability of $\frac{1}{5}$.

$$E(x) = 2\left(\frac{1}{5}\right) + 3\left(\frac{1}{5}\right) + 4\left(\frac{1}{5}\right) + 5\left(\frac{1}{5}\right) + 6\left(\frac{1}{5}\right)$$

$$= \frac{1}{5}(20) = 4$$

Thus, if a player rolls n dice the expected average score is

$$n \cdot E(x) = n \cdot 4 = 4n.$$

(d) If a player rolls n dice, a nonzero score will occur whenever each die rolls a number other than 1. For each die there are 5 possibilities so the possible scoring ways for n dice is 5^n. When rolling one die there are 6 possibilities so the possible outcomes for n dice is 6^n. The probability of rolling a scoring set of dice is $\frac{5^n}{6^n}$; thus the expected value of the player's score when rolling n dice is $E(x) = \frac{5^n(4n)}{6^n}$.

57. Let x represent the number of hits. Since $p = 0.342, 1 - p = 0.658$.

$$P(0) = C(4,0)(0.342)^0(0.658)^4 = 0.1875$$

$$P(1) = C(4,1)(0.342)^1(0.658)^3 = 0.3897$$

$$P(2) = C(4,2)(0.342)^2(0.658)^2 = 0.3038$$

$$P(3) = C(4,3)(0.342)^3(0.658)^1 = 0.1053$$

$$P(4) = C(4,4)(0.342)^4(0.658)^0 = 0.0137$$

The distribution is shown in the following table.

x	0	1	2	3	4
$P(x)$	0.1875	0.3897	0.3038	0.1053	0.0137

The expected number of hits is $np = (4)(0.342) = 1.368$.

Chapter 8 Review Exercises

1. True

2. True

3. True

4. True

5. False: The probability of at least two occurrences is the probability of two or more occurrences.

6. True

7. True

8. False: Binomial probability applies to trials with exactly two outcomes.

9. True

10. False: For example, the random variable that assigns 0 to a head and 1 to a tail has expected value 1/2 for a fair coin.

11. True

12. False: The expected value of a fair game is 0.

13. 6 shuttle vans can line up at the airport in

$$P(6,6) = 6! = 720$$

different ways.

15. 3 oranges can be taken from a bag of 12 in

$$C(12,3) = \frac{12!}{9!3!} = \frac{12 \cdot 11 \cdot 10}{3 \cdot 2 \cdot 1} = 220$$

different ways.

17. (a) The sample will include 1 of the 2 rotten oranges and 2 of the 10 good oranges. Using the multiplication principle, this can be done in

$$C(2,1)C(10,2) = 2 \cdot 45 = 90 \, \text{ways.}$$

(b) The sample will include both of the 2 rotten oranges and 1 of the 10 good oranges. This can be done in

$$C(2,2)C(10,1) = 1 \cdot 10 = 10 \text{ ways.}$$

(c) The sample will include 0 of the 2 rotten oranges and 3 of the 10 good oranges. This can be done in

$$C(2,0)C(10,3) = 1 \cdot 120 = 120 \text{ ways.}$$

(d) If the sample contains at most 2 rotten oranges, it must contain 0, 1, or 2 rotten oranges. Adding the results from parts (a), (b), and (c), this can be done in

$$90 + 10 + 120 = 220 \text{ ways.}$$

19. (a) $P(5,5) = 5! = 120$

(b) $P(4,4) = 4! = 24$

21. (a) The order within each list is not important. Use combinations and the multiplication principle. The choice of three items from column A can be made in $C(8, 3)$ ways, and the choice of two from column B can be made in $C(6, 2)$ ways. Thus, the number of possible dinners is

$$C(8,3)C(6,2) = 56 \cdot 15 = 840.$$

(b) There are

$$C(8,0) + C(8,1) + C(8,2) + C(8,3)$$

ways to pick up to 3 items from column A. Likewise, there are

$$C(6,0) + C(6,1) + C(6,2)$$

ways to pick up to 2 items from column B. We use the multiplication principle to obtain

$$\left[C(8,0) + C(8,1) + C(8,2) + C(8,3)\right]$$
$$\cdot \left[C(6,0) + C(6,1) + C(6,2)\right]$$
$$= (1 + 8 + 28 + 56)(1 + 6 + 15)$$
$$= 93(22) = 2046.$$

Since we are assuming that the diner will order at least one item, subtract 1 to exclude the dinner that would contain no items. Thus, the number of possible dinners is 2045.

25. There are $C(13, 3)$ ways to choose the 3 balls and $C(4, 3)$ ways to get all black balls. Thus,

$$P(\text{all black}) = \frac{C(4,3)}{C(13,3)} = \frac{4}{286}$$

$$= \frac{2}{143} \approx 0.0140.$$

27. There are $C(4, 2)$ ways to get 2 black balls and $C(7, 1)$ ways to get 1 green ball. Thus,

$$P(\text{2 black and 1 green}) = \frac{C(4,2)C(7,1)}{C(11,3)}$$

$$= \frac{(6.7)}{286} = \frac{42}{286} = \frac{21}{143} \approx 0.1469.$$

29. There are $C(2, 1)$ ways to get 1 blue ball and $C(11, 2)$ ways to get 2 nonblue balls. Thus,

$$P(\text{exactly 1 blue}) = \frac{C(2,1)C(11,2)}{C(13,3)}$$

$$= \frac{2 \cdot 55}{286} = \frac{110}{286} = \frac{5}{13} \approx 0.3846.$$

31. This is a Bernoulli trial problem with

$$P(\text{success}) = P(\text{girl}) = \tfrac{1}{2}. \text{ Here,}$$

$$n = 6, p = \tfrac{1}{2}, \text{ and } x = 3.$$

$$P(\text{exactly 3 girls}) = C(6,3)\left(\frac{1}{2}\right)^3\left(\frac{1}{2}\right)^3$$

$$= \frac{20}{64} = \frac{5}{16} \approx 0.313$$

33. $P(\text{at least 4 girls})$

$$= P(4 \text{ girls}) + P(5 \text{ girls}) + P(6 \text{ girls})$$

$$= C(6,4)\left(\frac{1}{2}\right)^4\left(\frac{1}{2}\right)^2 + C(6,5)\left(\frac{1}{2}\right)^5\left(\frac{1}{2}\right)^1$$

$$+ C(6,6)\left(\frac{1}{2}\right)^6\left(\frac{1}{2}\right)^0$$

$$= \frac{22}{64} = \frac{11}{32} \approx 0.344$$

35. $P(\text{both red})$

$$= \frac{C(26,2)}{C(52,2)} = \frac{325}{1326} = \frac{25}{102} \approx 0.245$$

37. $P(\text{at least 1 card is a spade})$

$$= 1 - P(\text{neither is a spade})$$

$$= 1 - \frac{C(39,2)}{C(52,2)} = 1 - \frac{741}{1326}$$

$$= \frac{585}{1326} = \frac{15}{34} \approx 0.441$$

39. There are 12 face cards and 40 nonface cards in an ordinary deck.

$$P(\text{at least 1 face card})$$

$$= P(1 \text{ face card}) + P(2 \text{ face cards})$$

$$= \frac{C(12,1)C(40,1)}{C(52,2)} + \frac{C(12,2)}{C(52,2)}$$

$$= \frac{480}{1326} + \frac{66}{1326}$$

$$= \frac{546}{1326} \approx 0.4118$$

41. This is a Bernoulli trial problem.

(a) $P(\text{success}) = P(\text{head}) = \tfrac{1}{2}$. Hence,

$$n = 3 \text{ and } p = \tfrac{1}{2}.$$

Number of Heads	Probability
0	$C(3,0)\left(\frac{1}{2}\right)^0\left(\frac{1}{2}\right)^3 = 0.125$
1	$C(3,1)\left(\frac{1}{2}\right)^1\left(\frac{1}{2}\right)^2 = 0.375$
2	$C(3,2)\left(\frac{1}{2}\right)^2\left(\frac{1}{2}\right)^1 = 0.375$
3	$C(3,3)\left(\frac{1}{2}\right)^3\left(\frac{1}{2}\right)^0 = 0.125$

(b)

(c) $E(x) = 0(0.125) + 1(0.375) + 2(0.375)$
$$+ 3(0.125)$$
$$= 1.5$$

43. The probability that corresponds to the shaded region of the histogram is the total of the shaded areas, that is,

$$1(0.3) + 1(0.2) + 1(0.1) = 0.6.$$

45. The probability of rolling a 6 is $\frac{1}{6}$, and your net winnings would be \$2. The probability of rolling a 5 is $\frac{1}{6}$, and your net winnings would be \$1.

The probability of rolling something else is $\frac{4}{6}$, and your net winnings would be $-\$2$. Let x represent your winnings. The expected value is

$$E(x) = 2\left(\frac{1}{6}\right) + 1\left(\frac{1}{6}\right) + (-2)\left(\frac{4}{6}\right)$$
$$= -\frac{5}{6}$$
$$\approx -\$0.833 \text{ or } -83.3¢.$$

This is not a fair game since the expected value is not 0.

47. (a)

Number of Aces	Probability	
0	$\dfrac{C(4,0)\,C(48,3)}{C(52,3)}$	$= \dfrac{17,296}{22,100}$
1	$\dfrac{C(4,1)\,C(48,2)}{C(52,3)}$	$= \dfrac{4512}{22,100}$
2	$\dfrac{C(4,2)\,C(48,1)}{C(52,3)}$	$= \dfrac{288}{22,100}$
3	$\dfrac{C(4,3)\,C(48,0)}{C(52,3)}$	$= \dfrac{4}{22,100}$

$$E(x) = 0\left(\frac{17,296}{22,100}\right) + 1\left(\frac{4512}{22,100}\right) + 2\left(\frac{288}{22,100}\right)$$
$$+ 3\left(\frac{4}{22,100}\right)$$
$$= \frac{5100}{22,100} = \frac{51}{221} = \frac{3}{13} \approx 0.231$$

(b)

Number of Clubs	Probability	
0	$\dfrac{C(13,0)\,C(39,3)}{C(52,3)}$	$= \dfrac{9139}{22,100}$
1	$\dfrac{C(13,1)\,C(39,2)}{C(52,3)}$	$= \dfrac{9633}{22,100}$
2	$\dfrac{C(13,2)\,C(39,1)}{C(52,3)}$	$= \dfrac{3042}{22,100}$
3	$\dfrac{C(13,3)\,C(39,0)}{C(52,3)}$	$= \dfrac{286}{22,100}$

$$E(x) = 0\left(\frac{9139}{22,100}\right) + 1\left(\frac{9633}{22,100}\right) + 2\left(\frac{3042}{22,100}\right)$$
$$+ 3\left(\frac{286}{22,100}\right)$$
$$= \frac{16,575}{22,100} = \frac{3}{4} = 0.75$$

49. We define a success to be the event that a student flips heads and is on the committee. In this situation, $n = 6$; $x = 1, 2, 3, 4,$ or 5; $p = \frac{1}{2}$; and $1 - p = \frac{1}{2}$.

$P(x = 1, 2, 3, 4, \text{ or } 5)$
$= 1 - P(x = 6) - P(x = 0)$
$= 1 - C(6,6)\left(\dfrac{1}{2}\right)^6\left(\dfrac{1}{2}\right)^0 - C(6,0)\left(\dfrac{1}{2}\right)^0\left(\dfrac{1}{2}\right)^6$
$= 1 - \dfrac{1}{64} - \dfrac{1}{64} = \dfrac{62}{64} = \dfrac{31}{32}$

51. (a) Given a set with n elements, the number of subsets of size

0 is $C(n, 0) = 1,$

1 is $C(n, 1) = n,$

2 is $C(n, 2) = \dfrac{n(n-1)}{2}$, and

n is $C(n, n) = 1.$

(b) The total number of subsets is

$$C(n, 0) + C(n, 1) + C(n, 2) + \cdots + C(n, n).$$

(d) Let $n = 4$.

$$C(4,0) + C(4,1) + C(4,2) + C(4,3) + C(4,4)$$
$$= 1 + 4 + 6 + 4 + 1 = 16 = 2^4 = 2^n$$

Let $n = 5$.

$$C(5,0) + C(5,1) + C(5,2) + C(5,3) + C(5,4) + C(5,5)$$
$$= 1 + 5 + 10 + 10 + 5 + 1 = 32 = 2^5 = 2^n$$

(e) The sum of the elements in row n of Pascal's triangle is 2^n.

53. Use the multiplication principle.

$$3 \cdot 8 \cdot 2 = 48$$

55. $n = 12, x = 0, p = \dfrac{1}{6}$

$$P(0) = C(12,0)\left(\frac{1}{6}\right)^0 \left(\frac{5}{6}\right)^{12} \approx 0.1122$$

57. $n = 12, x = 10, p = \dfrac{1}{6}$

$$P(10) = C(12,10)\left(\frac{1}{6}\right)^{10} \left(\frac{5}{6}\right)^{2}$$
$$\approx 7.580 \times 10^{-7}$$

59. $n = 12, p = \dfrac{1}{6}$

$$P(\text{at least } 2) = 1 - P(\text{at most } 1)$$
$$= 1 - P(0) - P(1)$$
$$= 1 - C(12,0)\left(\frac{1}{6}\right)^0 \left(\frac{5}{6}\right)^{12}$$
$$\quad - C(12,1)\left(\frac{1}{6}\right)^1 \left(\frac{5}{6}\right)^{11}$$
$$\approx 0.6187$$

61. The expected value is $\frac{1}{6}(12) = 2$.

63. Observe that for $a + b = 7$,

$$P(a)P(b) = \left(\frac{1}{2^{a+1}}\right)\left(\frac{1}{2^{b+1}}\right) = \frac{1}{2^{a+b+2}} = \frac{1}{2^9}.$$

The probability that exactly seven claims will be received during a given two-week period is

$$P(0)P(7) + P(1)P(6) + P(2)P(5) + P(3)P(4) + P(4)P(3)$$
$$+ P(5)P(2) + P(6)P(1) + P(7)P(0)$$
$$= 8\left(\frac{1}{2^9}\right) = \frac{1}{64}.$$

The answer is d.

65. Denote by S the event that a product is successful.

Denote by U the event that a product is unsuccessful.

Denote by Q the event of passing quality control.

We must calculate the conditional probabilities $P(S|Q)$ and $P(U|Q)$ using Bayes' Theorem in order to calculate the expected net profit (in millions).

$$E = 40P(S|Q) - 15P(U|Q).$$
$$P(S) = P(U) = 0.5$$
$$P(Q|S) = 0.8, P(Q|U) = 0.25$$

$$P(S|Q) = \frac{P(S) \cdot P(Q|S)}{P(S) \cdot P(Q|S) + P(U) \cdot P(Q|U)}$$
$$= \frac{0.5(0.8)}{0.5(0.8) + 0.5(0.25)}$$
$$= \frac{0.4}{0.4 + 0.125} = 0.762$$

$$P(U|Q) = \frac{P(U) \cdot P(Q|U)}{P(U) \cdot P(Q|U) + P(S) \cdot P(Q|S)}$$
$$= \frac{0.125}{0.525} = 0.238$$

Therefore,

$$E = 40P(S|Q) - 15P(U|Q)$$
$$= 40(0.762) - 15(0.238)$$
$$\approx 27.$$

So the expected net profit is $27 million, or the correct answer is e.

67. Let $I(x)$ represent the airline's net income if x people show up.

$$I(0) = 0$$
$$I(1) = 400$$
$$I(2) = 2(400) = 800$$
$$I(3) = 3(400) = 1200$$
$$I(4) = 3(400) - 400 = 800$$
$$I(5) = 3(400) - 2(400) = 400$$
$$I(6) = 3(400) - 3(400) = 0$$

Let $P(x)$ represent the probability that x people will show up. Use the binomial probability formula to find the values of $P(x)$.

$$P(0) = C(6,0)(0.6)^0(0.4)^6 = 0.0041$$
$$P(1) = C(6,1)(0.6)^1(0.4)^5 = 0.0369$$
$$P(2) = C(6,2)(0.6)^2(0.4)^4 = 0.1382$$
$$P(3) = C(6,3)(0.6)^3(0.4)^3 = 0.2765$$
$$P(4) = C(6,4)(0.6)^4(0.4)^2 = 0.3110$$
$$P(5) = C(6,5)(0.6)^5(0.4)^1 = 0.1866$$
$$P(6) = C(6,6)(0.6)^6(0.4)^0 = 0.0467$$

(a) $E(I) = 0(0.0041) + 400(0.0369)$
$$+ 800(0.1382) + 1200(0.2765)$$
$$+ 800(0.3110) + 400(0.1866)$$
$$+ 0(0.0467)$$
$$= \$780.56$$

(b) $n = 3$

x	0	1	2	3
Income	0	100	200	300
$P(x)$	0.064	0.288	0.432	0.216

$E(I) = 0(0.064) + 400(0.288) + 800(0.432)$
$$+ 1200(0.216)$$
$$= \$720$$

On the basis of all the calculations, the table given in the exercise is completed as follows.

x	Income	$P(x)$
0	0	0.004
1	400	0.037
2	800	0.038
3	1200	0.276
4	800	0.311
5	400	0.187
6	0	0.047

$n = 4$

x	1	1	2	3	4
Income	0	400	800	1200	800
$P(x)$	0.0256	0.1536	0.3456	0.3456	0.1296

$E(I) = 0(0.0256) + 400(0.1536)$
$$+ 800(0.3456) + 1200(0.3456)$$
$$+ 800(0.1296)$$
$$= \$856.32$$

$n = 5$

x	Income	$\cdot P(x)$
0	0	0.01024
1	400	0.0768
2	800	0.2304
3	1200	0.3456
4	800	0.2592
5	400	0.07776

$E(I) = 0(0.01024) + 400(0.0768)$
$$+ 800(0.2304) + 1200(0.3456)$$
$$+ 800(0.2596) + 400(0.07776)$$
$$= \$868.22$$

Since $E(I)$ is greatest when $n = 5$, the airlines should book 5 reservations to maximize revenue.

69. $C(40,5)\left(\dfrac{1}{8}\right)^5\left(\dfrac{7}{8}\right)^{35} \approx 0.1875$

71. This is a set of binomial trials with $n = 5, p = 0.48$, and $1 - p = 0.52$.

(a) $P(0\,\text{women}) = C(5,0)(0.48)^0(0.52)^5$
$$\approx 0.0380$$
$$P(1\,\text{women}) = C(5,1)(0.48)^1(0.52)^4$$
$$\approx 0.1755$$

$P(2\,\text{women}) = C(5,2)(0.48)^2(0.52)^3$
$$\approx 0.3240$$
$$P(3\,\text{women}) = C(5,3)(0.48)^3(0.52)^2$$
$$\approx 0.2990$$
$$P(4\,\text{women}) = C(5,4)(0.48)^4(0.52)^1$$
$$\approx 0.1380$$
$$P(5\,\text{women}) = C(5,5)(0.48)^5(0.52)^0$$
$$\approx 0.0255$$

The distribution is shown in the following table.

Number of women	Probability
0	0.0380
1	0.1755
2	0.3240
3	0.2990
4	0.1380
5	0.0255

(b)

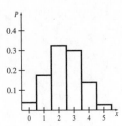

(c) Expected number of women
$= np = 5(0.48) = 2.4.$

73. (a)

Number Who Did Not Do Homework	Probability
0	$\dfrac{C(3,0)C(7,5)}{C(10,5)} = \dfrac{21}{252} = \dfrac{1}{12}$
1	$\dfrac{C(3,1)C(7,4)}{C(10,5)} = \dfrac{105}{252} = \dfrac{5}{12}$
2	$\dfrac{C(3,2)C(7,3)}{C(10,5)} = \dfrac{105}{252} = \dfrac{5}{12}$
3	$\dfrac{C(3,3)C(7,2)}{C(10,5)} = \dfrac{21}{252} = \dfrac{1}{12}$

(b) Draw a histogram with four rectangles.

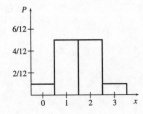

(c) Expected number who did not do homework

$$= 0\left(\frac{1}{12}\right) + 1\left(\frac{5}{12}\right) + 2\left(\frac{5}{12}\right) + 3\left(\frac{1}{12}\right)$$

$$= \frac{18}{12} = \frac{3}{2}$$

75. It costs $2(0.44 + 0.04) = 0.96$ to play the game.

x	\$1999.18	–\$0.96
$P(x)$	$\dfrac{1}{8000}$	$\dfrac{7999}{8000}$

$$E(x) = \$1999.18\left(\frac{1}{8000}\right) - \$0.96\left(\frac{7999}{8000}\right)$$

$$= -\$0.71$$

77. (a) The probability of the outcome 000 is 0.001, so the expected number of occurrences of this outcome in 30 years of play is $(30)(365)(0.001) = 10.95.$

(b) In 7 years of play the expected number of wins for 000 is $(7)(365)(0.001) \approx 2.56.$

79. (a) (i) When 5 socks are selected, we could get 1 matching pair and 3 odd socks or 2 matching pairs and 1 odd sock.

First consider 1 matching pair and 3 odd socks. The number of ways this could be done is

$C(10,1)\big[C(18,3) - C(9,1)C(16,1)\big] = 6720.$

$C(10,1)$ gives the number of ways for 1 pair, while $\big[C(18,3) - C(9,1)C(16,1)\big]$ gives the number of ways for choosing the remaining 3 socks from the 18 socks left. We must subtract the number of ways the last 3 socks could contain a pair from the 9 pairs remaining.

Next consider 2 matching pairs and 1 odd sock. The number of ways this could be done is

$$C(10,2)C(16,1) = 720.$$

$C(10,2)$ gives the number of ways for choosing 2 pairs, while $C(16,1)$ gives the number of ways for choosing the 1 odd sock.

The total number of ways is

$$6720 + 720 = 7440.$$

Then

$$P(\text{matching pair}) = \frac{7440}{C(20,5)} \approx 0.4799.$$

(ii) When 6 socks are selected, we could get 3 matching pairs and no odd socks or 2 matching pairs and 2 odd socks or 1 matching pair and 4 odd socks. The

number of ways of obtaining 3 matching
pairs is $C(10, 3) = 120$. The number of
ways of obtaining 2 matching pairs and
2 odd socks is

$$C(10,2)\big[C(16,2) - C(8,1)\big] = 5040.$$

The 2 odd socks must come from the
16 socks remaining but cannot be one of
the 8 remaining pairs.

The number of ways of obtaining
1 matching pair and 4 odd socks is

$$C(10,1)\big[C(18,4) - C(9,2) - C(9\,1)[C(16,2) - 8]\big]$$
$$= 20,160.$$

The 4 odd socks must come from the
18 socks remaining but cannot be 2 pairs
and cannot be 1 pair and 2 odd socks.

The total number of ways is

$$120 + 5040 + 20,160 - 25,320.$$

Thus,

$$P(\text{matching pair}) = \frac{25,320}{C(20,6)} \approx 0.6533.$$

(c)   Suppose 6 socks are lost at random. The
worst case is they are 6 odd socks. The best
case is they are 3 matching pairs.

First find the number of ways of selecting
6 odd socks. This is

$$C(10,6)\,C(2,1)\,C(2,1)\,C(2,1)\,C(2,1)\,C(2,1)\,C(2,1)$$
$$= 13,440.$$

The $C(10,6)$ gives the number of ways of
choosing 6 different socks from the 10 pairs.
But with each pair, $C(2,1)$ gives the number
of ways of selecting 1 sock. Then

$$P(6\,\text{odd socks}) = \frac{13,440}{C(20,6)}$$
$$\approx 0.3467.$$

Next find the number of ways of selecting
three matching pairs. This is
$C(10,3) = 120$. Then

$$P(3\,\text{matching pairs}) = \frac{120}{C(20,6)}$$
$$\approx 0.003096.$$

STATISTICS

9.1 Frequency Distributions; Measures of Central Tendency

Your Turn 1

$$\bar{x} = \frac{(12 + 17 + 21 + 25 + 27 + 38 + 49)}{7}$$

$$= \frac{189}{7} = 27$$

Your Turn 2

Interval	Midpoint, x	Frequency, f	Product, xf
0–6	3	2	6
7–13	10	4	40
14–20	17	7	119
21–27	24	10	240
28–34	31	3	93
35–41	38	1	38
Totals:		$n = 27$	536

$$\bar{x} = \frac{\sum xf}{n} = \frac{536}{27} \approx 19.85$$

Your Turn 3

The data are given in order: 12, 17, 21, 25, 27, 38, 49. The middle number is 25 so this is the median.

9.1 Exercises

1. (a)-(b) Since 0–24 is to be the first interval and there are 25 numbers between 0 and 24 inclusive, we will let all six intervals be of size 25. The other five intervals are 25–49, 50–74, 75–99, 100–124, and 125–149. Making a tally of how many data values lie in each interval leads to the following frequency distribution.

Interval	Frequency
0–24	4
25–49	8
50–74	5
75–99	10
100–124	4
125–149	5

(c) Draw the histogram. It consists of 6 bars of equal width having heights as determined by the frequency of each interval. See the histogram in part (d).

(d) To construct the frequency polygon, join consecutive midpoints of the tops of the histogram bars with line segments.

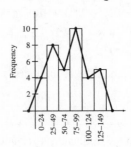

3. (a)-(b) There are eight intervals starting with 0–19. Making a tally of how many data values lie in each interval leads to the following frequency distribution.

Interval	Frequency
0–19	4
20–39	5
40–59	4
60–79	5
80–99	9
100–119	3
120–139	4
140–159	2

(c) Draw the histogram. It consists of 8 rectangles of equal width having heights as determined by the frequency of each interval. See the histogram in part (d).

(d) To construct the frequency polygon, join consecutive midpoints of the tops of the histogram bars with line segments.

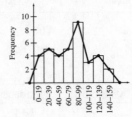

7. $$\bar{x} = \frac{\sum x}{n}$$

$$= \frac{8 + 10 + 16 + 21 + 25}{5}$$

$$= \frac{80}{5} = 16$$

9. $\sum x = 30,200 + 23,700 + 33,320$
$+ 29,410 + 24,600 + 27,750$
$+ 27,300 + 32,680$
$= 228,960$

The mean of the 8 numbers is

$$\bar{x} = \frac{\sum x}{n} = \frac{228,960}{8} = 28,620.$$

11. $\sum x = 9.4 + 11.3 + 10.5 + 7.4 + 9.1$
$+ 8.4 + 9.7 + 5.2 + 1.1 + 4.7$
$= 76.8$

The mean of the 10 numbers is

$$\bar{x} = \frac{\sum x}{n} = \frac{76.8}{10} = 7.68.$$

13. Add to the frequency distribution a new column, "Value × Frequency."

Value	Frequency	Value × Frequency
4	6	$4 \cdot 6 = 24$
6	1	$6 \cdot 1 = 6$
9	3	$9 \cdot 3 = 27$
15	2	$15 \cdot 2 = 30$
Totals:	12	87

The mean is

$$\bar{x} = \frac{\sum xf}{n} = \frac{87}{12} = 7.25.$$

15. (a)

Interval	Midpoint, x	Frequency, f	Product, xf
0–24	12	4	48
25–49	37	8	296
50–74	62	5	310
75–99	87	10	870
100–124	112	4	448
125–149	137	5	685
Totals:		$n = 36$	2657

$$\bar{x} = \frac{\sum xf}{n} = \frac{2657}{36} \approx 73.81$$

(b)

Interval	Midpoint, x	Frequency, f	Product, xf
0–19	9.5	4	38
20–39	29.5	5	147.5
40–59	49.5	4	198
60–79	69.5	5	347.5
80–99	89.5	9	805.5
100–119	109.5	3	328.5
120–139	129.5	4	518
140–159	149.5	2	299
Totals:		$n = 36$	2682

$$\bar{x} = \frac{\sum xf}{n} = \frac{2682}{36} = 74.5$$

17. 27, 35, 39, 42, 47, 51, 54

The numbers are already arranged in numerical order, from smallest to largest. The median is the middle number, 42.

19. 100, 114, 125, 135, 150, 172

The median is the mean of the two middle numbers, which is

$$\frac{125 + 135}{2} = \frac{260}{2} = 130.$$

21. Arrange the numbers in numerical order, from smallest to largest.

$$3.4, 9.1, 27.6, 28.4, 29.8, 32.1, 47.6, 59.8$$

There are eight numbers here; the median is the mean of the two middle numbers, which is

$$\frac{28.4 + 29.8}{2} = \frac{58.2}{2} = 29.1.$$

23. Using a graphing calculator, $\bar{x} \approx 73.861$ and the median is 80.5.

25. 4, 9, 8, 6, 9, 2, 1, 3

The mode is the number that occurs most often. Here, the mode is 9.

27. 55, 62, 62, 71, 62, 55, 73, 55, 71

The mode is the number that occurs most often. Here, there are two modes, 55 and 62, since they both appear three times.

29. 6.8, 6.3, 6.3, 6.9, 6.7, 6.4, 6.1, 6.0

The mode is 6.3.

33.

Interval	Midpoint, x	Frequency, f	Product, xf
0–24	12	4	48
25–49	37	8	296
50–74	62	5	310
75–99	87	10	870
100–124	112	4	448
125–149	137	5	685
Totals:		36	2657

The mean of this collection of grouped data is

$$\overline{x} = \frac{\sum xf}{n} = \frac{2657}{36} \approx 73.8.$$

The interval 75–99 contains the most data values, 10, so it is the modal class.

37. Find the mean of the numbers in the Production column.

$$\overline{x} = \frac{\sum x}{n} = \frac{20,959}{10} = 2095.9$$

The mean production of wheat is 2095.9 million bushels.

The middle two numbers in the Production column, when we list them in order, are 2103 and 2157. The average of these two numbers is the median.

$$\frac{2103 + 2157}{2} = \frac{4260}{2} = 2130$$

The median production of wheat is 2130 million bushels.

39. Find the mean for the grouped data. Note that the frequency is in thousands.

$$\overline{x} = \frac{\sum xf}{n} = \frac{664,840,000,000}{14,518,000} = 45,794$$

The estimated mean income for African American households in 2008 is $45,794.

41. **(a)** Find the mean of the numbers in the Complaints column.

$$\overline{x} = \frac{\sum x}{n} = \frac{3484}{10} = 348.4$$

The mean number of complaints is 348.4 complaints per airline.
The median is the mean of the two middle numbers.

$$\text{median} = \frac{350 + 149}{2} = 249.5$$

The median number of complaints is 249.5 complaints per airline.

(b) The averages found are not meaningful because not all airlines carry the same number of passengers.

(c) Find the mean of the numbers in the Complaints per 100,000 column.

$$\overline{x} = \frac{\sum x}{n} = \frac{11.87}{10} = 1.187$$

The mean number of complaints per 100,000 passengers boarding is 1.187.

The median is the mean of the two middle numbers once the values have been sorted.

$$\text{median} = \frac{0.87 + 1.56}{2} = 1.215$$

The median number of complaints per 100,000 passengers boarding is 1.215.

43. Find the mean.

$$\overline{x} = \frac{\sum x}{n} = \frac{16 + 12 + \ldots + 2}{13} = \frac{96}{13} \approx 7.38$$

The mean number of recognized blood types is 7.38.

Find the median.

The values are listed in order.

Since there are 13 values, the median is the seventh value, 7.

The median number of recognized blood types is 7.

The values 7, 5, and 4 each occur the greatest number of times, 2.

The modes are 7, 5, and 4.

45. **(a)** Find the mean of the numbers in the maximum temperature column.

$$\overline{x} = \frac{\sum x}{n} = \frac{666}{12} = 55.5$$

The mean of the maximum temperatures is 55.5°F. To find the median, list the 12 maximum temperatures from smallest to largest.

$$39, 39, 40, 44, 47, 50, 51, 60, 69, 70, 78, 79$$

The median is the mean of the two middle values.

$$\frac{50 + 51}{2} = 50.5°F$$

(b) Find the mean of the numbers in the minimum temperature column.

$$\overline{x} = \frac{\sum x}{n} = \frac{347}{12} \approx 28.9$$

The mean of the minimum temperatures is about 28.9°F.

To find the median, list the 12 minimum temperatures from smallest to largest.

$$16, 18, 20, 21, 24, 26, 31, 32, 37, 37, 42, 43$$

The median is the mean of the two middle values.

$$\frac{26 + 31}{2} = 28.5°F$$

47. (a) $\dfrac{5,700,000,000 \cdot 1 + 100 \cdot 80,000}{1 + 80,000}$

$$= \frac{5,708,000,000}{80,001}$$

$$\approx 71,349$$

The average worth of a citizen of Chukotka is $71,349.

49. (a) $\bar{x} = \dfrac{\sum x}{n} = \dfrac{206,333,389}{25} = 8,253,336$

The mean salary is $8,253,336.

Since the salaries are listed in order and there are 25 values, the median is the 13th value, which is $5,500,000.

The only value that occurs more than once is $5,500,000, which occurs twice. This is the mode.

(b) Most of the team earned below the mean salary, some well below. Either the mode or median makes a better description of the data.

9.2 Measures of Variation

Your Turn 1

The range is the difference of the largest and smallest values, or $35 - 7 = 28$.

For the variance, construct a table.

x	x^2
7	49
11	121
16	256
17	289
19	361
35	1225
Total: 105	2301

From the table we see that the mean is

$$\bar{x} = \frac{105}{6} = 17.5.$$

The variance is

$$s^2 = \frac{\sum x^2 - n\bar{x}^2}{n - 1} = \frac{2301 - (6)(17.5)^2}{6 - 1} = 92.7$$

The standard deviation is $\sqrt{92.7} \approx 9.628$.

Your Turn 2

Using the grouped data we first find the mean:

$$\bar{x} = \frac{\sum xf}{n} = \frac{536}{27} = 19.852$$

Interval	x	x^2	f	fx^2
0–6	3	9	2	18
7–13	10	100	4	400
14–20	17	289	7	2023
21–27	24	576	10	5760
28–34	31	961	3	2883
35–41	38	1444	1	1444
Total:			27	12,528

The variance is

$$s^2 = \frac{\sum fx^2 - n\bar{x}^2}{n - 1} = \frac{12,528 - (27)(19.852)^2}{27 - 1}$$

$$\approx 72.586$$

The standard deviation is $s = \sqrt{72.586} \approx 8.52$.

9.2 Exercises

1. The standard deviation of a sample of numbers is the square root of the variance of the sample.

3. The range is the difference of the highest and lowest numbers in the list, or $85 - 52 = 33$.

To find the standard deviation, first find the mean.

$$\bar{x} = \frac{72 + 61 + 57 + 83 + 52 + 66 + 85}{7}$$

$$= \frac{476}{7} = 68$$

To prepare for calculating the standard deviation, construct a table.

x	x^2
72	5184
61	3721
57	3249
83	6889
52	2704
66	4356
85	7225
Total:	33,328

The variance is

$$s^2 = \frac{\sum x^2 - n\bar{x}^2}{n-1}$$

$$= \frac{33{,}328 - 7(68)^2}{7-1}$$

$$= \frac{33{,}328 - 32{,}368}{6}$$

$$= 160$$

and the standard deviation is $s = \sqrt{160} \approx 12.6$.

5. The range is $287 - 241 = 46$. The mean is

$$\bar{x} = \frac{241 + 248 + 251 + 257 + 252 + 287}{6} = 256.$$

x	x^2
241	58,081
248	61,504
251	63,001
257	66,049
252	63,504
287	82,369
Total:	394,508

The standard deviation is

$$s = \sqrt{\frac{\sum x^2 - n\bar{x}^2}{n-1}}$$

$$= \sqrt{\frac{394{,}508 - 6(256)^2}{5}}$$

$$= \sqrt{258.4} \approx 16.1.$$

7. The range is $27 - 3 = 24$. The mean is

$$\bar{x} = \frac{\sum x}{n} = \frac{140}{10} = 14.$$

x	x^2
3	9
7	49
4	16
12	144
15	225
18	324
19	361
27	729
24	576
11	121
Total:	2554

The standard deviation is

$$s = \sqrt{\frac{\sum x^2 - n\bar{x}^2}{n-1}}$$

$$= \sqrt{\frac{2554 - 10(14)^2}{9}}$$

$$= \sqrt{66} \approx 8.1.$$

9. Using a graphing calculator, enter the 36 numbers into a list. Using the 1-Var Stats feature of a TI-84 Plus calculator, the standard deviation is found to be $Sx \approx 40.04793754$, or 40.05.

11. Expand the table to include columns for the midpoint x of each interval for xf, x^2, and fx^2.

Interval	f	x	xf	x^2	fx^2
30–39	4	12	48	144	576
40–49	8	37	296	1396	10,952
50–59	5	62	310	3844	19,220
60–69	10	87	870	7569	75,690
70–79	4	112	448	12,544	50,176
80–89	5	137	685	18,769	93,845
Total:	36		2657		250,459

The mean of the grouped data is

$$\bar{x} = \frac{\sum xf}{n} = \frac{2657}{36} \approx 73.8.$$

The standard deviation for the grouped data is

$$s = \sqrt{\frac{\sum fx^2 - n\bar{x}^2}{n-1}}$$

$$= \sqrt{\frac{250{,}459 - 36(73.8)^2}{35}}$$

$$\approx \sqrt{1554}$$

$$\approx 39.4.$$

13. Use $k = 3$ in Chebyshev's theorem.

$$1 - \frac{1}{k^2} = 1 - \frac{1}{3^2} = \frac{8}{9},$$

So at least $\frac{8}{9}$ of the distribution is within 3 standard deviations of the mean.

15. Use $k = 5$ in Chebyshev's theorem.

$$1 - \frac{1}{k^2} = 1 - \frac{1}{5^2} = \frac{24}{25},$$

so at least $\frac{24}{25}$ of the distribution is within 5 standard deviations of the mean.

17. We have $36 = 60 - 3 \cdot 8 = \bar{x} - 3s$ and $84 = 60 + 3 \cdot 8 = \bar{x} + 3s$, so Chebyshev's theorem applies with $k = 3$. Hence, at least

$$1 - \frac{1}{k^2} = 1 - \frac{1}{9} = \frac{8}{9}$$

of the numbers lie between 36 and 84.

19. The answer here is the complement of the answer to Exercise 16. It was found there that at least 8/9 of the distribution of the numbers are between 36 and 84, so at most $1 - 8/9 = 1/9$ of the numbers are less than 36 or more than 84.

23.
$$s^2 = \frac{\sum(x - \bar{x})^2}{n - 1}$$
$$= \frac{\sum\left(x^2 - 2x\bar{x} + \bar{x}^2\right)}{n - 1}$$
$$= \frac{\sum x^2 - 2\bar{x}\sum x + n\bar{x}^2}{n - 1}$$
$$= \frac{\sum x^2 - 2\bar{x}(n\bar{x}) + n\bar{x}^2}{n - 1}$$
$$= \frac{\sum x^2 - n\bar{x}^2}{n - 1}$$

25. $15, 18, 19, 23, 25, 25, 28, 30, 34, 38$

(a) $\bar{x} = \dfrac{1}{10}(15 + 18 + 19 + 23 + 25$
$\qquad\qquad + 25 + 28 + 30 + 34 + 38)$
$\qquad = \dfrac{1}{10}(255) = 25.5$

The mean life of the sample of Brand X batteries is 25.5 hr.

x	x^2
15	225
18	324
19	361
23	529
25	625
25	625
28	784
30	900
34	1156
38	1444
Total:	6973

$$s = \sqrt{\frac{\sum x^2 - n\bar{x}^2}{n - 1}}$$
$$= \sqrt{\frac{6973 - 10(25.5)^2}{9}}$$
$$\approx \sqrt{52.28} \approx 7.2$$

The standard deviation of the Brand X lives is 7.2 hr.

(b) Forever Power has a smaller standard deviation (4.1 hr, as opposed to 7.2 hr for Brand X), which indicates a more uniform life.

(c) Forever Power has a higher mean (26.2 hr, as opposed to 25.5 hr for Brand X), which indicates a longer average life.

27.

Sample Number	(a) $\bar{x}$	(b) s
1	$\frac{1}{3}$	2.1
2	2	2.6
3	$-\frac{1}{3}$	1.5
4	0	2.6
5	$\frac{5}{3}$	2.5
6	$\frac{7}{3}$	0.6
7	1	1.0
8	$\frac{4}{3}$	2.1
9	$\frac{7}{3}$	0.6
10	$\frac{2}{3}$	1.2

(c) $\bar{X} = \dfrac{\sum \bar{x}}{n} \approx \dfrac{11.3}{10} = 1.13$

(d) $\bar{s} = \dfrac{\sum s}{n} = \dfrac{16.8}{10} = 1.68$

(e) The upper control limit for the sample means is

$$\bar{X} + k_1\bar{s} = 1.13 + 1.954(1.68)$$
$$\approx 4.41.$$

The lower control limit for the sample means is

$$\bar{X} - k_1\bar{s} = 1.13 - 1.954(1.68)$$
$$\approx -2.15.$$

(f) The upper control limit for the sample standard deviations is

$$k_2\bar{s} = 2.568(1.68) \approx 4.31.$$

The lower control limit for the sample standard deviations is

$$k_3\bar{s} = 0(1.68) = 0.$$

29. This exercise should be solved using a calculator with a standard deviation key. The answers are $\bar{x} = 1.8158$ mm and $s = 0.4451$ mm.

31. (a) This exercise should be solved using a calculator with a standard deviation key. The answers are $\bar{x} = 7.3571$ and $s = 0.1326$.

(b) $\bar{x} + 2s = 7.3571 + 2(0.1326) = 7.6223$

$\bar{x} - 2s = 7.3571 - 2(0.1326) = 7.0919$

All the data, or 100%, are within these two values, that is, within 2 standard deviations of the mean.

33. (a) Find the mean.

$$\bar{x} = \frac{\sum x}{n} = \frac{84 + 91 + \cdots + 164}{7}$$

$$= \frac{894}{7} = 127.71$$

The mean is 127.71 days.

Find the standard deviation with a graphing calculator or spreadsheet.

$$s = 30.16$$

The standard deviation is 30.16 days.

(b) $\bar{x} + 2s = 127.71 + 2(30.16) = 188.03$

$\bar{x} - 2s = 127.71 - 2(30.16) = 67.39$

All seven of these cancers have doubling times that are within two standard deviations of the mean.

35. (a) Using a graphing calculator, the standard deviation is approximately $\$9,267,188$. In Section 9.2 we found the mean to be $\$8,253,336$.

(b) $\bar{x} - 2s \approx -7,239,488$

$\bar{x} + 2s \approx 25,773,864$

Note that the lower limit here is effectively 0. Only the highest-paid player, Alex Rodriguez, has a salary more than 2 standard deviations from the mean. This is 1/25 or 4% of the team.

9.3 The Normal Distribution

Your Turn 1

(a) Using the T1-84 Plus graphing calculator, type `normal cdf (-1E99,-0.76,0,1)`. Pressing ENTER gives an answer of 0.2236.

(b) Enter `1-normal cdf (-1E99,-1.36,0,1)` to get the answer, 0.9131.

(c) Enter `normal cdf (-1E99,1.33,0,1) - normal cdf (-1E99,-1.22,0,1)`. The answer is 0.7970.

Your Turn 2

(a) Using the Tl-84 Plus, type `invNorm (.025,0,1)`. Then pressing ENTER gives a z-score of -1.96.

(b) If 20.9% of the area is to the right, 79.1% is to the left. Enter `invNorm (.791,0,1)` to get a z-score of approximately 0.81.

Your Turn 3

$$z = \frac{x - \mu}{\sigma} = \frac{20 - 35}{20} = -0.75$$

Your Turn 4

For Example 4, $\mu = 1200$ and $\sigma = 150$. The calculator solution looks like this:

`normal cdf (-1EE99,1425,1200,150)-`
`normal cdf (-1EE99,1275,1200,150)`.
Pressing ENTER now gives an answer of 0.2417. To solve the problem using the normal curve table, first convert the given mileage values to z-scores:

$$z = \frac{x - \mu}{\sigma} = \frac{1275 - 1200}{150} = 0.5$$

$$z = \frac{x - \mu}{\sigma} = \frac{1425 - 1200}{150} = 1.5$$

Now subtract the area to the left of 0.5 from the area to the left of 1.5: $0.9332 - 0.6915 = 0.2417$

9.3 Exercises

1. The peak in a normal curve occurs directly above *the mean*.

3. For normal distributions where $\mu \neq 0$ or $\sigma \neq 1$, z-scores are found by using the formula

$$z = \frac{x - \mu}{\sigma}.$$

5. Use the table, "Area Under a Normal Curve to the Left of z", in the Appendix. To find the percent of the area under a normal curve between the mean and 1.70 standard deviations from the mean, subtract the table entry for $z = 0$ (representing the mean) from the table entry for $z = 1.7$.

$$0.9554 - 0.5000 = 0.4554$$

Therefore, 45.54% of the area lies between μ and $\mu + 1.7\sigma$.

7. Subtract the table entry for $z = -2.31$ from the table entry for $z = 0$.

$$0.5000 - 0.0104 = 0.4896$$

48.96% of the area lies between μ and $\mu - 2.31\sigma$.

9. $P(0.32 \leq z \leq 3.18)$

$= P(z \leq 3.18) - P(z \leq 0.32)$

$=$ (area to the left of 3.18)

$\quad -$ (area to the left of 0.32)

$= 0.9993 - 0.6255$

$= 0.3738$ or 37.38%

11. $P(-1.83 \leq z \leq -0.91)$

$= P(z \leq -0.91) - P(z \leq -1.83)$

$= 0.1814 - 0.0336$

$= 0.1478$ or 14.78%

13. $P(-2.95 \leq z \leq 2.03)$

$= P(z \leq 2.03) - P(z \leq -2.95)$

$= 0.9788 - 0.0016$

$= 0.9772$ or 97.72%

15. 5% of the total area is to the left of z.

Use the table backwards. Look in the body of the table for an area of 0.05, and find the corresponding z using the left column and top column of the table.

The closest values to 0.05 in the body of the table are 0.0505, which corresponds to $z = -1.64$, and 0.0495, which corresponds to $z = -1.65$.

17. 10% of the total area is to the right of z.

If 10% of the area is to the right of z, then 90% of the area is to the left of z. The closest value to 0.90 in the body of the table is 0.8997, which corresponds to $z = 1.28$.

19. For any normal distribution, the value of $P(x \leq \mu)$ is 0.5 since half of the distribution is less than the mean. Similarly, $P(x \geq \mu)$ is 0.5 since half of the distribution is greater than the mean.

21. According to Chebyshev's theorem, the probability that a number will lie within 3 standard deviations of the mean of a probability distribution is at least

$$1 - \frac{1}{3^2} = 1 - \frac{1}{9} = \frac{8}{9} \approx 0.8889.$$

Using the normal distribution, the probability that a number will lie within 3 standard deviations of the mean is 0.9974.

These values are not contradictory since "at least 0.8889" means 0.8889 or more. For the normal distribution, the value is more.

In Exercises 23–27, let x represent the life of a light bulb.

$$\mu = 500, \sigma = 100$$

23. Less than 500 hr

$$z = \frac{x - \mu}{\sigma} = \frac{500 - 500}{100} = 0, \text{ so}$$

$P(x < 500) = P(z < 0)$

$\qquad\qquad = $ area to the left of $z = 0$

$\qquad\qquad = 0.5000.$

Hence, $0.5000(10,000) = 5000$ bulbs can be expected to last less than 500 hr.

25. Between 350 and 550 hr

For $x = 350$,

$$z = \frac{350 - 500}{100} = -1.5,$$

and for $x = 550$,

$$z = \frac{550 - 500}{100} = 0.5.$$

Then

$P(350 < x < 550)$

$= P(-1.5 < z < 0.5)$

$= $ area between $z = -1.5$

$\qquad$ and $z = 0.5$

$= 0.6915 - 0.0668 = 0.6247.$

Hence, $0.6247(10,000) = 6247$ bulbs should last between 350 and 550 hr.

27. More than 440 hr

For $x = 440$,

$$z = \frac{440 - 500}{100} = -0.6.$$

Then

$$P(x > 440) = P(z > -0.6)$$
$$= \text{area to the right of } z = -0.6$$
$$= 1 - 0.2743$$
$$= 0.7257.$$

Hence, $0.7257(10,000) = 7257$ bulbs should last more than 440 hr.

29. Here, $\mu = 16.5, \sigma = 0.5$.

For $x = 16$,

$$z = \frac{16 - 16.5}{0.5} = -1.$$

$$P(x < 16) = P(z < -1) = 0.1587$$

The fraction of the boxes that are underweight is 0.1587.

31. Here, $\mu = 16.5, \sigma = 0.2$.

For $x = 16$,

$$z = \frac{16 - 16.5}{0.2} = -2.5.$$

$$P(x < 16) = P(z < -2.5) = 0.0062$$

The fraction of the boxes that are underweight is 0.0062.

In Exercises 33–37, let x represent the weight of a chicken.

33. More than 1700 g means $x > 1700$.

For $x = 1700$,

$$z = \frac{1700 - 1850}{150} = -1.0.$$

$$P(x > 1700) = 1 - P(x \le 1700)$$
$$= 1 - P(z \le -1.0)$$
$$= 1 - 0.1587$$
$$= 0.8413$$

Thus, 84.13% of the chickens will weigh more than 1700 g.

35. Between 1750 and 1900 g means $1750 \le x \le 1900$.

For $x = 1750$,

$$z = \frac{1750 - 1850}{150} = -0.67.$$

For $x = 1900$,

$$z = \frac{1900 - 1850}{150} = 0.33.$$

$$P(1750 \le x \le 1900)$$
$$= P(-0.67 \le z \le 0.33)$$
$$= P(z \le 0.33) - P(z \le -0.67)$$
$$= 0.6293 - 0.2514$$
$$= 0.3779$$

Thus, 37.79% of the chickens will weigh between 1750 and 1900 g.

37. More than 2100 g or less than 1550 g

$$P(x < 1550 \text{ or } x > 2100)$$
$$= 1 - P(1550 \le x \le 2100).$$

For $x = 1550$,

$$z = \frac{1550 - 1850}{150} = -2.00.$$

For $x = 2100$,

$$z = \frac{2100 - 1850}{150} = 1.67.$$

$$P(x < 1550 \text{ or } x > 2100)$$
$$= P(z \le -2.00) + [1 - P(z \le 1.67)]$$
$$= 0.0228 + (1 - 0.9525)$$
$$= 0.0228 + 0.0475$$
$$= 0.0703$$

Thus, 7.03% of the chickens will weigh more than 2100 g or less than 1550 g.

39. (a) In a normal distribution, 68.26% of the distribution is within one standard deviation of the mean. This leaves 31.74% for the two tails that are more than one standard deviation from the mean. Since the distribution is symmetrical, half of this tail area is on the left, so $\frac{0.3174}{2} = 0.1587$ or 15.87% of the population is more than one standard deviation below the mean.

(b) For a mean of \$169 million and a standard deviation of \$300, values more than one standard deviation below the mean would all be negative, so 0% of colleges and universities had expenditures in this range.

41. Let x represent the number of ounces of milk in a carton.

$$\mu = 32.2, \sigma = 1.2$$

Find the z-score for $x = 32$.

$$z = \frac{x - \mu}{\sigma} = \frac{32 - 32.2}{1.2} \approx -0.17$$

$$P(x < 32) = P(z < -0.17)$$
$$= \text{area to the left of } z = -0.17$$
$$= 0.4325$$

The probability that a filled carton will contain less than 32 oz is 0.4325.

43. Let x represent the weight of an egg (in ounces).

$$\mu = 1.5, \sigma = 0.4$$

Find the z-score for $x = 2.2$.

$$z = \frac{x - \mu}{\sigma} = \frac{2.2 - 1.5}{0.4} = 1.75,$$

so

$$P(x > 2.2)$$
$$= P(z > 1.75)$$
$$= \text{area to the right of } z = 1.75$$
$$= 1 - (\text{area to the left of } z = 1.75)$$
$$= 1 - 0.9599$$
$$= 0.0401.$$

Thus, out of 5 dozen eggs, we expect $0.0401(60) = 2.406$ eggs, or about 2, to be graded extra large.

45. The Recommended Daily Allowance is

$$\mu + 2.5\sigma = 1200 + 2.5(60)$$
$$= 1350 \text{ units.}$$

47. The Recommended Daily Allowance is

$$\mu + 2.5\sigma = 1200 + 2.5(92)$$
$$= 1430 \text{ units.}$$

49. Let x represent a driving speed.

$$\mu = 52, \sigma = 8$$

At the 85th percentile, the area to the left is 0.8500, which corresponds to about $z = 1.04$. Find the x-value that corresponds to this z-score.

$$z = \frac{x - \mu}{\sigma}$$
$$1.04 = \frac{x - 52}{8}$$
$$8.32 = x - 52$$
$$60.32 = x$$

The 85th percentile speed for this road is 60.32 mph.

51. $P\left(x \geq \mu + \frac{3}{2}\sigma\right) = P(z \geq 1.5)$

$$= 1 - P(z \leq 1.5)$$
$$= 1 - 0.9332 = 0.0668$$

Thus, 6.68% of the student receive A's.

53. $P\left(\mu - \frac{1}{2}\sigma \leq x \leq \mu + \frac{1}{2}\sigma\right)$

$$= P(-0.5 \leq z \leq 0.5)$$
$$= P(z \leq 0.5) - P(z \leq -0.5)$$
$$= 0.6915 - 0.3085$$
$$= 0.383$$

Thus, 38.3% of the students receive C's

In Exercises 55 and 57, let x represents a student's test score.

55. Since the top 8% get A's, we want to find the number a for which

$$P(x \geq a) = 0.08,$$
$$\text{or} \quad P(x \leq a) = 0.92.$$

Read the table backwards to find the z-score for an area of 0.92, which is 1.41. Find the value of x that corresponds to $z = 1.41$.

$$z = \frac{x - \mu}{\sigma}$$
$$1.41 = \frac{x - 76}{8}$$
$$11.28 = x - 76$$
$$87.28 = x$$

The bottom cutoff score for an A is 87.

57.   28% of the students will receive D's and F's, so to find the bottom cutoff score for a C we need to find the number c for which

$$P(x \le c) = 0.28.$$

Read the table backwards to find the z-score for an area of 0.28, which is -0.58. Find the value of x that corresponds to $z = -0.58$.

$$-0.58 = \frac{x - 76}{8}$$
$$-4.64 = x - 76$$
$$71.36 = x$$

The bottom cutoff score for a C is 71.

59.   **(a)**   The area above the 55th percentile is equal to the area below the 45th percentile.

$$2P(z > 0.55) = 2[1 - P(z \le 0.55)]$$
$$= 2(1 - 0.7088)$$
$$= 2(0.2912)$$
$$= 0.5824$$

$0.58 = 58\%$

   (b)   The area above the 60th percentile is equal to the area below the 40th percentile.

$$2P(z > 0.6) = 2[1 - P(z \le 0.6)]$$
$$= 2(1 - 0.7257)$$
$$= 2(0.2743)$$
$$= 0.5486$$

$0.55 = 55\%$

The probability that the student will be above the 60th percentile or below the 40th percentile is 55%.

61.   **(a)** $\mu = 93, \sigma = 16$

For $x = 130.5$,

$$z = \frac{130.5 - 93}{16} = 2.34.$$

Then,

$$P(x \ge 130.5) = P(z \ge 2.34)$$
$$= \text{area to the right of } 2.34$$
$$= 1 - 0.9904 = 0.0096$$

The probability is about 0.01 that a person from this time period would have a lead level of 130.5 ppm or higher. Yes, this provides evidence that Jackson suffered from lead poisoning during this time period.

   (b) $\mu = 10, \sigma = 5$

For $x = 130.5$,

$$z = \frac{130.5 - 10}{5} = 24.1.$$

Then,

$$P(x \ge 130.5) = P(z \ge 24.1)$$
$$= \text{area to the right of } 24.1$$
$$\approx 0$$

The probability is essentially 0 by these standards.

From this we can conclude that Andrew Jackson had lead poisioning.

63.

Reference	Models

(a) Head size: $z = \dfrac{55 - 55.3}{2.0} = -0.15$

$$P(x \geq 55) = P(z \geq -0.15)$$
$$= 1 - 0.4404$$
$$= 0.5596$$

$z = \dfrac{55 - 50.0}{2.4} = 2.08$

$$P(x \geq 55) = P(z \geq 2.08)$$
$$= 1 - 0.9812$$
$$= 0.0188$$

(b) Neck size: $z = \dfrac{23.9 - 32.7}{1.4} = -6.29$

$$P(x \leq 23.9) = P(z \leq -6.29)$$
$$\approx 0$$

$z = \dfrac{23.9 - 31.0}{1.0} = -7.1$

$$P(x \leq 23.9) = P(z \leq -7.1)$$
$$\approx 0$$

(c) Bust size: $z = \dfrac{82.3 - 90.3}{5.5} = -1.45$

$$P(x \geq 82.3) = P(z \geq -1.45)$$
$$= 1 - 0.0735$$
$$= 0.9265$$

$z = \dfrac{82.3 - 87.4}{3.0} = -1.70$

$$P(x \geq 82.3) = P(z \geq -1.70)$$
$$= 1 - 0.0446$$
$$= 0.9554$$

(d) Wrist size: $z = \dfrac{10.6 - 16.1}{0.8} = -6.88$

$$P(x \leq 10.6) = P(z \leq -6.88)$$
$$\approx 0$$

$z = \dfrac{10.6 - 15.0}{0.6} = -7.33$

$$P(x \leq 10.6) = P(z \leq -7.33)$$
$$\approx 0$$

(e) Waist size: $z = \dfrac{40.7 - 69.8}{4.7} = -6.19$

$$P(x \leq 40.7) = P(z \leq -6.19)$$
$$\approx 0$$

$z = \dfrac{40.7 - 65.7}{3.5} = -7.14$

$$P(x \leq 40.7) = P(z \leq -7.14)$$
$$\approx 0$$

9.4 Normal Approximation to the Binomial Distribution

Your Turn 1

For this experiment, $n = 12$ and $p = \frac{1}{6}$.

$$\mu = np = (12)\left(\frac{1}{6}\right) = 2$$

$$\sigma = \sqrt{np(1-p)} = \sqrt{(12)\left(\frac{1}{6}\right)\left(\frac{5}{6}\right)} = \sqrt{\frac{5}{3}} \approx 1.291$$

Your Turn 2

First find the mean and standard deviation using

$n = 120$ and $p = \dfrac{1}{5} = 0.2$.

$$\mu = np = (120)(0.2) = 24$$
$$\sigma = \sqrt{np(1-p)} = \sqrt{(120)(0.2)(0.8)}$$
$$= \sqrt{19.2} \approx 4.3818$$

Let x represent the number of correct answers. To find the probability of at least 32 correct answers using the normal distribution we must find $P(x > 31.5)$. Since the corresponding z-score is

$$z = \frac{x - \mu}{\sigma} = \frac{31.5 - 24}{4.3818} \approx 1.7116, \text{ we need to}$$

compute $P(z \geq 1.7116) = 1 - P(z \leq 1.7116)$. Using tables or a graphing calculator we find $P(z \leq 1.7116) \approx 0.9564$ so our answer is $1 - 0.9564 = 0.0436$.

The probability of getting at least 32 correct by random guessing is about 0.0436; depending on how you round intermediate results you may get a slightly different answer.

9.4 Exercises

1. In order to find the mean and standard deviation of a binomial distribution, you must know the number of trials and the probability of a success on each trial.

3. Let x represent the number of heads tossed. For this experiment, $n = 16$, $x = 4$, and $p = \frac{1}{2}$.

   (a) $P(x = 4) = C(16,4)\left(\frac{1}{2}\right)^4\left(1 - \frac{1}{2}\right)^4$

   ≈ 0.0278

   (b) $\mu = np = 16\left(\frac{1}{2}\right) = 8$

   $\sigma = \sqrt{np(1-p)}$

   $= \sqrt{16\left(\frac{1}{2}\right)\left(\frac{1}{2}\right)}$

   $= \sqrt{4} = 2$

   For $x = 3.5$,

   $z = \dfrac{3.5 - 8}{2} = -2.25.$

   For $x = 4.5$,

   $z = \dfrac{4.5 - 8}{2} = -1.75.$

   $P(z < -1.75) - P(z < -2.25)$

   $= 0.0401 - 0.0122 = 0.0279$

5. Let x represent the number of tails tossed. For this experiment, $n = 16$; $x = 13, 14, 15$, or 16; and $p = \frac{1}{2}$.

   (a) $P(x = 13, 14, 15,$ or $16)$

   $= C(16,13)\left(\frac{1}{2}\right)^{13}\left(1 - \frac{1}{2}\right)^3$

   $+ C(16,14)\left(\frac{1}{2}\right)^{14}\left(1 - \frac{1}{2}\right)^2$

   $+ C(16,15)\left(\frac{1}{2}\right)^{15}\left(1 - \frac{1}{2}\right)^1$

   $+ C(16,16)\left(\frac{1}{2}\right)^{16}\left(1 - \frac{1}{2}\right)^0$

   $\approx 0.00854 + 0.00183 + 0.00024$

   $+ 0.00001$

   ≈ 0.0106

   (b) $\mu = np = 16\left(\frac{1}{2}\right) = 8$

   $\sigma = \sqrt{np(1-p)}$

   $= \sqrt{16\left(\frac{1}{2}\right)\left(\frac{1}{2}\right)}$

   $= \sqrt{4} = 2$

For $x = 12.5$,

$z = \dfrac{12.5 - 8}{2} = 2.25.$

$P(z > 2.25) = 1 - P(z \le 2.25)$

$= 1 - 0.9878$

$= 0.0122$

In Exercises 7 and 9, let x represent the number of heads tossed. Since $n = 1000$ and $p = \frac{1}{2}$,

$$\mu = np = 1000\left(\frac{1}{2}\right) = 500$$

and

$$\sigma = \sqrt{np(1-p)}$$

$$= \sqrt{1000\left(\frac{1}{2}\right)\left(\frac{1}{2}\right)}$$

$$= \sqrt{250}$$

$$\approx 15.8.$$

7. To find P(exactly 500 heads), find the z-scores for $x = 499.5$ and $x = 500.5$.

   For $x = 499.5$,

   $z = \dfrac{499.5 - 500}{15.8} \approx -0.03.$

   For $x = 500.5$,

   $z = \dfrac{500.5 - 500}{15.8} \approx 0.03.$

   Using the table,

   P(exactly 500 heads) $= 0.5120 - 0.4880$

   $= 0.0240.$

9. Since we want 475 heads or more, we need to find the area to the right of $x = 474.5$. This will be $1 -$ (the area to the left of $x = 474.5$). Find the z-score for $x = 474.5$.

   $z = \dfrac{474.5 - 500}{15.8} \approx -1.61$

   The area to the left of 474.5 is 0.0537, so

   P(480 heads or more) $= 1 - 0.0537$

   $= 0.9463.$

11. Let x represent the number of 5's tossed.

$$n = 120, p = \frac{1}{6}$$

$$\mu = np = 120\left(\frac{1}{6}\right) = 20$$

$$\sigma = \sqrt{np(1 - p)}$$
$$= \sqrt{120\left(\frac{1}{6}\right)\left(\frac{5}{6}\right)}$$
$$\approx 4.08$$

Since we want the probability of getting exactly twenty 5's, we need to find the area between $x = 19.5$ and $x = 20.5$. Find the corresponding z-scores.

For $x = 19.5$,

$$z = \frac{19.5 - 20}{4.08} \approx -0.12.$$

For $x = 20.5$,

$$z = \frac{20.5 - 20}{4.08} \approx 0.12.$$

Using values from the table,

$$P(\text{exactly twenty 5's}) = 0.5478 - 0.4522$$
$$= 0.0956.$$

13. Let x represent the number of 3's tossed.

$$n = 120, p = \frac{1}{6}$$
$$\mu = 20, \sigma \approx 4.08$$

(These values for μ and σ are calculated in the solution for Exercise 11.)

Since

$$P(\text{more than fifteen 3's})$$
$$= 1 - P(\text{fifteen 3's or less}),$$

find the z-score for $x = 15.5$.

$$z = \frac{15.5 - 20}{4.08} \approx -1.10$$

From the table, $P(z < -1.10) = 0.1357$.
Thus,

$$P(\text{more than fifteen 3's}) = 1 - 0.1357$$
$$= 0.8643.$$

15. Let x represent the number of times the chosen number appears.

$$n = 130; x = 26, 27, 28, \ldots, 130; \text{ and } p = \frac{1}{6}$$

$$\mu = np = 130\left(\frac{1}{6}\right) = \frac{65}{3}$$

$$\sigma = \sqrt{np(1 - p)} = \sqrt{130\left(\frac{1}{6}\right)\left(\frac{5}{6}\right)} = \frac{5}{6}\sqrt{26}$$

For $x = 25.5$,

$$z = \frac{25.5 - \frac{65}{3}}{\frac{5}{6}\sqrt{26}} \approx 0.90.$$

$$P(z > 0.90) = 1 - P(z \leq 0.90)$$
$$= 1 - 0.8159$$
$$= 0.1841$$

17. **(a)** Let x represent the number of heaters that are defective.

$$n = 10,000, p = 0.02$$
$$\mu = np = 10,000(0.02)$$
$$= 200$$
$$\sigma = \sqrt{np(1 - p)} = \sqrt{10,000(0.02)(0.98)}$$
$$= 14$$

To find $P(\text{fewer than } 170)$, find the z-score for $x = 169.5$.

$$z = \frac{169.5 - 200}{14} \approx -2.18$$

$$P(\text{fewer than } 170) = 0.0146$$

(b) Let x represent the number of heaters that are defective.

$$n = 10,000, p = 0.02, \mu = np = 200$$

$$\sigma = \sqrt{np(1 - p)} = \sqrt{10,000(0.02)(0.98)}$$
$$= 14$$

We want the area to the right of $x = 222.5$. For $x = 222.5$,

$$z = \frac{222.5 - 200}{14} \approx 1.61.$$

$$P(\text{more than 222 defects})$$
$$= P(x \geq 222.5)$$
$$= P(z \geq 1.61)$$
$$= 1 - P(z \leq 1.61)$$
$$= 1 - 0.9463$$
$$= 0.0537$$

19. Use a calculator or computer to complete this exercise. The answers are given.

   (a) $P(\text{all 58 like it}) = 1.04 \times 10^{-9} \approx 0$

   (b) $P(\text{exactly 28, 29, or 30 like it}) = 0.0018$

21. Let x be the number of nests escaping predation.

   $n = 24, p = 0.3$

   $\mu = np = 24(0.3) = 7.2$

   $\sigma = \sqrt{np(1-p)} = \sqrt{24(0.3)(0.7)}$
   $= \sqrt{5.04} \approx 2.245$

   To find $P(\text{at least half escape predation})$, find the z-score for $x = 11.5$.

   $z = \dfrac{11.5 - 7.2}{2.245} \approx 1.92$

   $p(z > 1.92) = 1 - 0.9726 = 0.0274$

23. Let x represent the number of hospital patients struck by falling coconuts.

   (a) $n = 20; x = 0$ or 1; and $p = 0.025$

   $p(x = 0 \text{ or } 1) = C(20,0)(0.025)^0(0.975)^{20}$
   $+ (20,1)(0.025)^1(0.975)^{19}$
   $\approx 0.60269 + 0.30907$
   ≈ 0.9118

   (b) $n = 2000; x = 0, 1, 2, \ldots,$ or $70; p = 0.025$

   $\mu = np = 2000(0.025) = 50$

   $\sigma = \sqrt{np(1-p)} = \sqrt{2000(0.025)(0.975)}$
   $= \sqrt{48.75}$

   To find $P(70 \text{ or less})$, find the z-score for $x = 70.5$.

   $z = \dfrac{70.5 - 50}{\sqrt{48.75}} \approx 2.94$

   $P(x < 2.94) = 0.9984$

25. This exercise should be solved by calculator or computer methods. The answers, which may vary slightly, are

   (a) 0.0001,

   (b) 0.0002 and

   (c) 0.0000.

27. Let x represent the number of motorcyclists injured between 3 p.m. and 6 p.m. $n = 200$ and $p = 0.241$. To use the normal approximation we compute μ and σ.

   $\mu = np = (200)(0.241) = 48.2$

   $\sigma = \sqrt{np(1-p)}$
   $= \sqrt{(200)(0.241)(0.759)}$
   ≈ 6.0485

   To find $P(x < 51)$, find the z-score for 50.5.

   $z = \dfrac{50.5 - 48.2}{6.0485} \approx 0.3803$

   $P(z \le 0.3803) \approx 0.6480 \text{ (by table)}$

   The probability that at most 50 in the sample were injured between 3 p.m. and 6 p.m. is 0.6480.

29. Let x represent the number of ninth grade students who have tried cigarette smoking. $n = 500$ and $p = 0.463$. To use the normal approximation we compute μ and σ.

   $\mu = np = (500)(0.463) = 231.5$

   $\sigma = \sqrt{np(1-p)} = \sqrt{(500)(0.463)(0.537)}$
   ≈ 11.1497

   To find $P(x < 251)$ find the z-score for 250.5.

   $z = \dfrac{250.5 - 231.5}{11.1497} \approx 1.7041$

   $P(z \le 1.704) \approx 0.9554 \text{ (by table)}$

   The probability that at most half in the sample of 500 have tried cigarette smoking is approximately 0.9554.

31. **(a)** The numbers are too large for the calculator to handle.

   (b) $n = 5{,}825{,}043, p = 0.5$

   $\mu = np = 5{,}825{,}043(0.5) = 2{,}912{,}522$

   $\sigma = \sqrt{np(1-p)}$
   $= \sqrt{5{,}825{,}043(0.5)(0.5)}$
   ≈ 1206.8

   $z = \dfrac{2{,}912{,}253 - 2{,}912{,}522}{1206.8} = -0.22$

   $z = \dfrac{2{,}912{,}790 - 2{,}912{,}522}{1206.8} = 0.22$

$$P(2,912,253 \le x \le 2,912,790)$$
$$= P(-0.22 \le z \le 0.22)$$
$$= P(z \le 0.22) - P(z \le -0.22)$$
$$= 0.5871 - 0.4129$$
$$= 0.1742$$

33. Let x represent the number of questions.

$$n = 100; x = 60, 61, 62, \ldots, \text{ or } 100; p = \tfrac{1}{2}$$

$$\mu = np = 180\left(\frac{1}{2}\right) = 50$$

$$\sigma = \sqrt{np(1-p)} = \sqrt{100\left(\frac{1}{2}\right)\left(\frac{1}{2}\right)}$$

$$= \sqrt{25} = 5$$

To find $P(60 \text{ or more correct})$, find the z-score for $x = 59.5$.

$$z = \frac{59.5 - 50}{5} = 1.90$$

$$P(z > 1.90) = 1 - P(z \le 1.90)$$
$$= 1 - 0.9713$$
$$= 0.0287$$

Chapter 9 Review Exercises

1. True

2. False: Any symmetrical distribution which peaks in the middle will have equal mean and mode.

3. False: Variance measures spread

4. True

5. False: A large variance indicates that the data are widely spread.

6. False: The mode is the most frequently occurring value.

7. False: This will be true only if $\mu(x) = 0$.

8. True

9. True

10. True

11. False: The expected value of a sample variance is the population variance.

12. True

15. (a) since 450–474 is to be the first interval, let all the intervals be of size 25. The largest data value is 566, so the last interval that will be needed is 550–574. The frequency distribution is as follows.

Interval	Frequency
450–474	5
475–499	6
500–524	5
525–549	2
550–574	2

(b) Draw the histogram. It consists of equal width having heights as determined by the frequency of each interval. See the histogram in part (c).

(c) Construct the frequency polygon by joining consecutive midpoints of the tops of the histogram bars with line segments.

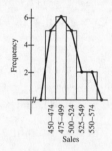

17. $\sum x = 30 + 24 + 34 + 30 + 29$
$$+ 28 + 30 + 29$$
$$= 234$$

The mean of the 8 numbers is

$$\overline{x} = \frac{\sum x}{n} = \frac{234}{8} = 29.25.$$

19.

Interval	Midpoint, x	Frequency, f	Product, xf
10–19	14.5	6	87
20–29	24.5	12	294
30–39	34.5	14	483
40–49	44.5	10	445
50–59	54.5	8	436
Totals:		50	1745

The mean of this collection of grouped data is

$$\overline{x} = \frac{\sum xf}{n} = \frac{1745}{50} = 34.9.$$

23. Arrange the numbers in numerical order, from smallest to largest.

$$35, 36, 36, 38, 38, 42, 44, 48$$

There are 8 numbers here; the median is the mean of the two middle numbers, which is

$$\frac{38 + 38}{2} = \frac{76}{2} = 38.$$

The mode is the number that occurs most often. Here, there are two modes, 36 and 38, since they both appear twice.

25. The modal class is the interval with the greatest frequency; in this case, the modal class is 30–39.

29. The range is $93 - 26 = 67$, the difference of the highest and lowest numbers in the distribution.

The mean is

$$\bar{x} = \frac{\sum x}{n} = \frac{520}{10} = 52.$$

Construct a table with the values of x and x^2.

x	x^2
26	676
43	1849
51	2601
29	841
37	1369
56	3136
29	841
82	6724
74	5476
93	8649
Total:	32,162

The standard deviation is:

$$s = \sqrt{\frac{\sum x^2 - n\bar{x}^2}{n-1}}$$

$$= \sqrt{\frac{32,162 - 10(52)^2}{9}}$$

$$\approx \sqrt{569.1} \approx 23.9.$$

31. Recall that when working with grouped data, x represents the midpoint of each interval. Complete the following table.

Interval	f	x	xf	x^2	fx^2
10–19	6	14.5	87.0	210.25	1261.50
20–29	12	24.5	294.0	600.25	7203.00
30–39	14	34.5	483.0	1190.25	16,663.50
40–49	10	44.5	445.0	1980.25	19,802.50
50–59	8	54.5	436.0	2970.25	23,762.00
Totals:	50		1745		68,692.50

Use the formulas for grouped frequency distribution to find the mean and then the standard deviation. (The mean was also calculated in Exercise 19.)

$$\bar{x} = \frac{\sum xf}{n} = \frac{1745}{50} = 34.9$$

$$s = \sqrt{\frac{\sum fx^2 - n\bar{x}^2}{n-1}}$$

$$= \sqrt{\frac{68,692.5 - 50(34.9)^2}{49}}$$

$$\approx 12.6$$

33. A skewed distribution has the largest frequency at one end rather than in the middle.

35. To the left of $z = 0.84$

Using the standard normal curve table,

$$P(z < 0.84) = 0.7995.$$

37. Between $z = 1.53$ and $z = 2.82$

$$P(1.53 \le z \le 2.82)$$
$$= P(z \le 2.82) - P(z \le 1.53)$$
$$= 0.9976 - 0.9370$$
$$= 0.0606$$

39. The normal distribution is not a good approximation of a binomial distribution that has a value of p close to 0 or 1 because the histogram of such a binomial distribution is skewed and therefore not close to the shape of a normal distribution.

41.

Number of Heads, x	Frequency, f	xf	fx^2
0	1	0	0
1	5	5	5
2	7	14	28
3	5	15	45
4	2	8	32
Totals:	20	42	110

(a) $\bar{x} = \frac{\sum xf}{n} = \frac{42}{20} = 2.1$

$$s = \sqrt{\frac{\sum fx^2 - n\bar{x}^2}{n-1}}$$

$$= \sqrt{\frac{110 - 20(2.1)^2}{20 - 1}}$$

$$\approx 1.07$$

(b) For this binomial experiment,

$$\mu = np = 4\left(\frac{1}{2}\right) = 2,$$

and

$$\sigma = \sqrt{np(1-p)}$$

$$= \sqrt{4\left(\frac{1}{2}\right)\left(\frac{1}{2}\right)}$$

$$= \sqrt{1} = 1.$$

(c) The answer to parts (a) and (b) should be close.

43. (a) For Stock I,

$$\overline{x} = \frac{11 + (-1) + 14}{3} = 8,$$

so, the mean (average return) is 8%.

$$s = \sqrt{\frac{\sum x^2 - n\overline{x}^2}{n-1}}$$

$$= \sqrt{\frac{318 - 3(8)^2}{2}}$$

$$= \sqrt{63} \approx 7.9$$

so the standard deviation is 7.9%.
For Stock II,

$$\overline{x} = \frac{9 + 5 + 10}{3} = 8,$$

so the mean is also 8%.

$$s = \sqrt{\frac{\sum x^2 - n\overline{x}^2}{n-1}}$$

$$= \sqrt{\frac{206 - 3(8)^2}{2}}$$

$$= \sqrt{7} \approx 2.6,$$

so the standard deviation is 2.6%.

(b) Both stocks offer an average (mean) return of 8%. The smaller standard deviation for Stock II indicates a more stable return and thus greater security.

45. Let x represents the number of overstuffed frankfurters.

$$n = 500, p = 0.04, 1 - p = 0.96$$

We also need the following results.

$$\mu = np = 500(0.04) = 20$$

$$\sigma = \sqrt{np(1-p)}$$

$$= \sqrt{500(0.04)(0.96)}$$

$$= \sqrt{19.2}$$

$$\approx 4.38$$

(a) P(twenty-five or fewer) or, equivalently, $P(x \le 25)$

First, using the binomial probability formula:

$$P(x \le 25) = C(500,0)(0.04)^0 (0.96)^{500}$$

$$+ C(500,1)(0.04)^1 (0.96)499$$

$$+ \cdots + C(500,25)(0.04)^{25} (0.96)^{475}$$

$$= (1.4 \times 10^{-9}) + (2.8 \times 10^{-8})$$

$$+ \cdots + 0.0446$$

$$\approx 0.8924$$

(To evaluate the sum, use a calculator or computer program. For example, using a TI-83/84 Plus calculator, enter the following:

sum(seq((500nCrX)(0.04^X)

(0.96^(500 − X)),X,0,25,1))

The displayed result is 0.8923644609.)

Second, using the normal approximation:

To find $P(x \le 25)$, first find the z-score for 25.5.

$$z = \frac{25.5 - 20}{4.38}$$

$$\approx 1.26$$

$$P(z < 1.26) = 0.8962$$

(b) P(exactly twenty-five) or $P(x = 25)$

Using the binomial probability formula:

$$P(x = 25)$$

$$= C(500,25)(0.04)^{25}(0.96)^{475}$$

$$\approx 0.0446$$

Using the normal approximation:

P(exactly twenty-five) corresponds to the area under the normal curve between $x = 24.5$ and $x = 25.5$. The correspond-ing z-scores are found as follows.

$$z = \frac{24.5 - 20}{4.38} \approx 1.03 \text{ and}$$

$$z = \frac{24.5 - 20}{4.38} \approx 1.26$$

$$P(x = 25)$$

$$= P(24.5 < x < 25.5)$$

$$= P(1.03 < z < 1.26)$$

$$= P(z < 1.26) - P(z < 1.03)$$

$$= 0.8962 - 0.8485$$

$$= 0.0477$$

(c) P(at least 30), or equivalently, $P(x \geq 30)$

Using the binomial probability formula:

This is the complementary event to "less than 30," which requires fewer calculations. We can use the results from Part (a) to reduce the amount of work even more.

$P(x < 30)$
$= P(x \leq 25) + P(x = 26)$
$\qquad + P(x = 27) + P(x = 28)$
$\qquad + P(x = 29)$
$= 0.8924 + C(500,26)(0.04)^{26}(0.96)^{474}$
$\qquad + \cdots + C(500,29)(0.04)^{29}(0.96)^{471}$
≈ 0.9804

Therefore,

$$P(x \geq 30) = 1 - P(x < 30)$$
$$= 1 - 0.9804$$
$$= 0.0196.$$

Using the normal approximation:

Again, use the complementary event. To find $P(x < 30)$, find the z-score for 29.5.

$$z = \frac{29.5 - 20}{4.38} \approx 2.17$$
$$P(z < 2.17) = 0.9850$$

Therefore,

$$P(x \geq 30) = 1 - P(x < 30)$$
$$= 1 - 0.9850$$
$$= 0.0150.$$

47. The table below records the mean and standard deviation for diet A and for diet B.

	$\bar{x}$	s
Diet A	2.7	2.26
Diet B	1.3	0.95

(a) Diet A had the greater mean gain, since the mean for diet A is larger.

(b) Diet B had a more consistent gain, since diet B has a smaller standard deviation.

49. Let x represents the number of flies that are killed.

$n = 1000; x = 0, 1, 2, \ldots, 986; p = 0.98$
$\mu = np = 1000(0.98) = 980$
$\sigma = \sqrt{np(1 - p)} = \sqrt{1000(0.98)(0.02)}$
$= \sqrt{19.6}$

To find P(no more that 986), find the z-score for $x = 986.5$.

$$z = \frac{986.5 - 980}{\sqrt{19.6}} \approx 1.47$$
$$P(z < 1.47) = 0.9292$$

51. Again, let x represents the number of flies that are killed.

$n = 1000; x = 973, 974, 975, \ldots, 993; p = 0.98$

As in Exercise 49, $\mu = 980$ and $\sigma = \sqrt{19.6}$. To find P(between 973 and 993), find the z-scores for $x = 972.5$ and $x = 993.5$.

For $x = 972.5$,

$$z = \frac{972.5 - 980}{\sqrt{19.6}} \approx -1.69.$$

For $x = 993.5$,

$$z = \frac{993.5 - 980}{\sqrt{19.6}} \approx 3.05.$$
$$P(-1.69 \leq z \leq 3.05)$$
$$= P(z \leq 3.05) - P(z \leq -1.69)$$
$$= 0.9989 - 0.0455$$
$$= 0.9534$$

53. No more that 40 min/day

$$\mu = 42, \sigma = 12$$

Find the z-score for $x = 40$.

$$z = \frac{40 - 42}{12} \approx -0.17$$
$$P(x \leq 40) = P(z \leq -0.17) = 0.4325$$

43.25% of the residents commute no more that 40 min/day.

55. Between 38 and 60 min/day

$$\mu = 42, \sigma = 12$$

Find the z-scores for $x = 38$ and $x = 60$.

For $x = 38$,

$$z = \frac{38 - 42}{12} \approx -0.33.$$

For $x = 60$,

$$z = \frac{60 - 42}{12} = 1.5.$$

$$P(38 \leq x \leq 60) = P(-0.33 \leq z \leq 1.5)$$
$$= P(z \leq 1.5) - P(z \leq -0.33)$$
$$= 0.9332 - 0.3707$$
$$= 0.5625$$

56.25% of the residents commute between 38 and 60 min/day.

57. **(a)** The mean is

$$\bar{x} = \frac{\sum x}{n} = \frac{1852}{13} \approx 142.46.$$

To find the median, arrange the values in order. The middle value is the 7th in the list, which is 140 (corresponding to Los Angeles), so this is the median. There are no repeated values so there is no mode.

(b) Find the standard deviation by using the 1-variable statistics available on a graphing calculator. Enter the data as a list in `L1` and then enter `1-Var Stats L1`. The standard deviation s is given as the value of Sx, which is approximately 48.67.

(c) $\bar{x} - s = 142.46 - 48.67 = 93.79$

$\bar{x} + s = 142.46 + 48.67 = 191.13$

Five of the data values, or 38.5%, fall within one standard deviation of the mean.

(d) $\bar{x} - 2s = 142.46 - (2)(48.67) = 45.12$

$\bar{x} + 2s = 142.46 + (2)(48.67) = 239.80$

All the data values, or 100%, fall within two standard deviations of the mean.

59. **(a)** $P(x \geq 35) = P\left(z \geq \dfrac{35 - 28.0}{5.5}\right)$

$$= P(z \geq 1.27)$$
$$= 1 - P(z < 1.27)$$
$$= 1 - 0.8980$$
$$= 0.1020$$

(b) $P(x \geq 35) = P\left(z \geq \dfrac{35 - 32.2}{8.4}\right)$

$$= P(z \geq 0.33)$$
$$= 1 - P(z < 0.33)$$
$$= 1 - 0.6293$$
$$= 0.3707$$

(d) $P(x \geq 1.4) = P\left(z \geq \dfrac{1.4 - 1.64}{0.08}\right)$

$$= P(z \geq -3)$$
$$= 1 - P(z < -3)$$
$$= 1 - 0.0013$$
$$= 0.9987$$

(e) $P(z < -1.5) + P(z > 1.5)$

$$= 2P(z < -1.5)$$
$$= 2(0.0668)$$
$$= 0.1336$$

MARKOV CHAINS

10.1 Basic Properties of Markov Chains

Your Turn 1

If the probability is 0.68 that a Johnson customer will stay with Johnson the next time, the probability that a Johnson customer will switch to NorthClean is $1 - 0.68 = 0.32$. If the probability is 0.21 that a North-Clean customer will switch to Johnson, the probability that this customer will stay with NorthClean next time is $1 - 0.21 = 0.79$. The transition matrix is as follows:

$$
\begin{array}{c}
 \text{Second batch} \\
 \text{Johnson} \quad \text{NorthClean} \\
\begin{array}{c} \text{First batch} \end{array}
\begin{array}{c} \text{Johnson} \\ \text{NorthClean} \end{array}
\begin{bmatrix} 0.68 & 0.32 \\ 0.21 & 0.79 \end{bmatrix}
\end{array}
$$

Your Turn 2

The probability that a lower-class person will have an upper-class great-grandchild can be found from P^3, where P is the income-class transition matrix for a single generation. P^3 is given in the text.

$$
P^3 = \begin{bmatrix} 0.3825 & 0.4446 & 0.1730 \\ 0.2513 & 0.5147 & 0.2322 \\ 0.2369 & 0.4877 & 0.2745 \end{bmatrix}
$$

The first row and column correspond to lower-class, the second row and column correspond to middle-class, and the third row and column correspond to upper-class. Thus the probability we want is the entry in row 1, column 3, which is 0.1730.

Your Turn 3

The probability that a person bringing his first batch to Johnson will bring his fifth batch to Johnson is found in row 1, column 2 of C^4. (As in Example 3, the batch number is one less than the exponent.)

$$
\text{For} \quad C = \begin{bmatrix} 0.8 & 0.2 \\ 0.35 & 0.65 \end{bmatrix}, \quad C^4 = \begin{bmatrix} 0.6513 & 0.3487 \\ 0.6103 & 0.3897 \end{bmatrix}
$$

(rounded to four places).

The probability is 0.3487.

For the income class distribution discussed in Example 4, the initial probability vector X_0 is [0.21 0.68 0.11] and the transition matrix is

$$
P = \begin{bmatrix} 0.65 & 0.28 & 0.07 \\ 0.15 & 0.67 & 0.18 \\ 0.12 & 0.36 & 0.52 \end{bmatrix}.
$$

The income distribution after three generations is equal to

$$
X_0 P^3 = \begin{bmatrix} 0.21 & 0.68 & 0.11 \end{bmatrix} \begin{bmatrix} 0.65 & 0.28 & 0.07 \\ 0.15 & 0.67 & 0.18 \\ 0.12 & 0.36 & 0.52 \end{bmatrix}^3
$$

$$
= \begin{bmatrix} 0.21 & 0.68 & 0.11 \end{bmatrix} \begin{bmatrix} 0.3825 & 0.4446 & 0.1730 \\ 0.2531 & 0.5147 & 0.2322 \\ 0.2369 & 0.4877 & 0.2754 \end{bmatrix}
$$

$$
= \begin{bmatrix} 0.2785 & 0.4970 & 0.2245 \end{bmatrix}.
$$

The vector [0.2785 0.4970 0.2245] gives the income distribution after three generations, where the first column represents lower, the second middle, and the third upper.

10.1 Exercises

1. $\begin{bmatrix} \frac{2}{3} & \frac{1}{2} \end{bmatrix}$ could not be a probability vector because the sum of the entries is not equal to 1.

3. [0 1] could be a probability vector since it is a matrix of only one row, having nonnegative entries whose sum is 1.

5. [0.4 0.2 0] could not be a probability vector because the sum of the entries in the row is not equal to 1.

7. [0.07 0.04 0.37 0.52] Could be a probability vector. It is a matrix of only one row, having nonnegative entries whose sum is 1.

Your Turn 4

9. $\begin{bmatrix} 0.6 & 0 \\ 0 & 0.6 \end{bmatrix}$

The entries in each row do not add up to 1, so this could not be a transition matrix.

11. $\begin{bmatrix} \frac{1}{3} & \frac{2}{3} \\ \frac{1}{2} & \frac{1}{2} \end{bmatrix}$

This could be a transition matrix since it is a square matrix, all entries are between 0 and 1, inclusive, and the sum of the entries in each row is 1.

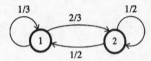

13. $\begin{bmatrix} \frac{1}{3} & \frac{1}{3} & \frac{1}{3} \\ 0 & 1 & 0 \\ \frac{1}{2} & 0 & \frac{1}{2} \end{bmatrix}$

This could be a transition matrix since it meets all the requirements given in the solution for Exercise 11.

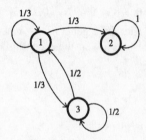

15. The transition diagram provides the information

$p_{AA} = 0.9$, $p_{AB} = 0.1$, $p_{AC} = 0$,

$p_{BA} = 0.1$, $p_{BB} = 0.7$, $p_{BC} = 0.2$,

$p_{CA} = 0$, $p_{CB} = 0.2$, and $p_{CC} = 0.8$.

The transition matrix associated with this diagram is

$$\begin{array}{c} \\ A \\ B \\ C \end{array} \begin{array}{ccc} A & B & C \\ \begin{bmatrix} 0.9 & 0.1 & 0 \\ 0.1 & 0.7 & 0.2 \\ 0 & 0.2 & 0.8 \end{bmatrix} \end{array}.$$

17. The given diagram does not yield a transition matrix because the sum of the probabilities from each of the three states is not 0. For example, the sum of the probabilities from state C is 2/3.

19. $A = \begin{bmatrix} 1 & 0 \\ 0.7 & 0.3 \end{bmatrix}$

$A^2 = \begin{bmatrix} 1 & 0 \\ 0.7 & 0.3 \end{bmatrix} \begin{bmatrix} 1 & 0 \\ 0.7 & 0.3 \end{bmatrix} = \begin{bmatrix} 1 & 0 \\ 0.91 & 0.09 \end{bmatrix}$

$A^3 = A \cdot A^2 = \begin{bmatrix} 1 & 0 \\ 0.7 & 0.3 \end{bmatrix} \begin{bmatrix} 1 & 0 \\ 0.91 & 0.09 \end{bmatrix}$

$= \begin{bmatrix} 1 & 0 \\ 0.973 & 0.027 \end{bmatrix}$

The entry in row 1, column 2 of A^3 gives the probability that state 1 changes to state 2 after 3 repetitions of the experiment. The probability is 0.

21. $C = \begin{bmatrix} 0 & 0 & 1 \\ 0.2 & 0.6 & 0.2 \\ 0.1 & 0.7 & 0.2 \end{bmatrix}$

$C^2 = \begin{bmatrix} 0.10 & 0.70 & 0.20 \\ 0.14 & 0.50 & 0.36 \\ 0.16 & 0.56 & 0.28 \end{bmatrix}$

$C^3 = \begin{bmatrix} 0.160 & 0.560 & 0.280 \\ 0.136 & 0.552 & 0.312 \\ 0.140 & 0.532 & 0.328 \end{bmatrix}$

The probability that state 1 changes to state 2 after three repetitions is the entry in row 1, column 2 of C^3, which is 0.560.

23. $E = \begin{bmatrix} 0.8 & 0.1 & 0.1 \\ 0.3 & 0.6 & 0.1 \\ 0 & 1 & 0 \end{bmatrix}$

$E^2 = \begin{bmatrix} 0.8 & 0.1 & 0.1 \\ 0.3 & 0.6 & 0.1 \\ 0 & 1 & 0 \end{bmatrix} \begin{bmatrix} 0.8 & 0.1 & 0.1 \\ 0.3 & 0.6 & 0.1 \\ 0 & 1 & 0 \end{bmatrix}$

$= \begin{bmatrix} 0.67 & 0.24 & 0.09 \\ 0.42 & 0.49 & 0.09 \\ 0.3 & 0.6 & 0.1 \end{bmatrix}$

$E^3 = \begin{bmatrix} 0.8 & 0.1 & 0.1 \\ 0.3 & 0.6 & 0.1 \\ 0 & 1 & 0 \end{bmatrix} \begin{bmatrix} 0.67 & 0.24 & 0.09 \\ 0.42 & 0.49 & 0.09 \\ 0.3 & 0.6 & 0.1 \end{bmatrix}$

$= \begin{bmatrix} 0.608 & 0.301 & 0.091 \\ 0.483 & 0.426 & 0.091 \\ 0.42 & 0.49 & 0.09 \end{bmatrix}$

The probability that state 1 changes to state 2 after 3 repetitions is 0.301, since that is the entry in row 1, column 2 of E^3.

25. The exercise should be solved by graphing calculator methods. The first five powers of the transition matrix are

$$A = \begin{bmatrix} 0.1 & 0.2 & 0.2 & 0.3 & 0.2 \\ 0.2 & 0.1 & 0.1 & 0.2 & 0.4 \\ 0.2 & 0.1 & 0.4 & 0.2 & 0.1 \\ 0.3 & 0.1 & 0.1 & 0.2 & 0.3 \\ 0.1 & 0.3 & 0.1 & 0.1 & 0.4 \end{bmatrix},$$

$$A^2 = \begin{bmatrix} 0.2 & 0.15 & 0.17 & 0.19 & 0.29 \\ 0.16 & 0.2 & 0.15 & 0.18 & 0.31 \\ 0.19 & 0.14 & 0.24 & 0.21 & 0.22 \\ 0.16 & 0.19 & 0.16 & 0.2 & 0.29 \\ 0.16 & 0.19 & 0.14 & 0.17 & 0.34 \end{bmatrix},$$

$$A^3 = \begin{bmatrix} 0.17 & 0.178 & 0.171 & 0.191 & 0.29 \\ 0.171 & 0.178 & 0.161 & 0.185 & 0.305 \\ 0.18 & 0.163 & 0.191 & 0.197 & 0.269 \\ 0.175 & 0.174 & 0.164 & 0.187 & 0.3 \\ 0.167 & 0.184 & 0.158 & 0.182 & 0.309 \end{bmatrix},$$

$$A^4 = \begin{bmatrix} 0.1731 & 0.175 & 0.1683 & 0.188 & 0.2956 \\ 0.1709 & 0.1781 & 0.1654 & 0.1866 & 0.299 \\ 0.1748 & 0.1718 & 0.1753 & 0.1911 & 0.287 \\ 0.1712 & 0.1775 & 0.1667 & 0.1875 & 0.2971 \\ 0.1706 & 0.1785 & 0.1641 & 0.1858 & 0.301 \end{bmatrix},$$

and

$$A^5 = \begin{bmatrix} 0.1719 & 0.1764 & 0.1678 & 0.1878 & 0.2961 \\ 0.1717 & 0.1769 & 0.1667 & 0.1872 & 0.2975 \\ 0.1729 & 0.1749 & 0.1701 & 0.1888 & 0.2933 \\ 0.1719 & 0.1765 & 0.1671 & 0.1874 & 0.297 \\ 0.1714 & 0.1773 & 0.1663 & 0.1870 & 0.2981 \end{bmatrix}.$$

The probability that state 2 changes to state 4 after 5 repetitions of the experiment is 0.1872, since that is the entry in row 2, column 4 of A^5.

27. (a) We are asked to show that

$$X_0(P^n) = (\cdots(((X_0P)P)P)\cdots P)$$

is true for any natural number n, where the expression on the right side of the equation has a total of n factors of P. This may be proven by mathematical induction on n.

When $n = 1$, the statement becomes $X_0P = (X_0P)$, which is obviously true. When $n = 2$, the statement becomes

$$X_0(P^2) = (X_0P)P, \text{ or } X_0(PP) = (X_0P)P,$$

which is true since matrix multiplication is associative. Next, assume the nth statement

is true in order to show that the $(n + 1)$st statement is true. That is, assume that

$$X_0(P^n) = (\cdots(((X_0P)P)P)\cdots P)$$

is true. Associativity plays a role here also.

$$\begin{aligned} X_0(P^{n+1}) &= X_0(P^n \cdot P) \\ &= (X_0P^n)P \\ &= (\cdots(((X_0P)P)P)\cdots P)P \end{aligned}$$

Conclude that

$$X_0(P^n) = (\cdots(((X_0P)P)P)\cdots P)$$

is true for any natural number n.

29. The probability vector is $[0.4 \quad 0.6]$.

(a) $[0.4 \quad 0.6]\begin{bmatrix} 0.8 & 0.2 \\ 0.35 & 0.65 \end{bmatrix} = [0.53 \quad 0.47]$

Thus, after 1 wk Johnson has a 53% market share, and NorthClean has a 47% share.

(b) $C^2 = \begin{bmatrix} 0.8 & 0.2 \\ 0.35 & 0.65 \end{bmatrix}\begin{bmatrix} 0.8 & 0.2 \\ 0.35 & 0.65 \end{bmatrix}$

$= \begin{bmatrix} 0.71 & 0.29 \\ 0.5075 & 0.4925 \end{bmatrix}$

$[0.4 \quad 0.6]\begin{bmatrix} 0.71 & 0.29 \\ 0.5075 & 0.4925 \end{bmatrix}$

$= [0.5885 \quad 0.4115]$

After 2 wk, Johnson has a 58.85% market share, and NorthClean has a 41.15% share.

(c) $C^3 = \begin{bmatrix} 0.8 & 0.2 \\ 0.35 & 0.65 \end{bmatrix}\begin{bmatrix} 0.71 & 0.29 \\ 0.5075 & 0.4925 \end{bmatrix}$

$= \begin{bmatrix} 0.6695 & 0.3305 \\ 0.5784 & 0.4216 \end{bmatrix}$

$[0.4 \quad 0.6]\begin{bmatrix} 0.6695 & 0.3305 \\ 0.5784 & 0.4216 \end{bmatrix}$

$= [0.6148 \quad 0.3852]$

After 3 wk, the shares are 61.48% and 38.52%, respectively

(d) $C^4 = \begin{bmatrix} 0.8 & 0.2 \\ 0.35 & 0.65 \end{bmatrix}\begin{bmatrix} 0.6695 & 0.3305 \\ 0.5784 & 0.4216 \end{bmatrix}$

$= \begin{bmatrix} 0.6513 & 0.3487 \\ 0.6103 & 0.3897 \end{bmatrix}$

$$[0.4 \quad 0.6] \begin{bmatrix} 0.6513 & 0.3487 \\ 0.6103 & 0.3897 \end{bmatrix}$$
$$= [0.6267 \quad 0.3733]$$

After 4 wk, the shares are 62.67% and 37.33%, respectively.

31. The transition matrix P is

$$\begin{array}{c} \\ G_0 \\ G_1 \\ G_2 \end{array} \begin{array}{ccc} G_0 & G_1 & G_2 \end{array} \\ \begin{bmatrix} 0.75 & 0.20 & 0.05 \\ 0 & 0.70 & 0.30 \\ 0 & 0 & 1 \end{bmatrix}$$

We have 50,000 new policyholders, all in G_0. The probability vector for these people is

$$\begin{array}{ccc} G_0 & G_1 & G_2 \end{array} \\ [1 \quad 0 \quad 0].$$

(a) After 1 year, the distribution of people in each group is

$$[1 \quad 0 \quad 0] \begin{bmatrix} 0.75 & 0.20 & 0.05 \\ 0 & 0.70 & 0.30 \\ 0 & 0 & 1 \end{bmatrix}$$
$$= [0.75 \quad 0.20 \quad 0.05].$$

There are

$$0.75(50,000) = 37,500 \text{ people in } G_0$$
$$0.2(50,000) = 10,000 \text{ people in } G_1$$
$$0.05(50,000) = 2500 \text{ people in } G_2.$$

(b) $P^2 = \begin{bmatrix} 0.75 & 0.20 & 0.05 \\ 0 & 0.70 & 0.30 \\ 0 & 0 & 1 \end{bmatrix} \begin{bmatrix} 0.75 & 0.20 & 0.05 \\ 0 & 0.70 & 0.30 \\ 0 & 0 & 1 \end{bmatrix}$

$$= \begin{bmatrix} 0.5625 & 0.29 & 0.1475 \\ 0 & 0.49 & 0.51 \\ 0 & 0 & 1 \end{bmatrix}$$

After 2 years, the distribution of people in each group is

$$[1 \quad 0 \quad 0] \begin{bmatrix} 0.5625 & 0.29 & 0.1475 \\ 0 & 0.49 & 0.51 \\ 0 & 0 & 1 \end{bmatrix}$$
$$= [0.5625 \quad 0.29 \quad 0.1475].$$

There are

$$0.5625(50,000) = 28,125 \text{ people in } G_0$$
$$0.29(50,000) = 14,500 \text{ people in } G_1$$
$$0.1475(50,000) = 7375 \text{ people in } G_2.$$

(c)

$$P^3 = \begin{bmatrix} 0.75 & 0.20 & 0.05 \\ 0 & 0.70 & 0.30 \\ 0 & 0 & 1 \end{bmatrix} \begin{bmatrix} 0.5625 & 0.29 & 0.1475 \\ 0 & 0.49 & 0.51 \\ 0 & 0 & 1 \end{bmatrix}$$

$$\approx \begin{bmatrix} 0.4219 & 0.3155 & 0.2626 \\ 0 & 0.343 & 0.657 \\ 0 & 0 & 1 \end{bmatrix}$$

After 3 year, the distribution of people in each group is

$$[1 \quad 0 \quad 0] \begin{bmatrix} 0.4219 & 0.3155 & 0.2626 \\ 0 & 0.343 & 0.657 \\ 0 & 0 & 1 \end{bmatrix}$$
$$\approx [0.4219 \quad 0.3155 \quad 0.2626].$$

There are

$$0.4219(50,000) = 21,094 \text{ people in } G_0$$
$$0.3155(50,000) = 15,775 \text{ people in } G_1$$
$$0.2626(50,000) = 13,131 \text{ peopele in } G_2.$$

(d)

$$P^4 = \begin{bmatrix} 0.75 & 0.20 & 0.05 \\ 0 & 0.70 & 0.30 \\ 0 & 0 & 1 \end{bmatrix} \begin{bmatrix} 0.4219 & 0.3155 & 0.2626 \\ 0 & 0.343 & 0.657 \\ 0 & 0 & 1 \end{bmatrix}$$

$$\approx \begin{bmatrix} 0.3164 & 0.3052 & 0.3784 \\ 0 & 0.2401 & 0.7599 \\ 0 & 0 & 1 \end{bmatrix}$$

After 4 years, the distribution of people in each group is

$$[1 \quad 0 \quad 0] \begin{bmatrix} 0.3164 & 0.3052 & 0.3784 \\ 0 & 0.2401 & 0.7599 \\ 0 & 0 & 1 \end{bmatrix}$$
$$\approx [0.3164 \quad 0.3052 \quad 0.3784].$$

There are

$$0.3164(50,000) = 15,820 \text{ people in } G_0$$
$$0.3052(50,000) = 15,261 \text{ people in } G_1$$
$$0.3784(50,000) = 18,918 \text{ people in } G_2.$$

33. (a) Letting e, f, and s stand for electric, fossil and solar, the transition matrix is

$$\begin{array}{c} \\ e \\ f \\ s \end{array} \begin{array}{ccc} e & f & s \end{array} \\ \begin{bmatrix} 0.9 & 0.05 & 0.05 \\ 0.15 & 0.75 & 0.1 \\ 0 & 0 & 1 \end{bmatrix}.$$

We will call this transition matrix P. The initial probability vector is

$$[0.35 \quad 0.60 \quad 0.05].$$

(b) The share held by each type after 1 year is given by

$$[0.35 \quad 0.60 \quad 0.05]P$$
$$= [0.4050 \quad 0.4675 \quad 0.1275].$$

(c) The share held by each type after 2 years is given by

$$[0.35 \quad 0.60 \quad 0.05]P^2$$
$$= [0.4346 \quad 0.3709 \quad 0.1945].$$

(d) The share held by each type after 3 years is given by

$$[0.35 \quad 0.60 \quad 0.05]P^3$$
$$= [0.4468 \quad 0.2999 \quad 0.2533]$$

(e) Continuing the computation, we find that

$$[0.35 \quad 0.60 \quad 0.05]P^9$$
$$= [0.3740 \quad 0.1221 \quad 0.5039],$$

so the goal is met after 9 years.

35. The transition matrix P is

$$\begin{array}{c} \\ \text{Small} \\ \text{Medium} \\ \text{Large} \end{array} \begin{array}{ccc} \text{Small} & \text{Medium} & \text{Large} \\ \begin{bmatrix} 0.9216 & 0.0780 & 0.0004 \\ 0.0460 & 0.8959 & 0.0581 \\ 0.0003 & 0.0301 & 0.9696 \end{bmatrix} \end{array}.$$

The initial distribution of businesses in 2011 is given by $[2094 \quad 2363 \quad 2378]$.

(a) The expected number of business of each type in 2012 is given by

$$[2094 \quad 2363 \quad 2378]P$$
$$= [2094 \quad 2363 \quad 2378] \begin{bmatrix} 0.9216 & 0.0780 & 0.0004 \\ 0.0460 & 0.8959 & 0.0581 \\ 0.0003 & 0.0301 & 0.9696 \end{bmatrix}$$
$$= [2039.24 \quad 2351.92 \quad 2443.84].$$

Rounding to whole numbers we get 2039 small, 2352 medium, and 2444 large businesses. Note that we have assumed that the total number of businesses stays constant at 6835.

(b) The expected number of business of each type in 2013 is given by

$$[2039.24 \quad 2351.92 \quad 244.84]P$$
$$= [2039.24 \quad 2351.92 \quad 2443.84] \begin{bmatrix} 0.9216 & 0.0780 & 0.0004 \\ 0.0460 & 0.8959 & 0.0581 \\ 0.0003 & 0.0301 & 0.9696 \end{bmatrix}$$
$$= [1988.29 \quad 2339.71 \quad 2507.01].$$

Rounding to whole numbers we get 1988 small, 2340 medium, and 2507 large businesses.

(c) For a two-year period the transition matrix is P^2.

$$P^2 = \begin{bmatrix} 0.8529 & 0.1418 & 0.0053 \\ 0.0836 & 0.8080 & 0.1084 \\ 0.0020 & 0.0562 & 0.9419 \end{bmatrix}$$

(d) The percentage of medium businesses that are small in two years is the entry in row 2, column 1 of P^2, which is 0.0836 or 8.36%. The percentage of medium businesses that are large in two years is the entry in row 2, column 3 of P^2, which is 0.1084 or 10.84%.

37. (a) Since the transition matrix

$$P = \begin{bmatrix} \frac{5}{7} & \frac{2}{7} & 0 & 0 \\ 0 & \frac{1}{2} & \frac{1}{3} & \frac{1}{6} \\ 0 & 0 & \frac{1}{2} & \frac{1}{2} \\ 0 & 0 & \frac{1}{4} & \frac{3}{4} \end{bmatrix}$$

describes the proportional change of rabbits among various immune response classifications from one week to the next, we need to consider the entry in the first row, first column of P^5. Because P has zero entries in rows two through four of column one, the proportion of the rabbits in group 1 that were still in group 1 five weeks later was $\frac{5^5}{7^5} \approx 0.1859$.

(b) Find the product $X_0 P^4$, where $X_0 = [9 \ 4 \ 0 \ 0]$.

$X_0 P^4 = [2.34 \quad 2.62 \quad 3.47 \quad 4.56]$, where the entries are approximate. Therefore, after 4 weeks there are approximately 2.34 rabbits in group 1, 2.62 rabbits in group 2, 3.47 in group 3, and 4.56 in group 4.

(c) Using a graphing calculator, when P is raised to a larger and larger power, the entries in the first row, first two columns, and the entries in the second row, first two columns are either zero or positive numbers that are getting smaller—that is, they are approaching zero. This leads to the conclusion that the long-range probability of rabbits in group 1 or 2 staying in group 1 or 2 is zero.

39. $P = \begin{bmatrix} 0.71 & 0.19 & 0.03 & 0.07 & 0.00 \\ 0.52 & 0.31 & 0.04 & 0.13 & 0.00 \\ 0.40 & 0.16 & 0.22 & 0.22 & 0.00 \\ 0.34 & 0.19 & 0.09 & 0.37 & 0.01 \\ 0.21 & 0.25 & 0.13 & 0.37 & 0.04 \end{bmatrix}$

The initial probability vector is
$X_0 = [0.2 \ 0.2 \ 0.2 \ 0.2 \ 0.2]$.

(a)   After 1 year,

$X_0 P = [0.2 \ 0.2 \ 0.2 \ 0.2 \ 0.2] \begin{bmatrix} 0.71 & 0.19 & 0.03 & 0.07 & 0.00 \\ 0.52 & 0.31 & 0.04 & 0.13 & 0.00 \\ 0.40 & 0.16 & 0.22 & 0.22 & 0.00 \\ 0.34 & 0.19 & 0.09 & 0.37 & 0.01 \\ 0.21 & 0.25 & 0.13 & 0.37 & 0.04 \end{bmatrix}$

$= [0.436 \ 0.220 \ 0.102 \ 0.232 \ 0.010]$

The percentages of the population that can be expected in each type of housing after one year are: 43.6% in type I, 22% in type II, 10.2% in type III, 23.2% in type IV, and 1% in type V.

(b)   After 2 years,

$[0.436 \ 0.220 \ 0.102 \ 0.232 \ 0.010]P$

$= [0.436 \ 0.220 \ 0.102 \ 0.232 \ 0.010] \begin{bmatrix} 0.71 & 0.19 & 0.03 & 0.07 & 0.00 \\ 0.52 & 0.31 & 0.04 & 0.13 & 0.00 \\ 0.40 & 0.16 & 0.22 & 0.22 & 0.00 \\ 0.34 & 0.19 & 0.09 & 0.37 & 0.01 \\ 0.21 & 0.25 & 0.13 & 0.37 & 0.04 \end{bmatrix}$

$\approx [0.5457 \ 0.2139 \ 0.0665 \ 0.1711 \ 0.0027]$.

The percentages of the population that can be expected in each type of housing after two years are: 54.57% in type I, 21.39% in type II, 6.65% in type III, 17.11% in type IV, and 0.27% in type V.

(c)

$P^2 = \begin{bmatrix} 0.71 & 0.19 & 0.03 & 0.07 & 0.00 \\ 0.52 & 0.31 & 0.04 & 0.13 & 0.00 \\ 0.40 & 0.16 & 0.22 & 0.22 & 0.00 \\ 0.34 & 0.19 & 0.09 & 0.37 & 0.01 \\ 0.21 & 0.25 & 0.13 & 0.37 & 0.04 \end{bmatrix} \begin{bmatrix} 0.71 & 0.19 & 0.03 & 0.07 & 0.00 \\ 0.52 & 0.31 & 0.04 & 0.13 & 0.00 \\ 0.40 & 0.16 & 0.22 & 0.22 & 0.00 \\ 0.34 & 0.19 & 0.09 & 0.37 & 0.01 \\ 0.21 & 0.25 & 0.13 & 0.37 & 0.04 \end{bmatrix}$

$= \begin{bmatrix} 0.6387 & 0.2119 & 0.0418 & 0.1069 & 0.0007 \\ 0.5906 & 0.2260 & 0.0485 & 0.1336 & 0.0013 \\ 0.5300 & 0.2026 & 0.0866 & 0.1786 & 0.0022 \\ 0.5041 & 0.2107 & 0.0722 & 0.2089 & 0.0041 \\ 0.4653 & 0.2185 & 0.0834 & 0.2275 & 0.0053 \end{bmatrix}$

(d)   The percent of those living in a commercial area (type V) that can be expected to be living in a middle class family household (type I) after 2 years is 0.4653, or 46.53%.

(e)   The percent of those living in a commercial area (type V) that can be expected to be living in an unsound rented multi-family dwelling (type IV) after 2 years is 0.2275, or 22.75%.

41. **(a)**

$\begin{bmatrix} 0.443 & 0.364 & 0.193 \\ 0.277 & 0.436 & 0.287 \\ 0.266 & 0.304 & 0.430 \end{bmatrix} \begin{bmatrix} 0.443 & 0.364 & 0.193 \\ 0.277 & 0.436 & 0.287 \\ 0.266 & 0.304 & 0.430 \end{bmatrix}$

$= \begin{bmatrix} 0.3484 & 0.3786 & 0.2730 \\ 0.3198 & 0.3782 & 0.3020 \\ 0.3164 & 0.3601 & 0.3235 \end{bmatrix}$

(b)   The desired probability is found in row 1, column 1. The probability that if England won the last game, England will win the game after the next one is 0.3484.

(c)   The desired probability is found in row 2, column 1. The probability that if Australia won the last game, England will win the game after the next one is 0.3198.

10.2   Regular Markov Chains

Your Turn 1

Since the matrix

$$C^2 = \begin{bmatrix} 0.21 & 0.09 & 0.70 \\ 0.27 & 0.52 & 0.21 \\ 0.20 & 0.02 & 0.78 \end{bmatrix}$$

has all positive entries, the matrix C is regular.

Your Turn 2

The transition matrix is

$$T = \begin{bmatrix} 0.7 & 0.3 \\ 0.2 & 0.8 \end{bmatrix},$$

we need to solve

$$[v_1 \quad v_2]\begin{bmatrix} 0.7 & 0.3 \\ 0.2 & 0.8 \end{bmatrix} = [v_1 \quad v_2],$$

with the condition $v_1 + v_2 = 1$.

As noted in Example 2, the two equations we get from the matrix equation $VP = V$ will always be dependent, so we only need to look at the first column of our product:

$$[v_1 \quad v_2]\begin{bmatrix} 0.7 & 0.3 \\ 0.2 & 0.8 \end{bmatrix} = [0.7v_1 + 0.2v_2 \quad ....]$$

This gives us

$$0.7v_1 + 0.2v_2 = v_1$$

or

$$-0.3v_1 + 0.2v_2 = 0$$

We now solve the following system:

$$-\frac{3}{10}v_1 + \frac{1}{5}v_2 = 0$$
$$v_1 + v_2 = 1$$

Multiply the first equation by 10/3 and add it to the second equation:

$$-\frac{3}{10}v_1 + \frac{1}{5}v_2 = 0$$
$$v_1 + \quad v_2 = 1$$
$$\frac{5}{3}v_2 = 1, \quad \frac{10}{3}R_1 + R_2 \to R_2$$

so $v_2 = \dfrac{3}{5} = 0.6.$

Since $v_1 + v_2 = 1, v_1 = 0.4.$ The long range trend is [0.4 0.6].

Your Turn 3

The transition matrix is

$$K = \begin{bmatrix} 0.2 & 0.1 & 0.7 \\ 0.3 & 0.7 & 0 \\ 0.2 & 0 & 0.8 \end{bmatrix}.$$

We first need to solve

$$[v_1 \quad v_2 \quad v_3]\begin{bmatrix} 0.2 & 0.1 & 0.7 \\ 0.3 & 0.7 & 0 \\ 0.2 & 0 & 0.8 \end{bmatrix} = [v_1 \quad v_2 \quad v_3].$$

Multiplying on the left, setting the first, second and third columns of the product equal to $v_1, v_2,$ and $v_3,$ and simplifying gives us three equations, and as a fourth equation we add the condition that the sum of the entries in the equilibrium vector must be 1. This produces the following system of four equations in three unknowns:

$$-0.8v_1 + 0.3v_2 + 0.2v_3 = 0$$
$$0.1v_1 - 0.3v_2 \qquad = 0$$
$$0.7v_1 \qquad - 0.2v_3 = 0$$
$$v_1 + v_2 + v_3 = 1$$

We know that the first three equations form a dependent system, and we can see that the first equation is the negative of the sum of the second and third equations, so to save ourselves some work we'll discard the first equation and solve this system:

$$0.1v_1 - 0.3v_2 \qquad = 0$$
$$0.7v_1 \qquad - 0.2v_3 = 0$$
$$v_1 + v_2 + v_3 = 1$$

We write the augmented matrix and transform it:

$$\begin{bmatrix} 0.1 & -0.3 & 0 & | & 0 \\ 0.7 & 0 & -0.2 & | & 0 \\ 1 & 1 & 1 & | & 1 \end{bmatrix}$$

$$\begin{matrix} 10R_1 \to R_1 \\ 10R_2 \to R_2 \end{matrix}\begin{bmatrix} 1 & -3 & 0 & | & 0 \\ 7 & 0 & -2 & | & 0 \\ 1 & 1 & 1 & | & 1 \end{bmatrix}$$

$$-7R_1 + R_2 \to R_2 \begin{bmatrix} 1 & -3 & 0 & | & 0 \\ 0 & 21 & -2 & | & 0 \\ 1 & 1 & 1 & | & 1 \end{bmatrix}$$

$$-R_1 + R_3 \to R_3 \begin{bmatrix} 1 & -3 & 0 & | & 0 \\ 0 & 21 & -2 & | & 0 \\ 1 & 4 & 1 & | & 1 \end{bmatrix}$$

$$\frac{1}{21}R_2 \to R_2 \begin{bmatrix} 1 & -3 & 0 & | & 0 \\ 0 & 1 & -\frac{2}{21} & | & 0 \\ 0 & 4 & 1 & | & 1 \end{bmatrix}$$

$$-4R_2 + R_3 \to R_3 \begin{bmatrix} 1 & -3 & 0 & | & 0 \\ 0 & 1 & -\frac{2}{21} & | & 0 \\ 0 & 0 & \frac{29}{21} & | & 1 \end{bmatrix}$$

$$\frac{21}{29}R_3 \to R_3 \begin{bmatrix} 1 & -3 & 0 & | & 0 \\ 0 & 1 & -\frac{2}{21} & | & 0 \\ 0 & 0 & 1 & | & \frac{21}{29} \end{bmatrix}$$

The matrix is now in echelon form, and we see

that $v_3 = \dfrac{21}{29}$. $v_2 = \left(\dfrac{2}{21}\right)\left(\dfrac{21}{29}\right) = \dfrac{2}{29}$, and using

the sum-to-1 condition, this means that $v_3 = \dfrac{6}{29}$.

The equilibrium vector is

$$\left[\dfrac{6}{29} \quad \dfrac{2}{29} \quad \dfrac{21}{29}\right] \approx [0.2069 \quad 0.0690 \quad 0.7241].$$

10.2 Exercises

1. Let $A = \begin{bmatrix} 0.2 & 0.8 \\ 0.9 & 0.1 \end{bmatrix}$.

    A is a regular transition matrix since $A^1 = A$
    contains all positive entries.

3. Let $B = \begin{bmatrix} 1 & 0 \\ 0.65 & 0.35 \end{bmatrix}$.

    $$B^2 = \begin{bmatrix} 1 & 0 \\ 0.65 & 0.35 \end{bmatrix}\begin{bmatrix} 1 & 0 \\ 0.65 & 0.35 \end{bmatrix}$$

    $$= \begin{bmatrix} 1 & 0 \\ 0.88 & 0.12 \end{bmatrix}$$

    B is not regular since any power of B will have
    $[1 \quad 0]$ as its first row and thus cannot have all
    positive entries.

5. Let $P = \begin{bmatrix} 0 & 1 & 0 \\ 0.4 & 0.2 & 0.4 \\ 1 & 0 & 0 \end{bmatrix}$.

    $$P^2 = \begin{bmatrix} 0 & 1 & 0 \\ 0.4 & 0.2 & 0.4 \\ 1 & 0 & 0 \end{bmatrix}\begin{bmatrix} 0 & 1 & 0 \\ 0.4 & 0.2 & 0.4 \\ 1 & 0 & 0 \end{bmatrix}$$

    $$= \begin{bmatrix} 0.4 & 0.2 & 0.4 \\ 0.48 & 0.44 & 0.08 \\ 0 & 1 & 0 \end{bmatrix}$$

    $$P^3 = \begin{bmatrix} 0 & 1 & 0 \\ 0.4 & 0.2 & 0.4 \\ 1 & 0 & 0 \end{bmatrix}\begin{bmatrix} 0.4 & 0.2 & 0.4 \\ 0.48 & 0.44 & 0.08 \\ 0 & 1 & 0 \end{bmatrix}$$

    $$= \begin{bmatrix} 0.48 & 0.44 & 0.08 \\ 0.256 & 0.568 & 0.176 \\ 0.4 & 0.2 & 0.4 \end{bmatrix}$$

    P is a regular transition matrix since P^3 contains all
    positive entries.

7. Let $P = \begin{bmatrix} \frac{1}{4} & \frac{3}{4} \\ \frac{1}{2} & \frac{1}{2} \end{bmatrix}$, and let V be the probability

    vector $[v_1 \quad v_2]$. We want to find V such that

    $$VP = V,$$

    or $[v_1 \quad v_2]\begin{bmatrix} \frac{1}{4} & \frac{3}{4} \\ \frac{1}{2} & \frac{1}{2} \end{bmatrix} = [v_1 \quad v_2].$

    Use matrix multiplication on the left to obtain

    $$\left[\tfrac{1}{4}v_1 + \tfrac{1}{2}v_2 \quad \tfrac{3}{4}v_1 + \tfrac{1}{2}v_2\right] = [v_1 \quad v_2].$$

    Set corresponding entries from the two matrices
    equal to get

    $$\tfrac{1}{4}v_1 + \tfrac{1}{2}v_2 = v_1$$
    $$\tfrac{3}{4}v_1 + \tfrac{1}{2}v_2 = v_2.$$

    Multiply both equations by 4 to eliminate fractions.

    $$v_1 + 2v_2 = 4v_1$$
    $$3v_1 + 2v_2 = 4v_2$$

    Simplify both equations.

    $$-3v_1 + 2v_2 = 0$$
    $$3v_1 - 2v_2 = 0$$

    This is a dependent system. To find the values of v_1
    and v_2, an additional equation is needed. Since
    $V = [v_1 \quad v_2]$ is a probability vector,

    $$v_1 + v_2 = 1.$$

    To find v_1 and v_2, solve the system

    $$-3v_1 + 2v_2 = 0 \qquad (1)$$
    $$v_1 + v_2 = 1. \qquad (2)$$

    From equation (2), $v_1 = 1 - v_2$. Substitute $1 - v_2$
    for v_1 in equation (1) to obtain

    $$-3(1 - v_2) + 2v_2 = 0$$
    $$-3 + 3v_2 + 2v_2 = 0$$
    $$-3 + 5v_2 = 5$$
    $$v_2 = \tfrac{3}{5}.$$

    Since $v_1 = 1 - v_2$, $v_1 = \tfrac{2}{5}$, and the equilibrium
    vector is

    $$V = \left[\tfrac{2}{5} \quad \tfrac{3}{5}\right].$$

9. Let $P = \begin{bmatrix} 0.4 & 0.6 \\ 0.3 & 0.7 \end{bmatrix}$, and let V be the probability

vector $[v_1 \quad v_2]$. We want to find V such that

$$VP = V$$

$$[v_1 \quad v_2] \begin{bmatrix} 0.4 & 0.6 \\ 0.3 & 0.7 \end{bmatrix} = [v_1 \quad v_2].$$

By matrix multiplication and equality of matrices,

$$0.4v_1 + 0.3v_2 = v_1$$
$$0.6v_1 + 0.7v_2 = v_2.$$

Simplify these equations to get the dependent system

$$-0.6v_1 + 0.3v_2 = 0$$
$$0.6v_1 - 0.3v_2 = 0.$$

Since V is a probability vector,

$$v_1 + v_2 = 1.$$

To find v_1 and v_2, solve the system

$$0.6v_1 - 0.3v_2 = 0$$
$$v_1 + v_2 = 1$$

by the substitution method. Observe that $v_2 = 1 - v_1$.

$$0.6v_1 - 0.3(1 - v_1) = 0$$
$$0.9v_1 - 0.3 = 0$$
$$v_1 = \frac{0.3}{0.9} = \frac{1}{3}$$
$$v_2 = 1 - \frac{1}{3} = \frac{2}{3}$$

The equilibrium vector is $\begin{bmatrix} \frac{1}{3} & \frac{2}{3} \end{bmatrix}$.

11. Let V be the probability vector $[v_1 \quad v_2 \quad v_3]$.

$$VP = V$$

$$[v_1 \ v_2 \ v_3] \begin{bmatrix} 0.1 & 0.1 & 0.8 \\ 0.4 & 0.3 & 0.3 \\ 0.1 & 0.2 & 0.7 \end{bmatrix} = [v_1 \ v_2 \ v_3]$$

$$0.1v_1 + 0.4v_2 + 0.1v_3 = v_1$$
$$0.1v_1 + 0.3v_2 + 0.2v_3 = v_2$$
$$0.8v_1 + 0.3v_2 + 0.7v_3 = v_3$$

Simplify the equations to get the dependent system

$$-0.9v_1 + 0.4v_2 + 0.1v_3 = 0$$
$$0.1v_1 - 0.7v_2 + 0.2v_3 = 0$$
$$0.8v_1 + 0.3v_2 - 0.3v_3 = 0.$$

Since V is a probability vector,

$$v_1 + v_2 + v_3 = 1.$$

Solving the above system of four equations using the Gauss-Jordan method, we obtain

$$v_1 = \frac{5}{31}, v_2 = \frac{19}{93}, v_3 = \frac{59}{93}.$$

The equilibrium vector is $\begin{bmatrix} \frac{5}{31} & \frac{19}{93} & \frac{59}{93} \end{bmatrix}$.

13. Let V be the probability vector $[v_1 \quad v_2 \quad v_3]$.

$$[v_1 \ v_2 \ v_3] \begin{bmatrix} 0.25 & 0.35 & 0.4 \\ 0.1 & 0.3 & 0.6 \\ 0.55 & 0.4 & 0.05 \end{bmatrix} = [v_1 \ v_2 \ v_3]$$

$$0.25v_1 + 0.1v_2 + 0.55v_3 = v_1$$
$$0.35v_1 + 0.3v_2 + 0.4v_3 = v_2$$
$$0.4v_1 + 0.6v_2 + 0.05v_3 = v_3$$

Simplify the equations to get the dependent system

$$-0.75v_1 + 0.1v_2 + 0.55v_3 = 0$$
$$0.35v_1 - 0.7v_2 + 0.4v_3 = 0$$
$$0.4v_1 + 0.6v_2 - 0.95v_3 = 0.$$

Since V is a probability vector,

$$v_1 + v_2 + v_3 = 1.$$

Solving this system we obtain

$$v_1 = \frac{170}{563}, v_2 = \frac{197}{563}, v_3 = \frac{196}{563}.$$

Thus, the equilibrium vector is

$$\begin{bmatrix} \frac{170}{563} & \frac{197}{563} & \frac{196}{563} \end{bmatrix}.$$

15. Let $P = \begin{bmatrix} 0.75 & 0.20 & 0.05 \\ 0.10 & 0.70 & 0.20 \\ 0.10 & 0.30 & 0.60 \end{bmatrix}$ and let V be the

probability vector $[v_1 \quad v_2 \quad v_3]$.

$$VP = V$$

$$[v_1 \ v_2 \ v_3] \begin{bmatrix} 0.75 & 0.20 & 0.05 \\ 0.10 & 0.70 & 0.20 \\ 0.10 & 0.30 & 0.60 \end{bmatrix} = [v_1 \ v_2 \ v_3].$$

$$0.75v_1 + 0.10v_2 + 0.10v_3 = v_1$$
$$0.20v_1 + 0.70v_2 + 0.30v_3 = v_2$$
$$0.05v_1 + 0.20v_2 + 0.60v_3 = v_3$$

Simplify the equations and use the equation $v_1 + v_2 + v_3 = 1$ to get the system

$$-0.25v_1 + 0.10v_2 + 0.10v_3 = 0$$
$$0.20v_1 - 0.30v_2 + 0.30v_3 = 0$$
$$0.05v_1 + 0.20v_2 - 0.40v_3 = 0$$
$$v_1 + \quad v_2 + \quad v_3 = 0$$

Solve the system using the Gauss-Jordan method to obtain

$$v_1 = \frac{2}{7}, v_2 = \frac{19}{42}, v_3 = \frac{11}{42}.$$

The equilibrium vector is $\begin{bmatrix} \frac{2}{7} & \frac{19}{42} & \frac{11}{42} \end{bmatrix}$.

17. $[v_1 \ v_2 \ v_3] \begin{bmatrix} 0.9 & 0.05 & 0.05 \\ 0.15 & 0.75 & 0.1 \\ 0 & 0 & 1 \end{bmatrix}$

$$= [v_1 \ v_2 \ v_3].$$

$$0.9v_1 + 0.15v_2 \qquad = v_1$$
$$0.05v_1 + 0.75v_2 \qquad = v_2$$
$$0.05v_1 + 0.1v_2 + v_3 = v_3$$

Also, $v_1 + v_2 + v_3 = 1$.

The second and third equations simplify to

$$0.05v_1 - 0.25v_2 = 0$$
$$0.05v_1 + 0.1v_2 = 0$$

This pair of equations implies that $v_2 = 0$ and hence $v_1 = 0$, which requires $v_3 = 1$. So the equilibrium vector is $[0 \ 0 \ 1]$.

19. $[v_1 \ v_2 \ v_3] \begin{bmatrix} 0.80 & 0.15 & 0.05 \\ 0.20 & 0.70 & 0.10 \\ 0.20 & 0.20 & 0.60 \end{bmatrix}$

$$= [v_1 \ v_2 \ v_3]$$

$$0.80v_1 + 0.20v_2 + 0.20v_3 = v_1$$
$$0.15v_1 + 0.70v_2 + 0.20v_3 = v_2$$
$$0.05v_1 + 0.10v_2 + 0.60v_3 = v_3$$

Solving this system, we obtain

$$v_1 = \frac{1}{2}, v_2 = \frac{7}{20}, v_3 = \frac{3}{20}.$$

The equilibrium vector is

$$\begin{bmatrix} \frac{1}{2} & \frac{7}{20} & \frac{3}{20} \end{bmatrix}.$$

21. Let V be the probability vector $[x_1 \ x_2]$. We want to find V such that

$$V \begin{bmatrix} p & 1-p \\ 1-q & q \end{bmatrix} = V.$$

The system of equations is

$$px_1 + (1-q)x_2 = x_1$$
$$(1-p)x_1 + qx_2 = x_2.$$

Collecting like terms and simplifying leads to

$$(p-1)x_1 + (1-q)x_2 = 0,$$

so

$$x_1 = \frac{1-q}{1-p}x_2.$$

Substituting this into $x_1 + x_2 = 1$, we obtain

$$\frac{1-q}{1-p}x_2 + x_2 = 1$$

or

$$\frac{2-p-q}{1-p}x_2 = 1;$$

therefore,

$$x_2 = \frac{1-p}{2-p-q}$$

and

$$x_1 = \frac{1-q}{2-p-q},$$

so

$$V = \begin{bmatrix} \dfrac{1-q}{2-p-q} & \dfrac{1-p}{2-p-q} \end{bmatrix}.$$

Since $0 < p < 1$ and $0 < q < 1$, the matrix is always regular.

23. Let V be the probability vector $[x_1 \ x_2]$.

We have $P = \begin{bmatrix} a_{11} & a_{12} \\ a_{21} & a_{22} \end{bmatrix}$,

where $a_{11} + a_{21} = 1$

and $a_{12} + a_{22} = 1$.

The resulting equations are

$$a_{11}x_1 + a_{21}x_2 = x_1$$
$$a_{12}x_1 + a_{22}x_2 = x_2,$$

which we simplify to

$$(a_{11} - 1)x_1 + a_{21}x_2 = 0$$
$$a_{12}x_1 + (a_{22} - 1)x_2 = 0.$$

Hence, $x_1 = \dfrac{a_{21}}{1-a_{11}}x_2 = \dfrac{a_{21}}{a_{21}}x_2 = x_2,$

which we substitute into $x_1 + x_2 = 1$, obtaining

$$x_2 + x_2 = 1$$
$$2x_2 = 1$$
$$x_2 = \frac{1}{2}$$

and, therefore,

$$x_1 = 1 - \frac{1}{2} = \frac{1}{2}.$$

The equilibrium vector is

$$\begin{bmatrix} \frac{1}{2} & \frac{1}{2} \end{bmatrix}.$$

25. The transition matrix for the given information is

$$\begin{array}{c} \\ \text{Works} \\ \text{Doesn't Work} \end{array} \begin{array}{c} \text{Works} \quad \text{Doesn't Work} \\ \begin{bmatrix} 0.9 & 0.1 \\ 0.8 & 0.2 \end{bmatrix}. \end{array}$$

Let V be the probability vector $[v_1 \quad v_2]$.

$$\begin{bmatrix} v_1 & v_2 \end{bmatrix} \begin{bmatrix} 0.9 & 0.1 \\ 0.8 & 0.2 \end{bmatrix} = \begin{bmatrix} v_1 & v_2 \end{bmatrix}$$

$$0.9v_1 + 0.8v_2 = v_1$$
$$0.1v_1 + 0.2v_2 = v_2$$

Simplify these equations to get the dependent system

$$-0.1v_1 + 0.8v_2 = 0$$
$$0.1v_1 - 0.8v_2 = 0.$$

Also, $v_1 + v_2 = 1$, so $v_1 = 1 - v_2$.

$$0.1(1 - v_2) - 0.8v_2 = 0$$
$$0.1 - 0.9v_2 = 0$$

$$v_2 = \frac{1}{9}, v_1 = \frac{8}{9}$$

The equilibrium vector is

$$\begin{bmatrix} \frac{8}{9} & \frac{1}{9} \end{bmatrix}.$$

The long-range probability that the line will work correctly is $\frac{8}{9}$.

27. This exercise should be solved by graphing calculator methods. The solution may vary. The answers are as follows.

(a) [0.4 0.6]; [0.53 0.47]; [0.5885 0.4115]; [0.614825 0.385175]; [0.626671 0.373329]; [0.632002 0.367998]; [0.634401 0.365599]; [0.635480 0.364520]; [0.635966 0.364034]; [0.636185 0.363815]

(b) 0.236364; 0.106364; 0.047864; 0.021539; 0.009693; 0.004362; 0.001963; 0.000884; 0.000398; 0.000179

(c) The ratio is roughly 0.45 for each week.

(d) Each week, the difference between the probability vector and the equilibrium vector is slightly less than half of what it was the previous week.

(e) [0.75 0.25]; [0.6875 0.3125]; [0.659375 0.340625]; [0.646719 0.353281]; [0.641023 0.358977]; [0.638461 0.361539]; [0.637307 0.362693]; [0.636788 0.363212]; [0.636555 0.363445]; [0.636450 0.363550];

0.113636; 0.051136; 0.023011; 0.010355; 0.004659; 0.002097; 0.000943; 0.000424; 0.000191; 0.000086

The ratio is roughly 0.45 for each week, which is the same conclusion as before.

29. Let V be the probability vector $[v_1 \quad v_2 \quad v_3]$.

$$VP = V$$

$$\begin{bmatrix} v_1 & v_2 & v_3 \end{bmatrix} \begin{bmatrix} 0.9216 & 0.0780 & 0.0004 \\ 0.0460 & 0.8959 & 0.0581 \\ 0.0003 & 0.0301 & 0.9696 \end{bmatrix} = \begin{bmatrix} v_1 & v_2 & v_3 \end{bmatrix}$$

$$0.921v_1 + 0.0460v_2 + 0.0003v_3 = v_1$$
$$0.0780v_1 + 0.8959v_2 + 0.0301v_3 = v_2$$
$$0.0004v_1 + 0.0581v_2 + 0.9696v_3 = v_3$$

Simplify the equations and use the equation $v_1 + v_2 + v_3 = 1$ to get the system

$$-0.0784v_1 + 0.0460v_2 + 0.0003v_3 = 0$$
$$0.0780v_1 - 0.1041v_2 + 0.0301v_3 = 0$$
$$0.0004v_1 + 0.0581v_2 - 0.0304v_3 = 0$$
$$v_1 + \quad v_2 + \quad v_3 = 1.$$

Using a graphing calculator, the solution to the system is

$$v_1 = 0.1691, v_2 = 0.2847, v_3 = 0.5462.$$

This means, if these trends continue, eventually 16.91% of the businesses will be small, 28.47% will be medium, and 54.62% will be large.

31. The family structure transition matrix F is as follows:

	C	M	F	R	O
Couple	0.566	0.309	0.083	0.023	0.019
Mother	0.392	0.453	0.066	0.061	0.028
Father	0.391	0.354	0.109	0.108	0.038
Relative	0.307	0.558	0.040	0.056	0.039
Other	0.337	0.320	0.025	0.252	0.066

Now we need to solve

$$[v_1 \quad v_2 \quad v_3 \quad v_4 \quad v_5]F = [v_1 \quad v_2 \quad v_3 \quad v_4 \quad v_5]$$

or

$$[v_1 \quad v_2 \quad v_3 \quad v_4 \quad v_5]F - [v_1 \quad v_2 \quad v_3 \quad v_4 \quad v_5]$$
$$= [0 \quad 0 \quad 0 \quad 0 \quad 0].$$

Simplifying the left side gives us the following system:

$$-0.434v_1 + 0.392v_2 + 0.391v_3 + 0.307v_4 + 0.337v_5 = 0$$
$$0.309v_1 - 0.547v_2 + 0.354v_3 + 0.558v_4 + 0.320v_5 = 0$$
$$0.083v_1 + 0.066v_2 - 0.891v_3 + 0.040v_4 + 0.025v_5 = 0$$
$$0.023v_1 + 0.061v_2 + 0.108v_3 - 0.944v_4 + 0.252v_5 = 0$$
$$0.019v_1 + 0.028v_2 + 0.038v_3 + 0.039v_4 - 0.934v_5 = 0$$

To this system we add the sum-to-1 condition on V, which gives the equation

$$v_1 + v_2 + v_3 + v_4 + v_5 = 1.$$

The augmented matrix for this system of six equations is the following.

$$\begin{bmatrix} -0.434 & 0.392 & 0.391 & 0.307 & 0.337 & 0 \\ 0.309 & -0.547 & 0.354 & 0.558 & 0.320 & 0 \\ 0.083 & 0.066 & -0.891 & 0.040 & 0.025 & 0 \\ 0.023 & 0.061 & 0.108 & -0.944 & 0.252 & 0 \\ 0.019 & 0.028 & 0.038 & 0.039 & -0.934 & 0 \\ 1 & 1 & 1 & 1 & 1 & 1 \end{bmatrix}$$

Solving with row operations using a graphing calculator, we get a row of 0's along the bottom; the first five rows contain the solution $v_1 = 0.4675$, $v_2 = 0.3802$, $v_3 = 0.0748$, $v_4 = 0.0515$, and $v_5 = 0.0261$. Thus the long-term trend will be 45.6% couples, 38.02% mother, 7.48% father, 5.15% relative, and 2.61% other.

33. Let V be the probability vector $[v_1 \quad v_2 \quad v_3 \quad v_4 \quad v_5]$.

$$VP = V$$

$$[v_1 \quad v_2 \quad v_3 \quad v_4 \quad v_5]\begin{bmatrix} 0.71 & 0.19 & 0.03 & 0.07 & 0.00 \\ 0.52 & 0.31 & 0.04 & 0.13 & 0.00 \\ 0.40 & 0.16 & 0.22 & 0.22 & 0.00 \\ 0.34 & 0.19 & 0.09 & 0.37 & 0.01 \\ 0.21 & 0.25 & 0.13 & 0.37 & 0.04 \end{bmatrix}$$
$$= [v_1 \quad v_2 \quad v_3 \quad v_4 \quad v_5]$$

$$0.71v_1 + 0.52v_2 + 0.40v_3 + 0.34v_4 + 0.21v_5 = v_1$$
$$0.19v_1 + 0.31v_2 + 0.16v_3 + 0.19v_4 + 0.25v_5 = v_2$$
$$0.03v_1 + 0.04v_2 + 0.22v_3 + 0.09v_4 + 0.13v_5 = v_3$$
$$0.07v_1 + 0.13v_2 + 0.22v_3 + 0.37v_4 + 0.37v_5 = v_4$$
$$0.01v_4 + 0.04v_5 = v_5$$

Simplify the equations and use the equation $v_1 + v_2 + v_3 + v_4 + v_5 = 1$ to get the following system.

$$-0.29v_1 + 0.52v_2 + 0.40v_3 + 0.34v_4 + 0.21v_5 = 0$$
$$0.19v_1 - 0.69v_2 + 0.16v_3 + 0.19v_4 + 0.25v_5 = 0$$
$$0.03v_1 + 0.04v_2 - 0.78v_3 + 0.09v_4 + 0.13v_5 = 0$$
$$0.07v_1 + 0.13v_2 + 0.22v_3 - 0.63v_4 + 0.37v_5 = 0$$
$$0.01v_4 - 0.96v_5 = 0$$
$$v_1 + v_2 + v_3 + v_4 + v_5 = 1$$

Using a graphing calculator, the solution to the system is

$$v_1 = 0.6053, v_2 = 0.2143, v_3 = 0.0494,$$
$$v_4 = 0.1295, v_5 = 0.0013.$$

This means, if these trends continue, the long-term probabilities of people living in each type of housing are: 0.6053 in type I, 0.2143 in type II, 0.0494 in type III, 0.1295 in type IV, and 0.0013 in type V.

35. **(a)** Let H, S, and U represent humanities, science, and undecided, respectively.

	H	S	U
H	0.35	0.2	0.45
S	0.15	0.5	0.35
U	0.5	0.3	0.2

Let $[v_1 \quad v_2 \quad v_3]$ be a probability vector. Then

$$[v_1 \quad v_2 \quad v_3]\begin{bmatrix} 0.35 & 0.2 & 0.45 \\ 0.15 & 0.5 & 0.35 \\ 0.5 & 0.3 & 0.2 \end{bmatrix} = [v_1 \quad v_2 \quad v_3].$$

We have the system

$$0.35v_1 + 0.15v_2 + 0.5v_3 = v_1$$
$$0.2v_1 + 0.5v_2 + 0.3v_3 = v_2$$
$$0.45v_1 + 0.35v_2 + 0.2v_3 = v_3$$
$$v_1 + v_2 + v_3 = 1,$$

which is equivalent to the system

$$-0.65v_1 + 0.15v_2 + 0.5v_3 = 0$$
$$0.2v_1 - 0.5v_2 + 0.3v_3 = 0$$
$$0.45v_1 + 0.35v_2 - 0.8v_3 = 0$$
$$v_1 + v_2 + v_3 = 1.$$

To solve this system, we use the augmented matrix

$$\begin{bmatrix} -0.65 & 0.15 & 0.5 & 0 \\ 0.2 & -0.5 & 0.3 & 0 \\ 0.45 & 0.35 & -0.8 & 0 \\ 1 & 1 & 1 & 1 \end{bmatrix}.$$

Solving by the Gauss-Jordan method, we obtain the matrix

$$\begin{bmatrix} 1 & 0 & 0 & \frac{1}{3} \\ 0 & 1 & 0 & \frac{1}{3} \\ 0 & 0 & 1 & \frac{1}{3} \\ 0 & 0 & 0 & 0 \end{bmatrix}.$$

Thus, $[v_1 \quad v_2 \quad v_3] = \begin{bmatrix} \frac{1}{3} & \frac{1}{3} & \frac{1}{3} \end{bmatrix}$.

The long-range prediction is that $\frac{1}{3}$ of the students will end up with each major.

37. (a) For

$$P = \begin{bmatrix} 0.065 & 0.585 & 0.34 & 0.01 \\ 0.042 & 0.44 & 0.42 & 0.098 \\ 0.018 & 0.276 & 0.452 & 0.254 \\ 0 & 0.044 & 0.292 & 0.664 \end{bmatrix},$$

since P^2 has only positive entries, P is a regular transition matrix. Therefore, there exists a unique probability vector $V = [v_1 \quad v_2 \quad v_3 \quad v_4]$ such that $VP = V$.

If we expand the matrix equation

$$[v_1 \quad v_2 \quad v_3 \quad v_4] \begin{bmatrix} 0.065 & 0.585 & 0.34 & 0.01 \\ 0.042 & 0.44 & 0.42 & 0.098 \\ 0.018 & 0.276 & 0.452 & 0.254 \\ 0 & 0.044 & 0.292 & 0.664 \end{bmatrix}$$
$$= [v_1 \quad v_2 \quad v_3 \quad v_4],$$

we get the system of equations

$$0.065v_1 + 0.042v_2 + 0.018v_3 = v_1$$
$$0.585v_1 + 0.44v_2 + 0.276v_3 + 0.044v_4 = v_2$$
$$0.34v_1 + 0.42v_2 + 0.452v_3 + 0.292v_4 = v_3$$
$$0.01v_1 + 0.098v_2 + 0.254v_3 + 0.664v_4 = v_4,$$

which along with the equation $v_1 + v_2 + v_3 + v_4 = 1$ simplifies to a system of equations, which can be written as the augmented matrix

$$\begin{bmatrix} -0.935 & 0.042 & 0.018 & 0 & 0 \\ 0.585 & -0.56 & 0.276 & 0.044 & 0 \\ 0.34 & 0.42 & -0.548 & 0.292 & 0 \\ 0.01 & 0.098 & 0.254 & -0.336 & 0 \\ 1 & 1 & 1 & 1 & 1 \end{bmatrix}.$$

Using the Gauss-Jordan method, this reduces to

$$\begin{bmatrix} 1 & 0 & 0 & 0 & 0.0180 \\ 0 & 1 & 0 & 0 & 0.2368 \\ 0 & 0 & 1 & 0 & 0.3847 \\ 0 & 0 & 0 & 1 & 0.3604 \\ 0 & 0 & 0 & 0 & 0 \end{bmatrix},$$

where entries in the fifth column are rounded to four decimal places. Therefore, the long-range prediction for the proportion of the students in each group is group 1, 1.80%; group 2, 23.68%; group 3, 38.47%; and group 4, 36.04%.

(b) Since we assume that students in group 4 will remain in group 4, and that everyone begins in group 1, we'll use the transition matrix

$$P = \begin{bmatrix} 0.065 & 0.585 & 0.34 & 0.01 \\ 0.042 & 0.44 & 0.42 & 0.098 \\ 0.018 & 0.276 & 0.452 & 0.254 \\ 0 & 0 & 0 & 1 \end{bmatrix}$$

and the initial probability vector $X_0 = [1 \quad 0 \quad 0 \quad 0]$.

Use a graphing calculator to calculate $X_0 P^n$ for increasing powers of n. Since

$$X_0 P^7 = [0.012 \quad 0.140 \quad 0.170 \quad 0.678],$$

and

$$X_0 P^8 = [0.010 \quad 0.115 \quad 0.140 \quad 0.735],$$

it will require 8 testing periods before 70% or more of the students will have mastered the material.

39. **(a)** Let's consider the possibilities for the 2 boxes. Suppose that you have some balls in box 1 (i balls, state i), and you want to get to state $j = i - 1$. There are n balls in total, and you would have to pick one of the i balls in box 1 and move it to box 2. The probability is

$$\frac{i}{n}. \quad \text{(Case 1)}$$

Suppose that you have at least 1 ball in box 1, but less than n balls, and you want to go to state $j = i + 1$. You would have to pick one of the balls in box 2, and put it in box 1. There are n total balls, and $(n - i)$ balls in box 2. The probability of this is

$$\frac{n - i}{n} = 1 - \frac{i}{n}. \quad \text{(Case 2)}$$

Suppose that box 1 is completely empty (0 balls) or completely full (n balls). To go to state 1 or state $n - 1$ respectively, you have only 1 choice in either case: move a ball from box 2 to box 1, or move a ball from box 1 to box 2 respectively. Since there is only 1 possibility in each case (only one thing that can be done), the probability is

$$\frac{n}{n} = 1. \quad \text{(Case 3)}$$

Since any other case would require you to go from state i to state j where $|i - j| \geq 2$, and you only can move 1 ball (that is, change a state by 1), these cases would be impossible and the probability would be 0. (Case 4).

(b) By the explanation in part (a), all entries will be 0 except those in row 0, column 1 which is 1, the entries where the row = column + 1 which is $\frac{\text{row}}{n}$, and the entries where column = row + 1 which is $1 - \frac{\text{row}}{n}$. Therefore, the transition matrix is

$$\begin{array}{c} \\ 0 \\ 1 \\ 2 \\ \vdots \\ n \end{array} \begin{array}{cccccc} 0 & 1 & 2 & \cdots & n \\ \left[\begin{array}{ccccc} 0 & 1 & 0 & \cdots & 0 \\ \frac{1}{n} & 0 & 1 - \frac{1}{n} & \cdots & 0 \\ 0 & \frac{2}{n} & 0 & \cdots & 0 \\ \vdots & \vdots & \vdots & \vdots & \vdots \\ 0 & 0 & 0 & \cdots & 0 \end{array}\right]. \end{array}$$

(c) $n = 2$

$p_{00} = 0; p_{10} = \frac{i}{n} = \frac{1}{2}; p_{20} = 0; p_{01} = 1; p_{11} = 0;$

$p_{21} = 1; p_{02} = 0; p_{12} = 1 - \frac{i}{n} = 1 - \frac{1}{2} = \frac{1}{2};$

$p_{22} = 0$

The transition matrix for the case $n = 2$ is

$$\begin{bmatrix} 0 & 1 & 0 \\ \frac{1}{2} & 0 & \frac{1}{2} \\ 0 & 1 & 0 \end{bmatrix}.$$

(d) $A^2 = \begin{bmatrix} \frac{1}{2} & 0 & \frac{1}{2} \\ 0 & 1 & 0 \\ \frac{1}{2} & 0 & \frac{1}{2} \end{bmatrix};$

$$A^3 = \begin{bmatrix} 0 & 1 & 0 \\ \frac{1}{2} & 0 & \frac{1}{2} \\ 0 & 1 & 0 \end{bmatrix}; A^4 = A^2$$

Therefore, the transition matrix is not regular.

(e) $\begin{bmatrix} v_1 & v_2 & v_3 \end{bmatrix} \begin{bmatrix} 0 & 1 & 0 \\ \frac{1}{2} & 0 & \frac{1}{2} \\ 0 & 1 & 0 \end{bmatrix} = \begin{bmatrix} v_1 & v_2 & v_3 \end{bmatrix}$

$$\frac{1}{2} v_2 = v_1$$
$$v_1 + v_3 = v_2$$
$$\frac{1}{2} v_2 = v_3$$

Therefore,

$$-v_1 + \frac{1}{2} v_2 \qquad = 0$$
$$v_1 - v_2 + v_3 = 0$$
$$\frac{1}{2} v_2 - v_3 = 0$$

and $v_1 + v_2 + v_3 = 1$.

Solve this system by the Gauss-Jordan method.

$$\begin{bmatrix} -1 & \frac{1}{2} & 0 & | & 0 \\ 1 & -1 & 1 & | & 0 \\ 0 & \frac{1}{2} & -1 & | & 0 \\ 1 & 1 & 1 & | & 1 \end{bmatrix}$$

$$\begin{bmatrix} 1 & -\frac{1}{2} & 0 & | & 0 \\ 0 & -\frac{1}{2} & 1 & | & 0 \\ 0 & \frac{1}{2} & -1 & | & 0 \\ 0 & \frac{3}{2} & 1 & | & 1 \end{bmatrix}$$

$$\begin{bmatrix} 1 & 0 & -1 & | & 0 \\ 0 & 1 & -2 & | & 0 \\ 0 & 0 & 0 & | & 0 \\ 0 & 0 & 4 & | & 1 \end{bmatrix}$$

$$\begin{bmatrix} 4 & 0 & 0 & | & 1 \\ 0 & 2 & 0 & | & 1 \\ 0 & 0 & 4 & | & 1 \\ 0 & 0 & 0 & | & 0 \end{bmatrix}$$

Therefore,

$$4v_1 = 1; v_1 = \frac{1}{4}$$

$$2v_2 = 1; v_2 = \frac{1}{2}$$

$$4v_3 = 1; v_3 = \frac{1}{4}.$$

The equilibrium vector is

$$\begin{bmatrix} \frac{1}{4} & \frac{1}{2} & \frac{1}{4} \end{bmatrix}.$$

41. **(a)** If the guard is at the middle door, he is equally likely to stay there or move to either door, so the second row must have all entries of $\frac{1}{3}$. If he is at the door at either end, he is equally likely to stay put or move to the middle door. Therefore, row 1 must be $\begin{bmatrix} \frac{1}{2} & \frac{1}{2} & 0 \end{bmatrix}$. (Note there is no chance, according to the problem, that the guard will go from door 1 to door 3.) Similarly, row 3 must be $\begin{bmatrix} 0 & \frac{1}{2} & \frac{1}{2} \end{bmatrix}$, since there is no chance that he can go from door 3 to door 1.

(b) $\begin{bmatrix} v_1 & v_2 & v_3 \end{bmatrix} \begin{bmatrix} \frac{1}{2} & \frac{1}{2} & 0 \\ \frac{1}{3} & \frac{1}{3} & \frac{1}{3} \\ 0 & \frac{1}{2} & \frac{1}{2} \end{bmatrix} = \begin{bmatrix} v_1 & v_2 & v_3 \end{bmatrix}$

$$\frac{1}{2}v_1 + \frac{1}{3}v_2 = v_1$$

$$\frac{1}{2}v_1 + \frac{1}{3}v_2 + \frac{1}{2}v_3 = v_2$$

$$\frac{1}{3}v_2 + \frac{1}{2}v_3 = v_3$$

Therefore, we have the system

$$-\frac{1}{2}v_1 + \frac{1}{3}v_2 \qquad\qquad = 0$$

$$\frac{1}{2}v_1 - \frac{2}{3}v_2 + \frac{1}{2}v_3 = 0$$

$$\frac{1}{3}v_2 - \frac{1}{2}v_3 = 0$$

$$v_2 + v_2 + v_3 = 1.$$

Solve this system by the Gauss-Jordan method to obtain $v_1 = \frac{2}{7}, v_2 = \frac{3}{7}, v_3 = \frac{2}{7}$.

The guard spends $\frac{2}{7}$ of the time in front of the first and third doors and $\frac{3}{7}$ of the time in front of the middle door.

10.3   Absorbing Markov Chains

Your Turn 1

For the transition matrix

$$\begin{array}{c c} & \begin{array}{c c c c} 1 & 2 & 3 & 4 \end{array} \\ \begin{array}{c} 1 \\ 2 \\ 3 \\ 4 \end{array} & \begin{bmatrix} 1 & 0 & 0 & 0 \\ \frac{1}{5} & \frac{2}{5} & 0 & \frac{2}{5} \\ 0 & 0 & 1 & 0 \\ \frac{1}{4} & \frac{3}{4} & 0 & 0 \end{bmatrix} \end{array}$$

states 1 and 3 are absorbing. Since it is possible to go from the other two states, 2 and 4, to the absorbing state 1, the Markov chain with this transition matrix is absorbing.

Your Turn 2

The transition matrix P is

$$\begin{array}{c c} & \begin{array}{c c c} 1 & 2 & 3 \end{array} \\ \begin{array}{c} 1 \\ 2 \\ 3 \end{array} & \begin{bmatrix} 1 & 0 & 0 \\ 0.1 & 0.6 & 0.3 \\ 0 & 0 & 1 \end{bmatrix}. \end{array}$$

First rewrite the matrix so that the absorbing states 1 and 3 are first. To do this, swap columns 1 and 2, then swap rows 2 and 3:

$$\begin{bmatrix} 1 & 0 & | & 0 \\ 0 & 1 & | & 0 \\ 0.1 & 0.3 & | & 0.6 \end{bmatrix}$$

Now we have $R = \begin{bmatrix} 0.1 & 0.3 \end{bmatrix}$ and $Q = \begin{bmatrix} 0.6 \end{bmatrix}$. Find the fundamental matrix F:

$$F = (I_1 - Q)^{-1} = [1 - 0.6]^{-1} = [1/0.4] = [5/2]$$

$$FR = [5/2][0.1 \quad 0.3] = \begin{bmatrix} \frac{1}{4} & \frac{3}{4} \end{bmatrix}$$

If the system starts in the nonabsorbing state 2, there is a $\frac{1}{4}$ chance of ending up in state 1 and a $\frac{3}{4}$ chance of ending up in state 3.

10.3   Exercises

1.

$$\begin{array}{c} \\ 1 \\ 2 \\ 3 \end{array} \begin{array}{ccc} 1 & 2 & 3 \\ \begin{bmatrix} 0.25 & 0.05 & 0.7 \\ 0.35 & 0 & 0.65 \\ 0 & 0 & 1 \end{bmatrix} \end{array}$$

Since $p_{33} = 1$, state 3 is absorbing. There is a probability of 0.7 of going from state 1 to state 3 and a probability of 0.65 of going from state 2 to state 3, so it is possible to go from each nonabsorbing state to the absorbing state. Thus, this is the transition matrix for an absorbing Markov chain.

3.

$$\begin{array}{c} \\ 1 \\ 2 \\ 3 \end{array} \begin{array}{ccc} 1 & 2 & 3 \\ \begin{bmatrix} 1 & 0 & 0 \\ 0 & 0.25 & 0.75 \\ 0 & 0.85 & 0.15 \end{bmatrix} \end{array}$$

Since $p_{11} = 1$, state 1 is absorbing. Since

$p_{21} = 0$ and $p_{31} = 0$, it is not possible to go from either of the nonabsorbing states (state 2 or state 3) to the absorbing state (state 1). Thus, this is not the transition matrix of an absorbing Markov chain.

5.

$$\begin{array}{c} \\ 1 \\ 2 \\ 3 \\ 4 \end{array} \begin{array}{cccc} 1 & 2 & 3 & 4 \\ \begin{bmatrix} 0.2 & 0.5 & 0.1 & 0.2 \\ 0 & 1 & 0 & 0 \\ 0.9 & 0.02 & 0.04 & 0.04 \\ 0 & 0 & 0 & 1 \end{bmatrix} \end{array}$$

Since $p_{22} = 1$ and $p_{44} = 1$, states 2 and 4 are absorbing. It is possible to get from state 1 to states 2 and 4, and from state 3 to states 2 and 4. Thus, this is the transition matrix of an absorbing Markov chain.

7. $P = \begin{bmatrix} 1 & 0 & 0 \\ 0 & 1 & 0 \\ \hline 0.15 & 0.35 & 0.5 \end{bmatrix}$

Here $R = [0.15 \quad 0.35]$ and $Q = [0.5]$. Find the fundamental matrix F.

$$F = (I_1 - Q)^{-1} = [1 - 0.5]^{-1}$$

$$= [0.5]^{-1} = \left[\frac{1}{0.5}\right] = [2]$$

The product FR is

$$FR = [2][0.15 \quad 0.35] = [0.3 \quad 0.7].$$

9.

$$\begin{bmatrix} 1 & 0 & 0 \\ 0 & 1 & 0 \\ \hline \frac{1}{2} & \frac{1}{6} & \frac{1}{3} \end{bmatrix} = P$$

$$R = \begin{bmatrix} \frac{1}{2} & \frac{1}{6} \end{bmatrix}, Q = \begin{bmatrix} \frac{1}{3} \end{bmatrix}$$

$$F = (I_1 - Q)^{-1} = \left[1 - \frac{1}{3}\right]^{-1} = \left[\frac{2}{3}\right]^{-1} = \left[\frac{3}{2}\right]$$

$$FR = \begin{bmatrix} \frac{3}{2} \end{bmatrix}\begin{bmatrix} \frac{1}{2} & \frac{1}{6} \end{bmatrix} = \begin{bmatrix} \frac{3}{4} & \frac{1}{4} \end{bmatrix}$$

11.

$$\begin{array}{c} \\ 1 \\ 2 \\ 3 \\ 4 \end{array} \begin{array}{cccc} 1 & 2 & 3 & 4 \\ \begin{bmatrix} 1 & 0 & 0 & 0 \\ \frac{1}{3} & 0 & \frac{2}{3} & 0 \\ 0 & 0 & 1 & 0 \\ \frac{1}{4} & \frac{1}{4} & \frac{1}{4} & \frac{1}{4} \end{bmatrix} \end{array} = P$$

Rearrange the rows and columns of P so that the absorbing states come first.

$$\begin{array}{c} \\ 1 \\ 3 \\ 2 \\ 4 \end{array} \begin{array}{cccc} 1 & 3 & 2 & 4 \\ \begin{bmatrix} 1 & 0 & 0 & 0 \\ 0 & 1 & 0 & 0 \\ \frac{1}{3} & \frac{2}{3} & 0 & 0 \\ \frac{1}{4} & \frac{1}{4} & \frac{1}{4} & \frac{1}{4} \end{bmatrix} \end{array}$$

$$R = \begin{bmatrix} \frac{1}{3} & \frac{2}{3} \\ \frac{1}{4} & \frac{1}{4} \end{bmatrix}; Q = \begin{bmatrix} 0 & 0 \\ \frac{1}{4} & \frac{1}{4} \end{bmatrix}$$

$$F = (I_2 - Q)^{-1} = \left(\begin{bmatrix} 1 & 0 \\ 0 & 1 \end{bmatrix} - \begin{bmatrix} 0 & 0 \\ \frac{1}{4} & \frac{1}{4} \end{bmatrix}\right)^{-1}$$

$$= \begin{bmatrix} 1 & 0 \\ -\frac{1}{4} & \frac{3}{4} \end{bmatrix}^{-1} = \begin{bmatrix} 1 & 0 \\ \frac{1}{3} & \frac{4}{3} \end{bmatrix}$$

$$FR = \begin{bmatrix} 1 & 0 \\ \frac{1}{3} & \frac{4}{3} \end{bmatrix}\begin{bmatrix} \frac{1}{3} & \frac{2}{3} \\ \frac{1}{4} & \frac{1}{4} \end{bmatrix} = \begin{bmatrix} \frac{1}{3} & \frac{2}{3} \\ \frac{4}{9} & \frac{5}{9} \end{bmatrix}$$

13.

$$\begin{array}{c} \\ 1 \\ 2 \\ 3 \\ 4 \\ 5 \end{array} \begin{array}{ccccc} 1 & 2 & 3 & 4 & 5 \\ \begin{bmatrix} 1 & 0 & 0 & 0 & 0 \\ 0 & 1 & 0 & 0 & 0 \\ 0.1 & 0.2 & 0.3 & 0.2 & 0.2 \\ 0.3 & 0.5 & 0.1 & 0 & 0.1 \\ 0 & 0 & 0 & 0 & 1 \end{bmatrix} \end{array} = P$$

Rearranging, we obtain the matrix

$$\begin{array}{c c} & \begin{array}{c c c c c} 1 & 2 & 3 & 4 & 5 \end{array} \\ \begin{array}{c} 1 \\ 2 \\ 5 \\ 3 \\ 4 \end{array} & \left[\begin{array}{c c c | c c} 1 & 0 & 0 & 0 & 0 \\ 0 & 1 & 0 & 0 & 0 \\ 0 & 0 & 1 & 0 & 0 \\ 0.1 & 0.2 & 0.2 & 0.3 & 0.2 \\ 0.3 & 0.5 & 0.1 & 0.1 & 0 \end{array} \right]. \end{array}$$

$$R = \begin{bmatrix} 0.1 & 0.2 & 0.2 \\ 0.3 & 0.5 & 0.1 \end{bmatrix}; Q = \begin{bmatrix} 0.3 & 0.2 \\ 0.1 & 0 \end{bmatrix}$$

$$F = (I_2 - Q)^{-1} = \left(\begin{bmatrix} 1 & 0 \\ 0 & 1 \end{bmatrix} - \begin{bmatrix} 0.3 & 0.2 \\ 0.1 & 0 \end{bmatrix} \right)^{-1}$$

$$= \begin{bmatrix} 0.7 & -0.2 \\ -0.1 & 1 \end{bmatrix}^{-1} = \begin{bmatrix} \frac{25}{17} & \frac{5}{17} \\ \frac{5}{34} & \frac{35}{34} \end{bmatrix}$$

$$FR = \begin{bmatrix} \frac{25}{17} & \frac{5}{17} \\ \frac{5}{34} & \frac{35}{34} \end{bmatrix} \begin{bmatrix} \frac{1}{10} & \frac{2}{10} & \frac{2}{10} \\ \frac{3}{10} & \frac{5}{10} & \frac{1}{10} \end{bmatrix}$$

$$= \begin{bmatrix} \frac{4}{17} & \frac{15}{34} & \frac{11}{34} \\ \frac{11}{34} & \frac{37}{68} & \frac{9}{68} \end{bmatrix}$$

15. (a) The transition matrix is

$$\begin{array}{c c} & \begin{array}{c c c c c} 0 & 1 & 2 & 3 & 4 \end{array} \\ \begin{array}{c} 0 \\ 1 \\ 2 \\ 3 \\ 4 \end{array} & \left[\begin{array}{c c c c c} 1 & 0 & 0 & 0 & 0 \\ \frac{1}{2} & 0 & \frac{1}{2} & 0 & 0 \\ 0 & \frac{1}{2} & 0 & \frac{1}{2} & 0 \\ 0 & 0 & \frac{1}{2} & 0 & \frac{1}{2} \\ 0 & 0 & 0 & 0 & 1 \end{array} \right]. \end{array}$$

(b) Rearranging, we have

$$\begin{array}{c c} & \begin{array}{c c c c c} 0 & 4 & 1 & 2 & 3 \end{array} \\ \begin{array}{c} 0 \\ 4 \\ 1 \\ 2 \\ 3 \end{array} & \left[\begin{array}{c c | c c c} 1 & 0 & 0 & 0 & 0 \\ 0 & 1 & 0 & 0 & 0 \\ \frac{1}{2} & 0 & 0 & \frac{1}{2} & 0 \\ 0 & 0 & \frac{1}{2} & 0 & \frac{1}{2} \\ 0 & \frac{1}{2} & 0 & \frac{1}{2} & 0 \end{array} \right]. \end{array}$$

$$R = \begin{bmatrix} \frac{1}{2} & 0 \\ 0 & 0 \\ 0 & \frac{1}{2} \end{bmatrix}; Q = \begin{bmatrix} 0 & \frac{1}{2} & 0 \\ \frac{1}{2} & 0 & \frac{1}{2} \\ 0 & \frac{1}{2} & 0 \end{bmatrix}$$

$$F = (I_3 - Q)^{-1} = \left(\begin{bmatrix} 1 & 0 & 0 \\ 0 & 1 & 0 \\ 0 & 0 & 1 \end{bmatrix} - \begin{bmatrix} 0 & \frac{1}{2} & 0 \\ \frac{1}{2} & 0 & \frac{1}{2} \\ 0 & \frac{1}{2} & 0 \end{bmatrix} \right)^{-1}$$

$$= \begin{bmatrix} 1 & -\frac{1}{2} & 0 \\ -\frac{1}{2} & 1 & -\frac{1}{2} \\ 0 & -\frac{1}{2} & 1 \end{bmatrix}^{-1} = \begin{bmatrix} \frac{3}{2} & 1 & \frac{1}{2} \\ 1 & 2 & 1 \\ \frac{1}{2} & 1 & \frac{3}{2} \end{bmatrix}$$

$$FR = \begin{bmatrix} \frac{3}{2} & 1 & \frac{1}{2} \\ 1 & 2 & 1 \\ \frac{1}{2} & 1 & \frac{3}{2} \end{bmatrix} \begin{bmatrix} \frac{1}{2} & 0 \\ 0 & 0 \\ 0 & \frac{1}{2} \end{bmatrix} = \begin{bmatrix} \frac{3}{4} & \frac{1}{4} \\ \frac{1}{2} & \frac{1}{2} \\ \frac{1}{4} & \frac{3}{4} \end{bmatrix}$$

(c) If player A starts with \$1, the probability of ruin for A is $\frac{3}{4}$, since that is the entry in row 1, column 1 of FR. The $\frac{3}{4}$ is the probability that the nonabsorbing state of starting with \$1 will lead to the absorbing state of ruin.

(d) If player A starts with \$3, the probability of ruin for A is $\frac{1}{4}$, since that is the entry in row 3, column 1 of FR.

17. Use the formulas given in the textbook to calculate r and then x_a if $a = 10$, $b = 30$, and $p = 0.49$.

$$r = \frac{1-p}{p} = \frac{1-0.49}{0.49} \approx 1.0408$$

The probability that A will be ruined in this situation is

$$x_a = \frac{r^a - r^{a+b}}{1 - r^{a+b}} = \frac{(1.0408)^{10} - (1.0408)^{40}}{1 - (1.0408)^{40}}$$

$$\approx 0.8756.$$

19. $a = 10, b = 10$

Complete the chart by using the formulas given in Exercise 17 for r and x_a.

p	r	x_a
0.1	9	0.9999999997
0.2	4	0.99999905
0.3	$\frac{7}{3}$	0.99979
0.4	1.5	0.98295
0.5	1	0.5
0.6	$\frac{2}{3}$	0.017046
0.7	$\frac{3}{7}$	0.000209
0.8	0.25	0.00000095
0.9	$\frac{1}{9}$	0.0000000003

21. Since every nonabsorbing state leads to the one absorbing state with probability 1, every entry in FR should be 1. For example, in the matrix

$$P = \begin{pmatrix} 1 & 0 & 0 \\ 0.5 & 0.5 & 0 \\ 0.5 & 0 & 0.5 \end{pmatrix}$$

the submatrix $Q = \begin{pmatrix} 0.5 & 0 \\ 0 & 0.5 \end{pmatrix}$, the fundamental

matrix $F = (I_2 - Q)^{-1} = \begin{pmatrix} 0.5 & 0 \\ 0 & 0.5 \end{pmatrix}^{-1} =$

$\begin{pmatrix} 2 & 0 \\ 0 & 2 \end{pmatrix}$, and the submatrix $R = \begin{pmatrix} 0.5 \\ 0.5 \end{pmatrix}$. So, the

product $FR = \begin{pmatrix} 1 \\ 1 \end{pmatrix}$ is a column matrix of all 1's.

23. The transition matrix is

$$\begin{array}{c c c c} & e & f & s \\ e & \begin{bmatrix} 0.9 & 0.05 & 0.05 \\ f & 0.15 & 0.75 & 0.1 \\ s & 0 & 0 & 1 \end{bmatrix} \end{array}.$$

Permuting rows and columns to put the absorbing state "solar" at the upper left gives

$$\begin{array}{c c c c} & s & e & f \\ s & \begin{bmatrix} 1 & 0 & 0 \\ e & 0.05 & 0.9 & 0.05 \\ f & 0.1 & 0.15 & 0.75 \end{bmatrix} \end{array}.$$

$R = \begin{bmatrix} 0.05 \\ 0.1 \end{bmatrix}$ and $Q = \begin{bmatrix} 0.9 & 0.05 \\ 0.15 & 0.75 \end{bmatrix}$.

(a) $F = (I_2 - Q)^{-1}$

$= \left(\begin{bmatrix} 1 & 0 \\ 0 & 1 \end{bmatrix} - \begin{bmatrix} 0.9 & 0.05 \\ 0.15 & 0.75 \end{bmatrix} \right)^{-1}$

$= \begin{bmatrix} 0.1 & -0.05 \\ -0.15 & 0.25 \end{bmatrix}^{-1}$

$= \begin{bmatrix} 14.2857 & 2.8571 \\ 8.5714 & 5.7143 \end{bmatrix}$

$FR = \begin{bmatrix} 14.2857 & 2.8571 \\ 8.5714 & 5.7143 \end{bmatrix} \begin{bmatrix} 0.05 \\ 0.1 \end{bmatrix}$

$= \begin{bmatrix} 1 \\ 1 \end{bmatrix}$

(b) Users of electric and users of fossil both end up as users of solar with probability 1.

(c) In the matrix F, row 1 corresponds to electric and row 2 to fossil. The expected number of years until an electric heat user becomes a solar user is the sum of the entries in the first row of F. $14.2857 + 2.8571 \approx 17.14$.

The expected transition time from electric to solar heat is 17.14 years.

25. (a) The transition matrix is

$$\begin{array}{c c c c} & 1 & 2 & 3 \\ 1 & \begin{bmatrix} 0.05 & 0.15 & 0.8 \\ 2 & 0.05 & 0.15 & 0.8 \\ 3 & 0 & 0 & 1 \end{bmatrix} \end{array}.$$

Rearranging, we obtain the matrix

$$\begin{array}{c c c c} & 3 & 1 & 2 \\ 3 & \begin{bmatrix} 1 & 0 & 0 \\ 1 & 0.8 & 0.05 & 0.15 \\ 2 & 0.8 & 0.05 & 0.15 \end{bmatrix} \end{array}.$$

$R = \begin{bmatrix} 0.8 \\ 0.8 \end{bmatrix}; Q = \begin{bmatrix} 0.05 & 0.15 \\ 0.05 & 0.15 \end{bmatrix}$

(b) $F = [I_2 - Q]^{-1}$

$= \left(\begin{bmatrix} 1 & 0 \\ 0 & 1 \end{bmatrix} - \begin{bmatrix} 0.05 & 0.15 \\ 0.05 & 0.15 \end{bmatrix} \right)^{-1}$

$= \begin{bmatrix} 0.95 & -0.15 \\ -0.05 & 0.85 \end{bmatrix}^{-1}$

$= \begin{bmatrix} 1.0625 & 0.1875 \\ 0.0625 & 1.1875 \end{bmatrix}$

$FR = \begin{bmatrix} 1.0625 & 0.1875 \\ 0.0625 & 1.1875 \end{bmatrix} \begin{bmatrix} 0.8 \\ 0.8 \end{bmatrix} = \begin{bmatrix} 1 \\ 1 \end{bmatrix}$

(c) The probability that the disease eventually disappears is 1, since that is the entry in row 2, column 1 of FR.

(d) The expected number of people is 1.25, since that is the sum of the entries in row 2 of F.

27. (a) The transition matrix is

$$\begin{array}{c c c c} & A & LR & SO \\ A & \begin{bmatrix} 0.80 & 0.15 & 0.05 \\ LR & 0.05 & 0.80 & 0.15 \\ SO & 0 & 0 & 1 \end{bmatrix} \end{array}.$$

Rearranging, we obtain the matrix

$$\begin{array}{c} \begin{array}{ccc} \text{SO} & \text{A} & \text{LR} \end{array} \\ \begin{array}{c} \text{SO} \\ \text{A} \\ \text{LR} \end{array}\left[\begin{array}{c|cc} 1 & 0 & 0 \\ \hline 0.05 & 0.80 & 0.15 \\ 0.15 & 0.05 & 0.80 \end{array}\right]. \end{array}$$

$$R = \begin{bmatrix} 0.05 \\ 0.15 \end{bmatrix}; Q = \begin{bmatrix} 0.80 & 0.15 \\ 0.05 & 0.80 \end{bmatrix}$$

$$F = (I_2 - Q)^{-1}$$

$$= \left(\begin{bmatrix} 1 & 0 \\ 0 & 1 \end{bmatrix} - \begin{bmatrix} 0.80 & 0.15 \\ 0.05 & 0.80 \end{bmatrix} \right)^{-1}$$

$$= \begin{bmatrix} 0.2 & -0.15 \\ -0.05 & 0.2 \end{bmatrix}^{-1}$$

$$= \begin{bmatrix} 6.154 & 4.615 \\ 1.538 & 6.154 \end{bmatrix}$$

$$FR = \begin{bmatrix} 6.154 & 4.615 \\ 1.538 & 6.154 \end{bmatrix}\begin{bmatrix} 0.05 \\ 0.15 \end{bmatrix}$$

$$= \begin{bmatrix} 1.000 \\ 1.000 \end{bmatrix}$$

(b) The probability that a person who commuted by car ends up avoiding the downtown area is 1, since that is the entry in row 1, column 1 of FR.

(c) The expected number of years until a person who commutes by automobile this year ends up avoiding the downtown area is $10.769 \approx 10.77$ yr since that is the sum of the entries in row 1 of F.

29. (a)

$$\begin{array}{c} \begin{array}{ccccc} 1 & 2 & 3 & 4 & 5 \end{array} \\ \begin{array}{c} 1 \\ 2 \\ 3 \\ 4 \\ 5 \end{array}\begin{bmatrix} 0.4 & 0.3 & 0.2 & 0.1 & 0 \\ 0.2 & 0.1 & 0 & 0.6 & 0.1 \\ 0 & 0 & 1 & 0 & 0 \\ 0.1 & 0.1 & 0.4 & 0.1 & 0.3 \\ 0 & 0 & 0 & 0 & 1 \end{bmatrix} \end{array}$$

is the transition matrix. States 3 and 5 are absorbing. Upon rearranging, we obtain

$$\begin{array}{c} \begin{array}{ccccc} 3 & 5 & 1 & 2 & 4 \end{array} \\ \begin{array}{c} 3 \\ 5 \\ 1 \\ 2 \\ 4 \end{array}\left[\begin{array}{cc|ccc} 1 & 0 & 0 & 0 & 0 \\ 0 & 1 & 0 & 0 & 0 \\ \hline 0.2 & 0 & 0.4 & 0.3 & 0.1 \\ 0 & 0.1 & 0.2 & 0.1 & 0.6 \\ 0.4 & 0.3 & 0.1 & 0.1 & 0.1 \end{array}\right]. \end{array}$$

Then

$$F = \begin{bmatrix} 0.6 & -0.3 & -0.1 \\ -0.2 & 0.9 & -0.6 \\ -0.1 & -0.1 & 0.9 \end{bmatrix}^{-1}$$

$$= \begin{bmatrix} 2.0436 & 0.7629 & 0.7357 \\ 0.6540 & 1.4441 & 1.0354 \\ 0.2997 & 0.2452 & 1.3079 \end{bmatrix},$$

and

$$\begin{array}{c} \begin{array}{cc} 3 & 5 \end{array} \\ FR = \begin{array}{c} 1 \\ 2 \\ 4 \end{array}\begin{bmatrix} 0.703 & 0.297 \\ 0.545 & 0.455 \\ 0.583 & 0.417 \end{bmatrix}. \end{array}$$

The second column of FR gives the probability of ending up in compartment 5 given the initial compartment.

(b) From compartment 1, the probability is 0.297.

(c) From compartment 2, the probability is 0.455.

(d) From compartment 3, the probability is 0 since state 3 is absorbing.

(e) From compartment 4, the probability is 0.417.

(f) To find the expected number of times that a rat in compartment 1 will be in compartment 1 before ending up in compartment 3 or 5, look at the entry in row 1, column 1 of F, $2.0436 \approx 2.04$.

(g) To find the expected number of times that a rat in compartment 4 will be in compartment 4 before ending up in compartment 3 or 5, look at the entry in row 3, column 3 of F, $1.3079 \approx 1.31$.

31. (a) Having 0 dollars is an absorbing state and having 7 dollars is an absorbing state, so in the transition matrix, P, the elements $p_{00} = 1$, $p_{77} = 1$, and the remaining entries in rows 0 and 7 are 0. Since for each of the other states the probability is $\frac{1}{2}$ of either gaining or losing a dollar, the transition matrix is

$$P = \begin{array}{c} \\ 0 \\ 1 \\ 2 \\ 3 \\ 4 \\ 5 \\ 6 \\ 7 \end{array} \begin{array}{c} \begin{matrix} 0 & 1 & 2 & 3 & 4 & 5 & 6 & 7 \end{matrix} \\ \begin{bmatrix} 1 & 0 & 0 & 0 & 0 & 0 & 0 & 0 \\ \frac{1}{2} & 0 & \frac{1}{2} & 0 & 0 & 0 & 0 & 0 \\ 0 & \frac{1}{2} & 0 & \frac{1}{2} & 0 & 0 & 0 & 0 \\ 0 & 0 & \frac{1}{2} & 0 & \frac{1}{2} & 0 & 0 & 0 \\ 0 & 0 & 0 & \frac{1}{2} & 0 & \frac{1}{2} & 0 & 0 \\ 0 & 0 & 0 & 0 & \frac{1}{2} & 0 & \frac{1}{2} & 0 \\ 0 & 0 & 0 & 0 & 0 & \frac{1}{2} & 0 & \frac{1}{2} \\ 0 & 0 & 0 & 0 & 0 & 0 & 0 & 1 \end{bmatrix} \end{array} .$$

To answer parts (b) and (c), we need to calculate the product matrix FR. To do this, first re-write P so that the absorbing states come first.

$$P = \begin{array}{c} \\ 0 \\ 7 \\ 1 \\ 2 \\ 3 \\ 4 \\ 5 \\ 6 \end{array} \begin{array}{c} \begin{matrix} 0 & 7 & 1 & 2 & 3 & 4 & 5 & 6 \end{matrix} \\ \left[\begin{array}{cc|cccccc} 1 & 0 & 0 & 0 & 0 & 0 & 0 & 0 \\ 0 & 1 & 0 & 0 & 0 & 0 & 0 & 0 \\ \hline \frac{1}{2} & 0 & 0 & \frac{1}{2} & 0 & 0 & 0 & 0 \\ 0 & 0 & \frac{1}{2} & 0 & \frac{1}{2} & 0 & 0 & 0 \\ 0 & 0 & 0 & \frac{1}{2} & 0 & \frac{1}{2} & 0 & 0 \\ 0 & 0 & 0 & 0 & \frac{1}{2} & 0 & \frac{1}{2} & 0 \\ 0 & 0 & 0 & 0 & 0 & \frac{1}{2} & 0 & \frac{1}{2} \\ 0 & \frac{1}{2} & 0 & 0 & 0 & 0 & \frac{1}{2} & 0 \end{array} \right] \end{array} ,$$

so $R = \begin{array}{c} 1 \\ 2 \\ 3 \\ 4 \\ 5 \\ 6 \end{array} \begin{array}{c} \begin{matrix} 0 & \;\;7 \end{matrix} \\ \begin{bmatrix} \frac{1}{2} & 0 \\ 0 & 0 \\ 0 & 0 \\ 0 & 0 \\ 0 & 0 \\ 0 & \frac{1}{2} \end{bmatrix} \end{array}$ and

$$Q = \begin{bmatrix} 0 & \frac{1}{2} & 0 & 0 & 0 & 0 \\ \frac{1}{2} & 0 & \frac{1}{2} & 0 & 0 & 0 \\ 0 & \frac{1}{2} & 0 & \frac{1}{2} & 0 & 0 \\ 0 & 0 & \frac{1}{2} & 0 & \frac{1}{2} & 0 \\ 0 & 0 & 0 & \frac{1}{2} & 0 & \frac{1}{2} \\ 0 & 0 & 0 & 0 & \frac{1}{2} & 0 \end{bmatrix}$$

Therefore,

$$F = (I_6 - Q)^{-1} = \begin{bmatrix} 1 & -\frac{1}{2} & 0 & 0 & 0 & 0 \\ -\frac{1}{2} & 1 & -\frac{1}{2} & 0 & 0 & 0 \\ 0 & -\frac{1}{2} & 1 & -\frac{1}{2} & 0 & 0 \\ 0 & 0 & -\frac{1}{2} & 1 & -\frac{1}{2} & 0 \\ 0 & 0 & 0 & -\frac{1}{2} & 1 & -\frac{1}{2} \\ 0 & 0 & 0 & 0 & -\frac{1}{2} & 1 \end{bmatrix}^{-1}$$

$$= \begin{bmatrix} \frac{12}{7} & \frac{10}{7} & \frac{8}{7} & \frac{6}{7} & \frac{4}{7} & \frac{2}{7} \\ \frac{10}{7} & \frac{20}{7} & \frac{16}{7} & \frac{12}{7} & \frac{8}{7} & \frac{4}{7} \\ \frac{8}{7} & \frac{16}{7} & \frac{24}{7} & \frac{18}{7} & \frac{12}{7} & \frac{6}{7} \\ \frac{6}{7} & \frac{12}{7} & \frac{18}{7} & \frac{24}{7} & \frac{16}{7} & \frac{8}{7} \\ \frac{4}{7} & \frac{8}{7} & \frac{12}{7} & \frac{16}{7} & \frac{20}{7} & \frac{10}{7} \\ \frac{2}{7} & \frac{4}{7} & \frac{6}{7} & \frac{8}{7} & \frac{10}{7} & \frac{12}{7} \end{bmatrix},$$

so $FR = \begin{array}{c} 1 \\ 2 \\ 3 \\ 4 \\ 5 \\ 6 \end{array} \begin{array}{c} \begin{matrix} 0 & \;\;7 \end{matrix} \\ \begin{bmatrix} \frac{6}{7} & \frac{1}{7} \\ \frac{5}{7} & \frac{2}{7} \\ \frac{4}{7} & \frac{3}{7} \\ \frac{3}{7} & \frac{4}{7} \\ \frac{2}{7} & \frac{5}{7} \\ \frac{1}{7} & \frac{6}{7} \end{bmatrix} \end{array} .$

(b) The probability of ruin for player A if A starts with \$4 is the entry in the fourth row, first column of FR, which is $\frac{3}{7}$.

(c) The probability of ruin for player A if A starts with \$5 is the entry in the fifth row, first column, which is $\frac{2}{7}$.

Chapter 10 Review Exercises

1. False: In a Markov chain, the outcome of an experiment depends only on the current state.

2. False: The transition matrix is the same for any transition.

3. True

4. False: One should take the kth power of the transition matrix.

5. False: A transition matrix is regular if some power of the matrix has all positive entries.

6. True

7. False: $VP = V$ will have infinitely many solutions.

8. False: If successive powers have a 0 in some location, this 0 will persist.

9. False: The statement is true for a regular transition matrix but not necessarily true for a non-regular transition matrix

10. True

11. True

12. True

15. $\begin{bmatrix} -0.2 & 1.2 \\ 0.8 & 0.2 \end{bmatrix}$

This could not be a transition matrix because it contains a negative entry.

17. $\begin{bmatrix} 0.8 & 0.2 & 0 \\ 0 & 1 & 0 \\ 0.1 & 0.4 & 0.5 \end{bmatrix}$

This could be a transition matrix for the same reasons stated in Exercise 3.

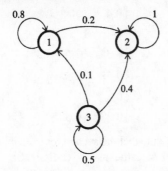

19. **(a)** $C = \begin{bmatrix} 0.7 & 0.3 \\ 1 & 0 \end{bmatrix}$

$C^2 = \begin{bmatrix} 0.7 & 0.3 \\ 1 & 0 \end{bmatrix}\begin{bmatrix} 0.7 & 0.3 \\ 1 & 0 \end{bmatrix}$

$= \begin{bmatrix} 0.79 & 0.21 \\ 0.7 & 0.3 \end{bmatrix}$

$C^3 = \begin{bmatrix} 0.7 & 0.3 \\ 1 & 0 \end{bmatrix}\begin{bmatrix} 0.79 & 0.21 \\ 0.7 & 0.3 \end{bmatrix}$

$= \begin{bmatrix} 0.763 & 0.237 \\ 0.79 & 0.21 \end{bmatrix}$

(b) The probability that state 2 changes to state 1 after three repetitions of the experiment is the entry in row 2, column 1 of C^3, or 0.79.

21. **(a)** $E = \begin{bmatrix} 0.2 & 0.5 & 0.3 \\ 0.3 & 0.4 & 0.3 \\ 0 & 1 & 0 \end{bmatrix}$

$E^2 = \begin{bmatrix} 0.2 & 0.5 & 0.3 \\ 0.3 & 0.4 & 0.3 \\ 0 & 1 & 0 \end{bmatrix}\begin{bmatrix} 0.2 & 0.5 & 0.3 \\ 0.3 & 0.4 & 0.3 \\ 0 & 1 & 0 \end{bmatrix}$

$= \begin{bmatrix} 0.19 & 0.6 & 0.21 \\ 0.18 & 0.61 & 0.21 \\ 0.3 & 0.4 & 0.3 \end{bmatrix}$

$E^3 = \begin{bmatrix} 0.2 & 0.5 & 0.3 \\ 0.3 & 0.4 & 0.3 \\ 0 & 1 & 0 \end{bmatrix}\begin{bmatrix} 0.19 & 0.6 & 0.21 \\ 0.18 & 0.61 & 0.21 \\ 0.3 & 0.4 & 0.3 \end{bmatrix}$

$= \begin{bmatrix} 0.218 & 0.545 & 0.237 \\ 0.219 & 0.544 & 0.237 \\ 0.18 & 0.61 & 0.21 \end{bmatrix}$

(b) The probability that state 2 changes to state 1 after 3 repetitions of the experiment is the entry in row 2, column 1 of E^3, or 0.219.

23. $T^2 = \begin{bmatrix} 0.2 & 0.8 \\ 0.5 & 0.5 \end{bmatrix}\begin{bmatrix} 0.2 & 0.8 \\ 0.5 & 0.5 \end{bmatrix} = \begin{bmatrix} 0.44 & 0.56 \\ 0.35 & 0.65 \end{bmatrix}$

The distribution after two repetitions is

$$DT^2 = [0.3 \quad 0.7]\begin{bmatrix} 0.44 & 0.56 \\ 0.35 & 0.65 \end{bmatrix}$$

$$= [0.377 \quad 0.623].$$

To predict the long-range distribution, let V be the probability vector $[v_1 \quad v_2]$. We want to find V such that

$$VT = V$$

$$[v_1 \quad v_2]\begin{bmatrix} 0.2 & 0.8 \\ 0.5 & 0.5 \end{bmatrix} = [v_1 \quad v_2].$$

By matrix multiplication and equality of matrices,

$$0.2v_1 + 0.5v_2 = v_1$$
$$0.8v_1 + 0.5v_2 = v_2.$$

Simplify the equations to get the dependent system

$$-0.8v_1 + 0.5v_2 = 0$$
$$0.8v_1 - 0.5v_2 = 0.$$

Since V is a probability vector,

$$v_1 + v_2 = 1.$$

To find v_1 and v_2, solve the system

$$0.8v_1 - 0.5v_2 = 0$$
$$v_1 + v_2 = 1$$

by the substitution method. Observe that
$v_2 = 1 - v_1$.

$$0.8v_1 - 0.5(1 - v_1) = 0$$
$$1.3v_1 - 0.5 = 0$$
$$v_1 = \frac{0.5}{1.3} = \frac{5}{13}$$
$$v_2 = 1 - \frac{5}{13} = \frac{8}{13}$$

The long-range distribution is $\left[\frac{5}{13} \quad \frac{8}{13}\right]$.

25. $T^2 = \begin{bmatrix} 0.7 & 0.1 & 0.2 \\ 0.3 & 0.3 & 0.4 \\ 0.4 & 0.5 & 0.1 \end{bmatrix} \begin{bmatrix} 0.7 & 0.1 & 0.2 \\ 0.3 & 0.3 & 0.4 \\ 0.4 & 0.5 & 0.1 \end{bmatrix}$

$= \begin{bmatrix} 0.6 & 0.2 & 0.2 \\ 0.46 & 0.32 & 0.22 \\ 0.47 & 0.24 & 0.29 \end{bmatrix}$

The distribution after two repetitions is

$$DT^2 = [0.2 \quad 0.4 \quad 0.4] \begin{bmatrix} 0.6 & 0.2 & 0.3 \\ 0.46 & 0.32 & 0.22 \\ 0.47 & 0.24 & 0.29 \end{bmatrix}$$

$$= [0.492 \quad 0.264 \quad 0.244].$$

To predict the long-range distribution, let V be the probability vector $[v_1 \quad v_2 \quad v_3]$. We want to find V such that

$$VT = V$$

$$[v_1 \quad v_2 \quad v_3] \begin{bmatrix} 0.7 & 0.1 & 0.2 \\ 0.3 & 0.3 & 0.4 \\ 0.4 & 0.5 & 0.1 \end{bmatrix} = [v_1 \quad v_2 \quad v_3].$$

By matrix multiplication and equality of matrices,

$$0.7v_1 + 0.3v_2 + 0.4v_3 = v_1$$
$$0.1v_1 + 0.3v_2 + 0.5v_3 = v_2$$
$$0.2v_1 + 0.4v_2 + 0.1v_3 = v_3.$$

Simplify the equations to get the dependent system

$$-0.3v_1 + 0.3v_2 + 0.4v_3 = 0$$
$$0.1v_1 - 0.7v_2 + 0.5v_3 = 0$$
$$0.2v_1 + 0.4v_2 - 0.9v_3 = 0.$$

Since V is a probability vector,

$$v_1 + v_2 + v_3 = 1.$$

To find v_1, v_2 and v_3, solve the system

$$-0.3v_1 + 0.3v_2 + 0.4v_3 = 0$$
$$0.1v_1 - 0.7v_2 + 0.5v_3 = 0$$
$$0.2v_1 + 0.4v_2 - 0.9v_3 = 0$$
$$v_1 + v_2 + v_3 = 1$$

by the Gauss-Jordan method. The solution is

$$v_1 = \frac{43}{80}, v_2 = \frac{19}{80}, v_3 = \frac{9}{40}.$$

The long-range distribution is $\left[\frac{43}{80} \quad \frac{19}{80} \quad \frac{9}{40}\right]$.

27. $A = \begin{bmatrix} 0 & 1 \\ 0.2 & 0.8 \end{bmatrix}$

$A^2 = \begin{bmatrix} 0 & 1 \\ 0.2 & 0.8 \end{bmatrix} \begin{bmatrix} 0 & 1 \\ 0.2 & 0.8 \end{bmatrix}$

$= \begin{bmatrix} 0.2 & 0.8 \\ 0.16 & 0.84 \end{bmatrix}$

A^2 has all positive entries, so A is regular.

29. $A = \begin{bmatrix} 0.4 & 0.2 & 0.4 \\ 0 & 1 & 0 \\ 0.6 & 0.3 & 0.1 \end{bmatrix}$

$A^2 = \begin{bmatrix} 0.4 & 0.4 & 0.2 \\ 0 & 1 & 0 \\ 0.3 & 0.45 & 0.25 \end{bmatrix}$

$A^3 = \begin{bmatrix} 0.28 & 0.54 & 0.18 \\ 0 & 1 & 0 \\ 0.27 & 0.585 & 0.145 \end{bmatrix}$

Note that the second row will always have zeros; hence, the matrix is not regular.

35. $\begin{bmatrix} 0 & 1 & 0 \\ 0.5 & 0.1 & 0.4 \\ 0 & 0 & 1 \end{bmatrix}$

Since $p_{33} = 1$, state 3 is absorbing. Since $p_{23} \neq 0$, it is possible to go from state 2 to the absorbing state. Since $p_{12} = 1$ and $p_{23} = 0.4$, it is possible to go from state 1 to the absorbing state in two steps. This is the transition matrix of an absorbing Markov chain.

37.
$$\begin{bmatrix} 0.5 & 0.1 & 0.1 & 0.3 \\ 0 & 0 & 1 & 0 \\ 1 & 0 & 0 & 0 \\ 0.1 & 0.8 & 0.05 & 0.05 \end{bmatrix}$$

There are no absorbing states. Hence, this is not the transition matrix for an absorbing Markov chain.

39.

$$P = \begin{array}{c} \\ 1 \\ 2 \\ 3 \end{array} \begin{array}{ccc} 1 & 2 & 3 \\ \begin{bmatrix} 0.2 & 0.45 & 0.35 \\ 0 & 1 & 0 \\ 0 & 0 & 1 \end{bmatrix} \end{array}$$

Arrange the rows and columns of P so the absorbing states come first.

$$\begin{array}{c} \\ 1 \\ 2 \\ 3 \end{array} \begin{array}{ccc} 2 & 3 & 1 \\ \left[\begin{array}{cc|c} 1 & 0 & 0 \\ 0 & 1 & 0 \\ \hline 0.45 & 0.35 & 0.2 \end{array} \right] \end{array}.$$

Thus, $Q = [0.2]$. Now find F.

$$F = (I_1 - Q)^{-1}$$
$$= ([1] - [0.2])^{-1}$$
$$= [0.8]^{-1} = \left[\frac{4}{5}\right]^{-1}$$
$$= \left[\frac{5}{4}\right]$$

$$FR = \left[\frac{5}{4}\right][0.45 \quad 0.35]$$
$$= \left[\frac{5}{4}\right]\left[\frac{9}{20} \quad \frac{7}{20}\right]$$
$$= \left[\frac{9}{16} \quad \frac{7}{16}\right]$$

41.

$$P = \begin{array}{c} \\ 1 \\ 2 \\ 3 \\ 4 \end{array} \begin{array}{cccc} 1 & 2 & 3 & 4 \\ \begin{bmatrix} \frac{1}{5} & \frac{1}{5} & \frac{2}{5} & \frac{1}{5} \\ 0 & 1 & 0 & 0 \\ \frac{1}{2} & \frac{1}{4} & \frac{1}{8} & \frac{1}{8} \\ 0 & 0 & 0 & 1 \end{bmatrix} \end{array}$$

Rearranging, we have

$$\begin{array}{c} \\ 2 \\ 4 \\ 1 \\ 3 \end{array} \begin{array}{cccc} 2 & 4 & 1 & 3 \\ \left[\begin{array}{cc|cc} 1 & 0 & 0 & 0 \\ 0 & 1 & 0 & 0 \\ \hline \frac{1}{5} & \frac{1}{5} & \frac{1}{5} & \frac{2}{5} \\ \frac{1}{4} & \frac{1}{8} & \frac{1}{2} & \frac{1}{8} \end{array} \right] \end{array}.$$

$$R = \begin{bmatrix} \frac{1}{5} & \frac{1}{5} \\ \frac{1}{4} & \frac{1}{8} \end{bmatrix} \text{ and } Q = \begin{bmatrix} \frac{1}{5} & \frac{2}{5} \\ \frac{1}{2} & \frac{1}{8} \end{bmatrix}$$

$$F = (I_2 - Q)^{-1} = \left(\begin{bmatrix} 1 & 0 \\ 0 & 1 \end{bmatrix} - \begin{bmatrix} \frac{1}{5} & \frac{2}{5} \\ \frac{1}{2} & \frac{1}{8} \end{bmatrix} \right)^{-1}$$

$$= \begin{bmatrix} \frac{4}{5} & -\frac{2}{5} \\ -\frac{1}{2} & \frac{7}{8} \end{bmatrix}^{-1} = \begin{bmatrix} \frac{7}{4} & \frac{4}{5} \\ 1 & \frac{8}{5} \end{bmatrix}$$

$$FR = \begin{bmatrix} \frac{7}{4} & \frac{4}{5} \\ 1 & \frac{8}{5} \end{bmatrix} \begin{bmatrix} \frac{1}{5} & \frac{1}{5} \\ \frac{1}{4} & \frac{1}{8} \end{bmatrix} = \begin{bmatrix} \frac{11}{20} & \frac{9}{20} \\ \frac{3}{5} & \frac{2}{5} \end{bmatrix} \text{ or } \begin{bmatrix} 0.55 & 0.45 \\ 0.6 & 0.4 \end{bmatrix}$$

43. The initial distribution matrix $D = [0.35 \quad 0.65]$.

(a) The distribution after one campaign is

$$DP = [0.35 \quad 0.65]\begin{bmatrix} 0.8 & 0.2 \\ 0.45 & 0.55 \end{bmatrix}$$
$$= [0.5725 \quad 0.4275].$$

(b) The distribution after three campaigns is

$$DP^3 = [0.35 \quad 0.65]\begin{bmatrix} 0.8 & 0.2 \\ 0.45 & 0.55 \end{bmatrix}^3$$

$$\approx [0.35 \quad 0.65]\begin{bmatrix} 0.7055 & 0.2945 \\ 0.6626 & 0.3374 \end{bmatrix}$$

$$\approx [0.6776 \quad 0.3224].$$

45. The distribution after 1 mo is

$$[0.4 \quad 0.4 \quad 0.2]\begin{bmatrix} 0.8 & 0.15 & 0.05 \\ 0.25 & 0.55 & 0.2 \\ 0.04 & 0.21 & 0.75 \end{bmatrix}$$
$$= [0.428 \quad 0.322 \quad 0.25].$$

47. $P^2 = \begin{bmatrix} 0.8 & 0.15 & 0.05 \\ 0.25 & 0.55 & 0.2 \\ 0.04 & 0.21 & 0.75 \end{bmatrix}\begin{bmatrix} 0.8 & 0.15 & 0.05 \\ 0.25 & 0.55 & 0.2 \\ 0.04 & 0.21 & 0.75 \end{bmatrix}$

$$= \begin{bmatrix} 0.6795 & 0.213 & 0.1075 \\ 0.3455 & 0.382 & 0.2725 \\ 0.1145 & 0.279 & 0.6065 \end{bmatrix}$$

$$P^3 = \begin{bmatrix} 0.8 & 0.15 & 0.05 \\ 0.25 & 0.55 & 0.2 \\ 0.04 & 0.21 & 0.75 \end{bmatrix} \begin{bmatrix} 0.6795 & 0.213 & 0.1075 \\ 0.3455 & 0.382 & 0.2725 \\ 0.1145 & 0.279 & 0.6065 \end{bmatrix}$$

$$= \begin{bmatrix} 0.60115 & 0.24165 & 0.1572 \\ 0.3828 & 0.31915 & 0.29805 \\ 0.18561 & 0.29799 & 0.5164 \end{bmatrix}$$

The distribution after 3 mo is

$$[0.4 \quad 0.4 \quad 0.2] \begin{bmatrix} 0.60115 & 0.24165 & 0.1572 \\ 0.3828 & 0.31915 & 0.29805 \\ 0.18561 & 0.29799 & 0.5164 \end{bmatrix}$$

$$= [0.4307 \quad 0.2839 \quad 0.2854].$$

49. **(a)** To find the fundamental matrix F associated with the transition matrix

$$\begin{array}{c} \\ 0 \\ 1 \\ 2 \end{array} \begin{array}{ccc} 0 & 1 & 2 \\ \left[\begin{array}{c|cc} 1 & 0 & 0 \\ \hline 0.085 & 0.779 & 0.136 \\ 0.017 & 0.017 & 0.966 \end{array} \right], \end{array}$$

First note that $Q = \begin{bmatrix} 0.779 & 0.136 \\ 0.017 & 0.966 \end{bmatrix}$.

Therefore,

$$F = (I_2 - Q)^{-1} = \begin{bmatrix} 0.221 & -0.136 \\ -0.017 & 0.034 \end{bmatrix}^{-1}$$

$$= \begin{array}{c} 1 \\ 2 \end{array} \begin{array}{cc} 1 & 2 \\ \begin{bmatrix} 6.536 & 26.144 \\ 3.268 & 42.484 \end{bmatrix}, \end{array}$$

where the entries of F are rounded to three decimal places.

(b) For a patient with favorable status, the expected number of 72-hr cycles that the patient will continue to have that status before dying is the entry located in the second row, second column of F, which is 42.484.

(c) For a patient with unfavorable status, the expected number of 72-hr cycles that the patient will have a favorable status before dying is the entry in the first row, second column of F, which is 26.144.

In Exercises 51–59 the original transition matrix is

$$\begin{array}{c} \\ A = \begin{array}{c} T \\ N \\ O \end{array} \end{array} \begin{array}{ccc} T & N & O \\ \begin{bmatrix} 0.3 & 0.5 & 0.2 \\ 0.2 & 0.6 & 0.2 \\ 0.1 & 0.5 & 0.4 \end{bmatrix}. \end{array}$$

51. A grandson is 2 generations down, so we need to look at A^2.

$$\begin{array}{c} \\ A^2 = \begin{array}{c} T \\ N \\ O \end{array} \end{array} \begin{array}{ccc} T & N & O \\ \begin{bmatrix} 0.21 & 0.55 & 0.24 \\ 0.2 & 0.6 & 0.2 \\ 0.17 & 0.55 & 0.28 \end{bmatrix} \end{array}$$

The probability that the grandson of a normal man is thin is given by the entry in row 2, column 1 of A^2. This probability is 0.2.

53. We can read this probability directly from matrix A. The probability that an overweight man has an overweight son is the entry in row 3, column 3 of A. This probability is 0.4.

55.

$$\begin{array}{c} \\ A^3 = \begin{array}{c} T \\ N \\ O \end{array} \end{array} \begin{array}{ccc} T & N & O \\ \begin{bmatrix} 0.197 & 0.555 & 0.248 \\ 0.196 & 0.556 & 0.248 \\ 0.189 & 0.555 & 0.256 \end{bmatrix} \end{array}$$

is the transition matrix over 3 generations. The probability of an overweight man having an overweight great-grandson is the entry in row 3, column 3 of A^3. This probability is 0.256.

57. Let $D = \begin{bmatrix} 0.2 & 0.55 & 0.25 \end{bmatrix}$, the initial distribution.

$$A^2 = \begin{bmatrix} 0.21 & 0.55 & 0.24 \\ 0.2 & 0.56 & 0.24 \\ 0.17 & 0.55 & 0.28 \end{bmatrix}.$$

The distribution after 2 generations is

$$DA^2 = [0.2 \quad 0.55 \quad 0.25] \begin{bmatrix} 0.21 & 0.55 & 0.24 \\ 0.2 & 0.56 & 0.24 \\ 0.17 & 0.55 & 0.28 \end{bmatrix}$$

$$= [0.1945 \quad 0.5555 \quad 0.25].$$

59. $A = \begin{bmatrix} 0.3 & 0.5 & 0.2 \\ 0.2 & 0.6 & 0.2 \\ 0.1 & 0.5 & 0.4 \end{bmatrix}$

To find the long-term distribution, use the system

$$v_1 + v_2 + v_3 = 1$$
$$0.3v_1 + 0.2v_2 + 0.1v_3 = v_1$$
$$0.5v_1 + 0.6v_2 + 0.5v_3 = v_2$$
$$0.2v_1 + 0.2v_2 + 0.4v_3 = v_3.$$

Simplify these equations to obtain the system

$$v_1 + v_2 + v_3 = 1$$
$$-0.7v_1 + 0.2v_2 + 0.1v_3 = 0$$
$$0.5v_1 - 0.4v_2 + 0.5v_3 = 0$$
$$0.2v_1 + 0.2v_2 - 0.6v_3 = 0.$$

Solve this system by the Gauss-Jordan method to obtain $v_1 = \frac{7}{36}$, $v_2 = \frac{5}{9}$, and $v_3 = \frac{1}{4}$. The long-range distribution is

$$\begin{bmatrix} \frac{7}{36} & \frac{5}{9} & \frac{1}{4} \end{bmatrix}.$$

61. The absorbing states are states 1 and 6 since $a_{11} = 1$ and $a_{66} = 1$.

63. $F = (I_4 - Q)^{-1}$

$$= \left(\begin{bmatrix} 1 & 0 & 0 & 0 \\ 0 & 1 & 0 & 0 \\ 0 & 0 & 1 & 0 \\ 0 & 0 & 0 & 1 \end{bmatrix} - \begin{bmatrix} \frac{1}{2} & 0 & \frac{1}{4} & 0 \\ 0 & 0 & 1 & 0 \\ \frac{1}{4} & \frac{1}{8} & \frac{1}{4} & \frac{1}{4} \\ 0 & 0 & \frac{1}{4} & \frac{1}{2} \end{bmatrix} \right)^{-1}$$

$$= \begin{bmatrix} \frac{1}{2} & 0 & -\frac{1}{4} & 0 \\ 0 & 1 & -1 & 0 \\ -\frac{1}{4} & -\frac{1}{8} & \frac{3}{4} & -\frac{1}{4} \\ 0 & 0 & -\frac{1}{4} & \frac{1}{2} \end{bmatrix}^{-1}$$

Use row operations to find this inverse.

$$\begin{bmatrix} \frac{1}{2} & 0 & -\frac{1}{4} & 0 & 1 & 0 & 0 & 0 \\ 0 & 1 & -1 & 0 & 0 & 1 & 0 & 0 \\ -\frac{1}{4} & -\frac{1}{8} & \frac{3}{4} & -\frac{1}{4} & 0 & 0 & 1 & 0 \\ 0 & 0 & -\frac{1}{4} & \frac{1}{2} & 0 & 0 & 0 & 1 \end{bmatrix}$$

$$\begin{bmatrix} 1 & 0 & -\frac{1}{2} & 0 & 2 & 0 & 0 & 0 \\ 0 & 1 & -1 & 0 & 0 & 1 & 0 & 0 \\ 0 & -\frac{1}{8} & \frac{5}{8} & -\frac{1}{4} & \frac{1}{2} & 0 & 1 & 0 \\ 0 & 0 & -\frac{1}{4} & \frac{1}{2} & 0 & 0 & 0 & 1 \end{bmatrix}$$

$$\begin{bmatrix} 1 & 0 & -\frac{1}{2} & 0 & 2 & 0 & 0 & 0 \\ 0 & 1 & -1 & 0 & 0 & 1 & 0 & 0 \\ 0 & 0 & \frac{1}{2} & -\frac{1}{4} & \frac{1}{2} & \frac{1}{8} & 1 & 0 \\ 0 & 0 & -\frac{1}{4} & \frac{1}{2} & 0 & 0 & 0 & 1 \end{bmatrix}$$

$$\begin{bmatrix} 1 & 0 & 0 & -\frac{1}{4} & \frac{5}{2} & \frac{1}{8} & 1 & 0 \\ 0 & 1 & 0 & -\frac{1}{2} & 1 & \frac{5}{4} & 2 & 0 \\ 0 & 0 & 1 & -\frac{1}{2} & 1 & \frac{1}{4} & 2 & 0 \\ 0 & 0 & 0 & \frac{3}{8} & \frac{1}{4} & \frac{1}{16} & \frac{1}{2} & 1 \end{bmatrix}$$

$$\begin{bmatrix} 1 & 0 & 0 & 0 & \frac{8}{3} & \frac{1}{6} & \frac{4}{3} & \frac{2}{3} \\ 0 & 1 & 0 & 0 & \frac{4}{3} & \frac{4}{3} & \frac{8}{3} & \frac{4}{3} \\ 0 & 0 & 1 & 0 & \frac{4}{3} & \frac{1}{3} & \frac{8}{3} & \frac{4}{3} \\ 0 & 0 & 0 & 1 & \frac{2}{3} & \frac{1}{6} & \frac{4}{3} & \frac{8}{3} \end{bmatrix}$$

$$F = \begin{bmatrix} \frac{8}{3} & \frac{1}{6} & \frac{4}{3} & \frac{2}{3} \\ \frac{4}{3} & \frac{4}{3} & \frac{8}{3} & \frac{4}{3} \\ \frac{4}{3} & \frac{1}{3} & \frac{8}{3} & \frac{4}{3} \\ \frac{2}{3} & \frac{1}{6} & \frac{4}{3} & \frac{8}{3} \end{bmatrix}$$

$$FR = \begin{bmatrix} \frac{8}{3} & \frac{1}{6} & \frac{4}{3} & \frac{2}{3} \\ \frac{4}{3} & \frac{4}{3} & \frac{8}{3} & \frac{4}{3} \\ \frac{4}{3} & \frac{1}{3} & \frac{8}{3} & \frac{4}{3} \\ \frac{2}{3} & \frac{1}{6} & \frac{4}{3} & \frac{8}{3} \end{bmatrix} \begin{bmatrix} \frac{1}{4} & 0 \\ 0 & 0 \\ \frac{1}{16} & \frac{1}{16} \\ 0 & \frac{1}{4} \end{bmatrix} = \begin{bmatrix} \frac{3}{4} & \frac{1}{4} \\ \frac{1}{2} & \frac{1}{2} \\ \frac{1}{2} & \frac{1}{2} \\ \frac{1}{4} & \frac{3}{4} \end{bmatrix}$$

65. If two parents with the genes Aa are mated, the probability that the recessive gene will eventually disappear is $\frac{1}{2}$. This is found in row 3, column 1 of the matrix FR, because this is the probability of state 4 becoming state 1.

67. The class mobility transition matrix M is as follows:

	B	S	M	F	T
Bottom	0.42	0.25	0.15	0.10	0.08
Second	0.19	0.28	0.21	0.18	0.14
Middle	0.19	0.19	0.26	0.20	0.16
Fourth	0.13	0.18	0.20	0.25	0.24
Top	0.10	0.12	0.19	0.23	0.36

For information about grandparent to grandson transitions, we need to look at M^2.

$$M^2 = \begin{bmatrix} 0.2734 & 0.2311 & 0.1897 & 0.1064 & 0.1454 \\ 0.2103 & 0.2150 & 0.2045 & 0.1886 & 0.1816 \\ 0.2073 & 0.2053 & 0.2064 & 0.1920 & 0.1890 \\ 0.1833 & 0.1947 & 0.2049 & 0.2031 & 0.2140 \\ 0.1668 & 0.1793 & 0.2040 & 0.2099 & 0.2400 \end{bmatrix}$$

(a) The probability that the grandson of someone in the bottom income group is in the top income group is the entry in row 1, column 5, which is 0.1454.

(b) The probability that the grandson of someone in the top income group is in the top income group is the entry in row 5, column 5, which is 0.2400.

(c) We need to solve

$$[v_1 \quad v_2 \quad v_3 \quad v_4 \quad v_5]\, M = [v_1 \quad v_2 \quad v_3 \quad v_4 \quad v_5]$$

or

$$[v_1 \quad v_2 \quad v_3 \quad v_4 \quad v_5]\, M - [v_1 \quad v_2 \quad v_3 \quad v_4 \quad v_5]$$
$$= [0 \quad 0 \quad 0 \quad 0 \quad 0].$$

Simplifying the left side gives us the following system:

$$-0.58v_1 + 0.19v_2 + 0.19v_3 + 0.13v_4 + 0.10v_5 = 0$$
$$0.25v_1 - 0.72v_2 + 0.19v_3 + 0.18v_4 + 0.12v_5 = 0$$
$$0.15v_1 + 0.21v_2 - 0.74v_3 + 0.20v_4 + 0.19v_5 = 0$$
$$0.10v_1 + 0.18v_2 + 0.20v_3 - 0.75v_4 + 0.23v_5 = 0$$
$$0.08v_1 + 0.14v_2 + 0.16v_3 + 0.24v_4 - 0.64v_5 = 0$$

To this system we add the sum-to-1 condition on V, which gives the equation

$$v_1 + v_2 + v_3 + v_4 + v_5 = 1.$$

The augmented matrix for this system of six equations is the following.

$$\begin{bmatrix} -0.58 & 0.19 & 0.19 & 0.13 & 0.10 & 0 \\ 0.25 & -0.72 & 0.19 & 0.18 & 0.12 & 0 \\ 0.15 & 0.21 & -0.74 & 0.20 & 0.19 & 0 \\ 0.10 & 0.18 & 0.20 & -0.75 & 0.23 & 0 \\ 0.08 & 0.14 & 0.16 & 0.24 & -0.64 & 0 \\ \hline 1 & 1 & 1 & 1 & 1 & 1 \end{bmatrix}$$

Solving with row operations using a graphing calculator, we get a row of 0's along the bottom; the first five rows contain the solution $v_1 = 0.2094$, $v_2 = 0.2057$,

$v_3 = 0.2018$, $v_4 = 0.1902$, and

$v_5 = 0.1929$. These are the probabilities that a male is in each income group if the trends given by the transition matrix continue.

Chapter 11

GAME THEORY

11.1   Strictly Determined Games

Your Turn 1

The payoff matrix is

$$\begin{array}{c} & \text{B} \\ & \begin{array}{cc} 1 & 2 \end{array} \\ \text{A} \begin{array}{c} 1 \\ 2 \end{array} & \begin{bmatrix} 2 & -1 \\ -3 & 4 \end{bmatrix}. \end{array}$$

If A chooses row 1 and B chooses column 1, the payoff is 2, which means that B pays A $2.

Your Turn 2

The payoff matrix is

$$\begin{bmatrix} 4 & -2 & -3 \\ 0 & -3 & 4 \\ -2 & -4 & -5 \end{bmatrix}.$$

Each entry in row 1 is larger than the corresponding entry in row 3. (Also each entry in row 2 is larger than the corresponding entry in row 3. Thus row 3 is a dominated strategy and we can remove this row, giving

$$\begin{bmatrix} 4 & -2 & -3 \\ 0 & -3 & 4 \end{bmatrix}.$$

Each entry in column 2 is less than the corresponding entry in column 1, so column 1 can be removed, giving

$$\begin{bmatrix} -2 & -3 \\ -3 & 4 \end{bmatrix}.$$

Your Turn 3

The payoff matrix for the game is

$$\begin{bmatrix} -2 & 1 & 0 \\ -3 & 2 & -5 \\ -1 & 0 & 2 \end{bmatrix}.$$

Underline the smallest number in each row and box the smallest number in each column:

$$\begin{bmatrix} \underline{-2} & 1 & 0 \\ -3 & \boxed{2} & \underline{-5} \\ \boxed{\underline{-1}} & 0 & \boxed{2} \end{bmatrix}$$

The number -1 is both underlined and boxed, so it is a saddle point. The corresponding strategy is $(3, 1)$, that is, A plays row 3 and B plays column 1. The value of the game is -1.

11.1   Excercises

For Exercises 1–7, use the following game.

$$\begin{array}{c} & \text{B} \\ & \begin{array}{ccc} 1 & 2 & 3 \end{array} \\ \text{A} \begin{array}{c} 1 \\ 2 \\ 3 \end{array} & \begin{bmatrix} 7 & -5 & 0 \\ 3 & -3 & 8 \\ -1 & 5 & 11 \end{bmatrix} \end{array}$$

1. Consider the strategy $(1, 1)$. Player A chooses row 1, and player B chooses column 1. A positive number represents a payoff from B to A. The first row, first column entry is 6, indicating a payoff of $7 from B to A.

3. Consider the strategy $(2, 2)$. Player A chooses row 2, and player B chooses column 2. A negative number represents a payoff from A to B. The second row, second column entry is -3, indicating a payoff of $3 from A to B.

5. Consider the strategy $(3, 1)$. Player A chooses row 3, and player B chooses column 1. The third row, first column entry is -1, indicating a payoff of $1 from A to B.

7. Yes, each entry in column 2 is smaller than the corresponding entry in column 3, so column 2 dominates column 3.

9. $$\begin{bmatrix} 0 & 9 & -3 \\ 3 & -8 & -2 \end{bmatrix}$$

Column 3 dominates column 1, so remove column 1 to obtain

$$\begin{bmatrix} 9 & -3 \\ -8 & -2 \end{bmatrix}.$$

11. $\begin{bmatrix} 3 & 6 \\ 1 & 4 \\ 4 & -2 \\ -4 & 0 \end{bmatrix}$

Row 1 dominates row 2 and row 4, so remove row 2 and row 4 to obtain

$$\begin{bmatrix} 3 & 6 \\ 4 & -2 \end{bmatrix}.$$

13. $\begin{bmatrix} 8 & 12 & -7 \\ -2 & 1 & 4 \end{bmatrix}$

Column 1 dominates column 2, so remove column 2 to obtain

$$\begin{bmatrix} 8 & -7 \\ -2 & 4 \end{bmatrix}.$$

15. $\begin{bmatrix} \underline{-5} & 2 \\ \boxed{5} & \boxed{\underline{4}} \end{bmatrix}$

Underline the smallest number in each row and draw a box around the largest number in each column. The 4 at (2, 2) is the smallest number in its row and the largest number in its column, so the saddle point is 4 at (2, 2). This game is strictly determined and its value is 4.

17. $\begin{bmatrix} 3 & \underline{-4} & \boxed{1} \\ \boxed{5} & \boxed{3} & \underline{-2} \end{bmatrix}$

Underline the smallest number in each row, and box the largest number in each column; in this matrix, the two categorizations do not overlap. There is no saddle point. This game is not strictly determined.

19. $\begin{bmatrix} \underline{-6} & 2 \\ -1 & \underline{-10} \\ \boxed{\underline{3}} & \boxed{5} \end{bmatrix}$

The 3 at (3, 1) is the smallest number in its row and the largest number in its column, so the saddle point is 3 at (3, 1). This game is strictly determined, and its value is 3.

21. $\begin{bmatrix} 2 & 3 & \boxed{\underline{1}} \\ -1 & \boxed{4} & \underline{-7} \\ 5 & 2 & \underline{0} \\ \boxed{8} & \underline{-4} & -1 \end{bmatrix}$

The 1 at (1, 3) is the smallest number in its row and the largest number in its column, so the saddle point is 1 at (1, 3). This game is strictly determined, and its value is 1.

23. $\begin{bmatrix} \underline{-6} & 1 & \boxed{4} & \boxed{2} \\ \boxed{9} & \boxed{3} & \underline{-8} & -7 \end{bmatrix}$

There is no saddle point. This game is not strictly determined.

25. $\begin{bmatrix} 2 & 3 & 1 \\ -1 & 4 & -7 \\ 5 & 2 & 0 \\ 8 & -4 & -1 \end{bmatrix}$

(a) Every entry in column 3 is smaller than the corresponding entry in column 1. Thus, column 3 dominates column 1. So, remove column 1 to obtain

$$\begin{bmatrix} 3 & 1 \\ 4 & -7 \\ 2 & 0 \\ -4 & -1 \end{bmatrix}.$$

(b) Every entry in row 1 is greater than the corresponding entries in rows 3 and 4. Thus, row 1 dominates row 3 and row 4. So, remove rows 3 and 4 to obtain

$$\begin{bmatrix} 3 & 1 \\ 4 & -7 \end{bmatrix}.$$

(c) Every entry in column 2 is smaller than the corresponding entry in column 1. Thus, column 2 dominates column 1. So, remove column 1 to obtain

$$\begin{bmatrix} 1 \\ -7 \end{bmatrix}.$$

(d) The entry in row 1 is larger than the entry in row 2. Thus, row 1 dominates row 2. So, remove row 2 to obtain

$$[1].$$

To verify that this is the saddle point, underline the smallest number in each row and box the largest number in each column.

$$\begin{bmatrix} 2 & 3 & \boxed{\underline{1}} \\ -1 & \boxed{4} & -7 \\ 5 & 2 & 0 \\ \boxed{8} & \underline{-4} & -1 \end{bmatrix}$$

Thus, 1 is the saddle point.

27. $\begin{bmatrix} -3 & -2 & 6 \\ 2 & 0 & 2 \\ 5 & -2 & -4 \end{bmatrix}$

(a) There are no dominated columns.

(b) There are no dominated rows.

29.

	Rain	No Rain
Stadium	−$1800	$2400
Gym	$1500	$1500
Both	$1200	$2100

(a) If the dean is an optimist, she doesn't think it will rain. $2400 is the largest possible net profit in that column, so she should set up in the stadium.

(b) If the dean is a pessimist, she thinks it will rain. $1500 is the largest possible net profit in that column, so she should set up in the gym.

(c) If there is a 0.6 probability of rain, there is a 0.4 probability of no rain. Find the dean's expected profit for each strategy.

Stadium: $-\$1800(0.6) + \$2400(0.4) = -\$120$

Gym: $\$1500(0.6) + \$1500(0.4) = \$1500$

Both: $\$1200(0.6) + \$2100(0.4) = \$1560$

She should set up both the stadium and the gym for a maximum profit of $1560.

31. (a) The payoff matrix is as follows.

	Better	Not Better
Market	$50,000	−$25,000
Don't Market	−$40,000	−$10,000

(b) Find the expected profit under the 2 strategies.

Market product:

$0.4(50,000) + 0.6(-25,000) = \$5000$

Don't market:

$0.4(-40,000) + 0.6(-10,000)$
$= -\$22,000$

They should market the product and make a profit of $5000 since that is better than losing $22,000.

(c) There are no dominated strategies. There is no saddle point.

33.

		B		
		City 1	City 2	City 3
	City 1	15	−2	6
A	City 2	7	15	9
	City 3	3	−3	15

To get the entries in the above matrix, look, for example, at the entry in row 1, column 2. If merchant A locates in city 1 and merchant B in city 2, then merchant A will get 80% of the business in city 1, 20% in city 2, and 60% in city 3. Taking into account the fraction of the population living in each city, we get

$0.80(0.30) + 0.20(0.45) + 0.60(0.25) = 0.48.$

Thus, merchant A gets 48% of the total business. However, this is 2 percentage points below 50%. Hence the entry in row 1, column 2 is −2.

For row 1, column 3, we have

$0.80(0.30) + 0.20(0.25) + 0.60(0.45) = 0.56,$

or 56% of the total business for merchant A. However, this is 6 percentage points above 50%.

The remaining entries are found in a similar manner. This game is not strictly determined because there is no saddle point.

37. $\begin{bmatrix} -8 & \underline{-10} & 4 \\ 0 & \underline{-12} & 6 \\ \boxed{3} & \boxed{\underline{-7}} & \boxed{8} \end{bmatrix}$

Underline the smallest number in each row, and box the largest number in each column. −7 is the smallest number in its row and the largest number in its column. The saddle point is −7 at (3, 2) and the value of the game is −7.

39. The payoff matrix is as follows.

	Rock	Paper	Scissors
Rock	0	−1	1
Paper	1	0	−1
Scissors	−1	1	0

Underline the smallest number in each row, and box the largest number in each column; in this matrix, the two categorizations do not overlap. The game is not strictly determined since it does not have a saddle point.

11.2   Mixed Strategies

Your Turn 1

The game in Examples 1 and 2 has the payoff matrix

$$M = \begin{bmatrix} -1 & 2 \\ 1 & 0 \end{bmatrix},$$

where the numbers are in dollars.

If player A chooses row 1 with probability 0.4, player A's probabilities are

$$A = \begin{bmatrix} 0.4 & 0.6 \end{bmatrix}.$$

If player B chooses column 1 with probability 0.8, player B's probabilities are

$$\begin{bmatrix} 0.8 \\ 0.2 \end{bmatrix}.$$

The expected value of the game is

$$AMB = \begin{bmatrix} 0.4 & 0.6 \end{bmatrix} \begin{bmatrix} -1 & 2 \\ 1 & 0 \end{bmatrix} \begin{bmatrix} 0.8 \\ 0.2 \end{bmatrix}$$

$$= \begin{bmatrix} 0.4 & 0.6 \end{bmatrix} \begin{bmatrix} -0.4 \\ 0.8 \end{bmatrix} = \begin{bmatrix} 0.32 \end{bmatrix}.$$

The expected value of the game is 32 cents.

Your Turn 2

Brian's payoff matrix is as follows:

	Attack	No attack
Spray	$8000	$2000
Don't	−$9000	$15,000

First note that there is no saddle point for this payoff matrix, so a mixed strategy is required. Let p_1 be the probability with which Brian chooses to spray; then

$1 - p_1$ is the probability that he does not spray. If nature attacks, his expected payoff is

$$E_1 = 8000p_1 - 9000(1 - p_1)$$
$$= 17{,}000p_1 - 9000.$$

If there is no attack by nature, his expected payoff is

$$E_2 = 2000p_1 + 15{,}000(1 - p_1)$$
$$= -13{,}000p_1 + 15{,}000.$$

Set E_1 equal to E_2.

$$17{,}000p_1 - 9000 = -13{,}000p_1 + 15{,}000$$
$$30{,}000p_1 = 24{,}000$$
$$p_1 = \frac{4}{5}$$

So Brian should spray with probability $4/5$ and not spray with probability $1 - 4/5 = 1/5$.

Your Turn 3

The payoff matrix is as follows:

$$\begin{bmatrix} a_{11} & a_{12} \\ a_{21} & a_{22} \end{bmatrix} = \begin{bmatrix} -4 & 2 \\ 1 & 0 \end{bmatrix}$$

First note that the game is not strictly determined. Apply the formulas used in Example 4.

$$d = a_{11} - a_{21} - a_{12} + a_{22}$$
$$= -4 - 1 - 2 + 0 = -7$$

The optimum strategy for A is

$$p_1 = \frac{a_{22} - a_{21}}{d} = \frac{0 - 1}{-7} = \frac{1}{7}.$$

So A should play row 1 with probability $1/7$ and row 2 with probability $1 - 1/7 = 6/7$.

The optimum strategy for B is

$$q_1 = \frac{a_{22} - a_{12}}{d} = \frac{0 - 2}{-7} = \frac{2}{7}.$$

So B should play column 1 with probability $2/7$ and column 2 with probability $1 - 2/7 = 5/7$.

The value of the game is

$$g = \frac{a_{11}a_{22} - a_{12}a_{21}}{d}$$
$$= \frac{(-4)(0) - (2)(1)}{-7}$$
$$= \frac{2}{7}.$$

11.2 Exercises

1. (a) $AMB = [0.5 \quad 0.5] \begin{bmatrix} 3 & -4 \\ -5 & 2 \end{bmatrix} \begin{bmatrix} 0.4 \\ 0.6 \end{bmatrix}$

$= [0.5 \quad 0.5] \begin{bmatrix} -1.2 \\ -0.8 \end{bmatrix}$

$= [-0.6 - 0.4] = [-1]$

The expected value is -1.

(b) $AMB = [0.1 \quad 0.9] \begin{bmatrix} 3 & -4 \\ -5 & 2 \end{bmatrix} \begin{bmatrix} 0.4 \\ 0.6 \end{bmatrix}$

$= [0.1 \quad 0.9] \begin{bmatrix} -1.2 \\ -0.8 \end{bmatrix} = [-0.84]$

The expected value is -0.84.

(c) $AMB = [0.8 \quad 0.2] \begin{bmatrix} 3 & -4 \\ -5 & 2 \end{bmatrix} \begin{bmatrix} 0.4 \\ 0.6 \end{bmatrix}$

$= [0.8 \quad 0.2] \begin{bmatrix} -1.2 \\ -0.8 \end{bmatrix} = [-1.12]$

The expected value is -1.12.

(d) $AMB = [0.2 \quad 0.8] \begin{bmatrix} 3 & -4 \\ -5 & 2 \end{bmatrix} \begin{bmatrix} 0.4 \\ 0.6 \end{bmatrix}$

$= [0.2 \quad 0.8] \begin{bmatrix} -1.2 \\ -0.8 \end{bmatrix} = [-0.88]$

The expected value is -0.88.

3. $\begin{bmatrix} 7 & 1 \\ 3 & 4 \end{bmatrix}$

There are no saddle points so the game is not strictly determined, and mixed strategies must be used. Here $a_{11} = 7, a_{12} = 1, a_{21} = 3$, and $a_{22} = 4$. For player A, the optimum strategy is

$p_1 = \dfrac{a_{22} - a_{21}}{a_{11} - a_{21} - a_{12} + a_{22}}$

$= \dfrac{4 - 3}{7 - 3 - 1 + 4} = \dfrac{1}{7},$

$p_2 = 1 - p_1 = 1 - \dfrac{1}{7} = \dfrac{6}{7}.$

For player B, the optimum strategy is

$q_1 = \dfrac{a_{22} - a_{12}}{a_{11} - a_{21} - a_{12} + a_{22}}$

$= \dfrac{4 - 1}{7 - 3 - 1 + 4} = \dfrac{3}{7},$

$q_2 = 1 - q_1 = 1 - \dfrac{3}{7} = \dfrac{4}{7}.$

The value of the game is

$\dfrac{a_{11}a_{22} - a_{12}a_{21}}{a_{11} - a_{21} - a_{12} + a_{22}} = \dfrac{7(4) - 1(3)}{7 - 3 - 1 + 4}$

$= \dfrac{28 - 3}{7} = \dfrac{25}{7}.$

5. $\begin{bmatrix} -2 & 0 \\ 5 & -4 \end{bmatrix}$

There are no saddle points so the game is not strictly determined, and mixed strategies must be used. Here $a_{11} = -2, a_{12} = 0, a_{21} = 5$, and $a_{22} = -4$. For player A, the optimum strategy is

$p_1 = \dfrac{-4 - 5}{-2 - 5 - 0 + (-4)} = \dfrac{-9}{-11} = \dfrac{9}{11},$

$p_2 = 1 - \dfrac{9}{11} = \dfrac{2}{11}.$

For player B, the optimum strategy is

$q_1 = \dfrac{-4 - 0}{-2 - 5 - 0 + (-4)} = \dfrac{-4}{-11} = \dfrac{4}{11},$

$q_2 = 1 - \dfrac{4}{11} = \dfrac{7}{11}.$

The value of the game is

$\dfrac{-2(-4) - 0(5)}{-2 - 5 - 0 + (-4)} = -\dfrac{8}{11}.$

7. $\begin{bmatrix} 4 & -3 \\ -1 & 9 \end{bmatrix}$

There are no saddle points. For player A, the optimum strategy is

$p_1 = \dfrac{9 - (-1)}{4 - (-1) - (-3) + 9} = \dfrac{10}{17},$

$p_2 = 1 - \dfrac{10}{17} = \dfrac{7}{17}.$

For player B, the optimum strategy is

$q_1 = \dfrac{9 - (-3)}{4 - (-1) - (-3) + 9} = \dfrac{12}{17},$

$q_2 = 1 - \dfrac{12}{17} = \dfrac{5}{17}.$

The value of the game is

$\dfrac{4(9) - (-3)(-1)}{4 - (-1) - (-3) + 9} = \dfrac{33}{17}.$

9. $\begin{bmatrix} \boxed{-1} & 2 \\ \boxed{3} & \boxed{5} \end{bmatrix}$

The game is strictly dominated since it has a saddle point. Thus, pure strategies can be used. The saddle point is at (2, 1) and the value of the game is 3.

11. $\begin{bmatrix} \frac{8}{3} & -\frac{1}{2} \\ \frac{3}{4} & -\frac{5}{12} \end{bmatrix}$

The game is strictly determined since it has a saddle point at (2, 2). The value of the game is $-\frac{5}{12}$.

13. $\begin{bmatrix} -2 & \frac{1}{2} \\ 0 & -3 \end{bmatrix}$

There are no saddle points. For player A, the optimum strategy is

$$p_1 = \frac{-3 - 0}{-2 - 0 - \frac{1}{2} + (-3)} = \frac{-3}{\frac{-11}{2}} = \frac{6}{11},$$

$$p_2 = 1 - \frac{6}{11} = \frac{5}{11}.$$

For player B, the optimum strategy is

$$q_1 = \frac{-3 - \frac{1}{2}}{-2 - 0 - \frac{1}{2} + (-3)} = \frac{-\frac{7}{2}}{\frac{-11}{2}} = \frac{7}{11},$$

$$q_2 = 1 - \frac{7}{11} = \frac{4}{11}.$$

The value of the game is

$$\frac{-2(-3) - \frac{1}{2}(0)}{-2 - 0 - \frac{1}{2} + (-3)} = \frac{6}{\frac{-11}{2}} = -\frac{12}{11}.$$

15. $\begin{bmatrix} -4 & 9 \\ 3 & -5 \\ 8 & 7 \end{bmatrix}$

Row 3 dominates row 2, so remove row 2. This gives the matrix

$$\begin{bmatrix} -4 & 9 \\ 8 & 7 \end{bmatrix}.$$

For player A, the optimum strategy is

$$p_1 = \frac{7 - 8}{-4 - 8 - 9 + 7} = \frac{-1}{-14} = \frac{1}{14},$$

$$p_2 = 0 (\text{row 2 was removed}),$$

$$p_3 = 1 - (p_1 + p_2) = 1 - \frac{1}{14} = \frac{13}{14}.$$

For player B, the optimum strategy is

$$q_1 = \frac{7 - 9}{-4 - 8 - 9 + 7} = \frac{-2}{-14} = \frac{1}{7},$$

$$q_2 = 1 - \frac{1}{7} = \frac{6}{7}.$$

The value of the game is

$$\frac{-4(7) - 9(8)}{-4 - 8 - 9 + 7} = \frac{-100}{-14} = \frac{50}{7}.$$

17. $\begin{bmatrix} 8 & 6 & 3 \\ -1 & -2 & 4 \end{bmatrix}$

Column 2 dominates column 1, so remove column 1. This gives the matrix

$$\begin{bmatrix} 6 & 3 \\ -2 & 4 \end{bmatrix}.$$

For player A, the optimum strategy is

$$p_1 = \frac{4 - (-2)}{6 - (-2) - 3 + 4} = \frac{6}{9} = \frac{2}{3},$$

$$p_2 = 1 - \frac{2}{3} = \frac{1}{3}.$$

For player B, the optimum strategy is

$$q_1 = 0 \text{ (column 1 was removed)},$$

$$q_2 = \frac{4 - 3}{6 - (-2) - 3 + 4} = \frac{1}{9},$$

$$q_3 = 1 - (q_1 + q_2) = 1 - \frac{1}{9} = \frac{8}{9}.$$

The value of the game is

$$\frac{6(4) - 3(-2)}{6 - (-2) - 3 + 4} = \frac{30}{9} = \frac{10}{3}.$$

19. $\begin{bmatrix} 9 & -1 & 6 \\ 13 & 11 & 8 \\ 6 & 0 & 9 \end{bmatrix}$

Row 2 dominates row 1, so remove row 1. This gives the matrix

$$\begin{bmatrix} 13 & 11 & 8 \\ 6 & 0 & 9 \end{bmatrix}.$$

Now, column 2 dominates column 1, so remove column 1. This gives the matrix

$$\begin{bmatrix} 11 & 8 \\ 0 & 9 \end{bmatrix}.$$

For player A, the optimum strategy is

$p_1 = 0$ (row 1 was removed),

$$p_2 = \frac{9-0}{11-0-8+9} = \frac{9}{12} = \frac{3}{4},$$

$$p_3 = 1 - \frac{3}{4} = \frac{1}{4}.$$

For player B, the optimum strategy is

$q_1 = 0$ (column 1 was removed),

$$q_2 = \frac{9-8}{11-0-8+9} = \frac{1}{12},$$

$$q_3 = 1 - \frac{1}{12} = \frac{11}{12}.$$

The value of the game is

$$\frac{11(9)-8(0)}{11-0-8+9} = \frac{99}{12} = \frac{33}{4}.$$

21. In a non-strictly-determined game, there is no saddle point. Let

$$M = \begin{bmatrix} a_{11} & a_{12} \\ a_{21} & a_{22} \end{bmatrix}$$

be the payoff matrix of the game. Assume that player B chooses column 1 with probability q_1. The expected value for B, assuming A plays row 1, is E_1, where

$$E_1 = a_{11}q_1 + a_{12}(1-q_1).$$

The expected value for B if A plays row 2 is E_2, where

$$E_2 = a_{21}q_1 + a_{22}(1-q_1).$$

The optimum strategy for player B is found by letting $E_1 = E_2$.

$a_{11}q_1 + a_{12}(1-q_1) = a_{21}q_1 + a_{22}(1-q_1)$

$a_{11}q_1 + a_{12} - a_{12}q_1 = a_{21}q_1 + a_{22} - a_{22}q_1$

$a_{11}q_1 - a_{21}q_1 - a_{12}q_1 + a_{22}q_1 = a_{22} - a_{12}$

$q_1(a_{11} - a_{21} - a_{12} + a_{22}) = a_{22} - a_{12}$

$$q_1 = \frac{a_{22} - a_{12}}{a_{11} - a_{21} - a_{12} + a_{22}}$$

Since $q_2 = 1 - q_1$,

$$q_2 = 1 - \frac{a_{22} - a_{12}}{a_{11} - a_{21} - a_{12} + a_{22}}$$

$$= \frac{a_{11} - a_{21} - a_{12} + a_{22} - (a_{22} - a_{12})}{a_{11} - a_{21} - a_{12} + a_{22}}$$

$$= \frac{a_{11} - a_{21}}{a_{11} - a_{21} - a_{12} + a_{22}}.$$

These are the formulas given in the text for q_1 and q_2.

27. $\begin{bmatrix} 1 & 2 & 3 \\ 4 & 3 & 1 \end{bmatrix}$

(a) $E_1 = 1p_1 + 4(1 - p_1)$

$= p_1 + 4 - 4p_1$

$= 4 - 3p_1$

$E_2 = 2p_1 + 3(1 - p_1)$

$= 2p_1 + 3 - 3p_1$

$= 3 - p_1$

$E_3 = 3p_1 + 1(1 - p_1)$

$= 3p_1 + 1 - p_1$

$= 1 + 2p_1$

(b)

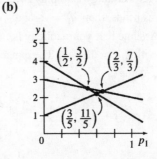

(c) From the graph, $p_1 = \frac{3}{5}$ maximizes the minimum expected value the row player receives.

29. The payoff matrix for this game is

$$\begin{array}{c c} & \begin{array}{c c} H & T \end{array} \\ \begin{array}{c} H \\ T \end{array} & \begin{bmatrix} 3 & -2 \\ -2 & 1 \end{bmatrix} \end{array}$$

with $a_{11} = 3, a_{12} = -2, a_{21} = -2, a_{22} = 1$ and

$d = a_{11} - a_{21} - a_{12} + a_{22}$

$= 3 - (-2) - (-2) + 1$

$= 8.$

(a) Your optimum strategy is

$$H: p_1 = \frac{a_{22} - a_{21}}{d} = \frac{1 - (-2)}{d} = \frac{3}{8}$$

$$T: p_2 = 1 - p_1 = 1 - \frac{3}{8} = \frac{5}{8}.$$

You should play heads with probability $\frac{3}{8}$ and tails with probability $\frac{5}{8}$.

Your opponent's optimum strategy is

$$H: q_1 = \frac{a_{22} - a_{21}}{d} = \frac{1 - (-2)}{d} = \frac{3}{8}$$

$$T: q_2 = 1 - q_1 = 1 - \frac{3}{8} = \frac{5}{8}.$$

Your opponent should also play heads with probability $\frac{3}{8}$ and tails with probability $\frac{5}{8}$.

The value of the game is

$$g = \frac{a_{11}a_{22} - a_{12}a_{21}}{d}$$

$$= \frac{(3)(1) - (-2)(-2)}{d}$$

$$= \frac{3 - 4}{8} = -\frac{1}{8}.$$

(b) If the stranger's strategy is to play twice as many tails as heads, then $q_1 = \frac{1}{3}$ and $q_2 = \frac{2}{3}$. Assume that your strategy is $[p_1 \quad p_2]$ with $p_2 = 1 - p_1$. The expected value of the game is

$$AMB = [p_1 \quad 1 - p_1] \begin{bmatrix} 3 & -2 \\ -2 & 1 \end{bmatrix} \begin{bmatrix} \frac{1}{3} \\ \frac{2}{3} \end{bmatrix}$$

$$= [5p_1 - 2 \quad 1 - 3p_1] \begin{bmatrix} \frac{1}{3} \\ \frac{2}{3} \end{bmatrix}$$

$$= \left[-\frac{p_1}{3} \right].$$

Marilyn claims the stranger will win \$1 for every six plays, so

$$6\left(-\frac{p_1}{3} \right) = -1$$

$$-2p_1 = -1$$

$$p_1 = \frac{1}{2}.$$

Thus, Marilyn is assuming your strategy is to play heads and tails with probability $\frac{1}{2}$ each.

(c) From part (b), the expected value of the game with the given strategy for your opponent is

$-p_1/3$. Since $0 \le p_1 \le 1$, no matter how you choose p_1, $-p_1/3$ will always be nonpositive—i.e., zero or negative. The best you can hope for is an expected value of zero, which will occur when $p_1 = 0$. In other words you should always play tails. The value of the game is then 0.

31.

		Competitor's Strategy	
		4.9	4.75
Boeing's	4.9	$\begin{bmatrix} 2 & -4 \\ 0 & 2 \end{bmatrix}$	
Strategy	4.75		

There is no saddle point. The optimum strategy for pricing is

$$p_1 = \frac{0 - 2}{-4 - 2 - 2 + 0} = \frac{-2}{-8} = \frac{1}{4},$$

$$p_2 = 1 - p_1 = \frac{3}{4}.$$

(q_1 and q_2 are not of interest here.) This means that Boeing's price strategy should be to aim for the \$4.9 million profit $\frac{1}{4}$ of the time and the \$4.75 million profit $\frac{3}{4}$ of the time. The value is

$$\frac{-4(0) - 2(2)}{-4 - 2 - 2 + 0} = \frac{-4}{-8} = \frac{1}{2},$$

which means this strategy will increase Boeing's profit by $\frac{1}{2}$ million dollars.

33. (a)

		Strain	
		1	2
Medicine	1	$\begin{bmatrix} 0.75 & 0.4 \\ 0 & 1 \end{bmatrix}$	
	2		

(b) There is no saddle point. The optimum strategy for prescribing medicine is

$$p_1 = \frac{1 - 0}{0.75 - 0 - 0.4 + 1} = \frac{1}{1.35} = \frac{20}{27},$$

$$p_2 = 1 - p_1 = \frac{7}{27}.$$

(q_1 and q_2 are not of interest here.) This means that Dr. Goedeker should prescribe medicine 1 about $\frac{20}{27}$ of the time and medicine 2 about $\frac{7}{27}$ of the time. The result (value) will be

$$\frac{0.75(1) - 0.4(0)}{0.75 - 0 - 0.4 + 1} = \frac{0.75}{1.35} = \frac{5}{9},$$

which indicates an effectiveness of 55.56%.

35. **(a)** $\begin{bmatrix} 1 & -1 \\ -1 & 1 \end{bmatrix}$

(b) There is no saddle point. For player A, the optimum strategy is

$$p_1 = \frac{1 - (-1)}{1 - (-1) - (-1) + 1} = \frac{2}{4} = \frac{1}{2},$$

$$p_2 = 1 - p_1 = \frac{1}{2}.$$

For player B, the optimum strategy is

$$q_1 = \frac{1 - (-1)}{1 - (-1) - (-1) + 1} = \frac{2}{4} = \frac{1}{2},$$

$$q_2 = 1 - q_1 = \frac{1}{2}.$$

The value of the game is

$$\frac{1(1) - (-1)(-1)}{1 - (-1) - (-1) + 1} = \frac{0}{4} = 0,$$

so this is a fair game.

37. **(a)** The playoff matrix for the game is

Number of Fingers

$$\begin{array}{c} \text{Number of} \\ \text{Fingers} \end{array} \begin{array}{cc} & 0 \qquad\quad 2 \\ \begin{array}{c} 0 \\ 2 \end{array} & \begin{bmatrix} 0 & -2 \\ -2 & 4 \end{bmatrix}. \end{array}$$

(b) For player A, the optimum strategy is

$$p_1 = \frac{4 - (-2)}{0 - (-2) - (-2) + 4} = \frac{6}{8} = \frac{3}{4},$$

$$p_2 = 1 - p_1 = \frac{1}{4}.$$

For player B, the optimum strategy is

$$q_1 = \frac{4 - (-2)}{0 - (-2) - (-2) + 4} = \frac{6}{8} = \frac{3}{4},$$

$$q_2 = 1 - q_1 = \frac{1}{4}.$$

Both players should choose 0 fingers with probability $\frac{3}{4}$ and 2 fingers with probability $\frac{1}{4}$.

The value of the game is

$$\frac{0(4) - (-2)(-2)}{0 - (-2) - (-2) + 4} = \frac{-4}{8} = -\frac{1}{2}.$$

11.3 Game Theory and Linear Programming

Your Turn 1

First add 4 to each element of the payoff matrix to obtain nonnegative entries:

$$\begin{bmatrix} -2 & 1 & 0 \\ 1 & -4 & -2 \end{bmatrix} \text{ becomes } \begin{bmatrix} 2 & 5 & 4 \\ 5 & 0 & 2 \end{bmatrix}$$

In linear programming terms our problem is now the following:

$$\begin{aligned} \text{Maximize} \quad & z = x_1 + x_2 + x_3 \\ \text{Subject to:} \quad & 2x_1 + 5x_2 + 4x_3 \leq 1 \\ & 5x_1 \qquad\quad + 2x_3 \leq 1 \\ \text{with} \quad & x_1 \geq 0, \ x_2 \geq 0, \ x_3 \geq 0. \end{aligned}$$

The initial tableau is

$$\begin{array}{cccccc} x_1 & x_2 & x_3 & s_2 & s_3 & z \\ \end{array}$$
$$\left[\begin{array}{cccccc|c} 2 & 5 & 4 & 1 & 0 & 0 & 1 \\ \boxed{5} & 0 & 2 & 0 & 1 & 0 & 1 \\ \hline -1 & -1 & -1 & 0 & 0 & 1 & 0 \end{array}\right]$$

Pivot on the indicated 5.

$$\begin{array}{cccccc} & x_1 & x_2 & x_3 & s_2 & s_3 & z \\ \end{array}$$
$$\begin{array}{c} -2R_2 + 5R_1 \to R_1 \\ \\ R_2 + 5R_3 \to R_3 \end{array} \left[\begin{array}{cccccc|c} 0 & \boxed{25} & 16 & 5 & -2 & 0 & 3 \\ 5 & 0 & 2 & 0 & 1 & 0 & 1 \\ 0 & -5 & -3 & 0 & 1 & 5 & 1 \end{array}\right]$$

Next pivot on the indicated 25.

$$\begin{array}{cccccc} x_1 & x_2 & x_3 & s_2 & s_3 & z \\ \end{array}$$
$$\begin{array}{c} \\ \\ \frac{1}{5}R_1 + R_3 \to R_3 \end{array}\left[\begin{array}{cccccc|c} 0 & 25 & 16 & 5 & -2 & 0 & 3 \\ 5 & 0 & 2 & 0 & 1 & 0 & 1 \\ 0 & 0 & \frac{1}{5} & 1 & \frac{3}{5} & 5 & \frac{8}{5} \end{array}\right]$$

Finally get coefficients of 1 for the x variables and for z.

$$\begin{array}{cccccc} x_1 & x_2 & x_3 & s_2 & s_3 & z \\ \end{array}$$
$$\begin{array}{c} \frac{1}{25}R_1 \to R_1 \\ \frac{1}{5}R_2 \to R_2 \\ \frac{1}{5}R_3 \to R_3 \end{array}\left[\begin{array}{cccccc|c} 0 & 1 & \frac{16}{25} & \frac{1}{5} & -\frac{2}{25} & 0 & \frac{3}{25} \\ 1 & 0 & \frac{2}{5} & 0 & \frac{1}{5} & 0 & \frac{1}{5} \\ 0 & 0 & \frac{1}{25} & \frac{1}{5} & \frac{3}{25} & 1 & \frac{8}{25} \end{array}\right]$$

From the bottom right entry we obtain

$$z = \frac{8}{25}; g = \frac{1}{z} = \frac{25}{8}.$$

The fourth and fifth entries in the last row give the values of the y's.

$$y_1 = \frac{1}{5}, y_2 = \frac{3}{35}$$

The values of the p's for player A's strategy are found by multiplying the y's by g.

$$p_1 = y_1 g = \left(\frac{1}{5}\right)\left(\frac{25}{8}\right) = \frac{5}{8}$$

$$p_2 = y_2 g = \left(\frac{3}{25}\right)\left(\frac{25}{8}\right) = \frac{3}{8}$$

The values of the x's are found from the first two rows and first three columns

$$x_1 = \frac{1}{5}, x_2 = \frac{3}{25}, x_3 = 0$$

The values of the q's for player B's strategy are found by multiplying the x's by g.

$$q_1 = x_1 g = \left(\frac{1}{5}\right)\left(\frac{25}{8}\right) = \frac{5}{8}$$

$$q_2 = x_2 g = \left(\frac{3}{25}\right)\left(\frac{25}{8}\right) = \frac{3}{8}$$

$$q_3 = x_3 g = (0)\left(\frac{25}{8}\right) = 0$$

Thus the strategy for A is $[p_1 \quad p_2] = [5/8 \quad 3/8]$ and the strategy for B is $[q_1 \quad q_2 \quad q_3] = [5/8 \quad 3/8 \quad 0]$. To find the value of the game, we subtract the 4 that we initially added from the value of g to get

$$\frac{25}{8} - 4 = \frac{25}{8} - \frac{32}{8} = -\frac{7}{8}.$$

11.3 Exercises

1. $\begin{bmatrix} 1 & 2 \\ 4 & 1 \end{bmatrix}$

To find the optimum strategy for player A, use the following linear programming problem.

Minimize $w = x + y$
subject to: $x + 4y \geq 1$
$2x + y \geq 1$
with $x \geq 0, y \geq 0$.

Solve this linear programming problem by the graphical method. Sketch the feasible region.

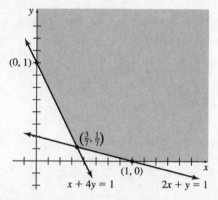

The region is unbounded, with corner points $(0, 1)$, $\left(\frac{3}{7}, \frac{1}{7}\right)$, and $(1, 0)$.

Corner Point	Value of $w = x + y$
$(0, 1)$	$0 + 1 = 1$
$\left(\frac{3}{7}, \frac{1}{7}\right)$	$\frac{3}{7} + \frac{1}{7} = \frac{4}{7}$
$(1, 0)$	$1 + 0 = 1$

The minimum value is $w = \frac{4}{7}$ at the point where $x = \frac{3}{7}$, $y = \frac{1}{7}$. Thus, the value of the game is $g = \frac{1}{w} = \frac{7}{4}$, and the optimum strategy for A is

$$p_1 = gx = \frac{7}{4}\left(\frac{3}{7}\right) = \frac{3}{4},$$

$$p_2 = gy = \frac{7}{4}\left(\frac{1}{7}\right) = \frac{1}{4}.$$

To find the optimum strategy for player B, use the following linear programming problem.

Maximize $z = x + y$
Subject to: $x + 2y \leq 1$
$4x + y \leq 1$
with $x \geq 0, y \geq 0$.

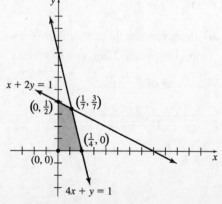

The region is bounded, with corner points $\left(0, \frac{1}{2}\right)$, $\left(\frac{1}{7}, \frac{3}{7}\right)$, $\left(\frac{1}{4}, 0\right)$, and $(0, 0)$.

Corner Point	Value of $z = x + y$
$\left(0, \frac{1}{2}\right)$	$0 + \frac{1}{2} = \frac{1}{2}$
$\left(\frac{1}{7}, \frac{3}{7}\right)$	$\frac{1}{7} + \frac{3}{7} = \frac{4}{7}$
$\left(\frac{1}{4}, 0\right)$	$\frac{1}{4} + 0 = \frac{1}{4}$
$(0, 0)$	$0 + 0 = 0$

The maximum value is $z = \frac{4}{7}$ at the point where $x = \frac{1}{7}, y = \frac{3}{7}$. Thus, the value of the game is

$g = \frac{1}{z} = \frac{4}{7}$ (agreeing with our earlier finding), and the optimum strategy for B is

$$q_1 = gx = \frac{7}{4}\left(\frac{1}{7}\right) = \frac{1}{4},$$

$$q_2 = gy = \frac{7}{4}\left(\frac{3}{7}\right) = \frac{3}{4}.$$

3. $\begin{bmatrix} 2 & -2 \\ -1 & 6 \end{bmatrix}$

To guarantee that the value of the game is positive, add 2 to each of the entries to eliminate any negative numbers. The 2 will be subtracted later, after the calculations have been performed.

$$\begin{bmatrix} 4 & 0 \\ 1 & 8 \end{bmatrix}$$

To find the optimum strategy for player A, use the following linear programming problem.

$$\text{Minimize} \quad w = x + y$$
$$\text{subject to:} \quad 4x + y \geq 1$$
$$8y \geq 1$$
$$\text{with} \quad x \geq 0, \ y \geq 0.$$

Solve this linear programming problem by the graphical method. Sketch the feasible region.

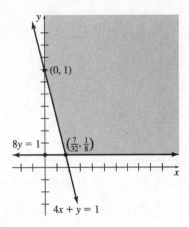

The region is unbounded, with corner points $(0, 1)$ and $\left(\frac{7}{32}, \frac{1}{8}\right)$.

Corner Point	Value of $w = x + y$
$(0, 1)$	$0 + 1 = 1$
$\left(\frac{7}{32}, \frac{1}{8}\right)$	$\frac{7}{32} + \frac{1}{8} = \frac{11}{32}$

The minimum value is $w = \frac{11}{32}$ at the point where $x = \frac{7}{32}, y = \frac{1}{8}$. Thus, the value of the game is $g = \frac{1}{w} = \frac{32}{11}$, and the optimum strategy for A is

$$p_1 = gx = \frac{32}{11}\left(\frac{7}{32}\right) = \frac{7}{11},$$

$$p_2 = gy = \frac{32}{11}\left(\frac{1}{8}\right) = \frac{4}{11}.$$

The value of the game is

$$g - 2 = \frac{32}{11} - 2 = \frac{10}{11}.$$

To find the optimum strategy for player B, use the following linear programming problem.

$$\text{Maximize} \quad z = x + y$$
$$\text{subject to:} \quad 4x \leq 1$$
$$x + 8y \leq 1$$
$$\text{with} \quad x \geq 0, y \geq 0.$$

The region is bounded, with corner points $\left(0, \frac{1}{8}\right)$, $\left(\frac{1}{4}, \frac{3}{32}\right), \left(\frac{1}{4}, 0\right)$, and $(0, 0)$.

Corner Point	Value of $z = x + y$
$\left(0, \frac{1}{8}\right)$	$0 + \frac{1}{8} = \frac{1}{8}$
$\left(\frac{1}{4}, \frac{3}{32}\right)$	$\frac{1}{4} + \frac{3}{32} = \frac{11}{32}$
$\left(\frac{1}{4}, 0\right)$	$\frac{1}{4} + 0 = \frac{1}{4}$
$(0, 0)$	$0 + 0 = 0$

The maximum value is $z = \frac{11}{32}$ at the point where
$x = \frac{1}{4}, y = \frac{3}{32}$. Thus, the value of the game is
$g = \frac{1}{z} = \frac{32}{11}$ (agreeing with our earlier finding),
and the optimum strategy for B is

$$q_1 = gx = \frac{32}{11}\left(\frac{1}{4}\right) = \frac{8}{11},$$

$$q_2 = gy = \frac{32}{11}\left(\frac{3}{32}\right) = \frac{3}{11}.$$

The value of the game is

$$g - 2 = \frac{32}{11} - 2 = \frac{10}{11}$$

which agrees with the earlier value.

5. $\begin{bmatrix} 7 & -8 \\ -3 & 3 \end{bmatrix}$

Add 8 to each entry to obtain

$$\begin{bmatrix} 15 & 0 \\ 5 & 11 \end{bmatrix}.$$

To find the optimum strategy for player A,

Minimize $w = x + y$
subject to: $15x + 5y \geq 1$
$\qquad\qquad\quad 11y \geq 1$
with $x \geq 0, y \geq 0.$

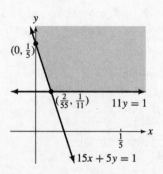

Corner Point	Value of $w = x + y$
$\left(0, \frac{1}{5}\right)$	$0 + \frac{1}{5} = \frac{1}{5}$
$\left(\frac{2}{55}, \frac{1}{11}\right)$	$\frac{2}{55} + \frac{1}{11} = \frac{7}{55}$

The minimum value is $w = \frac{7}{55}$ at $\left(\frac{2}{55}, \frac{1}{11}\right)$.
Thus, $g = \frac{1}{w} = \frac{55}{7}$, and the optimum strategy
for A is

$$p_1 = gx = \frac{55}{7}\left(\frac{2}{55}\right) = \frac{2}{7},$$

$$p_2 = gy = \frac{55}{7}\left(\frac{1}{11}\right) = \frac{5}{7}.$$

The value of the game is

$$\frac{55}{7} - 8 = -\frac{1}{7}.$$

To find the optimum strategy for player B,

Maximize $z = x + y$
subject to: $15x \qquad\quad \leq 1$
$\qquad\qquad\quad 5x + 11y \leq 1$
with $x \geq 0, \ y \geq 0.$

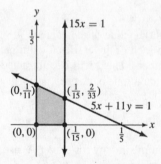

Corner Point	Value of $z = x + y$
$\left(0, \frac{1}{11}\right)$	$0 + \frac{1}{11} = \frac{1}{11}$
$\left(\frac{1}{15}, \frac{2}{33}\right)$	$\frac{1}{15} + \frac{2}{33} = \frac{21}{165} = \frac{7}{55}$
$\left(\frac{1}{15}, 0\right)$	$\frac{1}{15} + 0 = \frac{1}{15}$
$(0, 0)$	$0 + 0 = 0$

The maximum value is $z = \frac{7}{55}$ at $\left(\frac{1}{15}, \frac{2}{33}\right)$.
The optimum strategy for B is

$$q_1 = gx = \frac{55}{7}\left(\frac{1}{15}\right) = \frac{11}{21},$$

$$q_2 = gy = \frac{55}{7}\left(\frac{2}{33}\right) = \frac{10}{21}.$$

7. $\begin{bmatrix} 3 & -4 & 1 \\ 5 & 3 & -2 \end{bmatrix}$

Column 1 is dominated by the other two columns, so remove it.

$$\begin{bmatrix} -4 & 1 \\ 3 & -2 \end{bmatrix}$$

Add 4 to each entry to obtain

$$\begin{bmatrix} 0 & 5 \\ 7 & 2 \end{bmatrix}.$$

The linear programming problem to be solved is as follows.

$$\text{Maximize} \quad z = x_2 + x_3$$
$$\text{subject to:} \quad 5x_3 \le 1$$
$$7x_2 + 2x_3 \le 1$$
$$\text{with} \quad x_2 \ge 0, x_3 \ge 0.$$

Use the simplex method to solve the problem. The initial tableau is

$$
\begin{array}{ccccc}
x_2 & x_3 & s_1 & s_2 & z \\
\end{array}
$$
$$
\left[\begin{array}{ccccc|c}
0 & 5 & 1 & 0 & 0 & 1 \\
\boxed{7} & 2 & 0 & 1 & 0 & 1 \\
\hline
-1 & -1 & 0 & 0 & 1 & 0
\end{array} \right].
$$

Pivot on the indicated entry.

$$
\begin{array}{ccccc}
x_2 & x_3 & s_1 & s_2 & z \\
\end{array}
$$
$$
R_2 + 7R_3 \rightarrow R_3 \quad
\left[\begin{array}{ccccc|c}
0 & \boxed{5} & 1 & 0 & 0 & 1 \\
\boxed{7} & 2 & 0 & 1 & 0 & 1 \\
\hline
0 & -5 & 0 & 1 & 7 & 1
\end{array} \right]
$$

Pivot again.

$$
\begin{array}{ccccc}
x_2 & x_3 & s_1 & s_2 & z \\
\end{array}
$$
$$
\begin{array}{c}
-2R_1 + 5R_2 \rightarrow R_2 \\
R_1 + R_3 \rightarrow R_3
\end{array}
\left[\begin{array}{ccccc|c}
0 & 5 & 1 & 0 & 0 & 1 \\
35 & 0 & -2 & 5 & 0 & 3 \\
\hline
0 & 0 & 1 & 1 & 7 & 2
\end{array} \right]
$$

Create a 1 in the columns corresponding to $x_2, x_3,$ and z.

$$
\begin{array}{ccccc}
x_2 & x_3 & s_1 & s_2 & z \\
\end{array}
$$
$$
\begin{array}{c}
\frac{1}{5}R_1 \rightarrow R_1 \\[4pt]
\frac{1}{35}R_2 \rightarrow R_2 \\[4pt]
\frac{1}{7}R_3 \rightarrow R_3
\end{array}
\left[\begin{array}{ccccc|c}
0 & 1 & \frac{1}{5} & 0 & 0 & \frac{1}{5} \\[4pt]
1 & 0 & -\frac{2}{35} & \frac{1}{7} & 0 & \frac{3}{35} \\[4pt]
0 & 0 & \frac{1}{7} & \frac{1}{7} & 1 & \frac{2}{7}
\end{array} \right]
$$

From this final tableau, we have $x_2 = \frac{3}{35}$, $x_3 = \frac{1}{5}$, $y_1 = \frac{1}{7}$, $y_2 = \frac{1}{7}$, $z = \frac{2}{7}$. Note that $g = \frac{1}{z} = \frac{7}{2}$.

The optimum strategy for player A is

$$p_1 = gy_1 = \frac{7}{2}\left(\frac{1}{7}\right) = \frac{1}{2},$$

$$p_2 = gy_2 = \frac{7}{2}\left(\frac{1}{7}\right) = \frac{1}{2}.$$

The optimum strategy for player B is

$$q_1 = 0 \text{ (column 1 was removed)},$$

$$q_2 = gx_2 = \frac{7}{2}\left(\frac{3}{35}\right) = \frac{3}{10},$$

$$q_3 = gx_3 = \frac{7}{2}\left(\frac{1}{5}\right) = \frac{7}{10}.$$

The value of the game is

$$\frac{7}{2} - 4 = -\frac{1}{2}.$$

9. $\begin{bmatrix} -1 & 1 & 4 \\ 3 & -2 & -3 \end{bmatrix}$

Because of negative entries, add 3 to all entries. The resulting payoff matrix is

$$\begin{bmatrix} 2 & 4 & 7 \\ 6 & 1 & 0 \end{bmatrix}.$$

The linear programming problem to be solved is

$$\text{Maximize} \quad z = x_1 + x_2 + x_3$$
$$\text{subject to:} \quad 2x_1 + 4x_2 + 7x_3 \le 1$$
$$6x_1 + x_2 \le 1$$
$$\text{with} \quad x_1 \ge 0, x_2 \ge 0, x_3 \ge 0.$$

We will solve this problem using the simplex method. The initial tableau is

$$
\begin{array}{cccccc}
x_1 & x_2 & x_3 & s_1 & s_2 & z \\
\end{array}
$$
$$
\left[\begin{array}{cccccc|c}
2 & 4 & 7 & 1 & 0 & 0 & 1 \\
\boxed{6} & 1 & 0 & 0 & 1 & 0 & 1 \\
\hline
-1 & -1 & -1 & 0 & 0 & 1 & 0
\end{array} \right].
$$

We arbitrarily choose the first column. The smallest ratio is formed by the 6 in row 2. We make this the pivot and arrive at the following matrix.

$$-R_2 + 3R_1 \to R_1 \quad \begin{array}{cccccc} x_1 & x_2 & x_3 & s_1 & s_2 & z \\ \left[\begin{array}{cccccc|c} 0 & 11 & \boxed{21} & 3 & -1 & 0 & 2 \\ 6 & 1 & 0 & 0 & 1 & 0 & 1 \\ \hline 0 & -5 & -6 & 0 & 1 & 6 & 1 \end{array} \right] \end{array}$$

$$R_2 + 6R_3 \to R_3$$

The next pivot is the 21 in row 1, column 3.

$$\begin{array}{cccccc} x_1 & x_2 & x_3 & s_1 & s_2 & z \\ \left[\begin{array}{cccccc|c} 0 & \boxed{11} & 21 & 3 & -1 & 0 & 2 \\ 6 & 1 & 0 & 0 & 1 & 0 & 1 \\ \hline 0 & -13 & 0 & 6 & 5 & 42 & 11 \end{array} \right] \end{array}$$

$$2R_1 + 7R_3 \to R_3$$

The next pivot is the 11 in row 1, column 2.

$$\begin{array}{cccccc} x_1 & x_2 & x_3 & s_1 & s_2 & z \\ \left[\begin{array}{cccccc|c} 0 & 11 & 21 & 3 & -1 & 0 & 2 \\ 66 & 0 & -21 & -3 & 12 & 0 & 9 \\ \hline 0 & 0 & 273 & 105 & 42 & 462 & 147 \end{array} \right] \end{array}$$

$$-R_1 + 11R_2 \to R_2$$
$$13R_1 + 11R_3 \to R_3$$

We have the final tableau. Dividing the bottom row by 462 gives a z-value of $z = \frac{147}{462} = \frac{7}{22}$ so $g = \frac{1}{z} = \frac{22}{7}$.

The values of y_1 and y_2 are read from the bottom of the columns for the two slack variables after dividing the entries by 462.

$$y_1 = \frac{105}{462} = \frac{5}{22}, \quad y_2 = \frac{42}{462} = \frac{1}{11}$$

We find the values of p_1 and p_2 by multiplying the values of y_1 and y_2 by g.

$$p_1 = gy_1 = \left(\frac{22}{7}\right)\left(\frac{5}{22}\right) = \frac{5}{7},$$

$$p_2 = gy_2 = \left(\frac{22}{7}\right)\left(\frac{1}{11}\right) = \frac{2}{7}$$

Next, we find the values of x_1, x_2, and x_3 by using the first four columns combined with the last column.

$$x_1 = \frac{9}{66} = \frac{3}{22}, \quad x_2 = \frac{2}{11}, \quad x_3 = 0$$

We find the values of q_1, q_2, and q_3 by multiplying the values of x_1, x_2, and x_3 by g.

$$q_1 = gx_1 = \left(\frac{22}{7}\right)\left(\frac{3}{22}\right) = \frac{3}{7},$$

$$q_2 = \left(\frac{22}{7}\right)\left(\frac{2}{11}\right) = \frac{4}{7}, q_3 = 0$$

Finally, the value of the game is found by subtracting from g the 3 that was added at the beginning, yielding $\frac{22}{7} - 3 = \frac{1}{7}$.

To summarize, the optimum strategy for player A is $p_1 = \frac{5}{7}$ and $p_2 = \frac{2}{7}$. The optimum strategy for player B is $q_1 = \frac{3}{7}, q_2 = \frac{4}{7}$, and $q_3 = 0$. When these strategies are used, the value of the game is $\frac{1}{7}$.

11. $\begin{bmatrix} 1 & 0 & -1 \\ -1 & 0 & 1 \\ 2 & -1 & 2 \end{bmatrix}$

Add 1 to each entry to obtain

$$\begin{bmatrix} 2 & 1 & 0 \\ 0 & 1 & 2 \\ 3 & 0 & 3 \end{bmatrix}.$$

The linear programming problem to be solved is as follows.

Maximize $z = x_1 + x_2 + x_3$

subject to: $2x_1 + x_2 \quad\quad \le 1$

$\quad\quad\quad x_2 + 2x_3 \le 1$

$\quad 3x_1 \quad\quad + 3x_3 \le 1$

with $\quad x_1 \ge 0, x_2 \ge 0, x_3 \ge 0.$

The initial tableau is

$$
\begin{array}{c}
\begin{array}{ccccccc} x_1 & x_2 & x_3 & s_1 & s_2 & s_3 & z \end{array} \\
\left[\begin{array}{ccccccc|c}
2 & 1 & 0 & 1 & 0 & 0 & 0 & 1 \\
0 & 1 & 2 & 0 & 1 & 0 & 0 & 1 \\
\boxed{3} & 0 & 3 & 0 & 0 & 1 & 0 & 1 \\
\hline
-1 & -1 & -1 & 0 & 0 & 0 & 1 & 0
\end{array} \right].
\end{array}
$$

Pivot on each indicated entry.

$$
\begin{array}{c}
\begin{array}{ccccccc} \quad x_1 & x_2 & x_3 & s_1 & s_2 & s_3 & z \end{array} \\
\begin{array}{r}
-2R_3 + 3R_1 \rightarrow R_1 \\
\\
\\
R_3 + 3R_4 \rightarrow R_4
\end{array}
\left[\begin{array}{ccccccc|c}
0 & \boxed{3} & -6 & 3 & 0 & -2 & 0 & 1 \\
0 & 1 & 2 & 0 & 1 & 0 & 0 & 1 \\
3 & 0 & 3 & 0 & 0 & 1 & 0 & 1 \\
\hline
0 & -3 & 0 & 0 & 0 & 1 & 3 & 1
\end{array} \right]
\end{array}
$$

$$
\begin{array}{c}
\begin{array}{ccccccc} x_1 & x_2 & x_3 & s_1 & s_2 & s_3 & z \end{array} \\
\begin{array}{r}
\\
-R_1 + 3R_2 \rightarrow R_2 \\
\\
R_1 + R_4 \rightarrow R_4
\end{array}
\left[\begin{array}{ccccccc|c}
0 & 3 & -6 & 3 & 0 & -2 & 0 & 1 \\
0 & 0 & \boxed{12} & -3 & 3 & 2 & 0 & 2 \\
3 & 0 & 3 & 0 & 0 & 1 & 0 & 1 \\
\hline
0 & 0 & -6 & 3 & 0 & -1 & 3 & 2
\end{array} \right]
\end{array}
$$

$$
\begin{array}{c}
\begin{array}{ccccccc} x_1 & x_2 & x_3 & s_1 & s_2 & s_3 & z \end{array} \\
\begin{array}{r}
R_2 + 2R_1 \rightarrow R_1 \\
\\
-R_2 + 4R_3 \rightarrow R_3 \\
R_2 + 2R_4 \rightarrow R_4
\end{array}
\left[\begin{array}{ccccccc|c}
0 & 6 & 0 & 3 & 3 & -2 & 0 & 4 \\
0 & 0 & 12 & -3 & 3 & 2 & 0 & 2 \\
12 & 0 & 0 & 3 & -3 & 2 & 0 & 2 \\
\hline
0 & 0 & 0 & 3 & 3 & 0 & 6 & 6
\end{array} \right]
\end{array}
$$

Create a 1 in the columns corresponding to $x_1, x_2,\ x_3,$ and z.

$$
\begin{array}{c}
\begin{array}{ccccccc} x_1 & x_2 & x_3 & s_1 & s_2 & s_3 & z \end{array} \\
\begin{array}{r}
\frac{1}{6}R_1 \rightarrow R_1 \\[2pt]
\frac{1}{12}R_2 \rightarrow R_2 \\[2pt]
\frac{1}{12}R_3 \rightarrow R_3 \\[2pt]
\frac{1}{6}R_4 \rightarrow R_4
\end{array}
\left[\begin{array}{ccccccc|c}
0 & 1 & 0 & \frac{1}{2} & \frac{1}{2} & -\frac{1}{3} & 0 & \frac{2}{3} \\[2pt]
0 & 0 & 1 & -\frac{1}{4} & \frac{1}{4} & \frac{1}{6} & 0 & \frac{1}{6} \\[2pt]
1 & 0 & 0 & \frac{1}{4} & -\frac{1}{4} & \frac{1}{6} & 0 & \frac{1}{6} \\[2pt]
\hline
0 & 0 & 0 & \frac{1}{2} & \frac{1}{2} & 0 & 1 & 1
\end{array} \right]
\end{array}
$$

From this final tableau, we have $x_1 = \frac{1}{6}, x_2 = \frac{2}{3}, x_3 = \frac{1}{6}, y_1 = \frac{1}{2}, y_2 = \frac{1}{2}, y_3 = 0,$ and $z = 1$. Note that

$g = \frac{1}{z} = 1$. The optimum strategy for player A is

$$
p_1 = gy_1 = 1\left(\frac{1}{2} \right) = \frac{1}{2},
$$

$$
p_2 = gy_2 = 1\left(\frac{1}{2} \right) = \frac{1}{2},
$$

$$
p_3 = gy_3 = 1(0) = 0.
$$

The optimum strategy for player B is

$$q_1 = gx_1 = 1\left(\frac{1}{6}\right) = \frac{1}{6},$$

$$q_2 = gx_2 = 1\left(\frac{2}{3}\right) = \frac{2}{3},$$

$$q_3 = gx_3 = 1\left(\frac{1}{6}\right) = \frac{1}{6}.$$

The value of the game is $1 - 1 = 0$.

13. The payoff matrix is as follows.

$$
\begin{array}{c c}
 & \begin{array}{c c} \text{Strike} & \text{No Strike} \end{array} \\
\begin{array}{c} \text{Bid } \$30,000 \\ \text{Bid } \$40,000 \end{array} &
\left[\begin{array}{c c} -\$5500 & \$4500 \\ \$4500 & \$0 \end{array} \right]
\end{array}
$$

Add $5500 to each entry to obtain

$$
\begin{bmatrix} 0 & 10,000 \\ 10,000 & 5500 \end{bmatrix}.
$$

To find the optimum strategy for the contractor,

$$\text{Minimize} \quad w = x + y$$
$$\text{subject to:} \qquad 10,000y \geq 1$$
$$10,000x + 5500y \geq 1$$
$$\text{with} \qquad x \geq 0, y \geq 0.$$

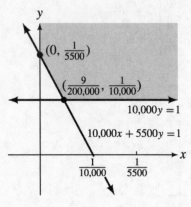

Corner Point	Value of $w = x + y$
$\left(0, \frac{1}{5500}\right)$	$0 + \frac{1}{5500} = \frac{1}{5500}$
$\left(\frac{9}{200,000}, \frac{1}{10,000}\right)$	$\frac{9}{200,000} + \frac{1}{10,000} = \frac{29}{200,000}$

The minimum value is $w = \frac{29}{200,000}$ at $\left(\frac{9}{200,000}, \frac{1}{10,000}\right)$. Thus, $g = \frac{1}{w} = \frac{200,000}{29}$, and the optimum strategy for

the contractor is

$$p_1 = gx = \frac{200{,}000}{29}\left(\frac{9}{200{,}000}\right) = \frac{9}{29},$$

$$p_2 = gy = \frac{200{,}000}{29}\left(\frac{1}{10{,}000}\right) = \frac{20}{29}.$$

That is, the contractor should bid $30,000$ with probability $\frac{9}{29}$ and $40,000$ with probability $\frac{20}{29}$. The value of the game is

$$\frac{200{,}000}{29} - 5500 \approx \$1396.55.$$

15. (a) The payoff matrix is as follows.

$$
\begin{array}{cc}
 & OI \\
 & \begin{array}{ccc} A & B & C \end{array} \\
GI\ \begin{array}{c} A \\ B \\ C \end{array} & \left[\begin{array}{ccc} 5000 & 10{,}000 & 10{,}000 \\ 8000 & 4000 & 8000 \\ 6000 & 6000 & 3000 \end{array}\right]
\end{array}
$$

Note that $5000 = \frac{1}{2}(10{,}000)$, $4000 = \frac{1}{2}(8000)$, and $3000 = \frac{1}{2}(6000)$ are the reduced profits for General Items when the two companies run ads in the same city.

(b) The linear programming problem to be solved is as follows.

$$\text{Maximize} \quad z = x_1 + x_2 + x_3$$
$$\text{subject to:} \quad 5000x_1 + 10{,}000x_2 + 10{,}000x_3 \le 1$$
$$8000x_1 + \quad 4000x_2 + \quad 8000x_3 \le 1$$
$$6000x_1 + \quad 6000x_2 + \quad 3000x_3 \le 1$$
$$\text{with} \quad x_1 \ge 0,\, x_2 \ge 0,\, x_3 \ge 0.$$

The initial tableau is

$$
\begin{array}{ccccccc}
x_1 & x_2 & x_3 & s_1 & s_2 & s_3 & z \\
\end{array}
$$
$$
\left[\begin{array}{ccccccc|c}
5000 & 10{,}000 & 10{,}000 & 1 & 0 & 0 & 0 & 1 \\
\boxed{8000} & 4000 & 8000 & 0 & 1 & 0 & 0 & 1 \\
6000 & 6000 & 3000 & 0 & 0 & 1 & 0 & 1 \\
\hline
-1 & -1 & -1 & 0 & 0 & 0 & 1 & 0
\end{array}\right].
$$

Pivot on each indicated entry.

$$
\begin{array}{ccccccc}
 & x_1 & x_2 & x_3 & s_1 & s_2 & s_3 & z \\
\end{array}
$$

$$
\begin{array}{l}
-5R_2 + 8R_1 \rightarrow R_1 \\
\\
-3R_2 + 4R_3 \rightarrow R_3 \\
R_2 + 8000R_4 \rightarrow R_4
\end{array}
\left[\begin{array}{ccccccc|c}
0 & \boxed{60{,}000} & 40{,}000 & 8 & -5 & 0 & 0 & 3 \\
8000 & 4000 & 8000 & 0 & 1 & 0 & 0 & 1 \\
0 & 12{,}000 & -12{,}000 & 0 & -3 & 4 & 0 & 1 \\
0 & -4000 & & 0 & 0 & 1 & 1 & 8000 & 1
\end{array}\right]
$$

$$
\begin{array}{ccccccc}
 & x_1 & x_2 & x_3 & s_1 & s_2 & s_3 & z \\
\end{array}
$$

$$
\begin{array}{l}
\\
-R_1 + 15R_2 \rightarrow R_2 \\
-R_1 + 5R_3 \rightarrow R_3 \\
R_1 + 15R_4 \rightarrow R_4
\end{array}
\left[\begin{array}{ccccccc|c}
0 & 60{,}000 & 40{,}000 & 8 & -5 & 0 & 0 & 3 \\
120{,}000 & 0 & 80{,}000 & -8 & 20 & 0 & 0 & 12 \\
0 & 0 & -100{,}000 & -8 & -10 & 20 & 0 & 2 \\
0 & 0 & 40{,}000 & 8 & 10 & 0 & 120{,}000 & 18
\end{array}\right]
$$

Create a 1 in the columns corresponding to x_1, x_2, s_3, and z.

$$
\begin{array}{c}
\frac{1}{60,000}R_1 \to R_1 \\[4pt]
\frac{1}{120,000}R_2 \to R_2 \\[4pt]
\frac{1}{20}R_3 \to R_3 \\[4pt]
\frac{1}{120,000}R_4 \to R_4
\end{array}
\quad
\begin{array}{ccccccc|c}
x_1 & x_2 & x_3 & s_1 & s_2 & s_3 & z & \\
\hline
0 & 1 & \frac{2}{3} & \frac{1}{7500} & -\frac{1}{12,000} & 0 & 0 & \frac{1}{20,000} \\[4pt]
1 & 0 & \frac{2}{3} & -\frac{1}{15,000} & \frac{1}{6000} & 0 & 0 & \frac{1}{10,000} \\[4pt]
0 & 0 & -5000 & -\frac{2}{5} & -\frac{1}{2} & 1 & 0 & \frac{1}{10} \\[4pt]
0 & 0 & \frac{1}{3} & \frac{1}{15,000} & \frac{1}{12,000} & 0 & 1 & \frac{3}{20,000}
\end{array}
$$

From this final tableau, we have

$$x_1 = \frac{1}{10,000}, x_2 = \frac{1}{20,000}, x_3 = 0,$$

$$y_1 = \frac{1}{15,000}, y_2 = \frac{1}{12,000},$$

$$y_3 = 0, \text{ and } z = \frac{3}{20,000}.$$

Note that

$$g = \frac{1}{z} = \frac{20,000}{3} \approx 6666.67,$$

so the value of the game is $6666.67. The optimum strategy for General Items is

$$p_1 = gy_1 = \frac{20,000}{3}\left(\frac{1}{15,000}\right) = \frac{4}{9},$$

$$p_2 = gy_2 = \frac{20,000}{3}\left(\frac{1}{12,000}\right) = \frac{5}{9},$$

$$p_3 = gy_3 = \frac{20,000}{3}(0) = 0.$$

That is, General Items should advertise in Atlanta with probability $\frac{4}{9}$, in Boston with probability $\frac{5}{9}$, and never in Cleveland.

The optimum strategy for Original Imitators is

$$q_1 = gx_1 = \frac{20,000}{3}\left(\frac{1}{10,000}\right) = \frac{2}{3},$$

$$q_2 = gx_2 = \frac{20,000}{3}\left(\frac{1}{20,000}\right) = \frac{1}{3},$$

$$q_3 = gx_3 = \frac{20,000}{3}(0) = 0.$$

That is, Original Imitators should advertise in Atlanta with probability $\frac{2}{3}$, in Boston with probability $\frac{1}{3}$, and never in Cleveland.

17. This exercise should be solved by graphing calculator or computer methods. The solution may vary slightly. The answer is that merchant A should locate in cities 1, 2, and 3 with probabilities $\frac{27}{101}$, $\frac{129}{202}$, and $\frac{19}{202}$, respectively; merchant B should locate in cities 1, 2, and 3 with probabilities $\frac{39}{101}$, $\frac{9}{101}$, and $\frac{53}{101}$, respectively. The value of the game is $\frac{885}{101} \approx 8.76$ percentage points.

19. (a)
$$\begin{bmatrix} 17.3 & 11.5 \\ -4.4 & 20.6 \\ 5.2 & 17.0 \end{bmatrix}$$

Because of negative entries, add 4.4 to all entries. The resulting payoff matrix is

$$\begin{bmatrix} 21.7 & 15.9 \\ 0 & 25 \\ 9.6 & 21.4 \end{bmatrix}.$$

The linear programming problem to be solved is

$$\text{Maximize: } z = x_1 + x_2$$
$$\text{subject to: } 21.7x_1 + 15.9x_2 \le 1$$
$$25x_2 \le 1$$
$$9.6x_1 + 21.4x_2 \le 1$$
$$\text{with} \quad x_1 \ge 0, x_2 \ge 0.$$

We will solve this problem using the simplex method. The initial tableau is

x_1	x_2	s_1	s_2	s_3	z	
21.7	15.9	1	0	0	0	1
0	25	0	1	0	0	1
9.6	21.4	0	0	1	0	1
−1	−1	0	0	0	1	0

We arbitrarily choose the second column. The smallest ratio is formed by the 25 in row 2. We make this the pivot and arrive at the following matrix.

$$-15.9R_2 + 25R_1 \to R_1$$

	x_1	x_2	s_1	s_2	s_3	z	
	542.5	0	25	−15.9	0	0	9.1
	0	25	0	1	0	0	1
$-21.4R_2 + 25R_3 \to R_3$	240	0	0	−21.4	25	0	3.6
$R_2 + 25R_4 \to R_4$	−25	0	0	1	0	25	1

The next pivot is the 240 in row 3, column 1.

$$-542.5R_3 + 240R_1 \to R_1$$

	x_1	x_2	s_1	s_2	s_3	z	
	0	0	6000	7793.5	−13,562.5	0	231
	0	25	0	1	0	0	1
	240	0	0	−21.4	25	0	3.6
$5R_3 + 48R_4 \to R_4$	0	0	0	−59	125	1200	66

The next pivot is the 7793.5 in row 1, column 4.

	x_1	x_2	s_1	s_2	s_3	z	
	0	0	6000	7793.5	−13,562.5	0	231
$-R_1 + 7793.5R_2 \to R_2$	0	194,837.5	−6000	0	13,562.5	0	7562.5
$21.4R_1 + 7793.5R_3 \to R_3$	1,870,440	0	128,400	0	−95,400	0	33,000
$59R_1 + 7793.5R_4 \to R_4$	0	0	354,000	0	174,000	9,352,200	528,000

This is the final tableau. From the bottom row, we find the z-value.

$$z = \frac{528{,}000}{9{,}352{,}200} = \frac{80}{1417}$$

Therefore,

$$g = \frac{1}{z} = \frac{1417}{80}.$$

The values of x_1 and x_2 are obtained by using the first two columns combined with the last column.

$$x_1 = \frac{33{,}000}{1{,}870{,}440} = \frac{25}{1417},$$

$$x_2 = \frac{7562.5}{194{,}837.5} = \frac{55}{1417}$$

The values of y_1, y_2, and y_3 are read from the bottom of the columns for the three slack variables after dividing the entries by 9,352,000.

$$y_1 = \frac{354{,}000}{9{,}352{,}200} = \frac{590}{15{,}587}, y_2 = 0,$$

$$y_3 = \frac{174{,}000}{9{,}352{,}200} = \frac{290}{15{,}587}.$$

The optimum strategy for the fisherman is determined by finding the values of q_1, q_2, and q_3.

$$q_1 = gy_1 = \left(\frac{1417}{80}\right)\left(\frac{590}{15{,}587}\right) = \frac{59}{88} \approx 0.6705$$

$$q_2 = gy_2 = 0$$

$$q_3 = gy_3 = \left(\frac{1417}{80}\right)\left(\frac{290}{15{,}587}\right) = \frac{29}{88} \approx 0.3295$$

The fishermen should fish the inside banks with probability 0.6705 and fish a mixture of the inside and outside banks with probability 0.3295. They should never fish the outside banks exclusively.

The optimum strategy for the environment is determined by finding the values of p_1 and p_2.

$$p_1 = gx_1 = \left(\frac{1417}{80}\right)\left(\frac{25}{1417}\right)$$

$$= \frac{5}{16} = 0.3125$$

$$p_2 = gx_2 = \left(\frac{1417}{80}\right)\left(\frac{55}{1417}\right)$$

$$= \frac{11}{16} = 0.6875$$

There is a strong current with probability 0.3125 and no current with probability 0.6875.

Finally, the value of the game is found by subtracting from g the 4.4 that was added at the beginning, yielding

$$\frac{1417}{80} - 4.4 = 13.3125.$$

(b) Return to the original table and calculate the payoffs when fishing with the current has probability 0.25 and fishing with no current has probability 0.75.

$$\begin{aligned}
\text{Inside:} &\quad 0.25(17.3) + 0.75(11.5) = 12.95 \\
\text{Outside:} &\quad 0.25(-4.4) + 0.75(20.6) = 14.35 \\
\text{Mixture:} &\quad 0.25(5.2) + 0.75(17.0) = 14.05
\end{aligned}$$

The largest value is 14.35. The fishermen should fish the outside banks exclusively, with payoff 14.35.

21.
$$\begin{bmatrix} 50 & 0 \\ 10 & 40 \end{bmatrix}$$

To find the student's optimum strategy,

Minimize $w = x + y$
subject to: $50x + 10y \geq 1$
$40y \geq 1$
with $x \geq 0, \ y \geq 0.$

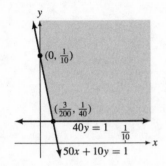

Corner Point	Value of $w = x + y$
$\left(0, \dfrac{1}{10}\right)$	$0 + \dfrac{1}{10} = \dfrac{1}{10}$
$\left(\dfrac{3}{200}, \dfrac{1}{40}\right)$	$\dfrac{3}{200} + \dfrac{1}{40} = \dfrac{1}{25}$

The minimum value is $w = \frac{1}{25}$ at $\left(\frac{3}{200}, \frac{1}{40}\right)$. Thus, the value of the game is $g = \frac{1}{w} = 25$ points, and the optimum strategy for the student is

$$p_1 = gx = 25\left(\frac{3}{200}\right) = \frac{3}{8},$$

$$p_2 = gy = 25\left(\frac{1}{40}\right) = \frac{5}{8}.$$

That is, the student should choose the calculator with probability $\frac{3}{8}$ and the book with probability $\frac{5}{8}$.

23.
$$\begin{bmatrix} 63.2 & 100 & 94.1 \\ 81.2 & 0 & 89.3 \\ 89.5 & 100 & 44.0 \end{bmatrix}$$

The linear programming problem to be solved is

Maximize $z = x_1 + \quad x_2 + \quad x_3$
subject to: $63.2x_1 + 100x_2 + 94.1x_3 \leq 1$
$81.2x_1 \qquad\qquad + 89.3x_3 \leq 1$
$89.5x_1 + 100x_2 + 44.0x_3 \leq 1$
with $x_1 \geq 0, \ x_2 \geq 0, \ x_3 \geq 0.$

The initial tableau is

$$\begin{array}{ccccccc|c}
x_1 & x_2 & x_3 & s_1 & s_2 & s_3 & z & \\
\hline
63.2 & 100 & 94.1 & 1 & 0 & 0 & 0 & 1 \\
81.2 & 0 & 89.3 & 0 & 1 & 0 & 0 & 1 \\
89.2 & 100 & 44.0 & 0 & 0 & 1 & 0 & 1 \\
\hline
-1 & -1 & -1 & 0 & 0 & 0 & 1 & 0
\end{array}$$

The final tableau is found using a calculator or computer program.

x_1	x_2	x_3	s_1	s_2	s_3	z	
0	1	0	0.006888	-0.008791	0.003112	0	0.001209
0	0	1	0.01265	0.004099	-0.01265	0	0.004099
1	0	0	-0.01392	0.007808	0.01392	0	0.007808
0	0	0	0.005625	0.003115	0.004375	1	0.013115

The values of z and g are as follows.

$$z \approx 0.013115, g = \frac{1}{z} \approx 76.25$$

The values of x_1, x_2, and x_3 are read directly from the last column.

$$x_1 \approx 0.007808, x_2 \approx 0.001209, x_3 \approx 0.004099$$

The values of y_1, y_2, and y_3 are read from the bottom of the columns for the three slack variables.

$$y_1 \approx 0.005625,$$
$$y_2 \approx 0.003115,$$
$$y_3 \approx 0.004375.$$

The optimum strategy for the kicker is determined by finding the values of q_1, q_2, and q_3.

$$q_1 = gy_1 \approx 76.25(0.005625) \approx 0.4289$$
$$q_2 = gy_2 \approx 76.25(0.003115) \approx 0.2375$$
$$q_3 = gy_3 \approx 76.25(0.004375) \approx 0.3336$$

The kicker should kick left with a probability of 0.4289, middle with a probability of 0.2375, and right with a probability of 0.3336.

The optimum strategy for the goalie is determined by finding the values of p_1, p_2, and p_3.

$$p_1 = gx_1 \approx 76.25(0.007808) \approx 0.5953$$
$$p_2 = gx_2 \approx 76.25(0.001209) \approx 0.0922$$
$$p_3 = gx_3 \approx 76.25(0.004099) \approx 0.3125$$

The goalie should move left with a probability of 0.5953, middle with a probability of 0.0922, and right with a probability of 0.3125. The payoff for the game is 76.25.

25. **(a)** Use the graphical method. The payoff matrix for the game is the following:

Receiver

$$\begin{array}{c} \\ \text{Server} \end{array} \begin{array}{c} \\ \text{L} \\ \text{R} \end{array} \begin{array}{cc} \text{L} & \text{R} \\ \begin{bmatrix} 0.58 & 0.79 \\ 0.73 & 0.49 \end{bmatrix} \end{array}$$

The linear programming statement of the server's strategy is

$$\text{Minimize} \quad w = x + y$$
$$\text{subject to:} \quad 0.58x + 0.73y \geq 1$$
$$0.79x + 0.49y \geq 1$$
$$\text{with} \quad x \geq 0, y \geq 0.$$

We plot the constraint boundaries in the first quadrant by plotting the two lines

$$0.58x + 0.73y = 1$$
$$0.79x + 0.49y = 1.$$

To find their point of intersection we solve these two equations as a system. A graphing calculator gives the approximate solution $x = 0.8205$, $y = 0.7179$.

We have marked this point on the graph. There are two other corner points, but they correspond to pure strategies so they won't be needed. (We can also see that the minimum cannot occur at these points on the axes because one line has a slope less than -1 and the other has a slope greater than -1 Since a graph of the objective function $x + y = $ constant has slope exactly -1, as we move this constraint line toward the origin looking for a minimum, it will leave the feasible region at the point $(0.8205, 0.7179)$.)

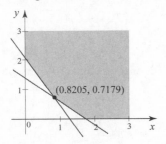

Now we can find the value of the game, which is

$$g = \frac{1}{x + y} = \frac{1}{0.8205 + 0.7179}$$

$$= \frac{1}{1.5384} \approx 0.650.$$

The probabilities for the server's strategy are found by multiplying x and y by g.

$$p_1 = gx = (0.650)(0.8205) = 0.533$$
$$p_2 = gy = (0.650)(0.7179) = 0.467$$

The server should serve left with probability 0.533 and right with probability 0.476.

The linear programming statement of the receiver's strategy is

$$\text{Maximize} \quad w = x + y$$
$$\text{subject to:} \quad 0.58x + 0.79y \le 1$$
$$0.73x + 0.49y \le 1$$
$$\text{with} \quad x \ge 0, y \ge 0.$$

We plot the constraint boundaries in the first quadrant by plotting the two lines

$$0.58x + 0.79y = 1$$
$$0.73x + 0.49y = 1.$$

To find their point of intersection we solve these two equations as a system. A graphing calculator gives the approximate solution $x = 1.0256$, $y = 0.5128$.

We have marked this point on the graph.

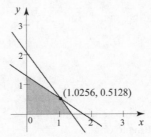

The probabilities for the server's strategy are found by multiplying x and y by g.

$$q_1 = gx = (0.650)(1.0256) = 0.667$$
$$q_2 = gy = (0.650)(0.5128) = 0.333$$

The receiver should move left with probability 0.667 and right with probability 0.333.

Chapter 11 Review Exercises

1. False: This chapter covers only games for two players.

2. True

3. True

4. True

5. False: The expected value of the game must be zero.

6. False: If the game has a saddle point, both players have optimum pure strategies; if not, both players will need to use a mixed strategy.

7. False: A payoff matrix may have neither a dominated row nor a dominated column.

8. True

9. True

10. True

For Exercises 13–17, use the following payoff matrix.

$$\begin{bmatrix} -3 & 5 & -6 & 2 \\ 0 & -1 & 9 & 5 \\ 2 & 6 & -4 & 3 \end{bmatrix}$$

13. The strategy $(1, 1)$ means that player A chooses row 1 and player B chooses column 1. A negative number represents a payoff from A to B. The entry at $(1, 1)$ is -3, indicating that the payoff is $3 from A to B.

15. The entry at $(2, 3)$ is 9. A positive number represents a payoff from B to A, indicating that the payoff is $9 from B to A.

17. Row 3 dominates row 1 and column 1 dominates column 4.

19. $\begin{bmatrix} -11 & 6 & 8 & 9 \\ -10 & -12 & 3 & 2 \end{bmatrix}$

Column 1 dominates both column 3 and column 4. Remove the dominated columns to obtain

$$\begin{bmatrix} -11 & 6 \\ -10 & -12 \end{bmatrix}.$$

21. $\begin{bmatrix} -2 & 4 & 1 \\ 3 & 2 & 7 \\ -8 & 1 & 6 \\ 0 & 3 & 9 \end{bmatrix}$

Row 2 dominates row 3. Remove row 3 to obtain

$$\begin{bmatrix} -2 & 4 & 1 \\ 3 & 2 & 7 \\ 0 & 3 & 9 \end{bmatrix}.$$

Column 1 dominates column 3. Remove column 3 to obtain

$$\begin{bmatrix} -2 & 4 \\ 3 & 2 \\ 0 & 3 \end{bmatrix}.$$

23. $\begin{bmatrix} \boxed{5} & \underline{-4} \\ 3 & \boxed{\underline{-3}} \end{bmatrix}$

Underline the smallest number in each row and draw a box around the largest number in each column. The -3 at $(2, 2)$ is the smallest number in its row and the largest number in its column, so the saddle point is -3 at $(2, 2)$. The value of the game is -3.

25. $\begin{bmatrix} \underline{-4} & -1 \\ 6 & \boxed{0} \\ \boxed{8} & \underline{-3} \end{bmatrix}$

The 0 at $(2, 2)$ is the smallest number in its row and the largest number in its column, so the saddle point is 0 at $(2, 2)$. The value of the game is 0, so it is a fair game.

27. $\begin{bmatrix} -1 & \boxed{4} & \boxed{3} & \boxed{\underline{-4}} \\ \boxed{8} & 1 & 2 & \underline{-7} \end{bmatrix}$

The -4 at $(1, 4)$ is the smallest number in its row and the largest number in its column, so the saddle point is -4 at $(1, 4)$. The value of the game is -4.

29. $\begin{bmatrix} 1 & 0 \\ -2 & 3 \end{bmatrix}$

The optimum strategy for player A is

$$p_1 = \frac{3 - (-2)}{1 - (-2) - 0 + 3} = \frac{5}{6},$$

$$p_2 = 1 - \frac{5}{6} = \frac{1}{6}.$$

The optimum strategy for player B is

$$q_1 = \frac{3 - 0}{1 - (-2) - 0 + 3} = \frac{3}{6} = \frac{1}{2},$$

$$q_2 = 1 - \frac{1}{2} = \frac{1}{2}.$$

The value of the game is

$$\frac{1(3) - 0(-2)}{1 - (-2) - 0 + 3} = \frac{1}{2}.$$

31. $\begin{bmatrix} -3 & 5 \\ 1 & 0 \end{bmatrix}$

The optimum strategy for player A is

$$p_1 = \frac{0 - 1}{-3 - 1 - 5 + 0} = \frac{-1}{-9} = \frac{1}{9},$$

$$p_2 = 1 - \frac{1}{9} = \frac{8}{9}.$$

The optimum strategy for player B is

$$q_1 = \frac{0 - 5}{-3 - 1 - 5 + 0} = \frac{-5}{-9} = \frac{5}{9},$$

$$q_2 = 1 - \frac{5}{9} = \frac{4}{9}.$$

The value of the game is

$$\frac{-3(0) - 5(1)}{-3 - 1 - 5 + 0} = \frac{-5}{-9} = \frac{5}{9}.$$

33. $\begin{bmatrix} -4 & 8 & 0 \\ -2 & 9 & -3 \end{bmatrix}$

Column 1 dominates column 2. Remove column 2 to obtain

$$\begin{bmatrix} -4 & 0 \\ -2 & -3 \end{bmatrix}.$$

The optimum strategy for player A is

$$p_1 = \frac{-3 - (-2)}{-4 - (-2) - 0 + (-3)} = \frac{-1}{-5} = \frac{1}{5},$$

$$p_2 = 1 - \frac{1}{5} = \frac{4}{5}.$$

The optimum strategy for player B is

$$q_1 = \frac{-3 - 0}{-4 - (-2) - 0 + (-3)} = \frac{-3}{-5} = \frac{3}{5},$$

$q_2 = 0$ (column 2 was removed),

$$q_3 = 1 - \frac{3}{5} = \frac{2}{5}.$$

The value of the game is

$$\frac{-4(-3) - 0(-2)}{-4 - (-2) - 0 + (-3)} = \frac{12}{-5} = -\frac{12}{5}.$$

35. $\begin{bmatrix} 2 & -1 \\ -4 & 5 \\ -1 & -2 \end{bmatrix}$

Row 1 dominates row 3. Remove row 3 to obtain

$$\begin{bmatrix} 2 & -1 \\ -4 & 5 \end{bmatrix}.$$

The optimum strategy for player A is

$$p_1 = \frac{5 - (-4)}{2 - (-4) - (-1) + 5} = \frac{9}{12} = \frac{3}{4},$$

$$p_2 = 1 - \frac{3}{4} = \frac{1}{4},$$

$p_3 = 0$ (row 3 was removed).

The optimum strategy for player B is

$$q_1 = \frac{5 - (-1)}{2 - (-4) - (-1) + 5} = \frac{6}{12} = \frac{1}{2},$$

$$q_2 = 1 - \frac{1}{2} = \frac{1}{2}.$$

The value of the game is

$$\frac{2(5) - (-1)(-4)}{2 - (-4) - (-1) + 5} = \frac{6}{12} = \frac{1}{2}.$$

37. $\begin{bmatrix} -4 & 2 \\ 3 & -5 \end{bmatrix}$

Get rid of the negative numbers by adding 5 to each entry to obtain

$$\begin{bmatrix} 1 & 7 \\ 8 & 0 \end{bmatrix}.$$

To find the optimum strategy for player A,

Minimize $\quad w = x + y$

subject to: $\quad x + 8y \geq 1$

$\qquad\qquad\quad 7x \qquad\ \geq 1$

with $\qquad\quad x \geq 0, y \geq 0.$

Solve this linear programming problem by the graphical method. Sketch the feasible region.

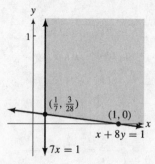

The region is unbounded, with corner points $\left(\frac{1}{7}, \frac{3}{28}\right)$ and $(1, 0)$.

Corner Point	Value of $w = x + y$
$\left(\frac{1}{7}, \frac{3}{28}\right)$	$\frac{1}{7} + \frac{3}{28} = \frac{1}{4}$
$(1, 0)$	$1 + 0 = 1$

The minimum value is $w = \frac{1}{4}$ at $\left(\frac{1}{7}, \frac{3}{28}\right)$. Thus, $g = \frac{1}{w} = 4$, and the optimum strategy for A is

$$p_1 = gx = 4\left(\frac{1}{7}\right) = \frac{4}{7},$$

$$p_2 = gy = 4\left(\frac{3}{28}\right) = \frac{3}{7}.$$

To find the optimum strategy for player B,

$$\text{Maximize} \quad z = x + y$$
$$\text{subject to:} \quad x + 7y \le 1$$
$$\qquad\qquad\qquad 8x \qquad \le 1$$
$$\text{with} \qquad\quad x \ge 0, y \ge 0.$$

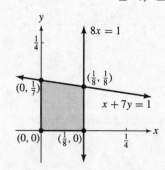

Corner Point	Value of $z = x + y$
$\left(0, \frac{1}{7}\right)$	$0 + \frac{1}{7} = \frac{1}{7}$
$\left(\frac{1}{8}, \frac{1}{8}\right)$	$\frac{1}{8} + \frac{1}{8} = \frac{1}{4}$
$\left(\frac{1}{8}, 0\right)$	$\frac{1}{8} + 0 = \frac{1}{8}$
$(0, 0)$	$0 + 0 = 0$

The maximum value is $z = \frac{1}{4}$ at $\left(\frac{1}{8}, \frac{1}{8}\right)$. The optimum strategy for B is

$$q_1 = gx = 4\left(\frac{1}{8}\right) = \frac{1}{2},$$

$$q_2 = gy = 4\left(\frac{1}{8}\right) = \frac{1}{2}.$$

The value of the game is $4 - 5 = -1$.

39. $\begin{bmatrix} 1 & 0 \\ -3 & 4 \end{bmatrix}$

Add 3 to each entry to obtain

$$\begin{bmatrix} 4 & 3 \\ 0 & 7 \end{bmatrix}.$$

To find the optimum strategy for player A,

$$\text{Minimize} \quad w = x + y$$
$$\text{subject to:} \quad 4x \qquad\quad \ge 1$$
$$\qquad\qquad\qquad 3x + 7y \ge 1$$
$$\text{with} \qquad\quad x \ge 0, y \ge 0.$$

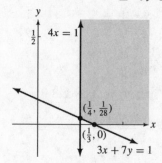

Corner Point	Value of $w = x + y$
$\left(\frac{1}{4}, \frac{1}{28}\right)$	$\frac{1}{4} + \frac{1}{28} = \frac{2}{7}$
$\left(\frac{1}{3}, 0\right)$	$\frac{1}{3} + 0 = \frac{1}{3}$

The minimum value is $w = \frac{2}{7}$ at $\left(\frac{1}{4}, \frac{1}{28}\right)$.

Thus, $g = \frac{1}{w} = \frac{7}{2}$, and the optimum strategy for A is

$$p_1 = gx = \frac{7}{2}\left(\frac{1}{4}\right) = \frac{7}{8},$$

$$p_2 = gy = \frac{7}{2}\left(\frac{1}{28}\right) = \frac{1}{8}.$$

To find the optimum strategy for player B,

Maximize $z = x + y$
subject to: $4x + 3y \leq 1$
$\quad\quad\quad\quad 7y \leq 1$
with $\quad x \geq 0, y \geq 0.$

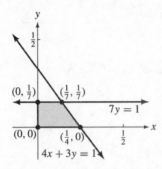

Corner Point	Value of $z = x + y$
$\left(0, \frac{1}{7}\right)$	$0 + \frac{1}{7} = \frac{1}{7}$
$\left(\frac{1}{7}, \frac{1}{7}\right)$	$\frac{1}{7} + \frac{1}{7} = \frac{2}{7}$
$\left(\frac{1}{4}, 0\right)$	$\frac{1}{4} + 0 = \frac{1}{4}$
$(0, 0)$	$0 + 0 = 0$

The maximum value is $z = \frac{2}{7}$ at $\left(\frac{1}{7}, \frac{1}{7}\right)$.
The optimum strategy for B is

$$q_1 = gx = \frac{7}{2}\left(\frac{1}{7}\right) = \frac{1}{2},$$

$$q_2 = gy = \frac{7}{2}\left(\frac{1}{7}\right) = \frac{1}{2}.$$

The value of the game is

$$\frac{7}{2} - 3 = \frac{1}{2}.$$

41. $\begin{bmatrix} 2 & 1 & -1 \\ -3 & -2 & 0 \end{bmatrix}$

Add 3 to each entry to obtain

$$\begin{bmatrix} 5 & 4 & 2 \\ 0 & 1 & 3 \end{bmatrix}.$$

The linear programming problem to be solved is as follows.

Maximize $z = x_1 + x_2 + x_3$
subject to: $5x_1 + 4x_2 + 2x_3 \leq 1$
$\quad\quad\quad\quad\quad x_2 + 3x_3 \leq 1$
with $\quad x_1 \geq 0, x_2 \geq 0, x_3 \geq 0.$

Use the simplex method to solve the problem.
The initial tableau is

$$\begin{array}{cccccc} x_1 & x_2 & x_3 & s_1 & s_2 & z \end{array}$$
$$\begin{bmatrix} 5 & \boxed{4} & 2 & 1 & 0 & 0 & 1 \\ 0 & 1 & 3 & 0 & 1 & 0 & 1 \\ \hline -1 & -1 & -1 & 0 & 0 & 1 & 0 \end{bmatrix}.$$

Pivot on each indicated entry.

$$\begin{array}{ccccccc} & x_1 & x_2 & x_3 & s_1 & s_2 & z \end{array}$$
$$\begin{bmatrix} 5 & 4 & 2 & 1 & 0 & 0 & 1 \\ 0 & 1 & \boxed{3} & 0 & 1 & 0 & 1 \\ \hline 0 & -1 & -3 & 1 & 0 & 5 & 1 \end{bmatrix}$$
$R_1 + 5R_3 \rightarrow R_3$

$$\begin{array}{cccccc} x_1 & x_2 & x_3 & s_1 & s_2 & z \end{array}$$
$-2R_2 + 3R_1 \rightarrow R_1 \begin{bmatrix} 15 & 10 & 0 & 3 & -2 & 0 & 1 \\ 0 & 1 & 3 & 0 & 1 & 0 & 1 \\ \hline 0 & 0 & 0 & 1 & 1 & 5 & 2 \end{bmatrix}$
$R_2 + R_3 \rightarrow R_3$

Create a 1 in the columns corresponding to x_1, x_3, and z.

$$\begin{array}{cccccc} x_1 & x_2 & x_3 & s_1 & s_2 & z \end{array}$$
$\frac{1}{15}R_1 \rightarrow R_1 \begin{bmatrix} 1 & \frac{2}{3} & 0 & \frac{1}{5} & -\frac{2}{15} & 0 & \frac{1}{15} \\ 0 & \frac{1}{3} & 1 & 0 & \frac{1}{3} & 0 & \frac{1}{3} \\ \hline 0 & 0 & 0 & \frac{1}{5} & \frac{1}{5} & 1 & \frac{2}{5} \end{bmatrix}$
$\frac{1}{3}R_2 \rightarrow R_2$
$\frac{1}{5}R_3 \rightarrow R_3$

From this final tableau, we have

$$x_1 = \frac{1}{15}, x_2 = 0, x_3 = \frac{1}{3},$$

$$y_2 = \frac{1}{5}, y_2 = \frac{1}{5}, \text{and } z = \frac{2}{5}.$$

Note that $g = \frac{1}{z} = \frac{5}{2}$. The optimum strategy for player A is

$$p_1 = gy_1 = \frac{5}{2}\left(\frac{1}{5}\right) = \frac{1}{2},$$

$$p_2 = gy_2 = \frac{5}{2}\left(\frac{1}{5}\right) = \frac{1}{2}.$$

The optimum strategy for player B is

$$q_1 = gx_1 = \frac{5}{2}\left(\frac{1}{15}\right) = \frac{1}{6},$$

$$q_2 = gx_2 = \frac{5}{2}(0) = 0,$$

$$q_3 = gx_3 = \frac{5}{2}\left(\frac{1}{3}\right) = \frac{5}{6}.$$

The value of the game is

$$\frac{5}{2} - 3 = -\frac{1}{2}.$$

43.
$$\begin{bmatrix} -2 & 1 & 0 \\ 2 & 0 & -2 \\ 0 & -1 & 3 \end{bmatrix}$$

Add 2 to each entry to obtain

$$\begin{bmatrix} 0 & 3 & 2 \\ 4 & 2 & 0 \\ 2 & 1 & 5 \end{bmatrix}.$$

The problem to be solved is as follows.

Maximize $z = x_1 + x_2 + x_3$

subject to:
$$3x_2 + 2x_3 \le 1$$
$$4x_1 + 2x_2 \qquad \le 1$$
$$2x_1 + x_2 + 5x_3 \le 1$$

with $x_1 \ge 0, x_2 \ge 0, x_3 \ge 0.$

The initial simplex tableau is

	x_1	x_2	x_3	s_1	s_2	s_3	z	
	0	3	2	1	0	0	0	1
	[4]	2	0	0	1	0	0	1
	2	1	5	0	0	1	0	1
	−1	−1	−1	0	0	0	1	0

Pivot on each indicated entry.

	x_1	x_2	x_3	s_1	s_2	s_3	z	
	0	3	2	1	0	0	0	1
	4	2	0	0	1	0	0	1
$-R_2 + 2R_3 \to R_3$	0	0	[10]	0	−1	2	0	1
$R_2 + 4R_4 \to R_4$	0	−2	−4	0	1	0	4	1

	x_1	x_2	x_3	s_1	s_2	s_3	z	
$-R_3 + 5R_1 \to R_1$	0	[15]	0	5	1	−2	0	4
	4	2	0	0	1	0	0	1
	0	0	10	0	−1	2	0	1
$2R_3 + 5R_4 \to R_4$	0	−10	0	0	3	4	20	7

	x_1	x_2	x_3	s_1	s_2	s_3	z	
	0	15	0	5	1	−2	0	4
$-2R_1 + 15R_2 \to R_2$	60	0	0	−10	13	4	0	7
	0	0	10	0	−1	2	0	1
$2R_1 + 3R_4 \to R_4$	0	0	0	10	11	8	60	29

Create a 1 in the columns corresponding to $x_1, x_2,$ $x_3,$ and z.

	x_1	x_2	x_3	s_1	s_2	s_3	z	
$\frac{1}{15}R_1 \to R_1$	0	1	0	$\frac{1}{3}$	$\frac{1}{15}$	$-\frac{2}{15}$	0	$\frac{4}{15}$
$\frac{1}{60}R_2 \to R_2$	1	0	0	$-\frac{1}{6}$	$\frac{13}{60}$	$\frac{1}{15}$	0	$\frac{7}{60}$
$\frac{1}{10}R_3 \to R_3$	0	0	1	0	$-\frac{1}{10}$	$\frac{1}{5}$	0	$\frac{1}{10}$
$\frac{1}{60}R_4 \to R_4$	0	0	0	$\frac{1}{6}$	$\frac{11}{60}$	$\frac{2}{15}$	1	$\frac{29}{60}$

From this final tableau, we have

$$x_1 = \frac{7}{60}, x_2 = \frac{4}{15}, x_3 = \frac{1}{10},$$

$$y_1 = \frac{1}{6}, y_2 = \frac{11}{60}, y_3 = \frac{2}{15}, \text{ and } z = \frac{29}{60}.$$

Note that $g = \frac{1}{z} = \frac{60}{29}$. The optimum strategy for player A is

$$p_1 = gy_1 = \frac{60}{29}\left(\frac{1}{6}\right) = \frac{10}{29},$$

$$p_2 = gy_2 = \frac{60}{29}\left(\frac{11}{60}\right) = \frac{11}{29},$$

$$p_3 = gy_3 = \frac{60}{29}\left(\frac{2}{15}\right) = \frac{8}{29}.$$

The optimum strategy for player B is

$$q_1 = gx_1 = \frac{60}{29}\left(\frac{7}{60}\right) = \frac{7}{29},$$

$$q_2 = gx_2 = \frac{60}{29}\left(\frac{4}{15}\right) = \frac{16}{29},$$

$$q_3 = gx_3 = \frac{60}{29}\left(\frac{1}{10}\right) = \frac{6}{29}.$$

The value of the game is

$$\frac{60}{29} - 2 = \frac{2}{29}.$$

In Exercises 47–51, use the following payoff matrix.

		Management	
		Friendly	Hostile
Labor	Friendly	\$700	\$ 900
	Hostile	\$500	\$1100

47. Be hostile; then he has a chance at the \$1100 wage increase, which is the largest possible increase.

49. If there is a 0.4 chance that the company will be hostile, then there is a 0.6 chance that it will be friendly. Find the expected payoff for each strategy.

Friendly: $\$700(0.6) + \$900(0.4) = \$780$

Hostile: $\$500(0.6) + \$1100(0.4) = \$740$

Therefore, he should be friendly. The expected payoff is $780.

51. The 700 at $(1, 1)$ is the smallest number in its row and the largest number in its column, so it is a saddle point, and the game is strictly determined. Labor and management should both always be friendly, and the value of the game is 700.

53. $\begin{bmatrix} 2800 & 3200 \\ 5000 & -2000 \end{bmatrix}$

To guarantee that the value of the game is positive, we add 2000 to all entries in the matrix to obtain

$$\begin{bmatrix} 4800 & 5200 \\ 7000 & 0 \end{bmatrix}.$$

Let Victor choose row 1 with probability p_1 and row 2 with probability p_2. Then,

$$E_1 = 4800p_1 + 7000p_2$$

and $E_2 = 5200p_1.$

Let g represent the minimum of the expected gains, so that

$$E_1 = 4800p_1 + 7000p_2 \geq g$$
$$E_2 = 5200p_1 \qquad\qquad \geq g.$$

Dividing by g yields

$$4800\left(\frac{p_1}{g}\right) + 7000\left(\frac{p_2}{g}\right) \geq 1$$

$$5200\left(\frac{p_1}{g}\right) \geq 1.$$

Let $x = \dfrac{p_1}{g}$ and $y = \dfrac{p_2}{g}.$

We have the following linear programming problem:

Minimize $w = x + y$

subject to: $4800x + 7000y \geq 1$
$$5200x \qquad\qquad \geq 1$$

with $x \geq 0, y \geq 0.$

Graph the feasible region.

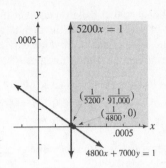

The corner points are $\left(\frac{1}{5200}, \frac{1}{91,000}\right)$ and $\left(\frac{1}{4800}, 0\right)$. The minimum value of $w = x + y$ is $\frac{37}{182,000}$ at $\left(\frac{1}{5200}, \frac{1}{91,000}\right)$. Thus, the value of the game is

$$g = \frac{1}{w} = \frac{182,000}{37} \approx 4918.92.$$

To find the value of the original game, we must subtract 2000:

$$4918.92 - 2000 = 2918.92.$$

The value of the original game is $2918.92. The optimum strategy for Victor is

$$p_1 = gx = \frac{182,000}{37}\left(\frac{1}{5200}\right) = \frac{35}{37},$$

$$p_2 = gy = \frac{182,000}{34}\left(\frac{1}{91,000}\right) = \frac{2}{37}.$$

Hector should invest in blue-chip stocks with probability $\frac{35}{37}$ and growth stocks with probability $\frac{2}{37}$.

For Exercises 55–59, use the following payoff matrix.

		Opponent		
		Favors	Waffles	Opposes
	Favors	0	1200	4200
Candidate	Waffles	−1000	0	600
	Opposes	−5000	−2000	0

55. As an optimist, Martha would look at the best possible outcome from each strategy, and choose the best of them. If she favors the factory, she could gain 4200 votes; if she waffles, she could gain 600 votes; and if she opposes the issue, she neither gains nor loses votes. The best of these options is to gain 4200 votes, so she should favor the factory.

57. If there is a 0.4 chance that Kevin will favor the plant, and 0.35 chance he will waffle, then there is a 0.25 chance he will oppose it. Find the expected payoffs in votes under the three strategies.

Favors: $0(0.4) + 1200(0.35) + 4200(0.25) = 1470$

Waffles: $-1000(0.4) + 0(0.35) + 600(0.25) = -250$

Opposes: $-5000(0.4) - 2000(0.35) + 0(0.25) = -2700$

Martha should favor the new factory and expect a gain of 1470 votes.

59. This game has a saddle point of 0 at $(1, 1)$. The value of the game is 0. The strategy $(1, 1)$ means that Martha favors the factory and her opponent also favors the factory. Therefore, each candidate should favor the factory.

For Exercise 61, use the following payoff matrix.

Rontovia

$$\begin{array}{c} \text{Ravogna} \end{array} \begin{array}{cc} & \text{Attack 1} \quad \text{Attack 2} \\ \begin{array}{c} \text{Defend } 1 \\ \text{Defend } 2 \end{array} & \left[\begin{array}{cc} 4 & 1 \\ 3 & 4 \end{array} \right] \end{array}$$

61. To find the optimum strategy for Ravogna,

Minimize $w = x + y$

Subject to: $4x + 3y \geq 1$

$x + 4y \geq 1$

with $x \geq 0,\ y \geq 0.$

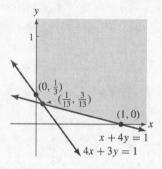

Corner Point	Value of $w = x + y$
$\left(0, \frac{1}{3}\right)$	$0 + \frac{1}{3} = \frac{1}{3}$
$\left(\frac{1}{13}, \frac{3}{13}\right)$	$\frac{1}{13} + \frac{3}{13} = \frac{4}{13}$
$(1, 0)$	$1 + 0 = 1$

The minimum value is $w = \frac{4}{13}$ at $\left(\frac{1}{13}, \frac{3}{13}\right)$.

Thus, $g = \frac{1}{w} = \frac{13}{4}$ is the value of the game.

The optimum strategy for Ravogna is

$$p_1 = gx = \frac{13}{4}\left(\frac{1}{13}\right) = \frac{1}{4},$$

$$p_2 = gy = \frac{13}{4}\left(\frac{3}{13}\right) = \frac{3}{4}.$$

That is, Ravogna should defend installation #1 with probability $\frac{1}{4}$ and installation #2 with probability $\frac{3}{4}$.

To find the optimum strategy for Rontovia,

Maximize $z = x + y$

subject to: $4x + y \leq 1$

$3x + 4y \leq 1$

with $x \geq 0,\ y \geq 0.$

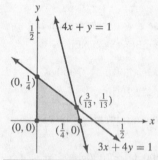

Corner Point	Value of $z = x + y$
$\left(0, \frac{1}{4}\right)$	$0 + \frac{1}{4} = \frac{1}{4}$
$\left(\frac{3}{13}, \frac{1}{13}\right)$	$\frac{3}{13} + \frac{1}{13} = \frac{4}{13}$
$\left(\frac{1}{4}, 0\right)$	$\frac{1}{4} + 0 = \frac{1}{4}$
$(0, 0)$	$0 + 0 = 0$

The maximum value is $z = \frac{4}{13}$ at $\left(\frac{3}{13}, \frac{1}{13}\right)$.

The optimum strategy for Rontovia is

$$q_1 = gx = \frac{13}{4}\left(\frac{3}{13}\right) = \frac{3}{4},$$

$$q_2 = gy = \frac{13}{4}\left(\frac{1}{13}\right) = \frac{1}{4}.$$

That is, Rontovia should attack installation #1 with probability $\frac{3}{4}$ and installation #2 with probability $\frac{1}{4}$.